Plant Breeding II

Plant Breeding II

EDITED BY

KENNETH J. FREY
C. F. CURTISS DISTINGUISHED PROFESSOR IN AGRICULTURE
IOWA STATE UNIVERSITY

THE IOWA STATE UNIVERSITY PRESS / AMES

Printed by
The Iowa State University Press
Ames, Iowa 50010

First edition, 1981

Library of Congress Cataloging in Publication Data

Plant Breeding Symposium, 2d, Iowa State University
1979.
Plant Breeding II.

Includes bibliographies and index.
1. Plant-breeding—Congresses. I. Frey, Kenneth J.
SB123.P56 1979 631.5'3 80-28879
ISBN 0-8138-1550-9

CONTENTS

Preface vii

List of Sustaining Members ix

1. Selection and Breeding Methods 3
A. R. Hallauer
Discussion Panel: *R. J. Baker*
R. R. Hill
J. J. MacKey

2. Germplasm Collection, Preservation, and Use 57
J. G. Hawkes
Discussion Panel: *E. E. Gerrish*
Q. Jones
C. M. Rick

3. Application of Tissue Culture and Somatic Hybridization to Plant Improvement 85
E. C. Cocking and R. Riley
Discussion Panel: *K. L. Giles*
C. E. Green
T. B. Rice

4. Chromosomal and Cytoplasmic Manipulations 117
S. J. Peloquin
Discussion Panel: *E. C. Bashaw*
J. James
J. R. Laughnan
R. L. Phillips

5. Breeding Plants for Stress Environments 151
C. F. Lewis and M. N. Christiansen
Discussion Panel: *M. N. Christiansen*
S. D. Jensen
M. C. Shannon

6. Development of Plant Genotypes for Multiple Cropping Systems 179
C. A. Francis
Discussion Panel: *R. K. Crookston*
R. M. Lantican

7. Breeding for Morphological and Physiological Traits 233
D. Wilson
Discussion Panel: *J. W. Hanover*
R. W. F. Hardy
D. C. Rasmusson

8. Breeding for Insect Resistance 291
J. N. Jenkins

9. Disease Resistance in Plants and Its Consequences for Plant Breeding 309
J. E. Parlevliet
Discussion Panel: *J. A. Browning*
W. D. Guthrie
F. G. Maxwell
R. R. Nelson

10. Breeding for Improved Nutritional Quality 365
J. D. Axtell
Discussion Panel: *W. H. Gabelman*
A. W. Hovin
W. Martinez

11. Meeting Human Needs Through Plant Breeding: Past Progress and Prospects for the Future 433
G. W. Burton
Discussion Panel: *W. L. Brown*
S. Fonseca Martinez
B. J. Zobel

12. Increasing and Stabilizing Food Production 467
N. E. Borlaug

Index 493

PREFACE

PLANT BREEDING SYMPOSIUM II was the third such conference sponsored by Iowa State University. Previous symposia in this series were "Heterosis" in 1950 and "Plant Breeding Symposium I" in 1965. These three conferences have provided the settings for periodic review of the accomplishments of plant breeding, for appraising the interactions of this discipline with related areas of science, and for charting the course that plant breeding should take in the future. Thus, Plant Breeding Symposium II was the latest in this series of international conferences that provided an in-depth look at the profession as a whole.

This book, *Plant Breeding II*, is a compendium of the lectures and discussions that made up the one-week program of Plant Breeding Symposium II. The symposium program and this publication have been divided in three sections. Chapters one through four evaluate or forecast the past, present, and future contributions of four major research areas to plant breeding accomplishments. Chapters five through ten review the past successes and predict future achievements from breeding for certain goals. And, chapters eleven and twelve concentrate on how plant breeding has and will continue to contribute to providing adequate food, clothing, and fuel for humankind.

Plant breeders, when conducting their research, work on specific field, forest, or horticultural crops that are allogamous or autogamous, annual or perennial, used for grain, fruit, or forage, and the like. Furthermore, for a given breeding project, a breeder may call upon one contributing scientific area for most of the techniques he or she uses, such as breeding for high protein content in sorghum grain. Plant Breeding Symposium II and this book have been organized to give appropriate recognition to the variations between plant species and among project objectives, but both deliberately focus on plant breeding in total and as a research discipline.

The symposium format was organized to engender audience participation. A session of the symposium, which occupied a half day, was built around one topic. Each session was initiated by a plenary speech that provided the substance for a chapter in this book. After the speech, a panel consisting of three to six scientists led a discussion in which the audience participated freely. The questions and comments from a discussion are included immediately after the chapter.

The success of Plant Breeding Symposium II was due to contributions from many persons and organizations. The symposium organizing committee, consisting of T. S. Cox, K. J. Frey, G. H. Ebert, D. G. Helsel, J. J. Mock, R. G. Palmer, W. A. Russell, M. D. Simons, and W. L. Summers, is especially indebted to the plenary speakers who prepared their scholarly contributions while engaged in other full-time assignments, and to the panelists who stimulated the audience to participate in the discussions. Moreover, without the generous grants from the sustaining members, the symposium would not have occurred. The advise and work contributed by numerous persons from Iowa State University were essential to the success of the symposium. A few who deserve special recognition are: T. L. Lund, for serving as executive assistant for the symposium; R. E. Atkins, T. B. Bailey, J. A. Browning, T. M Crosbie, W. R. Fehr, K. A. Kuenzel, J. P. Murphy, D. S. Robertson, R. M. Shibles, O. S. Smith, and J. Weigle for arranging the mechanics and amenities for the symposium; and Mary Jo Vivian for secretarial work associated with the symposium. And of signal importance were those who attended the symposium and provided loyal and intense support and participation in the symposium program.

List of Sustaining Members

Acco Seeds	*Belmond, Iowa*
Asgrow Seed Company	*Kalamazoo, Mich.*
Committee for Agricultural Development	*Ames, Iowa*
Callahan Enterprises, Inc.	*Westfield, Ind.*
Cal/West Seeds	*Woodland, Calif.*
Campbell Institute for Agricultural Research	*Camden, N. J.*
Coker's Pedigreed Seed Company	*Hartsville, S. C.*
College of Agriculture, Iowa State University	*Ames, Iowa*
DeKalb Ag Research, Inc.	*DeKalb, Ill.*
Delta and Pineland Company	*Scott, Miss.*
FFR Cooperative	*West Lafayette, Ind.*
FS Services, Inc.	*Bloomington, Ill.*
Funk Seeds International	*Bloomington, Ill.*
Golden Harvest Seeds	*Bloomington, Ill.*
Graduate College, Iowa State University	*Ames, Iowa*
Green Giant Company	*LeSueur, Minn.*
Hoegemeyer Hybrids, Inc.	*Hooper, Nebr.*
Illinois Foundation Seeds, Inc.	*Champaign, Ill.*
Iowa Crop Improvement Association	*Ames, Iowa*
McNair Seed Company	*Laurinburg, N.C.*
Mike Brayton Seeds	*Ames, Iowa*
Milling Oat Improvement Association	*Minneapolis, Minn.*
Moews Seed Company, Inc.	*Granville, Ill.*
National Science Foundation	*Washington, D.C.*
NC + Hybrids	*Lincoln, Nebr.*
Northrup King Company	*Minneapolis, Minn.*
O's Gold Seed Company	*Parkersberg, Iowa*

Pioneer Hi-Bred International, Inc.	*Des Moines, Iowa*
Pfister Hybrid Corn Company	*El Paso, Ill.*
Pfizer Genetics	*St. Louis, Mo.*
Quaker Oats Company	*Chicago, Ill.*
W. O. McCurdy and Sons	*Fremont, Iowa*
World Food Institute, Iowa State University	*Ames, Iowa*

Plant Breeding II

CHAPTER 1

Selection and Breeding Methods

A. R. HALLAUER

PLANT BREEDING was defined by Smith (1966) as the art and science of improving the genetic pattern of plants in relation to their economic use. Although selection is not included in Smith's definition of plant breeding, it is a primary activity in all plant breeding programs. Selection requires making a choice, and in a breeding program there are many choices to be made, such as choice of parental germplasm, choice of breeding methods, choice of genotypes for testing, choice of testing procedures, and choice of particular cultivar released for commercial use.

Improving the genetic pattern of plants in relation to their economic use in plant breeding programs emphasizes artificial selection. Natural selection is important in the evolution of crop species, but it is not of major benefit in most plant breeding programs. The plant breeder can use any of several selection methods, and the one chosen will depend upon the (1) objectives of the plant breeding program, (2) inheritance patterns of traits to be improved, (3) germplasm included in available breeding populations, and (4) selection history. The selection method used will also depend upon specific breeding goals in a given breeding program. Selection criteria change as economic conditions, crop management practices, and environmental conditions of the crop species change. But, in all instances, the primary goal of selection in plant breeding is to identify the desirable genotypes.

Contribution for Agricultural Research, USDA, and Journal Paper No. J-9425 of the Iowa Agricultural and Home Economics Experiment Station, Ames. Project No. 2194.

Research Geneticist, Agriculture Research, Science and Education Administration, USDA, and Professor of Plant Breeding, Iowa State University, Ames.

Herein, selection and breeding methods will be discussed relative to cyclical selection schemes for improvement of breeding populations. This does not imply, however, that conventional breeding methods are not important. Concerted effort to develop and improve plant populations started about 40 years ago. Sprague (1966) summarized the relation of quantitative genetics to population improvement and the early selection studies for different selection methods. But data upon responses from selection were limited at that time because cyclical selection studies are long-term. Additional data are now available with which to compare the rates of response from different selection methods in several crop species. Cyclical selection methods also have been used for autogamous species to supplement applied breeding programs. I will summarize the information related to cyclical selection and how this affects the choice of selection methods; the discussion will be restricted to artificial selection.

BREEDING OBJECTIVES

Breeding programs have sets of objectives to meet the short-, intermediate-, and long-term goals considered relevant for the particular crop species (Fig. 1.1). The objectives must not be rigid and

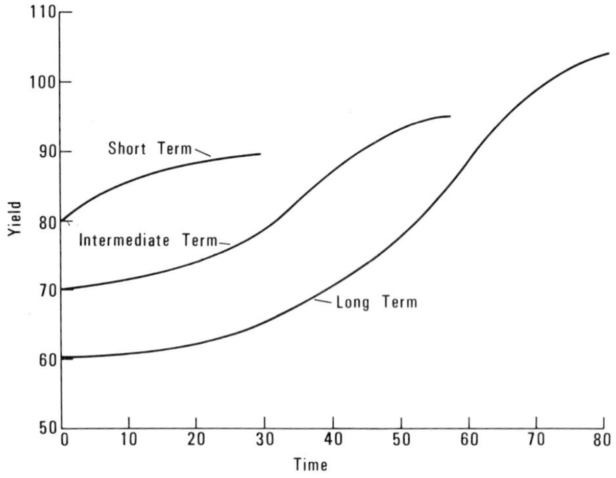

Fig. 1.1. Short-, intermediate-, and long-term goals for breeding programs.

should provide for the continuous flow of breeding materials in different stages of development. Short-term goals are designed to develop varieties and hybrids quickly and to meet any other immediate contingencies that arise. The intermediate- and long-term goals include development of materials that have potential in the breeding programs for the future. Breeders are concerned primarily with short-term goals, but the improvement of breeding populations and conservation of genetic variability are important long-term goals. Plant breeding has been successful in developing varieties and hybrids for environments in which they were grown. Frey (1971) concluded that there was no evidence that a yield plateau existed for any crop species receiving significant research attention. Russell (1974) and Duvick (1977) recently reported results comparing maize (*Zea mays* L.) hybrids developed and grown for the four decades since the introduction of hybrids in the USA Corn Belt. Both studies were conducted in Iowa. Duvick (1977), using two sets of hybrids, estimated the genetic gains were 57 and 60 percent. Russell (1974) calculated a genetic gain of 63 percent for his group of hybrids. Both researchers concluded that more than half of the increase achieved in Iowa maize yield during the past 40 years was due to improved hybrids. New hybrids produced these higher yields, however, only when modern cultural practices were used.

Russell (1974), from data collected at different plant densities, found that if we were still growing hybrids at 29,700 plants/ha, a common plant density used when hybrids were introduced, we would realize only small gains over the yields of the early double crosses. The genetic gains, therefore, were made in conjunction with improvements in the management and cultural practices used in growing maize. These estimates of genetic gains for maize hybrids emphasize that breeding goals must be flexible to incorporate the advances made in cultural techniques. Another example was the response of IR8 rice (*Oryza sativa* L.) cultivar to high rates of nitrogen; IR8 yielded only 20 percent more than the typical tropical cultivars at low nitrogen levels, but yielded 2 to 3 times more at higher nitrogen levels (Frey, 1971).

Breeding populations are continuously developed to meet short-, intermediate-, and long-term goals. The types of populations developed and used, however, can be quite different. Short-term goals may be met by use of F_2 and backcross populations either to develop new cultivars rapidly or to incorporate a simply inherited trait (such as Ht_1 in maize lines) into an otherwise desirable genotype. Efficient use of off-season nurseries and greenhouses may permit attainment of such goals in three to five years. Intermediate- and long-term goals usually do not include the expectation that new varieties and hybrids will be developed in the immediate future. Long-term goals may include the synthesis of composites, introduction of exotic germplasm, and mild selection pressure for adaptiveness. Because of the nature and unpredictibility of plant breeding, different types of breeding populations are essential, and the relative proportions of the different types of populations will vary according to the status of research and economic importance of the crop species, its mode of reproduction, and the type of cultivars used in commercial production.

In discussing breeding populations, I will assume they have a broad genetic base, that germplasm included in them has been based on selection goals, that cyclical selection is imposed to increase the frequency of favorable genes for the selected traits, and that the populations are possible sources of improved cultivars. Of course, development of superior cultivars is the ultimate goal of any breeding program, and to be useful, the cyclical selection procedures must contribute to this goal. In many instances, the long-term goals of cyclical selection programs must be based upon the goals of the breeding programs and not upon those of individual breeders. This is true because a long-term breeding program is likely to transcend the tenure of one breeder. Cyclical selection programs must be long-term projects, but to be acceptable for applied breeding programs they must meet the primary goal of developing superior cultivars. Because of their nature, cyclical selection programs can contribute to maintaining systematic genetic improvement of the populations for the extraction of the new cultivars.

GENETIC VARIABILITY

Effective selection is dependent on the existence of genetic variability. The extent of the genetic variability in a specific breeding population depends on the germplasm included in it and its selection history. When genetic variability in a breeding program is insufficient to permit attainment of a specific goal it will be necessary to increase the variability by using either mutagenic treatment or introduction of new germplasm. Collections of germplasm are available for most crop species and these can be used to increase the available genetic variability either by crossing them with the currently used germplasm or by selecting within them for local conditions.

Many studies have investigated the extent of genetic variability available in different types of maize populations. Most were conducted by use of mating designs suggested by Comstock and Robinson (1948, 1952). Development of mating designs for the estimation of the genetic components of variance was necessary to determine whether adequate additive genetic variance was available to permit progress from selection in maize populations. Hull (1945) believed that additive genetic effects in maize were inadequate to explain hybrid vigor, so he invoked overdominant effects to explain this phenomenon. He designed selection methods that emphasized selection for loci with overdominant effects. Divergent views of the causes of heterosis in maize also generated investigations of the relative magnitudes of additive and nonadditive genetic variances in maize populations.

Summaries of studies that estimated genetic variances in maize populations are given in Table 1.1 and Fig. 1.2 for grain yield.

Table 1.1. Summary[a] of estimates of additive genetic (σ^2_A) and dominance (σ^2_D) components of variance for yield for five types of maize populations.

Populations	σ^2_A	σ^2_D	σ^2_A/σ^2_D	No. of Estimates
F_2's	581 ± 338	451 ± 593	1.3	24
Synthetics	226 ± 59	129 ± 83	1.8	15
Open-pollinated	504 ± 179	246 ± 321	2.1	37
Variety crosses	306 ± 139	292 ± 32	1.1	13
Composites	722 ± 432	282 ...	2.6	10
Average	468 ± 229	280 ± 257	1.7	20

[a]Estimates were obtained for those reported in the literature since 1946 by use of different mating designs. No estimates were included from use of the diallel cross mating designs.

Estimates of additive genetic and dominance variances were summarized for five types of maize populations: F_2's from single crosses between inbred lines; synthetics formed by intercrossing elite lines; open-pollinated cultivars; and composites formed by intermating several cultivars and races. The average additive genetic variance estimate was greater than the average dominance estimate in all population types, and on the average, the additive genetic variance was 1.7 times greater than the dominance variance.

The estimates of additive genetic variance were plotted for the five types of maize populations to determine if any trends existed for

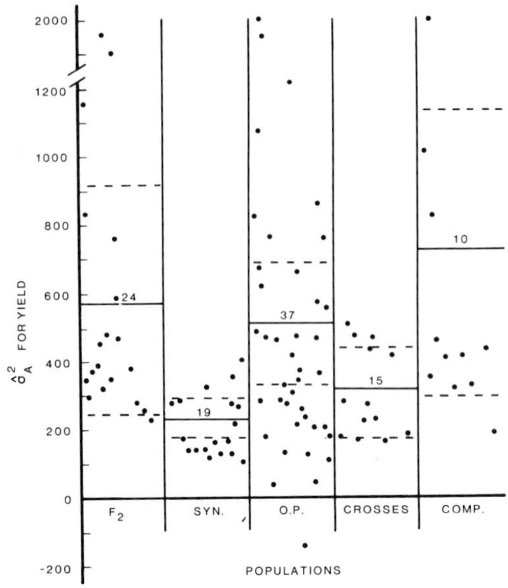

Fig. 1.2. Distribution of the estimates of additive genetic variance (σ^2_A) for yield for five types of maize populations. Each dot represents one estimate, solid line denotes average of σ^2_A estimates, slashed line denotes average standard error of σ^2_A and numbers indicate estimates for each population type.

the relative magnitude of the additive genetic variance for grain yield of maize (Fig. 1.2). No trends were evident, but average additive genetic variance estimates were greatest for the composites and F_2 populations and least for the synthetic cultivars. The average standard errors also were greater for the composite and F_2 populations and smaller for the synthetic cultivars (Table 1.1). Sampling and experimental technique would influence the standard errors of the estimates. Differential gene frequencies and number of segregating loci may have caused the differences in the estimates of additive genetic variance. The expected gene frequency for segregating loci in F_2 populations is 0.5, whereas gene frequency in the synthetic populations may be any value; hence, additive genetic variance could be at a maximum in F_2 populations and less than maximum in synthetic populations. Most synthetics were formed from elite lines, so on the average, their gene frequencies probably were greater than 0.5 for many loci. These studies showed that for grain yield in maize, additive genetic effects were more important than dominance effects and that this relationship occurred in all five population types. Several earlier studies that utilized F_2 populations indicated overdominance at some yield loci. It was recognized, however, that repulsion phase linkage would cause underestimates of additive genetic and overestimates of dominance variance. In later studies (Gardner, 1963), dominance variance was found to be 40 to 50 percent smaller in advanced random mating generations from a single cross than it was in the F_2's of the same crosses. Thus, linkage was a factor in giving evidence for overdominance in F_2 populations. The predominant type of genetic variation in maize seems to include genes with additive effects and partial to complete dominance.

Information on the genetic variability in other allogamous annual species is not as extensive as for maize. Quantitative genetic data for rye (*Secale cereale* L.) are restricted because of self-incompatibility in this species and required seed supplies can be produced only by use of cytoplasmic male sterility (Geiger and Schnell, 1975). In diallel experiments with unselected and selected inbred lines, general combining ability accounted for most of the genetic variability among crosses (Morgenstern and Geiger, 1975). Quantitative genetic information presently available in rye duplicates that from maize.

Limited estimates also are available for allogamous vegetable species. Estimates for onions (*Allium cepa* L.), carrots (*Daucus carota* L.), sweet potatoes (*Ipomoea batatas* L.), and pickling cucumbers (*Cucumis sativa* L.) indicate that adequate genetic variability exists for most traits in these species. McCollum (1971b), from an analysis of 10 onion populations, concluded that because of its high heritability, bulb shape could be changed by mass selection, but that difficulty would be expected for bulb size because its heritability was low. McCollum (1971a), working with a carrot population synthesized from diverse cultivars and breeding lines, found the heritability of greening on the shoulders of carrot roots was less than 25 percent, so he concluded that it would be difficult to eliminate greening genotypes by mass selection. Only intense inbreeding and progeny testing could reduce this undesirable greening trait.

Smith et al. (1978) estimated additive genetic and dominance variances for nine traits in a population of pickling cucumbers by use of the Design I. Only fruit color showed dominance variance, and the additive genetic variances for all traits except fruit weight were one to two times greater than its standard error. Heritability percentages were low for fruit number (9 percent), yield (12 percent), and fruit weight (1 percent); intermediate for fruit color (21 percent); and high for other traits. Significant progress from selection could be made for most traits, but either extensive testing or greater genetic variability would be needed to significantly increase fruit weight, fruit number, and yield.

Jones (1969) studied the inheritance of 10 vine traits in sweet potato. Heritabilities were sufficiently high for most traits to expect progress from selection, and the traits were uncorrelated, so simultaneous selection for any two or all of the traits would be possible. For traits with low heritabilities, response to selection was expected with use of greater selection intensities and more precise evaluation.

Estimates of genetic variability in autogamous crop species have been restricted mostly to F_2 populations developed by crossing two pure-line cultivars. Difficulties in making the cross pollinations and insufficient seed supplies have restricted the methods feasible for estimating genetic components of variance in these species. Methods developed by Mather (1949) and the diallel cross mating design have been used extensively to estimate genetic components of variance for several autogamous crop species. The estimates vary greatly for the same trait, which, of course, is a reflection of the specific parents and the environments (namely, space planted vs. solid stands) used in the studies. A survey by Matzinger (1963) showed that additive genetic variance predominated in autogamous species. In support of the predominance of additive genetic variance in these species is the common occurrence of pure lines that equal the best hybrids in yield. The continued progress in obtaining improved pure-line cultivars shows that adequate genetic variability exists in the autogamous species. But recently, breeders of these species have been developing breeding populations with a broader genetic base.

Studies on genetic variability in sorghum (*Sorghum bicolor* L. Moench) are not extensive, primarily because the development and rapid acceptance of hybrid sorghums have led to breeding methods that emphasized line and hybrid development. Studies generally show that additive gene effects (as measured by general combining ability) seem to be more important than nonadditive effects for this species (Maunder, 1969).

Jan-Orn et al. (1976) estimated genetic components of variance for nine traits for the 'NP3R' grain sorghum population. Additive genetic effects tended to be greater than dominance for all traits except yield; dominance effects were predominant for genes controlling grain yield. He used genetic variances to predict gain from mass, half-sib, full-sib, and S_1 selection and concluded that mass selection for days to flower and plant height and S_1 progeny selection for all traits suggested the greatest promise for improving the NP3R population. The improvement in sorghum yield that has occurred in the past 25 years shows that sorghum breeders are exploiting the genetic variability of the species, and they are using all types of genetic effects in the expression of heterosis in crosses between elite lines. Doggett and Eberhart (1968) indicated that there was no evidence that overdominance or overdominant types of epistasis were operating for grain yield in sorghum. They stated that breeding methods that emphasized selection for additive effects (that is, mass, ear-to-row, and S_1) would be efficient for improving grain yield of sorghum populations.

Estimates of genetic components of variance have been reported for alfalfa (*Medicago sativa* L.) and several forage grasses. Burton (1952) has emphasized the importance of quantitative inheritance of forage traits and how such parameters should be considered in breeding programs. Levings and Dudley (1963) illustrated how different mating designs could be used with autotetraploid species to estimate genetic components of variance. Studies that give estimates of genetic variability in forage grasses and legumes include those of Burton (1951) for pearl millet (*Pennisetum typhoides* (Burm.) Stapf and C. E. Hubb.), Burton and DeVane (1953) for tall fescue (*Festuca arundinacea* Schreb.), Hanson et al. (1956) for Korean Lespedeza (*Lespedeza stipulacea* Maxim.), Eberhart and Newell (1959) for switchgrass (*Panicum virgatum* L.), Kehr and Gardner (1960) and Dudley et al. (1963) for alfalfa, Potts and Holt (1967) for kleingrass (*Panicum coloratum* L.), and Asay et al. (1968) for reed canarygrass (*Phalaris arundinacea* L.). In nearly every instance, there was ample genetic variation for improving forage and seed yields of the populations studied. Because quantitative genetic data for the forage species were similar to those for maize, breeding methods that emphasize selection for additive genetic effects have been used.

RECURRENT SELECTION

Recurrent selection, in its broadest sense, is any cyclical scheme of plant selection by which frequencies of favorable genes are increased in plant populations. Some form of recurrent selection has always been used in plant improvement, but the methods of selection often were not used systematically. Recurrent selection methods were developed primarily for the improvement of quantitatively inherited traits, which involve a large number of genetic factor pairs, each with a small effect, and the genotypes cannot be classified into discrete classes; additionally, the environmental effects tend to obscure the genetic effects. Quantitative traits are measured metrically, and statistical parameters (means, variances, and covariances) are used to determine the effects of selection. Gene frequencies are not known, but if selection is effective, gene or genotypic frequencies must be changing in the desired direction. The basic objective of all recurrent selection methods is to increase systematically the frequency of desirable genes in a population so that the opportunities to extract superior genotypes are enhanced.

Success of recurrent selection depends, of course, on the genes present in the original breeding population. If frequencies for genes that control the trait under selection differ among populations, response to selection, even though realized, may occur at varying rates in the different populations. The potential of a recurrent selection program to meet the short-, intermediate-, and long-term goals will depend on the initial gene frequencies in the population.

A requisite for response is that the population under recurrent selection should have adequate genetic variability, which, as we have seen (Table 1.1), usually occurs. If genetic variances are equal in several populations, the best one to use is that with the higher mean.

A simplified example illustrating the objectives of recurrent selection is shown in Fig. 1.3. In this idealized example which assumes finite population size, the expected variabilities among single

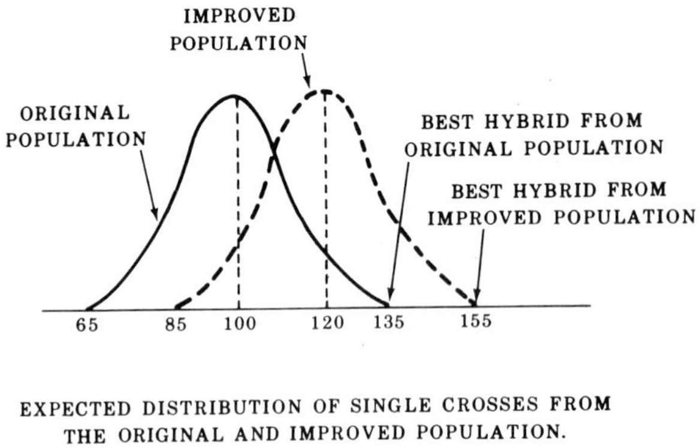

Fig. 1.3. Expected distribution of single crosses from the original and improved populations. (Eberhart, 1970a.)

crosses from the original and improved populations are the same, but the higher mean of hybrids and the best hybrid occurred in the improved population. Unless genetic variance decreases, improvement in the hybrids will parallel the improvement in the populations.

The results illustrated in Fig. 1.3 would not occur in one cycle of recurrent selection for most traits, but they illustrate the potential of recurrent selection after several cycles of selection and recombination. Recurrent selection methods, therefore, are designed to (1) improve the mean of the populations by increasing the frequencies of desirable genes for the traits under selection; and (2) maintain genetic variability to permit continued improvement and opportunity for selection of superior genotypes in any cycle.

Which of the several recurrent selection methods that is used for a given situation depends upon the mode of reproduction for the species, the primary type of gene action involved in the inheritance of the trait under selection, and the type of cultivar required for commercial production. Except for phenotypic (mass) selection, all recurrent selection methods involve three operations: (1) development of progenies, (2) evaluation of progenies, and (3) recombination of the superior progenies to synthesize the next cycle population. Variables involved in these three operations are: type of progeny structure, extent of progeny evaluation, selection intensity, and method of recombination. Adequate planning is essential for each operation because the rate of genetic gain from recurrent selection depends upon decisions made for each of the three operations (1), (2), and (3). Factors affecting genetic gain (Δ_G) are determined by the variables in the formula that predicts genetic gain (Eberhart, 1970a):

$$\Delta_G = kc\sigma_{g'}^2/(y\sigma_\rho)$$

where k is the standardized selection differential, c is a function of the parental control, $\sigma_{g'}^2$ is the additive genetic variance among progenies, y is the number of years required per cycle of recurrent selection, and σ_ρ is the phenotypic standard deviation among progenies. The factors in this formula can be adjusted to give more or less genetic gain, that is, k is a function of the proportion of evaluated progenies that are selected; c depends on the recombination unit; $\sigma_{g'}^2$ depends on the types of progenies evaluated; y depends on the years required to complete a cycle of selection; and σ_ρ is influenced by experimental precision. If possible, choices for these factors should be made before a recurrent selection program is begun. But, in an ongoing program, genetic gain can be increased by increasing parental control, reducing number of years per cycle, and improving the quality of evaluation trials. Increases in selection intensity, parental control, and additive genetic variance (assuming years per cycle and the phenotypic variance do not change) will increase the rate of genetic gain. A decrease in the number of years per cycle and in phenotypic variance (assuming the other factors are constant) also will increase the rate of genetic gain. Genetic gain predicted from this formula is expressed as gain per year, but often, genetic gain is expressed on a per cycle basis, which is valid, but to compare the relative efficiencies of different recurrent selection methods, it is preferable to express genetic gain on a per year basis because the various methods require different numbers of years or seasons per cycle.

Maize

The various recurrent selection schemes that have been conducted using maize are listed in Table 1.2. Maize is amenable to test different recurrent selection schemes because both cross and self-pollinations can be made easily in this species. Probably, the questions raised by Hull (1945) relative to overdominant genetic effects

Table 1.2. Different recurrent selection schemes for the cyclical improvement of breeding populations. (Adapted from Eberhart, 1970a.)

Selection Method	Years (y) per Cycle	Parental Control (c)	σ^2_G: σ^2_A,	σ^2_D
Phenotypic (mass)[a] with recombination				
One sex	1	1/2	1	1
Both sexes	1	1	1	1
Ear-to-row				
One sex	1	1/2	1/4[b]	0
Both sexes	2	1	1/4[b]	0
Half-sib				
Selfs recombined	2	2	1/4	0
Half-sibs recombined	2	1	1/4	0
Full-sib	2	1	1/2	1/4
S_1	2	1	1[a]	1/4[c]
S_2	3	1	3/2[a]	3/16[c]

[a]Not equal to σ^2_A unless p = q = 0.5 and dominance decreases to zero with inbreeding.
[b]If within plot selection is practiced, 3/8 σ^2_A, should be added.
[c]Coefficient difficult to define unless p = q = 0.5.

on yield of maize stimulated development of different methods of recurrent selection. If overdominant effects were important, recurrent selection that emphasized selection for specific combining ability would be most appropriate (Hull, 1945). After it became obvious that additive genetic variance was of major importance (Table 1.1), mass (Gardner, 1961) and S_1 or S_2 selection, which emphasized selection for additive gene effects, seemed appropriate methods for improving breeding populations. Reciprocal recurrent selection was designed to select for both types of gene action (Comstock et al., 1949).

Sprague and Eberhart (1977) summarized the observed genetic gains per cycle of selection for recurrent selection studies conducted to improve maize yield (Tables 1.3 and 1.4). Average genetic gain per cycle was about 3 to 5 percent for all recurrent selection methods and for intra- and interpopulation selection. The average gain of 2.9 percent for reciprocal recurrent selection includes both populations themselves and population crosses; average gain was 1.8 percent for populations themselves and 5.1 percent for population

Table 1.3. Average yield gains per cycle with different intrapopulation recurrent selection schemes in maize. (Adapted from Sprague and Eberhart, 1977.)

Selection Method	No. of Reports	No. of Cycles	Av. Gain per Cycle	Range in Gain per Cycle
			%	%
Mass	6	6.2	3.4	1.4 - 11.1
Ear-to-row	6	4.7	3.8	2.2 - 5.3
Full-sib	5	6.8	3.1	2.5 - 4.0
S_1	6	3.3	4.6	1.1 - 6.9
S_2	2	5.0	2.0	1.9 - 2.2
Testcross	4	4.6	2.8	−0.7 - 7.3
Average	4.8	4.9	3.5	1.4 - 6.1

crosses. Most evidence suggests that genes with additive effects and partial dominance are of primary importance for maize yield. The fact that all recurrent selection methods seem equally effective for improving maize populations for yield, indicates that selection is primarily for additive genetic effects.

Darrah et al. (1978) reported results from several recurrent selection methods in Kenya for 'Kitale Synthetic II' (KII), 'Ecuador 573' (Ec 573), and the syn-3 population, 'KCA', from the cross

Table 1.4. Average yield gains per cycle with different interpopulation recurrent selection schemes in maize. (Adapted from Sprague and Eberhart, 1977.)

Selection Scheme	No. of Reports	No. of Cycles	Av. Gain per Cycle	Range in Gain per Cycle
			%	%
Reciprocal				
Half-sib	3	4.7	5.1	3.5 - 7.4
Half-sib[a]	1	3.0	6.0	...
Half-sib[b]	1	3.0	3.5	...
Full-sib[c]	1	3.0	5.8	...
Testcross				
Broad-base tester	8	5.0	2.6	1.4 - 3.8
Narrow-base tester	7	4.8	4.5	1.8 - 7.4
Average	8.0	4.8	3.3	1.2 - 6.2

[a]Not included by Sprague and Eberhart (1977); see Gevers (1975).

[b]Not included by Sprague and Eberhart (1977); see Paterniani and Vencovsky (1978).

[c]Not included by Sprague and Eberhart (1977); see Hallauer (1978).

KII x Ec 573. Because the same populations were used for the different methods of recurrent selection, the rates of genetic gain could be compared directly. Three growing seasons could be obtained in two years in Kenya. One cycle per year was used for mass and ear-to-row selection, whereas two years were needed to complete one cycle in the other methods of selection. Rates of gain per cycle (Table 1.5) for the different selection methods were similar to average rates from other studies (Tables 1.3 and 1.4) except for reciprocal recurrent selection. Reciprocal recurrent selection in the KII x Ec 573 population cross had a genetic gain of 7 percent per cycle; this compares with the 3.5 percent for the 'Jarvis' x 'Indian Chief' and 4.5 percent for 'BSSS' x 'BSCB1' population crosses (Sprague and Eberhart, 1977). Gevers (1975) reported a gain of 6 percent per cycle for the 'Teko Yellow' x 'Natal Yellow Horsetooth' population cross in South Africa. Paterniani and Vencovsky (1978) reported a gain of 3.5 percent per cycle for three cycles of reciprocal recurrent selection in the 'Dent Composite' x 'Flint Composite' population cross in Brazil. Actual gains were less than predicted gains in most instances.

Table 1.5. Response to selection for yield in a breeding methods study in Kenya, East Africa. (Darrah et al., 1978.)

Selection Method	Gain per Cycle	
	q/ha	%
Mass		
KCA-M9C6	0.38	0.8
M10C6	0.93**[a]	1.6
M17C6	-0.70	...
Ear-to-row		
KII E1C6	0.83**	1.8
Ec573 E2C6	2.59**	6.6
KCA-E3C6	0.98**	2.1
E4C6	0.82**	1.4
E5C6	1.04**	2.2
E6C6	0.56*	1.2
E7C6	1.40**	2.5
Half-sib[b]		
KCA-H14C3	1.26	2.0
HL15C3	0.62	1.0
HI16C3	-3.26**	...
S_1^c: KCAC3	-0.50	...
Full-sib: KCAC3	1.60*	2.4
Reciprocal		
KIIC3	-0.04	...
Ec573C3	1.94**	2.5
(KIIxEc573)C3	4.18**	7.0

[a]* and ** indicate significance at P = 0.05 and 0.01, respectively, in tables throughout this book.

[b]Population by tester crosses and not populations themselves.

[c]Based on bulk S_1 lines rather than random mated populations.

Recurrent selection procedures also have been useful in the improvement of other traits of maize. Penny et al. (1967) reported on three cycles of recurrent selection for resistance to first-brood European corn borer (*Ostrinia nubilalis*, Hubner). Progress from selection was measured from the original (C0) and last cycle (C3) of selection for five populations. Based on a rating scale of 1 (resistant) to 9 (susceptible), recurrent selection improved the mean rating for the five populations from 5.5 in C0 to 2.5 in C3 (Fig. 1.4). Jinahyon and Russell (1969) used recurrent selection to develop resistance to

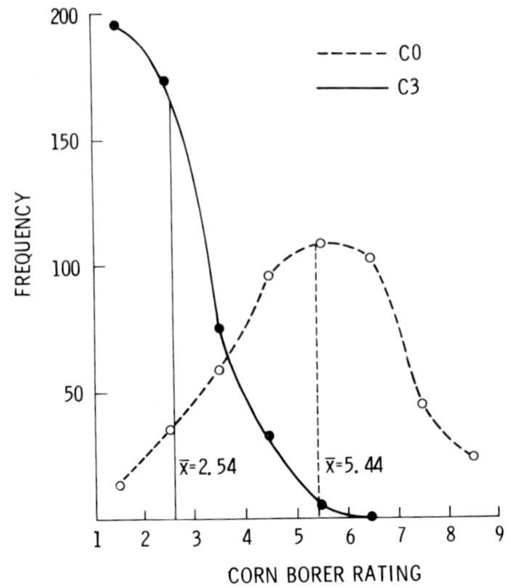

Fig. 1.4. Distribution of S_1 progenies for Eruopean corn borer resistance for the original (C0) and after three cycles (C3) of recurrent selection in maize. (Penny et al., 1967.)

Diplodia zeae in the open-pollinated cultivar, 'Lancaster Surecrop.' Average stalk rot ratings were 4.1 in C0 and 2.4 in C3 (Fig. 1.5). An important feature was the change in the distribution from the C0 to the C3 S_1 progenies. Few C3 S_1 progenies had a rating above 3 and few C0 S_1 progenies had a rating less than 3. Resistance to corn borer and *Diplodia zeae* was improved more rapidly than yield has been increased (Tables 1.3, 1.4, and 1.5) because the first two traits have higher heritabilities and good artificial infestation (corn borer) and infection (stalk rot) techniques that reduced the effects of environment.

Additional evidence for the value of recurrent selection to applied maize breeding programs is given by Harris et al. (1972), Suwantaradon and Eberhart (1974), Russell and Eberhart (1975), Moll et al. (1977), and Hallauer (1978). Each study reports upon hybrids among lines extracted from the original and improved populations. Harris et al. (1972), who practiced nine cycles of mass selection for yield improvement in the open-pollinated cultivar, 'Hays Golden,' reported that S_1 lines from the C9 population were significantly superior to S_1 lines from the C0 both when evaluations were based on S_1 lines themselves and testcrosses. There was evi-

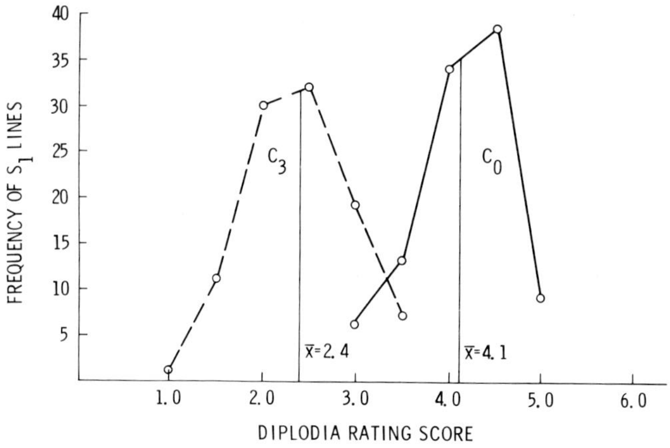

Fig. 1.5. Distribution of S_1 progenies for stalk-rot resistance for the original (C0) and after three cycles (C3) of recurrent selection in maize. (Jinahyon and Russell, 1969).

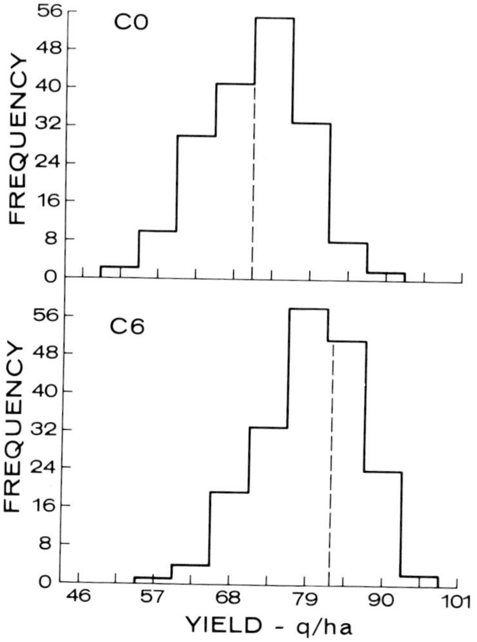

Fig. 1.6. Distribution of single crosses for yield for the original (C0) populations and after six cycles (C6) of reciprocal recurrent selection in maize. (Moll et al., 1977.)

dence that nine cycles of mass selection had reduced genetic variability for yield and combining ability.

Three studies have been reported on the effectiveness of reciprocal recurrent selection for improving yield of the crosses between two populations. Russell and Eberhart (1975) tested the population crosses and the crosses of five superior S_2 selections from 'BSSS' and 'BSCB1' after five cycles of reciprocal recurrent selection. Average yield of the S_2 line crosses exceeded that of the population crosses by 35 percent. Moll et al. (1977) examined the frequency distributions of single crosses derived from the C0 and C6 after reciprocal recurrent selection with Jarvis and Indian Chief maize cultivars (Fig. 1.6). Yield of the single crosses from the C6

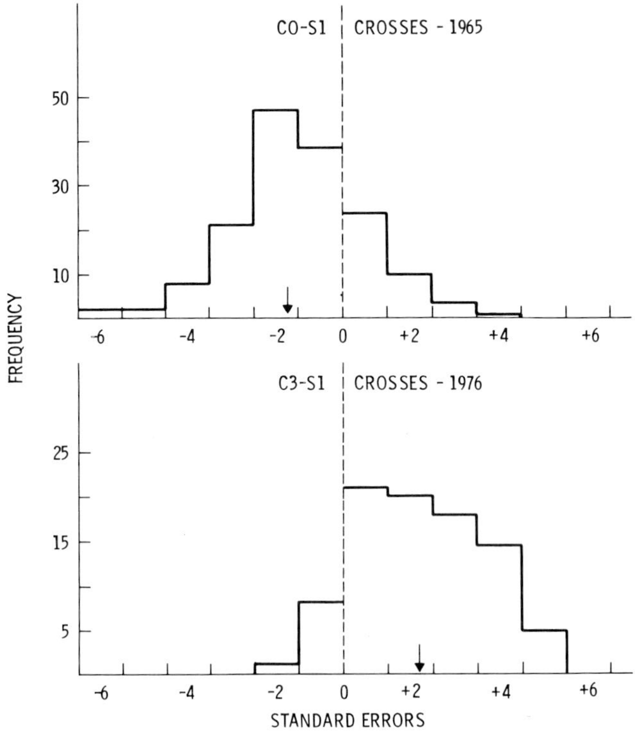

Fig. 1.7. Distribution of S_1 full-sib crosses for yield for the original (C0) and after three cycles (C3) of reciprocal full-sib selection in maize. (Hallauer, 1978). Vertical line is mean of six common check hybrids, and arrow mean of full-sib crosses.

populations averaged 12.5 percent greater than those from the C0 populations, but the distributions for the two sets of single crosses were similar. The 10 best yielding single crosses from the C6 were 8.6 percent greater in yield than the 10 best yielding single crosses from the C0. Hallauer (1978) summarized the yield distributions of the S_1 full-sib crosses for the C0 and C3 populations from reciprocal full-sib selection in 'BS10' and 'BS11' (Fig. 1.7). The means of the S_1 crosses were changed significantly from two standard errors below for the C0 to two standard errors above the mean of the checks for C3. The distributions of the C0 and C3 S_1 full-sib crosses were similar.

Self-Pollinating Species

Recurrent selection techniques have had limited testing with autogamous crop species because of the difficulties of producing enough seed for testing and necessary intercrossings in each cycle. The principles of recurrent selection are equally valid for autogamous species, but the technical problems of intermating and limited seed supplies with these species have discouraged use of recurrent selection. Some recent suggestions, such as using genetic male sterility to facilitate crossing, are causing greater interest and use of recurrent selection in the breeding programs for autogamous species. Gilmore (1964) outlined procedures for using reciprocal recurrent selection with autogamous species in which male-sterile plants (either genetic or cytoplasmic) of crops such as sorghum, barley (*Hordeum vulgare* L.), and wheat (*Triticum aestivum* (Vill., Host) MacKey) are wind pollinated; each of these species has the potential of producing hybrids for agricultural production. Compton (1968) discussed how the obstacle of inter-pollinations can be reduced in recurrent selection programs by use of the single-seed descent. Brim and Stuber (1973) outlined methods for conducting recurrent selection with soybeans (*Glycine max* (L.) Merr.) through the use of genetic male sterility to facilitate intercrossing. Cyclical selection methods suggested for use with autogamous species include: (1) phenotypic (mass) recurrent selection with and without recombination between cycles of selection; (2) progeny evaluation with or without the use of male sterility; (3) the diallel selective mating system for broadening the germplasm of breeding programs; (4) S_1 and half-sib progeny recurrent selection with or without the use of male sterility; and (5) use of the single-seed descent with cyclical selection procedures. Suggestions and modifications of recurrent selection methods for use with autogamous species generally are aimed at broadening the

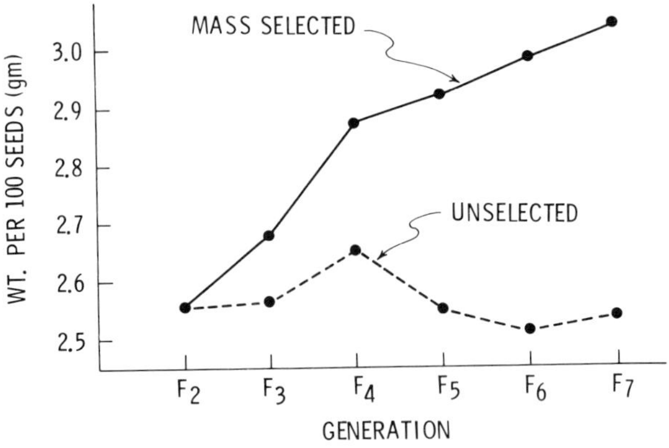

Fig. 1.8. Response of 100-seed weight with mass selection for seed width in oats. (Frey, 1967.)

genetic base of populations that are intended for extraction of improved cultivars.

Studies illustrating the effects of mass selection without recombination were reported by Frey (1967) and Romero and Frey (1966) in oats (*Avena sativa* L.) and by Fehr and Weber (1968) in soybeans. Frey (1967) conducted mass selection for seed width and measured the indirect effects of mass selection on 100-seed weight in a composite population originated by mixing 5-g samples of F_2 seeds from 250 oat crosses. Mass selection for seed width increased mean 100-seed weight 9 percent over five cycles (Fig. 1.8). Because heritability (35 percent) of seed width and the genotypic correlation ($r = 0.69$) between seed width and seed weight were both relatively high, indirect selection was effective for increasing 100-seed weight. Other correlated effects of mass selection for seed width were later heading dates and taller plants. In the same F_2 composite, Romero and Frey (1966) used mass selection effectively to reduce plant height (Fig. 1.9), but the mass selected populations were earlier and greater in grain yield also. Fehr and Weber (1968) found that mass selection for seed size and specific gravity in soybeans caused changes in protein and oil composition of the seeds, and maximum progress for high protein and low oil was obtained by mass selecting for large seed and low specific gravity. There was no intercrossing of selections between cycles of selection for these studies; hence genotypes that resulted from mass selection were already present in the original population or segregates from those genotypes.

Matzinger et al. (1977) studied the effects from phenotypic recurrent selection on a broad genetic-base population of tobacco

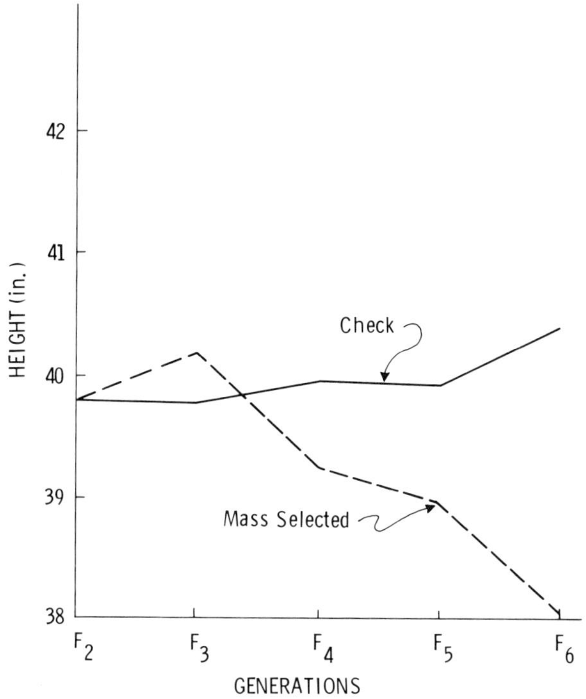

Fig. 1.9. Effects of mass selection for decreased plant height in oats. (Romero and Frey, 1966.)

(*Nicotiana tabacum* L.) (an autogamous species). Three independent studies with selection for decreased plant height, increased number of leaves, and an index for increased leaves on shorter plants were conducted. Selections were made before flowering, and the intermatings were made among selected individuals (parental control equal 1 in Table 1.2) to provide seed for the next cycle. Responses for direct and index selection were significant in all instances (Table 1.6). Response to direct selection was linear over the five cycles of selection for both traits, and there was little evidence that the genetic variability had decreased with selection. They concluded that phenotypic recurrent selection would be a useful breeding method for autogamous species because the intercrossing released genetic variability that would not be released via selection only.

The diallel selective mating system (DSM) outlined by Jensen (1970) was developed specifically for autogamous crop species that are difficult to cross and have few seeds per cross (Fig. 1.10). The DSM system was designed to broaden the genetic base of breeding

Table 1.6. Comparative selection response in plant height and number of leaves following direct and index selection (Matzinger et al., 1977.)

Selection Criteria	Plant Height		Number of Leaves	
	cm/cycle	% of CO	no./cycle	% of CO
		%		%
Short plants	-6.52**	-4.9	-0.46**	-2.5
Increased leaves	4.74**	3.6	1.28**	7.0
Index	-2.00*	-1.5	0.54**	2.9

populations and to supplement conventional breeding methods. It is a dynamic breeding method that permits screening of germplasm at different stages of development, intercrossing of elite germplasm, and infusing germplasm from outside the original populations. Conventional small grain breeding methods are used in conjunction with DSM. Jensen (1970) proposed establishing several breeding populations, each with a rather narrow objective. After the objective for a proposed breeding population is determined, the breeder selects the parental lines and crosses them in a diallel series (column 1, Fig. 1.10). The F_1's of the diallel series are used (1) to produce an F_2 population, and (2) for crossing to produce multiple-parent or

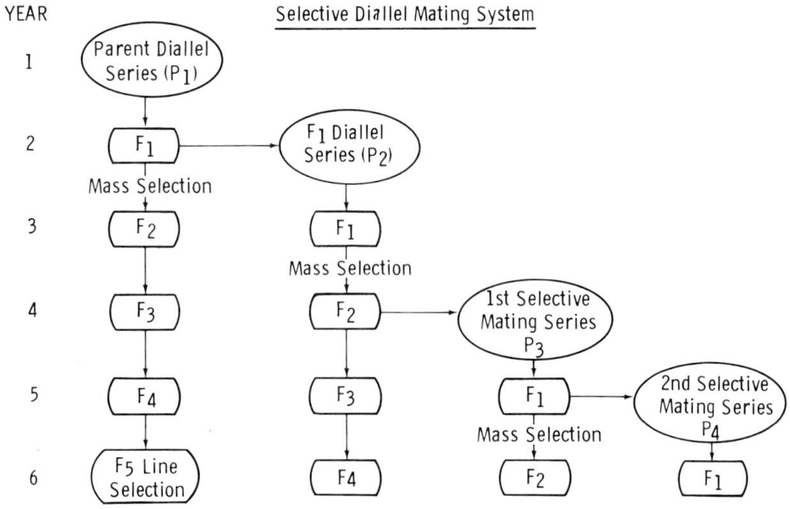

Fig. 1.10. Selective diallel mating system for self-fertilizing species. (Jensen, 1970.)

convergent crosses (column 2, Fig. 1.10). The F_1's from the multiple crosses are selfed to produce an F_2 population, and mass selection is practiced among the F_2 seeds to provide for the first selective series (column 3, Fig. 1.10); the remaining mass-selected F_2 seeds are selfed to produce a bulk F_3 generation. F_1's produced among the mass-selected F_2 plants (column 3, Fig. 1.10) are treated differently from the diallel F_1's from the single- and multiple-parent crosses, that is, some are selfed to produce a bulk F_2, some are selected for intermating to form the second selective series (column 4, Fig. 1.10), and others are crossed with parents not included in the original diallel.

The DSM system is unique because it permits the use of conventional breeding methods, but it also permits opportunities for additional recombination by intercrossing selected genotypes. DSM provides for: (1) conventional bulk-population breeding for the biparental diallel crosses (column 1, Fig. 1.10), (2) mass selection in each population series (all columns, Fig. 1.10), (3) recombination of selected genotypes (column 2, Fig. 1.10), (4) extraction of new cultivars at any stage, and (5) introgression of new germplasm into the breeding populations at any time. DSM seems formidable because of the large number of crosses required, but Jensen (1970) suggested two techniques that enhance its use: (1) using male sterility to facilitate crossing, and (2) growing the breeding populations in specialized environments to maximize genotypic expression of the traits under selection. Jensen (1978) reported that the DSM system was effective for developing new small grain cultivars.

Genetic improvement in soybeans generally has been restricted to pedigree and backcross breeding, but Brim (1978) and Burton and Brim (1978) also have included S_1 and half-sib recurrent selection in soybean breeding populations. To test the effectiveness of recurrent selection for population improvement of an autogamous species, Brim (1978) chose the trait percentage of seed protein: its heritability is high, and soybeans are a source of high-quality protein. Two populations were used: (1) population I was derived from a cross of two adapted lines, one high in seed protein percentage and the second high in seed oil percentage; and (2) population II was developed from a backcross of an adapted line to nine plant introductions. Response to selection was evaluated after six cycles of selection in population I and five cycles in population II. Percentage of seed protein was increased in population I at a linear rate of 0.33 percent per cycle, increasing from 46.3 to 48.4 percent. In population II, percentage of seed protein was changed 0.67 percent per cycle, increasing from 42.8 to 46.3 percent. The different rates of response in the two populations were attributed to the use of different selection intensities and to different levels of percentage of seed protein in the original populations. Seed oil percentage decreased, as expected, in both populations, whereas yield increased in population I but decreased in population II. Three cycles of recurrent selection caused a 16 percent increase in yield in population II, and the C3 composite yielded 20 percent more than the check cultivar. These advances expressed as percentages of the C0's are summarized in Table 1.7. Six cycles of recurrent selection in soybeans for percentage of seed protein in population I increased seed protein 5 percent, grain yield 9 percent, and protein yield 13 percent. In population II five cycles of selection for seed protein increased seed protein 8 percent and decreased yield 11 percent and protein yield 5 percent. Recurrent selection was successful for increasing the trait under selection in an autogamous species, in this case, soybeans.

Burton and Brim (1978) used three cycles of recurrent mass and half-sib family selection to increase seed oil percentage in a soybean population that was segregating for male sterility which facilitated intercrossing of selected individuals and the production of half-sib families. In composite populations, percentage of seed oil increased linearly from 18.4 to 19.5 percent at an average rate of 0.3 percent

Table 1.7. Summary of results from recurrent selection in soybeans as a percentage of the base population. (Brim, 1978.)

	Population		
		II[a]	
Trait	I[a]	A	B
Yield	109	116[b]	89
Protein, percent	105[b]	99	108[b]
Protein, total	113	115	95

[a] Population I was synthesized from a cross of two highly adapted experimental lines. Population II was synthesized from a backcross of a highly adapted experimental line to nine plant introductions. Six cycles of recurrent selection were completed for percentage of protein in population I. Five cycles of recurrent selection were completed for percentage of protein in population II-B and three cycles of recurrent selection for yield in population II-A.

[b] Selection criteria.

per cycle. They concluded that both mass and within-family selection contributed to the observed gain.

Single-seed descent is a method of sampling frequently used in autogamous species to assure that the range of genotypes in the original population also will be present in the future generations (Brim, 1966). The features of single-seed descent are: (1) to maintain the broadest possible representation of genotypes in the base population until selection is practiced in some inbred generation; and (2) to increase genetic variation among progenies as a result of inbreeding, namely, F_4 to F_6. Single-seed descent can be used as a supplement to cyclical selection in autogamous species.

Single-seed descent is being used by Fehr and Ortiz (1975) in conducting S_1 and S_4 recurrent selection for yield in soybeans. Yield evaluations are conducted in Iowa, and a winter nursery in Puerto Rico is used to advance the generations to the desired level of inbreeding and for recombination. Hence, one cycle of S_1 recurrent selection is completed in one year, and one cycle of S_4 recurrent selection is completed in two years. No actual gains from using this method are available, but Fehr and Ortiz (1975) did present expected genetic gain for the broad genetic-base soybean population, 'AP6CO' (Table 1.8). Expected genetic gain was greatest for S_4 selection on a per-cycle basis, but gain per year was greatest for S_1 selection because the other methods required two years to complete a cycle. Expected gain was least for half-sib selection because the genetic variability was less than for the other methods. Brown (C. M. Brown, pers. commun., Univ. of Ill.) has used single-seed descent successfully in his oat breeding program. He can

Table 1.8. Expected genetic gain for four methods of recurrent selection for yield in soybeans. (Fehr and Ortiz, 1975.)

Method of Selection	Years/ Cycle	Expected Genetic Gain (g/plot): Per cycle, 5%[a]	Per cycle, 10%[a]	Per year, 5%[a]	Per year, 10%[a]
S_1	1	...	54	...	54
S_1(ST)[b]	2	64	54	32	27
S_4	2	96	82	48	41
Half-sib	2	24	20	12	10

[a]Selection intensities.

[b]S_1 test with genetic male sterile.

complete two to three generations in winter greenhouses and have inbred progenies with sufficient seed for testing in two years. The potential for using the single-seed descent concept with methods of recurrent selection seems good, especially if greenhouse and winter nurseries are used for rapid generation advance. The methods proposed for soybeans would seem to be applicable for any autogamous species.

Forages

Recurrent selection methods have been used effectively to improve traits in forage crops. Past selection pressures, breeding systems, and modes of reproduction have contributed to the wealth of variability within the forage species of legumes and grasses, and breeders have been able to exploit this variability for the improvement of breeding populations and cultivars. The first evaluation of recurrent selection in forages was reported by Johnson (1952) and Johnson and Goforth (1953), who compared the response of S_1 progeny and mass selection for the improvement of forage yield in sweetclover (*Melilotis officinalis*) cultivar 'Madrid.' Both methods of selection rapidly increased forage yield per plant. Relative to Madrid, two cycles of S_1 progeny selection increased yield 52 percent, and four cycles of mass selection increased yield 11 percent.

An extensive mass selection program has been conducted in two broad genetic-base alfalfa populations (A and B) for pest resistance. Populations A and B were established in 1950 by intercrossing vigorous plants from three- and four-year-old broadcast stands (Hanson et al., 1972). In each cycle of selection, 1,800 to 10,000 plants were evaluated in each population, and from 80 to 500 plants were selected to propagate the next cycle for all traits except for spotted alfalfa aphid (*Therioaphis maculata*) resistance. Eighteen cycles of selection have been reported for resistance; resistant plants selected for intercrossing always were vigorous. Mass selection was conducted by evaluating seedlings and crossing among resistant plants, so parental control (c) would be equal to one (Table 1.2). Response to selection was realized in all instances, that is, 11 cycles of selection for resistance to leaf hopper were completed, and when expressed as a percentage of damage to 'Cherokee,' the values in the 11th cycle were 70 percent compared to 135 percent in C0 and seemingly additional improvement could be made (Fig. 1.11). Hill et al. (1969) evaluated the selected populations for leafhopper (*Empoasca fabae*) yellowing, spotted aphid damage, common leafspot (*Pseudopeziza medicaginis*) and bacterial

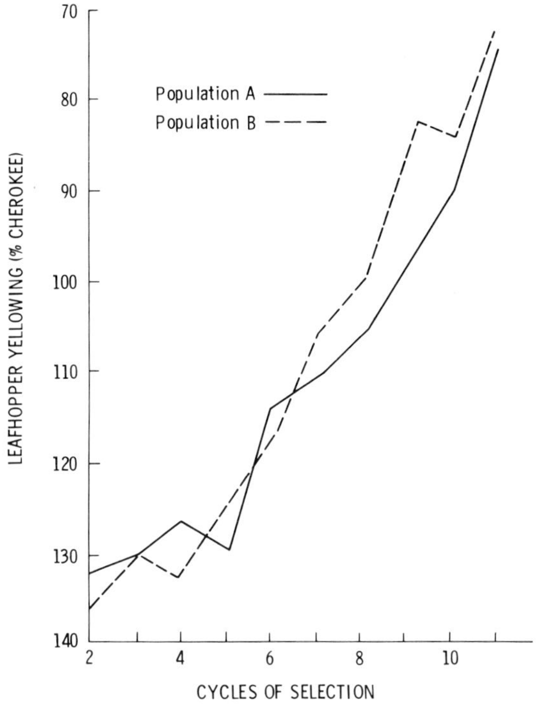

Fig. 1.11. Response of mass selection for leaf hopper yellowing in two populations of alfalfa. (Hanson et al., 1972.)

wilt (*Corynebacterium insidiosum*) infections, and for yield and other agronomic traits after 14 cycles of selection (Table 1.9). There were no consistent correlated changes for unselected traits. Changes in yield per plant were not consistent among the selected populations, and no important changes were noted for spring growth and fall height. Persistence did not change in population A, but it did increase with selection in population B.

According to Devine and Hanson (1973), three cycles of recurrent selection for anthracnose (*Colletotrichum trifolii* Bain) resistance in three alfalfa populations increased the level of resistance from 4.2 in the original to 1.7 after selection (1 equals highly resistant and 5 equals dead plants). More importantly, the stands for the selected populations were 85 to 95 percent alfalfa, whereas in stands of the original populations, 50 percent were alfalfa plants and 50 percent were weeds.

Recurrent restricted phenotypic selection was used by Burton

Table 1.9. **Agronomic traits and pest damage scores of alfalfa populations mass selected for pest resistance in populations A and B. (Hill et al., 1969.)**

	Leaf					Spring	Fall		Leaf Spot	
Population	Hopper	Spot	Wilt	Rust	Yield	Growth	Height	Persistance	Lepto.	Asco.
					g	*cm*	*cm*	%		
A8[a]	6.2	3.3	3.7	2.7	114	25.6	38.0	75	2.6	3.8
A-A3	6.4	3.5	3.0	2.4	91	24.2	37.3	80	3.0	4.2
A-L4	5.1[b]	3.4	3.3	1.8[b]	129	26.7	36.5	79	3.2	3.9
A-W4	5.4	3.9	2.3[b]	2.0	119	25.5	34.7	78	2.8	4.0
A-C4	5.3	1.4[b]	3.2	2.6	118	26.4	38.4	77	2.9	4.3
B8	6.0	2.8	3.3	2.3	106	25.2	38.4	58	2.7	4.0
B-A3	6.3	3.0	3.1	2.7	94	24.5	37.2	79	2.7	3.9
B-L4	4.8[b]	3.2	3.5	1.7[b]	127	26.2	39.8	75	3.5	3.7
B-W4	4.9	3.0	2.5[b]	1.6	110	24.8	38.6	73	2.7	3.7
B-C4	5.0	1.5[b]	4.0	1.8	110	24.5	34.6	67	3.4	4.0

[a]Cycle of selection for each pest.

[b]Population selected for the particular pest.

(1974) to increase the mean of the population under selection and the frequency of the superior plants in the advanced populations of Pensacola bahiagrass (*Paspalum notatum* var. *saure*). The mass-selection scheme was modified because selected plants were inter-crossed, so parental control (c) was increased from 0.5 to 1 (Table 1.2). As an offsetting factor, however, the method required two years per cycle. Winter or greenhouse nurseries could be used if winter injury and killing were ignored, but winter survival is an important trait, so to include selections for it, he extended cycle time to two years. Burton (1974) completed four cycles of recurrent restricted phenotypic selection in a narrow-gene population (NGP) formed by crossing F_1 plants to 'Tifhi-2,' and a wide-gene population (WGP) formed by compositing seed harvested from 39 south Georgia farms. The original and fourth cycle populations were evaluated under different cultural methods and the results are shown in Table 1.10. In the 1971 transplanted seedling test, the forage dry-matter yields of NGP-M4 exceeded those of the original population (NGP-M0) by 8.2 percent, and WGP-M4 yields exceeded the original population (WGP-M0) yields by 26.8 percent. Similar rates of improvement (9.2 and 24 percent, respectively) were obtained from the broadcast-seeded plots. Selection also increased the frequency of greater yielding plants, as shown in the last three columns of Table 1.10. Gain per cycle for forage dry-matter was about 2 percent in NGP and about 6 percent in WGP. There was no change in genetic variability in NGP, but in WGP it decreased 25 percent.

In summary, cyclical selection has been used extensively with forage grasses and legumes, and improvements have occurred in

Table 1.10. Effectiveness of recurrent restricted phenotypic selection for the improvement of forage dry-matter yields in two Pensacola bahia-grass populations. (Adapted from Burton, 1974.)

	Transplanted Seedlings 1971		Broadcast Seeding Av. 3 rates, 1973		Spaced Planted, 1973 Plants over 100 g		
Population[a]	T/ha	% original	T/ha	% original	no.	mean	highest
NGP-M0	1.39	100.0	0.57	100.0	13	117	143
NGP-M4	1.51	108.2	0.62	109.2	22	134	178
WGP-M0	1.16	100.0	0.50	100.0	4	123	160
WGP-M4	1.47	126.8	0.62	124.0	14	129	178

[a]NGP is narrow-genetic base population and WGP is wide-genetic base population.

nearly all instances. Phenotypic recurrent (mass) selection seems to be effective for most traits, especially when genetic control can be improved by vegetative propagation of the parents used for intercrossing. Mass selection for pest resistance (Hanson et al., 1972; Dudley et al., 1963, for alfalfa), forage dry-matter (Burton, 1974, for Pensacola bahiagrass), survival (Hayward, 1970, for perennial ryegrass, *Lolium perenne*), seed weight (Trupp and Carlson, 1971, for smooth bromegrass, *Bromus inermis*), and specific leaf weight (Topark-Ngarm et al., 1977, for reed canarygrass) has been successful.

Sorghum

Because of the wide diversity in germplasm within the sorghum species, the World Collection of sorghum was formed by assembling germplasm acquired from breeders throughout the world (Miller, 1968). Much of the germplasm could not be used by USA sorghum breeders who required temperate-zone types, so conversion programs were undertaken to make the material adapted to culture in the midwestern USA and other areas of the world. A conversion program entailed establishing appropriate maturity and plant height alleles into the genetic backgrounds of exotic sorghum lines, so these could be evaluated in environments where USA breeders wished to use them (Stephens et al., 1967). The success of the conversion programs was reported by Eberhart (1970b) and Webster (1975) for the material converted and released to breeders. Recurrent selection methods have been imposed in some populations for disease and insect resistance and yield improvement (Eberhart, 1970b).

Recurrent selection has only recently been initiated in sorghum populations, so data on its effectiveness are limited. Most recurrent selection methods with sorghum make use of a cytoplasmic-sterile-restorer system or the male-sterile gene, ms_3 to facilitate intercrossing. Gilmore (1964) outlined a reciprocal recurrent selection program for sorghum improvement, Maunder (1969) initiated a recurrent selection program, and Ross (1973) planned one, but generally, reciprocal recurrent selection is not used in many sorghum breeding programs. Doggett (1968, 1972), Doggett and Eberhart (1968), and Eberhart (1972) suggested recurrent selection methods that emphasized selection for additive genetic effects (mass, head-to-row, and S_1 selection) for sorghum.

Ross (1973) stated that the recurrent selection programs with grain sorghum populations at Nebraska were mass selection for yield, grain protein percentage, and drought tolerance; and family selection for yield, morphological traits, cold tolerance, and resistance to greenbug (*Schizaphis graminum*) and European corn borer. Most of these programs were in the early cycles of selection, but effective progress was being made for each selection method. S_1 selection seemed to be the most practical of the three family selection methods.

Horticultural Crops

Phenotypic recurrent selection has been used effectively in horticultural crops. Andrus and Bohn (1967) practiced mass selection for nine cycles in a heterozygous base population of cantaloupes (*Cucumis melo* L.) by use of an index representing 16 fruit characters. Parental control was 0.5. A unique feature of the experiment was that alternating selection was conducted at two locations in different seasons (high-stress environment at Charleston, S. C., in the spring and a low-stress environment at Brawley, Calif., in the fall); final evaluations were made in both environments after nine generations of mass selection. The nine mass-selected populations and nine randomly chosen inbred populations were evaluated to determine the response to selection, change in genetic variability with selection, and adaptiveness of the mass-selected populations. Mass selection resulted in improvement in most index-component traits without any evident loss of genetic variability. Phenotypic correlations among index-component traits across generations suggested that improved associations among traits paralleled individual trait improvement (Bohn and Andrus, 1969). Inbreds extracted from later cycles of mass selection performed better than those immediately following a cross, indicating that selection was effective for increasing the favorable-gene frequencies. Thus the populations improved by mass selection had greater potential, on inbreeding, to yield high-quality cultivars than did the original products of a cross. Data on the effects of alternating selection in stress and nonstress environments were not conclusive.

Jones et al. (1976) used six cycles of mass selection in sweet potato (*Ipomoea batatas*) to combine pest resistance with other desirable production and high market qualities. Because sweet potato is cross-pollinated by insects, selection was based only on the seed parents, so c = 0.5. Genetic advance was good for the traits selected, that is, selected roots with acceptable orange flesh color increased from 26 to 50 percent. The method of selection seemed successful because the plants in the sixth cycle of mass selection had higher frequencies of flowering and seed set, acceptable yield, orange flesh, and resistance to pests. The flexibility of mass selection was demonstrated by Jones et al. (1976), who introgressed other germplasm in cycles three to six and added new selection criteria as needed. They suggested mass selection in sweet potato would be more effective if a two-year selection cycle was used.

APPLICATION OF RECURRENT SELECTION

The primary goal of selection programs is to increase the frequency of desirable genotypes in the germplasm used in breeding programs. The choice of selection methods to attain this goal is nearly unlimited within the constraints of resources available in a particular breeding program. The selection method chosen should provide improved genotypes. The specific techniques used will differ among crop species because the reproductive system will determine the manner in which genetic variation is carried within a crop species. Most crop species have adequate genetic variability for reasonable progress to be made from selection. Plant breeders were successful in exploiting the variability within land-race cultivars, but after this source of variability was exploited, the pedigree, backcross, and other methods were developed to exploit variability of populations generated from controlled crosses of elite lines and cultivars. The conventional methods of plant breeding have been successful and are still important for plant improvement.

Recently, there has been increased interest in expanding the genetic variability available to breeders by infusing unadapted, exotic, and even weedy germplasm into breeding populations. According to Brown (1975), approximately 130 racial complexes make up the maize germplasm of the Western Hemisphere, but more than 90 percent of the breeding effort in USA breeding programs is devoted to germplasm whose origin traces to three of the 130 races; hence, USA maize improvement programs have largely ignored 98 percent of the maize germplasm. The example for maize may not be representative for all crop species because maize breeding methods have emphasized line development for use as parental stocks in hybrids. But Miller (1968) and the National Academy of Sciences (1972) have emphasized the restricted germplasm base of many crop species. Certainly, the increasing interest in the gene base of populations in many crop species indicates a concern that genetic variability is too limited in many instances.

Generally, infusion of unadapted or exotic germplasm is done by intercrossing it with adapted germplasm. Because new germplasm is intended to increase genetic variability, the breeder must grow large numbers of plants to enhance the opportunity to retain new genes and new gene combinations. One important part of introducing new germplasm to a breeding population is the selection for adaptiveness for the production environment. To prevent loss of potentially useful genes while selecting for adaptation, the breeder must use large populations with mild selection pressures. Recurrent selection is useful for this purpose because recombination on a regular basis is inherent in the method.

Different recurrent selection methods have been used with many crop species, but selection response was realized in nearly all instances. It seems that additive gene effects predominate and that all cyclical breeding methods select for them. Often, the selection method is chosen on factors other than its value in making good gains from selection. Rate of response to selection is determined by the trait under selection, mode of reproduction, experimental techniques, genetic variability, intensity of selection, and effective use of resources within the constraints of the breeding programs. Selection methods, therefore, should be chosen on the basis of how they contribute to the goals of a given breeding program.

Before recurrent selection methods can be accepted as an integral part of breeding programs, it is necessary to demonstrate that they are useful for developing new, improved cultivars. Recurrent selection insures continued genetic advance, but unless it is adaptable to integration with other selection methods currently used, it will not be accepted by the applied breeder. Many studies on recurrent selection were conducted to test a specific hypothesis. For example, a common concern in maize selection studies has been the type of tester to use for maximizing gain relative to thc types of genetic effects expressed in hybrids. Recent data indicate that response to selection is primarily for additive gene effects, regardless of the testers used (Tables 1.2 and 1.3). Therefore, the testers chosen should be based on the other goals of the breeding program.

Recurrent selection methods are not intended to be used alone in most breeding programs. For many crop species, the products of recurrent selection cannot be used directly for commercial production. Maize breeders, for example, need to exploit the heterosis in crosses between inbred lines. Populations improved by recurrent selection can serve only as source populations for the extraction of inbred lines. The primary goal for recurrent selection in maize breeding, therefore, is to develop populations that enhance the opportunity for extracting superior inbred lines for use in hybrids. Evidence by Andrus and Bohn (1967) and Russell and Eberhart (1975) shows this is an attainable goal. Jensen (1978) for oats, Brim (1978) and Fehr and Ortiz (1975) for soybeans, Andrus and Bohn

(1967) for cantaloupes, and Eberhart et al. (1967) for maize have emphasized that recurrent selection methods should supplement other breeding methods.

The phenotypic recurrent selection in alfalfa (Hanson et al., 1972) for pest resistance led to the release of the alfalfa cultivars 'Williams,' 'Cherokee,' 'Team,' and 'Arc' that were derived directly from recurrent selection (Busbice et al., 1972). The widely grown cantaloupe cultivar 'PMR45' was developed from a similar mass selection program (Andrus and Bohn, 1967). Jensen (1978) reported that 15 of the 21 small grain cultivars (oats, barley, and wheat) released in the past 25 years in New York were from the composite breeding program. The maize breeding program in Iowa has used recurrent selection extensively for the past 30 years. From these recurrent selection programs, maize inbred lines 'B14,' 'B37,' 'B73,' and 'B79' have been developed for use in commercial hybrids. 'B84,' extracted from a late cycle of selection, is a promising inbred line. In addition, the lines developed from recurrent selection have been used in pedigree and backcrossing programs, as evidenced by the 43 recovered strains of B14 and 21 recovered strains of B37 (Anonymous, 1976). Evidence of the benefits of recurrent selection also shows up in germplasm with greater pest resistance, improved adaptiveness from wide crosses, and the maintenance of genetic variability in breeding populations.

Nearly all recurrent selection methods can be integrated with other and more conventional methods used in a breeding program. Progenies selected for recombination to produce the next recurrent selection cycle are logical candidates to include in an applied breeding program. Of course, the selected progenies have had minimal testing; so further selection and testing are necessary before they are likely candidates for release. An example where recurrent selection has supplemented an applied breeding program is that of S_2 recurrent selection in maize. Selected S_0 plants are advanced to the S_1, and S_1 progenies are selected for pest resistance and agronomic traits and advanced to the S_2 when they are evaluated in replicated trials. Remnant seed of the superior progenies is planted for intermating to form the next cycle population for continued selection. Additionally, the S_2 progenies selected for recombination can be selected further in the breeding nursery. With continued selection, the progenies selected in successive cycles should have a better level of performance, as illustrated in Fig. 1.3.

Because of the nature of recurrent selection, maintenance of genetic variability in the breeding population is essential if continued genetic improvement is to be realized. Extensive data are not available for monitoring the changes in genetic variability in long-term plant selection experiments. Dudley (1977) summarized 76 generations of selection for high and low percentage of seed oil and protein in maize. He concluded that genetic variation was not exhausted for either trait in either direction of selection. Selection response for high percentage of protein and oil was 20 times the square root of the additive genetic variance (σ^2_A) in the base population 'Burr's White.' Dudley (1977) estimated that to have a response of $20\sigma^2_A$ would require 200 loci if p = q = 0.5 or between 50 and 100 loci if q is 0.25; the results, therefore, were well within the range of response expected for gene frequencies and number of loci accepted as operating on quantitatively inherited traits. Probably the number of segregating loci affecting yield is much greater than the number affecting the percentage of oil and protein, so there should be little concern about exhausting genetic variability for yield even in long-term breeding programs. Limited numbers of cycles of recurrent selection have been completed in the Iowa maize breeding program, but in none of six selection programs has there been a decrease in additive genetic variance with advanced cycles of selection (Table 1.11). Moll et al. (1977) had similar results from use of reciprocal recurrent selection in 'Jarvis' and 'Indian Chief' (Fig. 1.6).

Direct comparisons of the effectiveness of recurrent selection methods with that of conventional breeding methods generally are not available. Sprague (1952) compared the increase in oil percentage in maize grain from five years of pedigree selection (five generations) and two cycles of recurrent selection (Table 1.12). Recurrent selection was 1.3 to 3.0 times more effective than pedigree selection, depending upon the measure used. None of the comparisons, however, considered the residual genetic variability for percentage of oil with the recurrent selection method. Only by intercrossing the S_5 progenies from the pedigree selection would genetic variability be restored to the level expected in the recurrent selection population. Duvick (1977) also reported comparisons of recurrent and pedigree selection for rates of gain. Average rate of gain per year was 0.71 q/ha from recurrent selection and 0.68 q/ha from pedigree selection. Average cycle time was three years for recurrent selection and 13.3 years for pedigree selection. Duvick (1977) concluded that, because average rates of gain were the same, the methods of selection were equally effective. Comparisons of the rates of gain from recurrent selection and conventional methods of breeding, however, are not very meaningful for determining which method is preferable. The basic principles of the two categories of

Table 1.11. Estimates of additive genetic variance (g/plant) from six recurrent selection programs for yield improvement in maize in Iowa.

Cycles of	Half-sib Selection			S_1 Selection	Reciprocal Recurrent Selection	
Selection	BS13[a]	BS12	BSK	BSK	BSSS	BSCB1
C0	338	493	168	232	349	546
C1	110	417	341	255	291	443
C2	260	115	79	208	102	74
C3	...	420	94	259	110	178
C4	89	100	147	121	52	150
C5	270	517	260	211	346	289
C6	239	997	464	389	409	863
C7	258	...	454	255	352	545
C8	351	...	...	...	...	...
C9	218	...	...	...	...	...
Average	299	437	251	241	252	386

[a]S_2 recurrent selection conducted for the last three cycles.

Table 1.12. Increase in oil percentage in maize from five years of pedigree and recurrent selection. (Sprague, 1952.)

Generation	Pedigree	Cycle	Recurrent
S_1	5.0	C0	4.2
S_2	4.6	. . .	. . .
S_3	5.0	C1	5.2
S_4	5.2	. . .	. . .
S_5	5.6	C2	7.0
Gain per year	0.13	. . .	0.41

selection are quite different, and, to be useful, one should complement the other. For example, all the recurrent selection methods involve early testing, and progenies of the selected individuals may be placed in a pedigree breeding program to expedite the selection of a line for release. Also, pedigree selection can be considered in the broad context as one form of recurrent selection. Recurrent selection methods simply increase the frequency of favorable genes in broad gene-based populations and maintain genetic variability by intermating genotypes that possess favorable genes. Other selection methods are needed for refining traits and determining their productivity relative to check cultivars. Recurrent selection, therefore, is designed to supplement the other methods of selection used to develop improved cultivars and hybrids.

Recurrent selection contributes to the intermediate- and long-term goals of the breeding program (Fig. 1.1). Selection programs must be conducted continuously to provide viable sources of new genotypes, but because of the long-term nature of recurrent selection programs, they must be conducted as efficiently as possible. The parameter for determining the efficiency of selection is rate of genetic gain per year, and factors that determine the rate of genetic gain per year were given in the prediction equation. Breeders have more control over some of these factors than others. From Table 1.2, it was shown that the effectiveness of selection can be increased for mass and ear-to-row selection. But the increased gain from parental control can be offset if two years rather than one year are required to complete each cycle of selection. Development of small-plot planting and harvesting equipment, use of high-speed computers for handling large volumes of data quickly, and effective use of off-season nurseries have increased the efficiency of recurrent selection methods. These developments also have affected the other components of the prediction equation. The breeder can control the number of years required to complete each cycle, namely,

Table 1.13. Percentage of change in annual area, yield, and total production for eight crops in the United States for 1961-65 as compared with 1971-75.

Crop	Annual area (ha × 10^5)			Annual yield (q/ha)			Annual production (q × 10^6)		
	1961-65	1971-75	Change	1961-65	1971-75	Change	1961-65	1971-75	Change
			%			%			%
Maize	22.9	25.5	11.5	41.8	54.6	30.6	957.2	1392.3	45.4
Oats	8.6	5.7	-33.7	28.4	31.3	10.3	244.2	178.4	-26.9
Sorghum grain	4.9	6.1	23.9	28.3	33.6	18.7	138.7	205.0	47.8
Soybeans	12.0	20.2	68.3	15.2	16.9	11.3	182.4	341.4	87.2
Winter wheat	15.0	16.5	10.1	16.5	20.6	25.0	247.5	339.9	37.3
Spring wheat	3.6	5.1	40.5	13.4	17.2	28.6	48.2	87.7	82.0
Barley	4.5	3.8	-15.6	22.8	26.5	16.4	102.6	100.7	- 0.2
Hay	27.2	24.8	- 8.8	1.79[a]	2.13[a]	19.0	48.7[a]	52.8[a]	8.5

Source: USDA statistics, 1957 and 1976.

[a]Unit of measure in tons.

mechanical harvesting and computing facilities have reduced the cycle time of reciprocal recurrent selection for maize from four years to three years. These developments also have had an impact on the selection intensity (k) because mechanical planting and harvesting permit testing more progenies than was possible with labor-dependent programs. How the different components of the prediction equation affect genetic gain was illustrated by Eberhart (1970a) and by Fehr and Ortiz (1975) (Table 1.8). Economic restraints dictate efficiency, and the long-term selection programs will not be acceptable if genetic gain is not obtained efficiently. Recurrent selection methods should be considered as only one component of the matrix of breeding strategies available to the plant breeder. The evidence for the effectiveness of the different methods of recurrent selection shows they are effective for improving the means of quantitative traits and providing continued genetic variability for continued selection.

Plant breeders have successfully met the challenges of developing cultivars to meet the changing cultural and management practices (Table 1.13). Some adjustments have occurred in production area for the different crops, such as decreases for oats, barley, and hay and increases for maize, grain sorghum, soybeans, and wheat. Average annual area for the eight crops increased 9.2 percent from 1961-65 to 1971-75, total average production increased 37 percent for the same decade, and yield per hectare increased for all crops, ranging from 10.3 percent for oats and 11.3 percent for soybeans to 28.6 percent for spring wheat and 30.6 percent for maize. A large proportion of the increase in total production occurred because yield per hectare, for the eight crops listed in Table 1.13, increased 20 percent.

The challenges will continue in the future, and it is essential that all possible methods of selection be used to identify superior genotypes. Because of the great importance of a few crop species as sources of food and industrial products, the pressure for effective crop improvement will increase in the future. Neither recurrent selection nor any other method of selection can meet the challenge alone; they must be complementary to each other if they are to contribute to the overall goals of plant breeding. The appropriate recurrent selection method should be chosen on how it contributes to the overall goals of the breeding program.

Discussion

R. J. BAKER
R. R. HILL
J. J. MACKEY

1. R. R. HILL. Much of the published research information on selection and breeding methods in crop plants has been done with maize, but there are a number of examples in other crop species. Modifications in the various types of recurrent selection are adaptable to almost any mode of reproduction found in plants. Thus, the advantages of recurrent or cyclic breeding methods are usable for most crop species.

By averaging observed responses over a number of selection experiments, Dr. Hallauer presented evidence indicating that it is difficult to support the argument that one selection method is significantly better than another. This was a disappointment, but it is difficult to demonstrate that one method of selection is consistently superior to another. Selection methods should be chosen on the basis of how well they fit a particular breeding program, and little attention needs to be given to theoretical relative effectiveness.

Dr. Hallauer implied that short-, intermediate-, and long-term goals should be attempted simultaneously, with genetic material flowing smoothly from one goal or stage to another in the breeding program. It appears that the "art" of utilizing available resources so that the program operates smoothly and effectively is more important to the success of a plant breeding program than the "science" of choosing effective breeding methods.

2. R. J. BAKER. Dr. Hallauer implied that recurrent selection was necessary to meet long-term goals in plant breeding. Certainly, recurrent selection is a powerful breeding procedure. However, with many crops, breeding efforts are for short-term goals, and the

primary breeding procedure continues to be single-cycle selection. I have five concerns with regards to single-cycle selection:

1. Do breeders have the proper objectives for short-term breeding programs? The usual objective is to select the best possible genotype from a population. Perhaps a more realistic objective would be to maximize the probability for selecting a line that is "at least as good" as the best standard variety. According to W. G. Cochran (1951, Proc. Second Berkeley Symp. on Math. Stat. and Probab., pp. 449-70), such an objective would "lead to a different rule of selection."

2. Is the philosophical concept that "plant breeding is a numbers game" proper? Many feel that "bigger is better" in a breeding program, but there must be some point of diminishing returns as program size increases. In fact, there must be a point when it becomes more desirable to test each line more thoroughly than to increase the number of lines being tested. With intermediate heritabilities, testing 500 lines in any one cycle relative to one objective is probably near the optimum size.

3. Should most emphasis be put on selecting between crosses or within crosses? The best tools for deciding the allocation of resources to these two sources are the ratios of the heritabilities between and within crosses and the genetic variances within and between crosses. In most breeding programs, too little effort is spent on selecting between crosses.

4. In what generations should selection for qualitatively inherited traits, such as disease resistance, be practiced? Selection, when done in early generations, fails to consider undesirable linkages that may occur between qualitative genes and genes for quantitatively inherited traits. The increasing popularity of the modified pedigree or single-seed-descent method of breeding is a reflection of the attempt to reduce early selection for major genes.

5. For any one objective in a breeding program, is it better to carry overlapping generations in one season, or is it better to have all generations in phase? We need some plan for incorporating new genetic material into existing programs.

3. J. MACKEY. In plant breeding, crossing and selection are the fundamental operations, and different emphases can be placed on the two procedures. Crosses can be made solely to create genetic variation in which case the results depend entirely upon the success of selection. Or, crosses can be made with the intent to direct variability so as to avoid the need for a decisive selection procedure.

Alternating recombination and selection permits these two components of breeding to support each other.

The first case can be exemplified by the convergent cross, the second by the backcross, and the third by recurrent selection. Each approach has its advantages and disadvantages, mainly depending on degree of heritability of and efficiency in selection for the desired trait(s). The reproduction system of the crop plant, the level of progress from which to start, and the urgency of the breeding program are also important considerations.

For traits with simple oligogenic inheritance via major genes, a straight cross or a more or less complete backcross followed by some kind of plant progeny-selection method is most efficient.

When dealing with quantitatively inherited traits, it is essential to be aware of the possible degree of complexity that can exist. For example, if 30 gene pairs are segregating in a cross of wheat, it would require more space than the total land area of the world to grow a population in which all possible F_2 segregates would be represented. Even in generation infinity when all plants were homozygous due to inbreeding, more than 500 ha would be required to have a minimal chance that all segregates would be represented. Neither of these areas is the plot size used by breeders. Thirty segregating gene pairs must be considered as a modest number in crosses for improving agronomic traits if real progress is to be expected. As an illustration of the genetic complexity for even simple traits, the genetic controls for *erectoides* (shortened rachis length) and *eceriferum* (varied wax coating) mutants in barley occur at a minimum of 26 and 69 loci, respectively. These facts show that it is totally unrealistic in practice to expect a complete exploration of a polygenic recombination potential of a cross in a single cycle of crossing and selection. This situation is more true when a breeder selects for some trait that depends on breaking of linkages as part of potential variation.

Recurrent selection is a method that overcomes these restrictions and permits the complete exploration of a polygenic recombination potential. If time is not important and natural or mass selection can be sucessfully applied, the bulk method offers similar possibilities. Natural allogamy or outbreeding induced by male sterility provides a mechanism for alternation between crossing and selection.

To make greater progress by more complete exploration of a cross is, however, only one alternative. The other is to treat the population from each cross less intensively and make more crosses. Convergent backcross is such a method that effectively widens the gene pool. The conventional way of plant breeding based on a single cycle of crossing followed by selection works similarly in the long run because it is repeated over and over again with progressive selections.

The principle of successive, stepwise progress with some kind of alternation between crossing and selection apparently is basic to making improvements in polygenic traits. The major advantage of recurrent selection lies mainly in saving time and labor. When plant breeding has advanced closer to an ideotype, and a more conservative preserving method has to be developed, different versions of backcrossing will give the necessary narrowing of the genetic variation in the desired direction.

4. K. S. GILL. Using recurrent selection without consideration of the genetic background of the initial parental materials may limit the success of this breeding method. Perhaps the most important aspect of genetic background has to do with linkage phase, that is, are they in repulsion or coupling phase? If in repulsion phase, several cycles of random mating may be needed before the selection phases are initiated. Further, multiple crosses may provide good materials for initiating a recurrent selection program in autogamous crops.

5. A. R. HALLAUER. The diallel selective mating system is designed to incorporate the very points you make: (1) the objective with one population is generally a narrow one, such as improving test weight of wheat, and accordingly, all parents are selected to excel in this trait, and (2) mating does not stop at the single-cross level. Without exception this method results in convergent crosses.

6. F. MARQUEZ-SANCHEZ. Why is there a lack of agreement between predicted and actual responses with the various methods of selection? Generally they do not agree very well.

7. A. R. HALLAUER. There could be several causes for realized responses to be less than predicted ones. One cause could be insufficient sampling. A second could be a change in the mean gene frequency between the time when a population was initially sampled and when it was evaluated some years later. A third could be genotype-cultivar interaction, which is a type of response that cannot be fixed.

8. D. S. MURTY. If the objective of the breeding program is to produce a pure-line cultivar such as would be the case for an autogamous crop, recurrent selection may be disadvantageous because progeny selected from a composite that results from this breeding method will have much heterozygosity built into the selected progeny.

With recurrent selection, over time there are two conflicting objectives: (1) improvement of mean, and (2) maintenance of high genetic variability. With sorghum, few genes influence height and maturity; so is a composite is based on somewhat narrow height and maturity ranges, variation for these traits is apt to disappear after four or five cycles of recurrent selection.

9. A. R. HALLAUER. Yes, that possibility exists, at least theoretically. However, I know of no evidence with sorghum or any other crop where additive genetic variance has been reduced materially with only four or five cycles of recurrent selection.

10. A. JONES. The answer to this problem about fixation of major genes in the early cycles of recurrent selection is to delay selection for such traits for as long as possible. Hold the major genes at a low frequency while selecting for other traits with the result that major genes will not mask the effects of minor ones.

11. R. SWINDELL. There is an apparent contradiction between the theory which demonstrates that selection efficiency is reduced by intermating among F_2 plants as contrasted to one or more generations of selfing before selection, on the one hand, and the success from selection and intermating in segregating populations as demonstrated by Jensen and others. How can this be resolved?

12. R. J. BAKER. To assess the effect of intermating on a plant breeding program is a difficult problem. There is now considerable theoretical and experimental evidence to show that intermating does enhance recombination among linked genes. However, some workers will argue that enhanced recombination is detrimental because it will break up desirable linkages. Such a view is extremely shortsighted—the essential feature of plant breeding is the creation of new genetic combinations.

Theory suggesting that intermating among F_2's reduces selection efficiency is founded on the argument that to enhance recombination is undesirable. While such an argument may apply in certain situations, it is not generally applicable.

13. E. H. EVERSON. In our experience, the cultivars from diallel selection mating programs in other states are poorly adapted under extreme environmental conditions, such as severe winter weather. Probably the reason for this phenomenon is two-fold: (1) with the selective mating system, adapted parents occur in the diallel plan only once, so many crosses occur between unadapted cultivars which result in unadapted progenies; and (2) selection of segregates for recombination in the second, third, and subsequent cycles is done in a nonstress environment so these parental materials are not adapted to stress. Under ideal production conditions, the resultant cultivars are excellent, but in stress situations, their performance is disastrous.

In the wheat breeding program in Michigan, convergent breeding is used. With this type of recurrent selection a wide germplasm base is used. Several years of testing are done for yield and quality before the lines are chosen to intercross for the next cycle. In the 22-year period that this method has been used, genetic improvement for yield has been continuous. Also, the cultivars have been improved for winter hardiness and resistance to two or three races of Hessian fly, cereal leaf beetle, three or four races of powdery mildew, and three or four races of leaf rust, and baking quality has been maintained.

14. J. MACKEY. Natural evolution as well as plant breeding works with a balance between newly created elements and preservation of what has been gained already. Thus, there is a need for a more conservative program when a certain level of progress has been attained from past plant improvements. Segregates having their major gene contribution from a well-adapted and advanced parental cultivar are rare in straight crosses involving polygenic traits. Some kind of more or less complete backcross will enable the breeder to shift the cross population in the desired direction. Such a preadaptation should be made before any foreign germplasm is brought into a major breeding program to conserve what was gained earlier.

15. H. CORTEZ. Suppose a maize breeder has two plant populations that show a good heterosis when crossed. Now, if the method of S_1 progeny test is used instead of reciprocal recurrent selection to improve each population as well as the population cross, what will be the advantages and disadvantages of doing so?

16. A. R. HALLAUER. S_1's can be used for reciprocal recurrent selection. An advantage of using S_1 progenies for making the test crosses is that selection for agronomic traits would be more accurate than with S_0 plants. Remnant S_1 seed of those that had superior testcross performance would be recombined for each population. When averaged over all reciprocal recurrent selection programs with maize, heterosis between populations was about 20 percent greater in advanced cycles than in the C_0.

17. J. L. GEADELMANN. Dr. Baker, you stated that a sample of about 500 lines is a limit you would impose for any given phase of a breeding program. Would you care to expand on the reasons for that limit?

18. R. J. BAKER. The basis for that recommendation is the prediction equation given by Hallauer. With limited resources, there must be a balance between increased selection intensity that can be applied if a larger number of lines is tested and the increased heritability obtained by testing each line more thoroughly. Probably 500 is too high for traits that have low heritability; it's low for highly heritable traits.

19. J. O. GASKILL. For biennial crops, techniques for reducing the length of the life cycle may expedite breeding work materially. For example, the photothermal induction technique for seedlings has enabled breeders to reduce the length of the life cycle for sugar and garden beets to six months instead of the normal two years. So, for some plant characters in some crops, much can be gained in crop improvement, per unit of time, by refining plant propagation techniques.

20. R. J. BAKER. An analogous technique for annual seed crops would be the monoploid and haploid methods of producing inbred lines in maize and barley, respectively. Neither technique shortens the life cycle of the plant, but it shortens the breeding cycle.

21. C. E. LEWIS. The point was made that more yield progress was made per cycle of breeding twenty years ago than has been made per cycle in the past ten years, but genetic variability has not been reduced. This seems like a paradox, but what should breeders do to diminish this tendency for yield to increase at a slower rate, if in fact, it has occurred?

22. A. R. HALLAUER. Yes, the progress in improving yield per unit area of land in the last decade was only about half or less than that of the previous decade for all crops except hay and spinach, but there is no conflict here as you suggest. Actually, the seeming paradox may occur because estimates of crop yield are confounded from two different situations. The estimates of yield improvement that I gave for a number of crops are based upon agricultural production statistics, whereas the estimates of yield gains from selection and the size of the genotypic variances are based upon scientifically conducted experiments. Therefore, because they are estimated from different environments, they cannot be compared directly. Probably many of the yield increases of two decades ago were due to changes in management practices during that period, such as the rapid change to sorghum hybrids, and the changeover to using single crosses of maize. And, reporting changes in percent can lead to false impressions because as mean yields increase in absolute units these will become a smaller percentage value.

23. R. R. HILL. Also, maybe less progress is really being made today because there has been a decrease in absolute research level. The 1950s represented a very active period in plant breeding research, but in the 1960s, many projects were terminated.

24. D. R. WOOD. Dr. Baker, is your concern about overlapping generations in a breeding program based on genetic theory or resource allocation?

25. R. J. BAKER. My concern is: How can a breeder get total integration of new germplasm into the breeding program? In many breeding programs, a new set of crosses is made each year and there is no real attempt to integrate the germplasm from those crosses into breeding populations from previous years. Breeders have overlapping generations all directed toward the same goal, but with little integration between generations. Probably, we should look at the genetic theory involved to determine whether the present system is efficient.

26. J. D. MOLINA-GALAN. I have not heard any comment about mass selection, which has been an efficient and cheap breeding methodology for improving maize yields in Latin America. In six to ten cycles of visual mass selection to maize in Mexico (applying Gardner's principles), progress has averaged 3 percent per cycle.

27. A. R. HALLAUER. Mass selection is an effective breeding procedure, and its success depends, in large part, on the heritability of the trait under selection. Mass selection can be especially effective for improvement in breeding populations into which exotic or unadapted germplasm has been introgressed. Whether to use mass selection instead of another breeding procedure should depend upon such factors as stage of the program, the germplasm sources in the breeding populations, and the heritability of the traits under selection. For a trait such as yield, however, it seems some type of progeny evaluation is more effective than mass selection, particularly at higher levels of productivity.

28. A. GROBMAN. The discussion to now has centered upon cyclical selection as a breeding philosophy. But, in certain situations, might not more and easier progress be made by sampling additional germplasm that exists in a species without resorting to cyclical selection? Most of the literature on recurrent selection procedures comes from the USA, and the germplasm used in USA maize breeding programs may represent no more than 10 percent of what is available in this species. Years ago in Peru, reciprocal recurrent selection was interrupted, and instead, germplasm from additional races of maize was used to expand the genetic variability in the breeding populations. After 25 years with this approach, much greater gains have been made than could have been with recycling methods only. And, in the USA, there is evidence that exotic maize germplasm, when introgressed into local material, produces gains that are comparable with the best that can be obtained by recycling.

29. D. N. DUVICK. Have any studies been made on the amount of genetic variability (that is, genetic variance) that exists in ongoing breeding programs where the pedigree method is used vs. those where recurrent selection is used?

30. R. J. BAKER. Perhaps the best evidence on the importance of genetic variance in autogamous crops can be found in studies on small grains, where high x high yield and high x low yield crosses have been compared to see which give the best lines. In wide crosses, there is a counteracting effect whereby genetic variances are increased and overall means of populations are reduced. The evidence is fairly clear that high x high crosses, with low genetic variance and high cross mean, are more likely to give high lines in the short run. However, long-term breeding programs need more genetic

variability. I am not aware of any long-term studies in small grains that have addressed this latter point thoroughly.

31. S. T. KENNY. With forages, elite genotypes are selected and combined into a synthetic cultivar. In this case, would it be better to select genotypes for characteristics that contribute to population production as opposed to their performances as individual genotypes?

32. R. R. HILL. Burton, working with bahiagrass, improved mean productivity on the basis of selecting clones in a space-planted nursery, and this improvement also was present in a broadcast population. If the evaluations are done properly, selection in either space-planted or solid-seeded nurseries probably will work.

33. E. OMOLO. In developing countries, one of the limitations to progress from breeding is the limited availability of experienced scientific manpower. Oftentimes in these countries young workers with little experience handle breeding programs under the supervision of one or two senior scientists. Under those circumstances, use of recurrent selection may result in many mistakes; so progress in improving a crop is slow. For this reason, I recommend modified mass and ear-to-row selection as the most efficient breeding procedures for improving crops in developing countries. In East Africa, mass selection has resulted in Ecuador 573 and Kitale Synthetic 2 maize populations, and the Hybrid 611 made by crossing these populations yields 30 percent better than the best local check varieties.

REFERENCES

Andrus, C. F., and G. W. Bohn. 1967. Cantaloupe breeding: Shifts in population means and variability under mass selection. Proc. Am. Soc. Hortic. Sci. 90:209-22.

Anonymous. 1976. Seedman's Handbook—Corn and Sorghum. 3rd ed. Clyde Black & Sons and Mike Brayton Seeds, Ames, Ia.

Asay, K. H., I. T. Carlson, and C. P. Wilsie. 1968. Genetic variability in forage yield, crude protein percentage, and palatability in reed canarygrass, *Phalaris arundinacea* L. Crop Sci. 8:568-71.

Bohn, G. W., and C. F. Andrus. 1969. Cantaloupe breeding: Correlations among fruit characters under mass selection. USDA Tech. Bull. 1403.

Brim, C. A. 1966. A modified pedigree method of selection in soybeans. Crop Sci. 6:220.

Brim, C. A. 1978. Recurrent selection for percent protein in soybeans. Seminar presented at Ames, Iowa, sponsored by Dep. Agron.

Brim, C. A., and C. W. Stuber. 1973. Application of genetic male sterility to recurrent selection schemes in soybeans. Crop Sci. 13:528-30.

Brown, W. L. 1975. A broader germplasm base in corn and sorghum. Proc. Corn Sorghum Res. Conf. 30:81-89.

Burton, G. W. 1951. Quantitative inheritance in pearl millet (*Pennisetum glaucum*). Agron. J. 43:409-17.

Burton, G. W. 1952. Quantitative inheritance in grasses. Proc. Sixth Int. Grassl. Congr. 1:277-83.

Burton, G. W. 1974. Recurrent restricted phenotypic selection increase forage yield of Pensacola Bahiagrass. Crop Sci. 14:831-35.

Burton, G. W., and E. H. DeVane. 1953. Estimating heritability in tall fescue (*Festuca arundinacea*) from replicated clonal material. Agron. J. 45: 478-81.

Burton, J. W., and C. A. Brim. 1978. Recurrent mass and within half-sib family selection for high percent oil in soybeans. Agron. Abstr. p. 48.

Busbice, T. H., R. R. Hill, Jr., and H. L. Carnahan. 1972. Genetics and breeding procedures, pp. 283-318. *In* Hanson, C. H. (ed.), Alfalfa Sci. Tech. Agron. 15. Am. Soc. Agron., Madison, Wis.

Compton, W. A. 1968. Recurrent selection in self-pollinated crops without extensive crossing. Crop Sci. 8:773.

Comstock, R. E., and H. F. Robinson. 1948. The components of genetic variance in populations of biparental progenies and their use in estimating the average degree of dominance. Biometrics 4:254-66.

Comstock, R. E., and H. F. Robinson. 1952. Estimation of average dominance of genes, pp. 494-516. *In* Gowen, J. W. (ed.), Heterosis. Iowa State Univ. Press, Ames, Ia.

Comstock, R. E., H. F. Robinson, and P. H. Harvey. 1949. A breeding procedure designed to make maximum use of both general and specific combining ability. Agron. J. 41:360-67.

Darrah, L. L., S. A. Eberhart, and L. H. Penny. 1978. Six years of maize selection in 'Kitale Synthetic II,' 'Ecuador 573' and 'Kitale Composite' using methods of the comprehensive breeding system. Euphytica 27:191-204.

Devine, T. E., and C. H. Hanson. 1973. Hardy and resistant alfalfa. Agric. Res. 21(10):11.

Doggett, H. 1968. Mass selection schemes for sorghum, *Sorghum bicolor* (L.) Moench. Crop Sci. 8:391-92.

Doggett, H. 1972. Recurrent selection in sorghum populations. Heredity 28:9-29.

Doggett, H., and S. A. Eberhart. 1968. Recurrent selection in sorghum. Crop Sci. 8:119-21.

Dudley, J. W. 1977. Seventy-six generations of selection for oil and protein percentage in maize, pp. 459-73. *In* Pollak, E., O. Kempthorne, and T. B. Bailey (eds.), Proc. Int. Conf. Quant. Genet. Iowa State Univ. Press, Ames, Ia.

Dudley, J. W., R. R. Hill, Jr., and C. H. Hanson. 1963. Effects of seven cycles of recurrent phenotypic selection on means and genetic variance of several characters in two pools of alfalfa germplasm. Crop Sci. 3:543-46.

Duvick, D. N. 1977. Genetic rates of gain in hybrid maize yields during the past 40 years. Maydica 22:187-96.

Eberhart, S. A. 1970a. Factors affecting efficiencies of breeding methods. Afr. Soils 15:669-80.

Eberhart, S. A. 1970b. Progress report on the sorghum conversion program in Puerto Rico and plans for the future. Proc. Corn Sorghum Res. Conf. 25:41-54.

Eberhart, S. A. 1972. Techniques and methods for more efficient population improvement in sorghum, pp. 197-213. *In* Rao, N. G. P., and L. R. House (eds.). Sorghum in seventies. Oxford and IBH Publ. Co.

Eberhart, S. A., and L. C. Newell. 1959. Variation in domestic collections of switchgrass, *Panicum virgatum* L. Agron. J. 51:613-16.
Eberhart, S. A., M. N. Harrison, and F. Ogada. 1967. A comprehensive breeding system. Der Zuchter 37:169-74.
Fehr, W. R., and C. R. Weber. 1968. Mass selection by seed size and specific gravity in soybean populations. Crop Sci. 8:551-54.
Fehr, W. R., and L. B. Ortiz. 1975. Recurrent selection for yield in soybeans. J. Agric. Univ. P.R. 9:222-32.
Frey, K. J. 1967. Mass selection for seed width in oat populations. Euphytica 16:341-49.
Frey, K. J. 1971. Improving crop yields through plant breeding, pp. 15-58. *In* Eastin, J. D., and R. D. Munson (eds.), Moving off the yield plateau. Am. Soc. Agron., Madison, Wis.
Gardner, C. O. 1961. An evaluation of effects of mass selection and seed irradiation with thermal neutrons on yield of corn. Crop Sci. 1:241-45.
Gardner, C. O. 1963. Estimates of genetic parameters in cross-fertilizing plants and their implications in plant breeding, pp. 225-52. *In* Hanson, W. D., and H. F. Robinson (eds.), Statistical genetics and plant breeding. NAS-NRC Publ. 982.
Geiger, H. H., and F. W. Schnell. 1975. Experimental basis for breeding hybrid varieties in rye. Hodowla Roslin Aklimatyzacja 1 Nasiennictwo Tom 18. Zeszty 5,6.
Gevers, H. O. 1975. Three cycles of reciprocal recurrent selection in maize under two systems of parent selection. Agroplante 7:107-8.
Gilmore, E. C., Jr. 1964. Suggested method of using reciprocal recurrent selection in some naturally self-pollinated species. Crop Sci. 4:323-25.
Hallauer, A. R. 1978. Recurrent selection programs. Illinois Corn Breed. Sch. 14:28-45.
Hanson, C. E., H. F. Robinson, and R. E. Comstock. 1956. Biometrical studies of yield in segregating population of Korean lespedeza. Agron. J. 48: 268-72.
Hanson, C. H., T. H. Busbice, R. R. Hill, Jr., O. J. Hunt, and A. J. Oakes. 1972. Directed mass selection for developing multiple pest resistance and conserving germplasm in alfalfa. J. Environ. Qual. 1:106-11.
Harris, R. W., C. O. Gardner, and W. A. Compton. 1972. Effects of mass selection and irradiation in corn measured by random S_1 lines and their testcrosses. Crop Sci. 12:594-98.
Hayward, M. D. 1970. Selection and survival in *Lolium perenne.* Heredity 25:441-47.
Hill, R. R., Jr., C. H. Hanson, and T. H. Busbice. 1969. Effect of four recurrent selection programs on two alfalfa populations. Crop Sci. 9:363-65.
Hull, F. H. 1945. Recurrent selection and specific combining ability in corn. J. Am. Soc. Agron. 37:134-45.
Jan-Orn, Jinda, C. O. Gardner, and W. M. Ross. 1976. Quantitative genetic studies of the NP3R random mating grain sorghum population. Crop Sci. 16:489-96.
Jensen, N. F. 1970. A diallel selective mating system for cereal breeding. Crop Sci. 10:629-35.
Jensen, N. F. 1978. Composite breeding methods and the DSM system in cereals. Crop Sci. 18:622-26.
Jinahyon, S., and W. A. Russell. 1969. Evaluation of recurrent selection for stalk-rot resistance in an open-pollinated variety of maize. Iowa State J. Sci. 43:229-37.
Johnson, I. J. 1952. Effectiveness of recurrent selection for general combining ability in sweetclover, *Melilotus officinalis.* Agron. J. 44:476-81.

Johnson, I. J., and F. Goforth. 1953. Comparison of controlled mass selection and recurrent selection in sweetclover, *Melilotus officinalis*. Agron. J. 45: 535-39.

Jones, A. 1969. Quantitative inheritance of ten vine traits in sweet potatoes. J. Am. Soc. Hort. Sci. 94:408-11.

Jones, A., P. D. Dukes, and F. P. Cuthbert, Jr. 1976. Mass selection in sweet potato: Breeding resistance to insects and diseases and for horticultural characteristics. J. Am. Soc. Hort. Sci. 101:701-4.

Kehr, W. R., and C. O. Gardner. 1960. Genetic variability in Ranger alfalfa. Agron. J. 52:41-44.

Levings, C. S., III, and J. W. Dudley. 1963. Evaluation of certain mating designs for estimation of genetic variance in autotetraploid alfalfa. Crop Sci. 3: 532-35.

Mather, K. 1949. Biometrical Genetics. Methuen and Co., London.

Matzinger, D. F. 1963. Experimental estimates of genetic parameters and their applications in self-fertilizing species, pp. 253-79. *In* Hanson, W. D., and H. F. Robinson (eds.), Statistical genetics and plant breeding. NAS-NRC Publ. 982.

Matzinger, D. F., C. C. Cockerham, and E. A. Wernsman. 1977. Single character and index mass selection with random mating in a naturally self-fertilizing species, pp. 503-18. *In* Pollak, E., O. Kempthorne, and T. B. Bailey (eds.), Proc. Int. Conf. Quant. Genet. Iowa State Univ. Press, Ames, Ia.

Maunder, A. B. 1969. Meeting the challenge of sorghum improvement. Proc. Corn Sorghum Res. Conf. 24:135-51.

McCollum, G. D. 1971a. Greening of carrot roots (*Daucus carota* L.): Estimates of heritability and correlation. Euphytica 20:549-60.

McCollum, G. D. 1971b. Heritability of onion bulb shape and size. J. Heredity 62:101-4.

Miller, F. R. 1968. Genetic diversity in the world sorghum collection. Proc. Corn Sorghum Res. Conf. 23:120-28.

Moll, R. H., Abdul Raii, and C. W. Stuber. 1977. Frequency distributions of maize before and after reciprocal recurrent selection. Crop Sci. 17: 794-96.

Morgenstern, K., and H. Geiger. 1975. General and specific combining ability in test crosses between inbred lines of rye. Hodowla Roslin Aklimatyzacja I Nasiennictwo Tom 19. Zeszty 5, 6.

National Academy of Sciences. 1972. Genetic vulnerability of major crops. Washington, D.C. 307 pp.

Paterniani, E., and R. Vencovsky. 1978. Reciprocal recurrent selection based on half-sib progenies and prolific plants in maize (*Zea mays* L.) Maydica 23:209-19.

Penny, L. H., G. E. Scott, and W. D. Guthrie. 1967. Recurrent selection for European corn borer resistance in maize. Crop Sci. 7:407-9.

Potts, L. H. C., and E. C. Holt. 1967. Parent-offspring relationships in kleingrass, *Panicum coloratum* L. Crop Sci. 7:145-48.

Romero, G. E., and K. J. Frey. 1966. Mass selection for plant height in oat populations. Crop Sci. 6:283-87.

Ross, W. M. 1973. Use of population breeding in sorghum: Problems and progress. Proc. Corn Sorghum Res. Conf. 28:30-43.

Russell, W. A. 1974. Comparative performance for maize hybrids representing different eras of maize breeding. Proc. Corn Sorghum Res. Conf. 29: 81-101.

Russell, W. A., and S. A. Eberhart. 1975. Hybrid performance of selected maize lines from reciprocal recurrent selection and test-cross selection programs. Crop Sci. 15:1-4.

Smith, D. C. 1966. Plant breeding: Development and success, pp. 3-54. *In* Frey, K. J. (ed.), Plant breeding. Iowa State Univ. Press, Ames, Ia.

Smith, O. S., R. L. Lower, and R. H. Moll. 1978. Estimates of heritabilities and variance components in pickling cucumbers. J. Am. Hort. Sci. 103: 222-25.

Sprague, G. F. 1952. Additional studies of the relative effectiveness of two systems of selection for oil content of the corn kernel. Agron. J. 44: 329-31.

Sprague, G. F. 1966. Quantitative genetics in plant improvement, pp. 315-54. *In* Frey, K. J. (ed.), Plant breeding. Iowa State Univ. Press, Ames, Ia.

Sprague, G. F., and S. A. Eberhart. 1977. Corn breeding, pp. 305-62. *In* Sprague, G. F. (ed.), Corn and corn improvement. Am. Soc. Agron., Madison, Wis.

Stephens, J. C., F. R. Miller, and D. T. Rosenow. 1967. Conversion of alien sorghums to early combine genotypes. Crop Sci. 7:396.

Suwantaradon, K., and S. A. Eberhart. 1974. Developing hybrids from two improved maize populations. Theor. Appl. Genet. 44:206-10.

Topark-Ngarm, Anake, I. T. Carlson, and R. B. Pearce. 1977. Direct and correlated responses to selection for specific leaf weight in reed canarygrass. Crop Sci. 17:765-69.

Trupp, C. R., and I. T. Carlson. 1971. Improvement of seedling vigor of smooth bromegrass (*Bromus inermis* Leyss.) by recurrent selection for high seed weight. Crop Sci. 11:225-28.

Webster, O. J. 1975. Use of tropical germplasm in a sorghum breeding program for both tropical and temperate areas. Proc. Corn Sorghum Res. Conf. 30:1-12.

CHAPTER 2

Germplasm Collection, Preservation, and Use

J. G. HAWKES

IT IS PERHAPS unnecessary in this day and age to stress the need for the preservation of germplasm as basic material for plant breeding research. It is now generally agreed that a broad genetic base is essential, particularly as breeding objectives become more complex and more demanding. Breeders can no longer be content to use the basic stocks of cultivars which they inherited from their predecessors. Yields must be increased, adaptation to a wider range of environments must be sought, and more nutritious cultivars must be created with greater resistance to pests and diseases—to name only a few of the objectives in front of us—in our attempts to solve world problems of hunger and malnutrition.

It is clear that a much wider range of germplasm is needed by breeders today than in the past, including not only standard cultivars and breeding lines related to them, but also wild species and the primitive cultivars or land races that still exist under cultivation in remote areas of mountains and forest where ancient crop plants were first domesticated.

GENETIC DIVERSITY

N. I. Vavilov (1926), the Russian geneticist and plant breeder, was the first to realize the essential need for a really broad genetic base for crop plant improvement. Vavilov and his colleagues collected cultivated plants and their wild relatives from most parts of the world to provide basic germplasm for breeding cultivars adapted to the widely varying environmental conditions in the USSR. The

Department of Plant Biology, University of Birmingham, England.

results were astonishing in that far more genetic diversity was found than had ever been imagined possible. This variation had evolved through the natural processes of mutation, hybridization, and selection, both natural and artificial, over the periods that crops had been domesticated.

From the data collected on his germplasm collections, Vavilov postulated the existence of eight main "centers of diversity" and two or three subsidiary ones, all of which he at first considered to be the "centers of origin" for the crops found in them. Some modifications to this initial hypothesis were made by Vavilov and by others, but there is no doubt that centers of diversity do exist for most crop plants, and as such, they serve as invaluable sources of germplasm for plant breeders. Of course, genetic diversity exists in all biological organisms, wild or cultivated, plants or animals, throughout their distribution range. Thus, a search for genetic characters of value should not be limited to the Vavilovian centers of diversity.

In Vavilov's time the concept of active germplasm conservation had not occurred to breeders. True, Vavilov's collections and those assembled by the USDA and other government agencies were being preserved for screening and use by breeders, but if accessions were lost no one much cared. The material could always be re-collected. The only restraints to re-collecting were financial.

GENETIC EROSION

By the 1950s, concern developed that the natural reservoirs of germplasm resources were rapidly being destroyed, or to use a simile from soil science, genetic resources erosion was taking place. Indeed, genetic erosion has continued through the ensuing decades at such an increasing pace, that if no steps had been taken to prevent it, no natural germplasm reservoir would have remained by the end of the century.

Curiously enough, genetic erosion has occurred because plant breeders have been so successful. The new standard cultivars designed to give higher yields than those of the highly diverse land races in the gene centers have been replacing the land races at an amazing and ever-increasing pace, and once a land race has been replaced, it is lost forever. Thus, genetic diversity has been destroyed by the very cultivars created from it. A broad genetic base has been replaced by a narrow one, and the old genetic diversity is disappearing both inside and outside of the ancient gene centers. This trend is inevitable with the need for highly efficient and uniform cultivars in advanced and sophisticated farming systems.

The introduction of the Green Revolution wheat and rice cultivars into countries such as Pakistan and India is an example of the most recent and intensive stage in this process. With the virtual disappearance of many of the old wheat (*Triticum aestivum*) and barley (*Hordeum vulgare*) cultivars from Iraq, Iran, Pakistan, and India went the old "adaptive complexes," that is, the groups of genes conferring adaptation to the soil and climatic conditions of these regions that had evolved over several millenia.

Modern farming methods have brought about the near elimination of the attendant weedy forms of many crops, which, together with the crops themselves, formed the so-called crop/weed complexes. From these weed sources, introgression of genes from crop to weed and weed to crop takes place frequently, thus enriching the genetic diversity of both weeds and crops (Zohary, 1970).

The third process causing genetic erosion is the extension of farming and grazing land into wild habitats, with the ensuing destruction of the wild species in these habitats. Reduction in their ranges has diminished their genetic diversity also. Destruction of wild habitats in the humid tropics is especially serious where logging is destroying the forest ecosystems together with the wild relatives of fruit trees and other species that may have future value for medicinal or economic purposes. Overgrazing by cattle and particularly by goats has caused destruction of whole ecosystems on islands and in arid regions where the complete vegetation can be destroyed, never to return, if grazing pressures are too great. Wild species destruction also is taking place through the growth of towns and cities, roads, airports, and industrial areas, various types of environmental pollution, and the change of habitat wrought by poor land management.

How, then, can the genetic diversity of our crop plants and their wild relatives best be conserved? It is clearly impossible to impose restrictions on the farmers in the gene centers, because it would condemn them to growing the lower-yielding primitive cultivars, while their fellow countrymen outside the restriction area would enjoy the higher yields from newly bred cultivars. Neither is it possible to preserve all the areas where the wild species grow that may be of value for plant breeding in the future.

The answer to conserving genetic diversity lies in a series of necessary decisions and strategies for collecting and conserving genetic stocks, many of which are compromises, but all of which are workable in a real world, where absolute perfection is rarely attained. By doing nothing, all genetic diversity of our crop plants will be lost in a very short time indeed; by carrying out the activities that will be described herein, some genetic diversity will be lost, but certainly not all of it.

There are seven groups of activities needed for the conservation and utilization of genetic resources: (1) exploration, (2) conservation, (3) evaluation, (4) data storage and retrieval, (5) utilization, (6) training, and (7) global coordination. Each will be discussed in turn.

EXPLORATION AND COLLECTION

First, it is necessary to understand what genetic diversity still remains in our crop plants and to devise methods for collecting it. The Food and Agricultural Organization of the United Nations (FAO) has published a detailed work on *Crop Genetic Resources in Their Centres of Diversity* (Frankel, 1973), and individual crop committees and international research centers have brought parts of this work up-to-date and have added information on other crops not covered in Frankel's work (Grubben, 1977; IBPGR, 1978a; CIP, 1973; CIP, 1976). Apart from cultivated materials, related weed forms and wild species must be collected also, not always because they are under threat of erosion but often because they are needed by plant breeders. Priorities for collection have been established, and the letter "E" has been used by FAO to designate emergency situations, whereas at lower levels of priority, categories I, II, and III are used, listing species by species and region by region (see CIP, 1973, 1976).

In the past, priorities have been established first for crops and later for regions or countries. However, now that regional organizations are available, priorities are often set out by countries first and crops second, as for southeast Asian countries (see IBPGR, 1977d). Such priorities also can be established for conservation.

When working out plans for a collecting expedition, collectors should bear in mind that even though gene centers contain much germplasm of importance, they should not limit their collecting to gene centers only. Also, collections should be made from peripheral regions of the species distribution where the species is under environmental stress and where special mutations may have been selected for such conditions. Mutations also have arisen in regions where the crop was introduced in comparatively recent times, such as with tomatoes and peppers in Europe.

The germplasm available in the present world collections is very inadequate (Williams, 1978) because of inadequate sampling, loss of viability and genetic erosion during storage, and a number of other reasons. Few samples of wild or weed materials exist in the older collections, and there is much duplication of individual genotypes.

In the field, the objectives and methods of the collector of genetic resources are distinctly different from those of the botanical collector. The latter tries to select samples (generally dried herbarium materials with some seeds, roots, or bulbs) that are "typical specimens," that is, they are as uniform as possible. On the contrary, the genetic resources collector samples populations rather than individuals and looks for diversity rather than uniformity. His sampling strategy ideally should be based on a knowledge of the structure of a population of plants. Because this is rarely known in advance, population geneticists such as Allard (1979), Jain (1975a), and Marshall and Brown (1975), have proposed generalized sampling strategies for seed crops, which all collectors of genetic resources should follow.

Thus, in sampling populations, the collector aims to capture the maximum amount of diversity for the minimum number and size of samples. Ideally, gene pools should be sampled nonselectively to provide with at least 95 percent certainty one copy of each variant occurring in the target populations with a frequency greater than 0.05. This means that 50 to 100 individuals should be collected per site with some 50 seeds taken from each plant, and collecting sites should be scattered throughout an area either evenly or in clusters for cultivated or wild materials, respectively.

Selective or nonrandom sampling has been advocated by some, as has a mixture of selective and random samples (Bennett, 1970). With wholly selective sampling the collector may pick out morphological variation only, and as a result, he may miss entirely variation in disease resistance alleles or genes for adaptation to stress environments.

Equally unjustified are so-called mission-oriented expeditions, where collectors have looked for a limited range of traits that were discernible visually and collected only those plants that have such traits. For example, some foresters have preferred to collect seeds from "elite" trees only, and have completely ignored the fact that characters of value may reside in smaller trees that are not immediately suitable for timber.

Generally, nonselective sampling is designed to capture the greatest range of genetic diversity, some immediately useful to breeders and some not. Experience has shown that many characters regarded as useless several decades back are now considered of prime importance. Because collecting is often a "rescue operation" the complacent view that collecting should be done only for present-day breeding needs is intolerable. On the contrary, collecting must be done on the "now or never" principle.

The selection of sampling sites must take into account the ecological conditions of the region to be sampled (Williams, 1978). Clearly, patterns of genetic differentiation are strongly related to environmental heterogeneity. Thus, careful observations must be made in the field for crop differences; changes in ecological, agricultural and social conditions; soil patterns; and changes in agricultural practices. Sampling sites should be close together if there is not uniformity for such factors and farther apart if few changes are observed.

Particularly for wild species, the occurrence of topographical, geological, ecological, and altitudinal variation indicates the need for sampling at frequent intervals. Allard (1970) has outlined similar requirements for collecting wild grass species in California. So the collector must sample as wide a range of habitats as possible and always be alert for environmental and agricultural changes as he covers the target area.

Little attention has been given to the sampling of vegetatively propagated crops (Hawkes, 1975), but selective sampling would seem to be best for these species because of their nature and the strong artificial selection that has been applied to them.

Finally, conservation requirements cannot be ignored when germplasm is being collected in the field (Hawkes, 1976). The minimum numbers of seeds in a sample for conservation purposes should be from 2,500 to 5,000 seeds (Hawkes, 1980). This strategy cannot apply to plants with very large seeds, such as the coconut (*Cocos nucifera*); so clearly, for such species, expediency must triumph over theoretical perfection.

Keeping adequate field records is also of great importance, and most collectors have been remiss in the past in providing adequate collection information. On the other hand, too much information is equally undesirable (Bunting and Kuckuck, 1970). Minimum data sheets have been proposed (Hawkes, 1976) for general purposes, though for each crop some additions or modifications may be advisable (see Chang et al., 1975 for rice, and IBPGR, 1978 for bananas and plantains).

CONSERVATION AND STORAGE

Conservation can be carried out in two forms, namely, *in situ* and *ex situ*. Conservation *in situ* demands the establishment of nature or biosphere reserves, national parks, or special legislation to protect endangered or threatened species (Jain, 1975b). In these, the wild species and the complete natural or seminatural ecosystems are preserved together. For cultigens, biosphere reserves are not a solution because there are no natural ecosystems to support them. A controlled primitive or early farming system might provide a teaching or an experimental situation with value for other reasons, but in neither of these cases can the complete or even a broad genetic diversity of a species be conserved, because nature reserves can cover only a very small portion of the total distribution area of a major species.

By far the most practical method of preservation is to store germplasm *ex situ*, that is, in a gene bank, which normally means a seed bank. With tree fruits and vegetatively propagated crops, germplasm can be preserved in orchard or field plantings, but there are constant risks of loss of vegetatively propagated gene banks from disease and damage from man-made or natural disasters and human errors in handling of materials. Furthermore, the sheer expense and space involved with orchard plantings may be unacceptable.

Seed stores are relatively safe, easy to maintain, and they use minimum space for all but the largest seeds. Methods have been worked out for optimum storage under low moisture (5 ± 1 percent) and low temperature (−18°C, or even lower), in sealed glass containers, tinned steel cans, or plastic/foil bags (FAO, 1974; Roberts, 1975, 1978; IBPGR, 1976).

For efficiency and better management it has been recommended (FAO, 1973) that genetic resources centers should comprise base collections and active collections.

The base collections are for long-term conservation under optimum conditions, where the samples should not be disturbed until they are regenerated. The National Seed Storage Laboratory at Fort Collins, Colorado is a place where base collections are stored. The samples are regenerated when the germination capacity of the seeds falls below 95 percent of what it was when the material was first stored. Some authorities consider this requirement as too restrictive, and they would allow the germination percentage to fall even lower before regenerating the materials. Genotypes differ in the rates at which they die so if a seed sample represents a nonuniform seed population, the genetic integrity of the sample could change considerably if the viability was to drop to less than 95 percent before regeneration.

Roberts (1975, p. 277) describes the observation of a higher frequency of mutant phenotypes derived from old seed than that from young seed. In fact, in the materials studied (Roberts, 1978, p. 27), there was a close relationship between percentage loss of viability and the amount of genetic damage accumulated in the surviving seeds. In barley, loss of 50 percent viability was estimated to produce as many mutations in the surviving seeds as would be produced from treating fresh seeds with 10,000r of X-rays, and half the number of mutations were produced by a decrease in viability of only 20 percent. It is not known whether slow and rapid loss of viability both result in some mutation frequency for a given reduction. Most researchers assume they do, so the lesson is clear. Good storage conditions are useful first, because they lengthen the regeneration cycle, which cuts costs of storage and maintains the genetic integrity of the samples; and second, because they reduce the mutation rates in the stored samples.

There is one case where storage of seeds in low humidity may not decrease the risk of seed death. This was illustrated by Villiers (1975) who showed that seeds may be preserved under conditions of full imbibition providing germination is prevented. Also, seeds with membrane damage from irradiation can repair much of this damage when conserved in the fully imbibed condition.

Seeds that have been discussed until now, Roberts (1973) has designated "orthodox," that is, they respond in a normal way to drying and cooling (as described above). Another group of seeds, designated by Roberts (1973) as "recalcitrant," shows a very drastic loss in viability with a decrease in moisture content below 12 to 30 percent. Many forest trees; temperate and tropical fruit trees; and a number of tropical crops such as *Citrus*, cocoa (*Theobroma cacao*), coffee (*Coffea arabica*), rubber (*Hevea brasiliensis*), oil palm (*Elaeis guineensis*), and others seem to be recalcitrant. The recalcitrant nature of tropical plantation crops (see also Harrington, 1972) has encouraged research on careful drying and cooling as well as on conservation under fully imbibed conditions (see Mumford and Grout, 1979). Probably, future research will resolve this problem.

Active collections need not be stored under such stringent conditions as the base collections. Thus, active collections are used for medium-term storage, regeneration, multiplication and distribution, evaluation, and documentation. In any gene bank, base and active collections should be kept separate. Active collections may be stored at temperatures above 0°C and at moisture content above 5 percent. Usually, multiplication of samples in the active collection is done with seed from this collection directly, but to prevent too great a loss of genetic integrity in the active collections, base collection material should be used for regeneration from time to time (Ito, 1970).

Before leaving seed conservation, the problem of regeneration should be discussed. Routine germination tests are carried out every 5 to 10 years to ascertain the reduction in germination percentages of stored seeds, even though certain species have an estimated storage life of more than 100 years. For regeneration, samples sufficiently large to prevent genetic drift must be grown and allowed to self or sib pollinate. Generally, the number of plants grown to regenerate a sample should be the same as the number originally collected, namely, 50-100 (Frankel, pers. commun.). Some authorities, however, consider that a smaller number would be adequate. Care must be taken to prevent cross-fertilization among accessions of an allogamous crop, and this necessitates suitable isolation of samples.

Finally, samples should be regenerated under conditions isoclimatic to those in which they were collected. Such advice is easier to give than to carry out; actually, if the climatic conditions of the collecting and regeneration sites are not too dissimilar and if interplant competition is reduced by wide spacing, little loss of genetic integrity is likely to take place (Frankel, pers. commun.).

Another way of preserving genetic stocks is by meristem cultures (Morel, 1975; Henshaw, 1975; Henshaw and Grout, 1978). Establishment of meristem or shoot-tip banks on a large scale will occur soon because the basic techniques are already known. There are several good reasons for conserving stocks via meristem cultures. First, exact genotypes can be conserved indefinitely, free from virus or other pathogens, and without loss of genetic integrity. Second, meristem cultures of vegetatively propagating cultigens, such as potatoes (*Solanum tuberosum*), yams (*Dioscorea* spp.), cassava (*Manihot utilissima*), sweet potatoes (*Ipomoea batatas*), and a number of others, are particularly advantageous because their seed production is so poor. Third, loss of vegetatively propagated materials from natural disasters or pathogen attacks in the field can be avoided. Fourth, a reasonably long regeneration cycle can be envisaged for meristem cultures. Whereas potato meristems must be regenerated yearly under minimal growth conditions at reduced temperatures (Henshaw and Grout, 1978), potato meristems can be stored at the temperature of liquid nitrogen (−196°C) (Grout and Henshaw, 1978) with successful regeneration of plants from these cultures. Under such conditions, high genetic stability and very long regeneration cycles should be possible. Fifth, regeneration of forest and fruit tree germplasm via seeds (and methods of long-term seed storage of recalcitrant seeds, as we have seen, are not very well advanced at present) requires a very long time because trees require 10 to 20 years to reach sexual maturity, but meristems can be subcultured in a few minutes. Naturally, when mature plants are needed for breeding work, the 10-20 years are still needed, but regeneration of meristems is extremely easy when compared with seed regeneration for long-lived perennials.

EVALUATION

For germplasm samples to be of use to the breeder, they must be evaluated or screened. Many germplasm collections are inadequately evaluated, and therefore, are virtually of no use to the breeder.

Evaluation may consist of nothing more than the description of the place of origin and a morphological or phenological description, or it may consist of information on physiological, genetical, biochemical, plant pathological, or other characteristics (Frankel, 1978).

Oftentimes, evaluation may be regarded as an end in itself, but the best evaluation is one that relates the traits measured to plant breeders' needs. There clearly is no purpose to be served in making detailed descriptions of every morphological feature and screening for every known disease if these measurements do not serve breeders' needs. Yet, even if a sample when screened does not appear to be of value to breeders, it should not be discarded. It may be needed in the future.

Collaborative evaluation is essential for the future and is already being carried out in Europe through Eucarpia (European Association for Research and Plant Breeding) and other agencies (Hawkes and Lamberts, 1977, 1978; Scarascia-Mugnozza, FAO, 1977). More work of this nature is essential if the enormous stocks of genetic resources material now flowing into the gene banks are to be evaluated and utilized.

DATA STORAGE AND RETRIEVAL

Without a system for dissemination of evaluation data on the material held in the gene banks, none of it will be used by breeders. Hence, an efficient system for storage and retrieval of information about germplasm is the logical link between its evaluation and use. Ford-Lloyd (1978) has outlined the magnitude of the problem of handling data on genetic resources for the large collections of maize (*Zea mays*) from Mexico and South America which total some 40,000 accessions, and probably these represent only 40 percent of the global collection of some 100,000 accessions. Collections of rice (*Oryza sativa*), wheat, sorghum (*Sorghum bicolor*), and other major crops comprise large numbers of accessions also. The potato collection has 25,000 accessions from three principal sources, and this probably is only 75 percent of the number in the world. Already, collections exist for some 200 crop species with a total of ca. 7,300,000 accessions. And, of course, much collection is yet to be done. Consider that 10 to 50 pieces of descriptive data exist for each accession and it becomes clear that only electronic computers can deal with the massiveness of this information bank.

The problem is really twofold. First, the evaluation data must be transferred in an orderly and systematic manner into machine-readable form; and second, management systems must be devised to sort, store, and recall these data on command. The first problem involves the formulation of descriptors (characters) and descriptor states (character states), crop by crop. For this purpose the International Board for Plant Genetic Resources (IBPGR) has established international crop committees (IBPGR Annual Report 1976/77) to work out the best way to formulate the data. These formulations have been published for potatoes (*Solanum tuberosum*) (IBPGR, 1977c), wheat and *Aegilops* (IBPGR, 1978b), and bananas and plantains (IBPGR, 1978a), but many more are needed. Clearly, descriptors for morphological, physiological, pathological, and other aspects of each crop will differ, but those for field collecting of data should be somewhat similar. Especially valuable have been thesauri on forage crops, root and tuber crops, vegetable, oil and fiber crops, and pests and diseases published by Seidewitz (1973, 1974a, 1974b, 1975, 1976). As summary for this section, a useful start has been made on international standardization of the recording of descriptive and evaluation data on accessions in germplasm collections, but better coordination is still needed.

Management systems for data need to be standardized only to the point that the systems used are compatible. The group of information specialists at Colorado use EXIR (Executive Information Retrieval), a modified version of TAXIR (Taxonomic Information Retrieval) devised by Rogers (see Rogers, Snoad, and Seidewitz, 1975; Hersh and Rogers, 1975; IBPGR-GR/CIDS, 1976, 1977a, 1977b). This is the most widespread and internationally accepted information system to the present.

With the new generation of minicomputers and newer data management systems, certainly it will be easier to handle the immense volume of data now building up and to transmit it more readily to breeders on request. However, with 200 to 300 crops stored in germplasm banks all over the world and evaluation taking place in plant breeding stations, universities, and germplasm banks, the problem of information input, storage, and retrieval will never be simple—especially since collections are often held as populations from each of which conflicting data can emerge.

UTILIZATION OF GENETIC RESOURCES

The utilization aspect of genetic resources work encompasses the whole of plant breeding. Yet, to some, genetic resources are equated with exotic kinds of germplasm, collected from gene centers as primitive cultivars, old land races, or wild and weedy species, rather than the usual accessions contained in a breeder's working program.

When a plant breeder searches for a gene conferring resistance to a particular pathogen, as an example, he at first will screen his own materials and the national collections for the species. As an example of the order in which breeders proceed in their search for a desired gene(s) or trait, take the case of resistance to round cyst nematodes (*Globodera (Heterodera) rostochiensis* and *G. pallida*) in potatoes. Potato breeders first screened the European and North American collections. No resistance of value was found in them, so the breeders next screened the Andean cultivated species where they found resistance to the common race in *S. tuberosum* subsp. *andigena* and in several wild species. Recently, new races of the nematode have been found to which cultivated species have no resistance. This caused potato breeders to screen wild species in which resistance genes were found. However, inserting these resistance genes from a wild species into a cultigen involved the tedious process of backcrossing. Finally, this particular endeavor had a happier ending, because breeders found that the genes for resistance to the most aggressive races already existed in the cultigen due to naturally occurring hybridization and introgression in Bolivia where the resistant wild species occur (Astley, 1979).

The various ways in which genes from wild species can be transferred to cultivated ones have been outlined by Hawkes (1977). For each cultivated plant prebreeding or parental-line breeding is needed. Oftentimes this requires much basic research to obtain knowledge about the cytogenetics and biosystematics of the wild species and its genetic relationship to the cultigen. Thus, breeding lines that are similar to the cultivars are developed so as to get the desired characters from wild species into cultivated "backgrounds." These then can be used by breeders to create the cultivars for agricultural use. In this way the tedious work of transferring a gene from the wild species to a cultigen can be bridged.

TRAINING

An important aspect of this rapidly growing discipline of genetic resources is the training of young scientists in the various subdisciplines that make up the whole. At the University of Birmingham in England, more than 100 persons from 38 countries have trained to the Master's degree level in the past decade. Short courses on data management have been offered by the Information Sciences/Genetic Resources Program at the University of Colorado; three short courses have been given on plant exploration at Bogor, Indonesia, jointly by the Indonesian National Biological Institute and the University of Birmingham; and plans are under way for similar courses in India, Argentina, Nigeria (International Institute for Tropical Agriculture), and Colombia (Centro Internacional de Agricultura Tropical). Thus, a knowledgeable group of young scientists from many countries has been developed to play an important role in all aspects of genetic resources activities.

GLOBAL COORDINATION

The administrative and logistic needs of the plexus of activities associated with germplasm collection, preservation, and use on a worldwide basis will be the subject of this section. In the late 1950s and through the 1960s, FAO assumed the lead in developing the ideas necessary for the discipline of genetic resources. Three technical conferences were organized, one each in 1961, 1967, and 1973, with the last two being of greatest importance (see Frankel and Bennett, 1970; Frankel and Hawkes, 1975). A unit of crop ecology and genetic resources was organized at FAO in Rome in 1967 to be a clearing house for information on collections and expeditions and to encourage and coordinate initiatives. This unit was advised by a panel of experts which took scientific and organizational initiatives over the whole field. The panel or its individual members were responsible for the foundations and early development of the discipline.

Certain panel members, meeting in 1972 in Beltsville, Maryland, established a blueprint for a global network of genetic resources centers and activities under the auspices of the Consultative Group on International Agricultural Research (CGIAR). This latter body had previously established the network of international crop centers such as the International Potato Center (CIP), International Maize and Wheat Improvement Center (CIMMYT), International Rice Research Institute (IRRI), International Institute of Tropical Agriculture (IITA), and others. The blueprint was accepted in a modified form by CGIAR which in 1974 set up the International Board for Plant Genetic Resources (IBPGR).

The IBPGR, with generous funding from donor countries through the World Bank, has carried out much of the work which FAO, with its limited budget, could not do. Crop germplasm advisory committees were organized, and regional programs were established or planned, based geographically on the areas encompassed more or less by Vavilov's gene centers.

There is a thriving Southeast Asian program, with collaboration from Indonesia, Malaysia, Thailand, Philippines, and Papua-New Guinea; a Southwest Asian program comprising Turkey, Syria, Iraq, Iran, Afghanistan, and Pakistan; and a southern Asian program, just beginning, to include India, Bangladesh, Nepal, Bhutan, and Sri Lanka. There is a European network, which under the auspices of Eucarpia, includes all European countries, a Mediterranean program; and the beginnings of African and Latin American programs. Special bilaterally funded projects have been organized between the governments of West Germany and Ethiopia and Costa Rica respectively, while Brazil and India are developing interesting national programs. Additionally, the USA and the USSR have their own germplasm programs; and some individual countries which cooperate in regional programs make their own decisions to explore, conserve, evaluate, and utilize genetic resources.

This whole area of worldwide cooperation on germplasm activities is very heartening, and has the makings of a resounding success story. There have been some failures and some half-successful attempts that call for thought and possible revision. One saddening fact is the demise of the FAO panel of experts. Some point or focus for innovative thought is still needed, and perhaps Plant Breeding II will provide the incentive; but certainly genetic resources activities are flourishing and growing throughout the inhabited parts of the globe because it is understood by plant scientists in developing and developed countries alike that these are essential prerequisites for humankind's continued survival on this planet.

Discussion

E. E. GERRISH
Q. JONES
C. M. RICK

1. E. GERRISH. The most dramatic example of germplasm use has required the manipulation of photoperiod response to permit the exchange of germplasm between tropical and temperate crop races. In maize, this manipulation has involved crossing between germplasm sources with widely differing maturities, and the subsequent formation of new populations for recurrent selection schemes. An intermediate step, however, has entailed adapting exotic germplasm to a new area before introgressing it into adapted breeding populations much as Hallauer did with Eto. This type of conversion has been confined mostly to adapting tropical sorghums to temperate zone forms. It was accomplished by backcrossing several genes for reaction to photoperiod into tropical genotypes so they would mature in the temperate zone. The genetic mechanism for photoperiod response in maize may be more complicated than that for sorghum, but the method for inserting photoperiod-reacting genes into unadapted genotypes should work for maize or any other crop.

During the past 20 years our program has used this approach in maize. During the late 1950s, several tropical inbreds of maize, gathered from Latin American breeding programs, were crossed each to a different very early Corn Belt single cross, and the hybrids were backcrossed twice to the tropical forms. All have been adapted to central Iowa. Five of these inbreds were descended from Tuxpeno, a prominent, productive dent race found in lowland Mexico, and the sixth from the Mexican mid-elevation race, Celaya. Four additional inbreds derive from Caribbean coastal yellow flint backgrounds, a fifth inbred from Costeno, and a sixth from a Colombian population, Eto. The six converted Mexican inbreds have been intercrossed to form a six-line composite. Another composite has been made from the six converted Caribbean inbreds.

A third composite was made by intercrossing six unconverted inbreds from Argentine flint, a member of the Southern Cateto race. It is intended that these six-line composites, that is, from Mexican, Caribbean, and Argentine origin, together with three additional six-line composites made up from Corn Belt dent inbreds, will be crossed in an all-combination diallel, and that these crosses will be tested to obtain information about heterotic response among germplasm groups. This information should be helpful in showing what exotic germplasm sources would be most useful for supplying genes to incorporate into Corn Belt dent.

Another use that can be made of these converted lines of maize is to construct additional exotic populations which better represent exotic races for study and recurrent selection in the central Corn Belt. As an example, the adapted six-line Mexican composite mentioned previously was outcrossed to other sources of Tuxpeno germplasm, specifically, to seven inbreds, three improved synthetics, and three open-pollinated cultivars from Mexico and Brazil. Through mass selection a new adapted population has been formed. Similarly the six-line Caribbean composite was outcrossed to five Caribbean synthetics and to the race Costeno. The six-line Argentine composite was outcrossed to tropical synthetics of the Cateto race. These three representatives of tropical germplasm will be released to germplasm banks in 1980 or 1981.

An interesting population is one derived from intercrossing among five Caribbean inbreds and five Tuxpeno inbreds. It has undergone two cycles of recurrent selection, and at the University of Missouri, it was shown to be a good source of resistance to the southwestern corn borer and to aflatoxin problems.

It is not necessary to confine conversion programs to homozygous or homogeneous material. For example, race Cuzco, a large, white, flour type adapted to high elevations in Peru and Bolivia, was crossed to an early Corn Belt synthetic, Minnesota 3, and backcrossed three times to the Cuzco cultivar with a minimum of 30 plants being used in each backcross cycle to reasonably sample the parent cultivar. A fairly large population was maintained between backcrosses to ensure that the diversity of the Cuzco was maintained. Five subraces of Cuzco were carried this way concurrently to make a better sampling of this race. A similar project was carried out with the interlocking flour corn of the Amazon Basin, again making use of five subraces. The adapted versions of Cuzco and Amazon flour will be released in 1981 or 1982.

Unexpectedly, it has proved easy to convert tropical Cateto flint to temperate adaptation, and Caribbean and Mexican germplasm have converted without difficulty. The Amazon flour, however, has been difficult to convert because it is exceedingly late and somewhat intolerant of Corn Belt nursery conditions; and Cuzco, which is adapted to high altitudes, has been the most difficult.

The dramatic nature and results of this conversion program draw attention to the possibilities for even wider germplasm use. Other breeding approaches can be equally effective, but the ultimate goal, in any case, will be to elevate and stabilize future crop performance. To this end both the germplasm source and the breeding approach will affect success and the ease with which that success is gained. The possibilities may seem endless, but our experience suggests a few basic guidelines: (1) determine critically what qualities the germplasm source possesses in its own area of adaptation; (2) aside from photoperiodism, determine how closely donor and recipient regions compare climatically; and (3) decide how quickly some tangible gain from this work must be evident.

Where incorporation will be the ultimate fate of the introduced germplasm, the breeder might further consider: (1) whether it will be of advantage to evaluate and exploit the introduced sources per se, before incorporation; and (2) similarly, whether provision can be made for thorough evaluation of the source toward a future use of similar material.

2. C. M. RICK. I want to make four points about germplasm and its use.

1. A crisis has developed with respect to the rapidity with which primitive cultivars of many crops are becoming extinct—a point brought to my attention very forcefully during a recent trip to Peru to collect tomatoes. We found that land races are now extinct from the coastal region and western valleys, where the related wild species are also becoming eliminated. Abundant collections of these materials could be made as little as two or three decades ago. The cause of germplasm extinction, in the case of tomatoes, is the feeding of vast numbers of domesticated animals. The worst problem is posed by the rapidly increasing herds of goats, which systematically devour vegetation at the lower elevations.

2. There is a tremendous amount of genetic diversity in wild species. Among tomato species, one is cultivated and eight are wild. Usually the genetic variation contributed by a wild species has been assayed by crossing the garden tomato with one or two accessions of

the wild form. But we can now recognize at least 40 races in one of the species, *Lycopersicon peruvianum*, and within each race an immense amount of variability is to be found between and within populations. Our studies of allozymic variation have revealed that even within such largely self-pollinated species as *L. pimpinellifolium* the reserve of genetic variability is unexpectedly large.

3. Information useful to germplasm utilization can often be obtained from careful notes taken at the time of collecting; valuable clues may be gained by observations of the habitat. Thus, collections of wild species found in stress habitats may possess the genes sought in programs of breeding for resistance to drought, waterlogging, soil salinity, and various pests. As examples, our collections of *L. cheesmanii* biotypes discovered in a littoral habitat are so salt resistant that they will actually tolerate culture in 100 percent strength sea water. A few notes about such habitats could obviously have great usefulness, but they are often lacking.

4. Unexpected, unpredictable characteristics often appear in the progenies of hybrids between cultivated and wild forms. In this respect, the wild accessions cannot be evaluated directly because these novel traits are not observed in them. The genetic bases of such characteristics are various, including transgressive inheritance. DeWet and Harlan have uncovered many examples of this type in their studies of *Zea-Tripsacum* hybrids, including derivatives that have ears at every node. A case of an unexpected quantitatively inherited trait from interspecific hybridization is the increased grain yield obtained with oats. A 15-20 percent yield increase was realized from introgressing genes from the wild *Avena sterilis* into cultivated oats.

To exploit these unpredictable traits from species hybrids requires an investment in prebreeding. That is, some effort must be made to establish the genes responsible for these traits into desirable genotypic backgrounds before they can be utilized by plant breeders. This area is much neglected, but the National Plant Germplasm Resources Board (NPGS) has made a strong plea for its support.

3. Q. JONES. My mission will be to describe the U.S. National Plant Germplasm System and pose some questions and controversial concepts relative to germplasm activities.

The National Plant Germplasm System of the USDA has four Regional Plant Introduction Stations (RPIS). They are located at Ames, Iowa; Geneva, New York; Experiment, Georgia; and Pullman, Washington. The four RPISs have similar functions and responsibilities, but they differ in the crops they tend. All maintain active or working collections of germplasm for use by plant breeders. Each station has final responsibility for all germplasm collections within its region, including those maintained by off-station curators, such as potatoes at Sturgeon Bay, Wisconsin, and soybeans at Urbana, Illinois. The RPISs acquire their germplasm samples through the Plant Introduction Office in Beltsville, Maryland, which in turn, acquires samples through direct exploration overseas, by collection from the wild domestic flora, or from breeders' programs. The RPISs and their affiliated curators have responsibility for maintaining adequate seed stocks for all accessions held at the National Seed Storage Laboratory (NSSL) at Fort Collins, Colorado. The NSSL collection is called a "base collection," and a germplasm sample from it is not distributed to users unless it is unavailable elsewhere.

There are two new and developing elements within NPGS. The first is the development of a computer assisted information network, and the second is the funding of a set of national clonal repositories for fruit and nut crops. The information network will become the nerve system for the NPGS, providing channels for communication and exchange of information among all elements of the germplasm system and the users of germplasm collections. The objective, of course, is to make germplasm maximally accessible to users. The clonal repositories are being planned at 12 locations to handle seven crop groups—pome fruits, stone fruits, nuts, grapes, small fruits, citrus, and tropic and subtropical fruits. This is a very quick description of the NPGS and how it functions.

My second task will be to exemplify by these questions some of the pressing problems we now face in germplasm work. Who should bear the responsibility for increasing samples of exotic germplasm? Should NPGS be responsible for the maintenance and distribution of genetic marker stocks? Should NPGS be responsible for the maintenance and distribution of virus free stocks of germplasm? Should NPGS evaluate germplasm stocks in replicated tests? Does the integrity of individual stocks of germplasm need to be maintained or can germplasm pools or composites suffice?

As stewards of the world's plant germplasm, the scientists and administrators of the NPGS would like to use available funds for activities in addition to maintaining germplasm stocks, but to do so, resolution of these questions is necessary. Also, we need to have greater cooperation from germplasm users in reporting evaluation data back into the NPGS information system. Right now, much data that is collected either is not returned to us or it is reported in some nonstandard format or scaling system.

4. K. RAWAL. The Germplasm Resources Information Project (GRIP), located at Fort Collins, Colorado, is designed to support and complement the material exchange facilities within the NPGS. This information system has three basic components: data, methods, and environment. On data aspects, committees of crop experts develop uniform ways for recording reliable information. The methods and techniques used are well known in computer science, because they were developed for business applications. The environment refers to training people who work on germplasm in the storage and retrieval of information about germplasm.

5. S. JANA. It has been suggested by some that genetic resources can be managed best by continuously evolving gene pools. Such a system goes on continuously for plant species in the centers of genetic diversity.

However, the habitats in the centers of genetic diversity are being increasingly disturbed, with the consequent threatening of the evolution and existence of many species. In view of this real and immediate danger, would it not be worthwhile to have continuously evolving gene pools carried outside of these centers of diversity for cultivated species, primitive species, and maybe even wild species? These would be "gene parks" established outside of the centers of diversity.

6. J. G. HAWKES. Gene parks have value for experimental purposes, but not as repositories for the vast stocks of germplasm that must be stored for the future. True, when seed stocks are placed into a germplasm bank, evolution of the species stops. However, the evolutionary process is going to stop anyhow, because modern agriculture is rapidly destroying the natural habitats of the gene centers. By collecting, we are trying to capture the germplasm before it disappears. Further, a gene park, or for that matter, even several dozen or hundreds of gene parks, cannot conserve the whole of the genetic variability of the crop because the land areas that would be involved are small. On the other hand, as an experiment to understand what are the ongoing selective and evolutionary processes in ecosystems, gene parks certainly have a place. So from the scientific and experimental point of view, it would be quite interesting to establish a few gene parks here and there.

In Hungary, some uninhabited agricultural production areas are being reinhabited by educated middle-class people moving from the cities. The government has decreed that some of these "new farmers" should farm with traditional methods and traditional crops, and after a time the results will be carefully analyzed to see how the experiment works out.

The other thing that is done in many developed countries is to set up living museums on primitive farming. This has an educative purpose. Primitive farming methods are used, and primitive land race cultivars are grown. However, such museums could hardly be called gene parks.

7. L. HOUSE. Probably more effort should be put into collecting and preserving named cultivars. This is being done for sorghum at the International Crops Research Institute for the Semi Arid Tropics (ICRISAT). Named cultivars no longer useful in their areas of development may prove useful in different environments some day.

Also, plant collections probably should be evaluated initially in the location from which they were collected.

8. PATRICK GREGOIRE. What is the present status of the conservation of the northern flint maize from the USA?

9. E. GERRISH. There is no extensive effort to study and preserve New England flint maize in its original form throughout its former range. Many collections exist, however, and these do show differences, particularly in maturity and kernel traits. The race can be appreciated if only by its contribution to Corn Belt dent, but there is circumstantial evidence that it has combined additionally with the Southern Cateto flint of Uruguay and Argentina to form the modern Argentine flint type. Also, in its original and introgressed forms, it has been indispensable to the northern areas of Europe. At the time of its dispersal, the northern flint was itself relatively new, having appeared among the northern Indian tribes only two or three centuries prior to the arrival of the Europeans.

10. K. E. PRASADA RAO. Generally, much attention is given to the danger of extinction of land cultivars of a species as new and improved cultivars become available for agricultural production. But, an equally and perhaps more important factor leading to extinction is the speedy replacement of one crop species by another. Examples are the replacement of sorghum by maize in Tanzania, and replacement of sorghum by maize and mustard in central India.

Maybe seed increase and maintenance of germplasm need not be done in regions where a crop is not adapted, but the evaluation of germplasm certainly should be carried out in areas where the crop is adapted. Especially, the evaluation of germplasm collections for reactions to insect and disease pests needs to be done in areas where the pests are endemic. Wild and weedy types which are the sources for resistance are becoming extinct for various reasons. Perhaps there is need to establish gene parks in which they can be preserved for use in crop improvement programs. This would be a way to circumvent quarantine restrictions against their movement from one country to another.

11. R. ECOCHARD. Could the original variability in a germplasm collection be maintained by single-seed descent? That is, all samples would be sown in one bulk and one seed would be harvested from each plant to form the progeny bulk.

12. J. G. HAWKES. This approach to genetic variation maintenance has been advocated by many crop scientists, but on the whole, it is better to keep the lines separately because then a particular allele can be traced to its source. Also, in a large mixed population, rare alleles do have a fair probability of being lost. Certainly, if such bulking is done, lines should not be put into the bulk until they have been studied thoroughly.

13. W. L. BROWN. The ultimate purpose in collecting and preserving germplasm is to make that germplasm available to breeders. Yet, because of the great number of accessions available for certain species, the breeder finds it difficult to intelligently use this material until it has been evaluated. But with the vast number of accessions in the various germplasm collections, it seems nearly impossible that the bulk of this material can be evaluated in a systematic way within a reasonable period of time. What can be done to encourage the evaluation of all these materials so more of it is likely to be used by breeders?

14. J. G. HAWKES. One suggestion from EUCARPIA is to organize collaborative evaluation programs between countries with the breeders themselves establishing the priorities for evaluation. Clearly, a collaborative evaluation program does not make the task any more simple, but it does coordinate the task.

Evaluation has got to be done according to plant breeders'

needs; so there should be close communication between the director of the gene bank and the plant breeders who use the germplasm. Evaluation data are available, but the information management section needs to provide timely printouts of the traits that have been evaluated.

In summary, more coordination is needed. A germplasm bank should not be a museum where germplasm is just stored—genetic resources are to be used by breeders.

15. D. R. KNOTT. The real problem here is one of limited resources—personnel and money. There are not enough resources to carry out all germplasm activities simultaneously. Right now, emphasis is placed on collection, but maybe collection, at least for some crops, has reached the point of diminishing returns; that is, the point at which the chance of recovering valuable new germplasm does not justify the cost involved. If so, resources now used for collection should be shifted to where the next bottleneck occurs, namely, maintaining and evaluating current collections so they can be utilized. An example is wheat. The USDA has between 30,000 and 40,000 collections of this crop. Can additional collection really turn up any new and valuable traits?

16. J. G. HAWKES. It is important that one habitat should not be sampled over and over because the germplasm collected is likely to be redundant. But, collections from new habitats probably will provide much uncollected germplasm. For example, Dutch scientists, who recently collected wheat in remote parts of the Himalayas in Nepal and Pakistan, have found material that is new and considerably different from germplasm already in gene banks anywhere in the world.

Generally, past collectors have obtained many of their samples along roadsides or in markets; therefore, they have not sampled the full gene pool, and the full gene pool will not be sampled until the collectors traverse areas up in the high mountains and inaccessible regions. Probably no germplasm collection for any crop is yet adequate, but your concern about diminishing returns is justified.

17. C. M. RICK. I would argue in favor of overcollecting. We have measured substantially different allozyme gene frequencies in repeated collections of the same population of a wild tomato species. Also, in our recent experiences in Peru, we were impressed by the potential impact of annual climatic differences on the genetic make-

up of wild populations. The extremely dry conditions of early 1979 were clearly exerting extreme drought stress on the native tomato species as observed in the numbers of dead, weakened, and unfruitful plants. Almost certainly we were collecting genotypes that are the most drought resistant, yet in a year of moister conditions we might obtain different samples. These experiences impress us with the merits of re-collecting. At any rate, the probability of gaining valuable new material well outweighs the small price paid for over-collecting.

18. Q. JONES. Another argument for getting on with the evaluation phase of germplasm activities is that until the collection is thoroughly evaluated, there is no way to really know what traits the collection holds.

19. J. MACKEY. Genetic variation may be lost before sufficient collection can be made to preserve it. To what extent are we able to compensate such an erosion by induced mutagenesis and to what extent is natural mutation counteracting the loss? Mutation is both a new-creating and repetitive process.

20. J. G. HAWKES. Perhaps mutagenesis for creating genetic variation could be an alternative to collection and preservation of natural variation. However, naturally occurring genetic variation and that from recent mutagenesis may represent quite different arrays. Natural variation has passed through the sieves of natural selection and selection by agriculturalists, whereas the variation from mutagenesis has not undergone selection.

Of course, some genetic traits undoubtedly will get lost if genetic erosion continues, and it will continue, and new evolutionary progress in the original environment will not be entirely compensative. As an example, look at cassava, a crop that is normally propagated vegetatively and only seldom by seed. In Southeast Asia there are 300 to 400 cassava cultivars, so evolution must have been an important force in shaping the genetic variation of this crop.

21. T. GREGORI. Are collections and evaluations being made on plant species that are used as food in some niches of the world but are not major crop species?

22. Q. JONES. The USDA has supported explorations to the south-

west U.S. to collect beans and the buffalo gourd. These are locally important food crops to the indigenous people of that area, and they may have use in arid regions elsewhere in the world. This is a good point.

REFERENCES

Allard, R. W. 1970. Population structure and sampling methods, pp. 97-107. *In* Frankel, O. H., and E. Bennett (eds.), Genetic resources in plants: Their exploration and conservation. Blackwell, Oxford.

Astley, D. 1979. *Solanum sucrense* Hawkes: A classification, pp. 43-45. *In* Zeven, A. (ed.), Proc. Conf. Broadening Genet. Base Crops, Wageningen, 1978. Cent. Agric. Publ. and Doc., Wageningen.

Bennett, E. 1970. Tactics of plant exploration, pp. 157-59. *In* Frankel, O. H., and E. Bennett (eds.), Genetic resources in plants: Their exploration and conservation. Blackwell, Oxford.

Bunting, A. H., and H. Kuckuk. 1970. Ecological and agronomic studies related to plant exploration, pp. 181-88. *In* Frankel, O. H., and E. Bennett (eds.), Genetic resources in plants: Their exploration and conservation. Blackwell, Oxford.

Chang, T. T., et al. 1975. Manual for field collectors of rice. IRRI, Los Baños.

De Candolle, A. 1884. Origin of cultivated plants. Paul Kegan, translator. [Original published in French at Geneva, 1882.]

Food and Agric. Organ., UN. 1973. Report of the fifth session of the FAO panel of experts on plant exploration and introduction. FAO, Rome.

Food and Agric. Organ., UN. 1974. Report of the sixth session of the FAO panel of experts on plant exploration and introduction. FAO, Rome.

Ford-Lloyd, B. V. 1978. Data storage and retrieval systems in genetic resources information exchange. *In* Hughes, J. G. (ed)., Conservation of plant genetic resources, pp. 42-45. Proc. Sect. K., Br. Assoc. Adv. Sci., Univ. of Aston, Birmingham.

Frankel, O. H. (ed.). 1973. Survey of crop genetic resources in their centers of diversity. FAO/IBP, Rome.

Frankel, O. H. 1978. Conservation of crop genetic resources and their wild relatives: An overview, pp. 123-49. *In* Hawkes, J. G. (ed.), Conservation and agriculture. Duckworth, London; Allanheld, New York.

Frankel, O. H., and E. Bennett (eds.). 1970. Genetic resources in plants: Their exploration and conservation. Blackwell, Oxford.

Frankel, O. H., and J. G. Hawkes (eds.). 1975. Crop genetic resources for today and tomorrow. Cambridge Univ. Press.

Grout, B. W. W., and G. G. Henshaw. 1978. Freeze preservation of potato shoot-tip cultures. Ann. Bot. 42:1227-29.

Grubben, G. J. H. 1977. Tropical vegetables and their genetic resources. *In* Tindall, H. D. and J. T. Williams (eds.), Int. Board for Plant Genet. Resour., Rome.

Harrington, J. F. 1972. Seed storage and longevity. *In* Kozlowski, T. T. (ed.), Seed biology 3:145-245.

Hawkes, J. G. 1975. Vegetatively propagated crops. *In* Frankel, O. H., and J. G. Hawkes (eds.), Crop genetic resources for today and tomorrow, pp. 117-21. Cambridge Univ. Press.

Hawkes, J. G. 1976. Manual for field collectors (seed crops). FAO, Rome.

Hawkes, J. G. 1977. The importance of wild germplasm in plant breeding. Euphytica 26:615-21.

Hawkes, J. G. 1980. Crop genetic resources field collection manual. Eucarpia, Wageningen and IBPGR, Rome.

Hawkes, J. G., and H. Lamberts. 1977. Eucarpia's fifteen years of activities in genetic resources. Euphytica 26:1-3.

Hawkes, J. G., and and H. Lamberts. 1978. Eucarpia gene bank committee. Eucarpia Bull. 11, pp. 19-21.

Henshaw, G. G. 1975. Technical aspects of tissue culture storage for genetic conservation, pp. 399-405. *In* Frankel, O. H., and J. G. Hawkes (eds.), Crop genetic resources for today and tomorrow. Cambridge Univ. Press.

Henshaw, G. G., and B. W. W. Grout. 1978. The long-term storage of plant tissues by means of meristem culture and other *in vitro* techniques. *In* Hughes, J. G. (ed.), Conservation of plant genetic resources, pp. 36-41. Proc. Sect. K., Br. Assoc. Adv. Sci. Univ. of Aston, Birmingham.

Hersh, G. N., and D. J. Rogers. 1975. Documentation and information requirements for genetic resources application, pp. 407-46. *In* Frankel, O. H., and J. G. Hawkes (eds.), Crop genetic resources for today and tomorrow. Cambridge Univ. Press.

Int. Board Plant Genet. Resour. 1976. Report of IBPGR working group on engineering, design and cost aspects of long-term seed storage facilities. Int. Board Plant Genet. Resour., Rome.

Int. Board Plant Genet. Resour. 1976. Advisory committee on the genetic resources communication, information, and documentation system (GR/CIDS): First Report. Int. Board Plant Genet. Resour., Rome.

Int. Board Plant Genet. Resour. 1976/77. Annual report 1976. Int. Board Plant Genet. Resour., Rome.

Int. Board Plant Genet. Resour. 1977a. Advisory committee of the IBPGR on the genetic resources communication, information and documentation system (GR/CIDS): Second meeting. Int. Board Plant Genet. Resour., Rome.

Int. Board Plant Genet. Resour. 1977b. Advisory committee of the IBPGR on the genetic resources communication, information and documentation system (GR/CIDS): Third meeting. Int. Board Plant Genet. Resour., Rome.

Int. Board Plant Genet. Resour. 1977c. Descriptors for the cultivated potato. Huamán, Z., J. T. Williams, W. Salhuana, and L. Vincent (eds.). Int. Board Plant Genet. Resour., Rome.

Int. Board Plant Genet. Resour. 1977d. A cooperative regional program in southeast Asia. Int. Board Plant Genet. Resour., Rome.

Int. Board Plant Genet. Resour. 1978a. Genetic resources of bananas and plantains. Int. Board Plant Genet. Resour., Rome.

Int. Board Plant Genet. Resour., and information sciences/genetic resources program. 1978b. Descriptors for wheat and *Aegilops*. Int. Board Plant Genet. Resour., Rome.

Int. Potato Cent. (CIP). 1973. Germplasm exploration and taxonomy of potatos. Int. Potato Cent., Lima, Peru.

Int. Potato Cent., (CIP). 1976. Report of the planning conference on the exploration and maintenance of germplasm resources. Int. Potato Cent., Lima, Peru.

Ito, H. 1970. A new system of cereal breeding based on long-term seed storage. SABRAO Newsl. 2:65-70.

Jain, S. K. 1975a. Population structure and the effects of breeding system. *In* Frankel, O. H., and J. G. Hawkes (eds.), Crop genetic resources for today and tomorrow, pp. 15-36. Cambridge Univ. Press.

Jain, S. K. 1975b. Genetic reserves. *In* Frankel, O. H., and J. G. Hawkes (eds.), Crop genetic resources for today and tomorrow, pp. 379-96. Cambridge Univ. Press.

Marshall, D. R., and A. H. D. Brown. 1975. Optimum sampling strategies in genetic conservation. *In* Frankel, O. H., and J. G. Hawkes (eds.), Crop genetic resources for today and tomorrow, pp. 53-80. Cambridge Univ. Press.

Morel, G. 1975. Meristem culture techniques for the long-term storage of cultivated plants. *In* Frankel, O. H., and J. G. Hawkes (eds.), Crop genetic resources for today and tomorrow, pp. 327-32. Cambridge Univ. Press.

Mumford, P. M., and B. W. W. Grout. 1979. Desiccation and low temperature (−196° C) tolerance of *Citrus limon* seed. Seed Sci. and Technol. 7: 407-10.

Roberts, E. H. 1973. Predicting the storage life of seeds. Seed Sci. and Technol. pp. 499-514.

Roberts, E. H. 1975. Problems of long-term storage of seed and pollen for genetic resources conservation, pp. 269-96. *In* Frankel, O. H., and J. G. Hawkes (eds.), Crop genetic resources for today and tomorrow. Cambridge Univ. Press.

Roberts, E. H. 1978. The long-term storage of viable seeds. *In* Hughes, J. G. (ed.), Conservation of plant genetic resources, pp. 26-35. Proc. Sect. K., Br. Assoc. Adv. of Sci., 1977. Univ. of Aston, Birmingham.

Rogers, D. J., B. Snoad, and L. Seidewitz. 1975. Documentation for genetic resources centers. *In* Frankel, O. H., and J. G. Hawkes (eds.), Crop genetic resources for today and tommorrow, pp. 399-405. Cambridge Univ. Press.

Scarascia-Mugnozza, G. T. 1977. Report on the FAO European cooperative research network on durum wheat. FAO, Rome.

Seidewitz, L. 1976. Thesaurus for the international standardization of genebank documentation. Inst. Pflanzenbau u-Saatgutforschung, Braunschweig. 1st ed., 1973. Forage crops. 1974a. Root and tuber crops 1974b. Vegetable oil and fiber crop plants. 1975. Selection of common and scientific terms for plant pests and diseases.

Vavilov, N. I. 1926. Studies on the origins of cultivated plants. Bull. Appl. Bot. Plant Breed. 16:1-245.

Villiers, T. A. 1975. Genetic maintenance of seeds in imbibed storage. *In* Frankel, O. H., and J. G. Hawkes (eds.), Crop genetic resources for today and tomorrow, pp. 297-316. Cambridge Univ. Press.

Williams, J. T. 1978. Plant exploration: Capturing genetic variability. *In* Hughes, J. G. (ed.), Conservation of plant genetic resources, pp. 14-25. Proc. Sect. K. Br. Assoc. Adv. Sci., 1977. Univ. of Aston, Birmingham.

Zohary, D. 1970. Centers of diversity and centers of origin, pp. 32-42. *In* Frankel, O. H., and E. Bennett (eds.), Genetic resources in plants: Their exploration and conservation. Blackwell, Oxford.

CHAPTER 3

Application of Tissue Culture and Somatic Hybridization to Plant Improvement

E. C. COCKING
R. RILEY

RECENTLY, much has been written on the potential for using plant tissue culture and somatic hybridization for plant improvement, and it will not be useful to repeat most of what has been comprehensively reviewed by Scowcroft (1977) and Bhojwani et al. (1977). Most plant breeders are unfamiliar with the techniques of plant cell and tissue culture, but a few plant hybridizers are familiar with the techniques associated with *in vitro* fertilization and embryo culture; however, unless such techniques are essential, they are generally avoided. Plant tissue culture technology has been available for more than two decades; yet, its impact on plant improvement has been minimal.

This review will consider actual application of plant tissue culture technology and its present and foreseeable limitations, rather than focus on a discussion of its potentials. It has become clear in recent years that plant improvement through the application of cell and protoplast culture technology to plant tissue will require effort across a broad front of the new techniques. Different crops will vary markedly in their readiness to respond to such techniques; hence, judicious choice of the proper crop species for experimentation is essential at this stage in the development of the subject.

TISSUE CULTURE FOR PLANT PROPAGATION

Plant tissue culture methods can be useful to the breeder con-

E. C. Cocking, Agricultural Research Council Group, Department of Botany, University of Nottingham, Nottingham, England, and R. Riley, Agricultural Research Council, London.

cerned with propagation of certain economically valuable species. With horticultural species, tissue culture is valuable in the reduction of investment in conventional methods of propagation and in a reduced risk of infection by viruses and other diseases. It enables the marketing of new disease-indexed cultivars. In this type of *in vitro* propagation, meristems and calluses are most often used, but cell-suspension cultures can be utilized also. When a single-cell culture is desired, plant protoplasts (single cells with cell wall calluses removed) are very useful. It is much easier to utilize meristems or callus cultures than single-cell cultures for mass propagation; but even with callus tissues, and particularly with cell-suspension cultures formed from cultured protoplasts, regeneration of whole plants may be difficult to accomplish.

One reason why plant tissue culture methodology has not been used more extensively for plant propagation is that some cultures are unable to regenerate plants. Another requirement is that when tissue culture is used for vegetative propagation, the period of unorganized growth should be kept short to minimize the chances for genetic change in cells of the callus. Even though the basic requirements for growth of plant cells and tissue in culture are well established (Davey and Thomas, 1975), there is still a fundamental problem of how to manipulate the medium and environment to cause whole plant regeneration to occur. The present approach is largely empirical, and much work by tissue-culture workers on the fundamental requirements for regeneration is urgently needed. Recent studies by Ammirato (1979) on factors controlling embryogenesis in *Dioscorea* serve as an example of what is required to make the approach more systematic.

As far as actual plant improvement is concerned, there is no major advantage to be derived from propagation by tissue culture unless pressure for selection can be directed towards one particular characteristic.

SELECTION AT THE TISSUE CULTURE LEVEL

As discussed by Scowcroft (1977), the utilization of mutants for understanding biochemical and developmental processes in microorganogenesis is an obvious paradigm for the potential value of plant mutants. Clearly, if the plant could be handled in a manner similar to microorganisms, mutants could be selected for certain cases of disease resistance, adaptation to stress conditions, and improved nutritional quality. Superficially, selection *in vitro* would appear to be ideal for plant improvement. Numbers of mutations are induced and recovered in cell culture, and several of these have regenerated mutant plants. Technically it is possible to screen 10^7-10^8 cells easily, and if regeneration is possible, such mutants could be evaluated in whole plants. It may not be necessary to use cell suspensions or even callus tissues. Cultures derived from immature embryos of a maize (*Zea mays*) cultivar susceptible to race T of *Helminthosporium maydis* were treated with a pathotoxin produced by this fungus (Gengenbach et al., 1977). They were able to find resistant cell lines that gave rise to plants that were resistant also.

Nabors (1976) has described and assessed the relative merits of using either spontaneously occurring, or induced mutants, to obtain agriculturally useful plants from plant tissue cultures. From the practical aspect, he concluded that if little is known about the inheritance of the desired trait, it is better to search for spontaneous mutants in tissue cultures first, and if this fails, to induce mutations. It may be advantageous to use haploid cell lines for selecting recessive biochemical mutants.

There are several major problems confronting the plant breeder who wishes to utilize selection at the tissue-culture level. First, not all cultures will regenerate whole plants readily. Sometimes this difficulty can be bypassed by utilizing embryo culture; but in general, whole plant regeneration for cereals and legumes is not readily possible when small aggregates of cells and single protoplasts are utilized (Bhojwani et al., 1977; King et al., 1978). This difficulty does not preclude the ultimate application of such a selection procedure to the major crop plants; rather, more basic work is needed to discover ways to make culture and regeneration successful.

A second problem, even when regeneration is possible, is whether traits selected at the cell or callus level will be heritable in the regenerated whole plant. Increased synthesis of specific amino acids has been obtained by Widholm (1976) who selected lines resistant to 5-methyl tryptophan from carrot and tobacco suspension cultures. Resistance was stable against this analog of tryptophan, and resistant cultivated cells contained 20 times more free tryptophan than the corresponding parental cultures. These studies were extended to lysine and methionine analogs, and increased free pools of these amino acids were present in the analog resistant cell lines. Recently, valine-resistant tobacco plants have been regenerated from selected cells (Bourgin, 1978). Green and Phillips (1974) have questioned whether these feedback resistance mutants selected from among callus cells will result in increased lysine, threonine, and methionine in seeds. Remember that all increases were in soluble amino acids which is not the form of organic nitrogen stored in seeds. Polacco (see Boulter and Crocomo, 1977) selected cell lines of the soybean rich in the enzyme urease, a protein relatively rich in methionine. Clearly, soybean lines with high urease activity could be agronomically valuable if these cell lines regenerated plants with high levels of urease in the seeds, or even if there was an increased level of urease in the plant which facilitated the assimilation of urea fertilizer. Boulter and Crocomo (1977) stressed that nonexpression of a selected character in seeds of regenerated plants could be due to lack of transcription and/or translation of genetic information, or alternatively, activity of an enzyme (such as urease) could be inhibited by the internal environment of the plant. A possible analogous situation might exist for the genes that control cereal storage protein which are expressed only in the endosperm.

Selection for greater production of already existing enzymes in cells within a culture may be fruitful. For instance, selection of cells with high nitrate reductase activity (NRA) may be useful, particularly if regenerated whole plants from such cells carried the increased NRA. In some crops, reduction of nitrogen in the leaves by nitrate reductase may be the rate-limiting step in the plant's metabolism. Net photosynthesis might be increased in crops by lowering the amount of photorespiration that normally occurs, and cell lines that utilize glyoxylate as a source of carbon may have a lowered photorespiration; thus, selection among cells for this capability might provide cell lines which could regenerate plants with increased net photosynthesis. If more than one enzyme or factor was involved in improving a trait, selection would be very difficult, because this would involve a highly balanced system of metabolic and genetic interactions. The same considerations would apply to selection for salt-tolerant plants from cultures of cells *in vitro*. For disease resistance associated with the reaction to specific toxins at the biochemical level, selection of resistant cells, tissues, or embryos probably would be possible.

USE OF ANTHER CULTURE

Guha and Maheshwari (1964), who investigated *in vitro* culture of mature anthers of *Datura innoxia*, were attempting to obtain a better understanding of meiosis. However, they made the accidental but significant observation that when anthers were cultured in a suitable basal nutrient medium containing kinetin and coconut milk, numerous embryolike structures were formed. Later, these were shown to arise from pollen grains. As expected, the plantlets of pollen origin were haploid. As discussed by Bhojwani (1976), this report aroused a worldwide interest in developing the technique of anther culture for producing haploids from a wide array of crop plants because the haploid cell lines and plants regenerated from them were useful in basic and applied aspects for plant breeding. However, as pointed out by Nitzsche and Wenzel (1977), today's discussions on the utilization of haploids in plant breeding are frequently marked by exuberant optimism on the one side and total rejection on the other. (For an earlier review of basic research on haploidy see Kimber and Riley, 1963.)

Microspore culture, of course, is not the only method for obtaining haploid plants. Other methods are selection among twin seedlings and the use of interspecific crosses with associated chromosome elimination and embryo culture. It is difficult to generalize on this subject, but an excellent critical review of use of haploids in plant breeding has been published by Nitzsche and Wenzel (1977) and the reader is referred to this review for a more detailed consideration of the subject. Plant breeders would like to induce haploid sporophytes directly from gametophytes, because upon doubling the haploid chromosome complement, a homozygous pure line is produced instantaneously. That is, the first products of meiotic segregation and recombination can be fixed into pure-line cultivars in three generations. Consequently, this permits considerable acceleration in breeding programs. However, the production of haploids may constitute a bottleneck in routine breeding by limiting the range of genetic variation derived from intervarietal hybrids, especially when compared with that released from a conventional program where homozygosity is attained over several generations of selfing. So unless the haploid procedure can be undertaken on a very large scale, a price must be paid in terms of reduced variability in the progenies in return for the accelerated rate of attaining homozygosity. In a way, the progress, and lack of progress, with anther culture reflects the completely empirical nature of the procedures employed and the lack of knowledge about the interaction of genotype with cultural and microspore developmental stages. When success has been achieved (as by groups in China), it has taken enormous effort involving trial-and-error methods with a wide range of different media and genotypes of the chosen species.

APPLICATION OF *IN VITRO* CULTURE PROCEDURES TO PLANT HYBRIDIZATION

Tissue culture methodology can also be applied to higher plants to facilitate hybridization between species. Interspecific hybridization has seen only limited use as a source for deriving genetic variability because so few of these hybrids can be obtained through sexual reproduction. And, as discussed by Poehlman (1959), it is difficult to make generalizations about the breeding behavior in interspecific and intergeneric hybrids because the system of classifying plants into species was based mainly on their morphological and physiological characteristics and not on genomic characteristics or cross compatibility—this classification system, to a large extent, was worked out before the science of genetics developed. As a consequence, interspecific crossing may result in failure to obtain seed or in complete fertility of the F_1 plant. Amphiploids that occur in nature usually have fertility and their survival is related to this fertility. Artificially induced amphiploids, such as *Triticale*, vary from complete fertility to complete sterility.

The *in vitro* culture of excised embryos has been employed to grow whole plants from some sexually mediated interspecific crosses in cases where the embryo dies during later stages of seed development. Some instances of postzygotic embryo failure occur due to

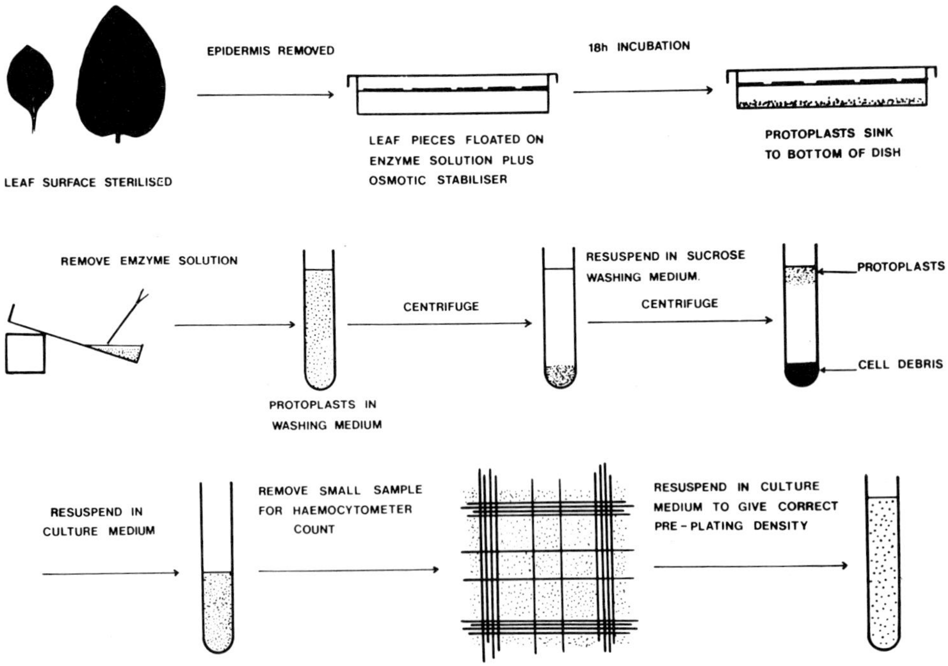

Fig. 3.1. Isolation of protoplasts (e.g. petunia and tobacco mesophyll).

breakdown in endosperm development, whereas in others, there is simple incompatability between embryo and endosperm. The recent sexual hybridization of barley and rye was achieved via embryo culture *in vitro* (Cooper et al., 1978). Sometimes during the early stages of embryo development in interspecific crosses, the chromosomes of one species are eliminated, resulting in the formation of a haploid embryo. In such cases, the endosperm degenerates; but if the embryos are excised and cultured *in vitro*, they give rise to haploid plantlets.

The culture of isolated ovules also provides a way to overcome barriers to fertilization imposed by the stigma, and *in vitro* fertil-

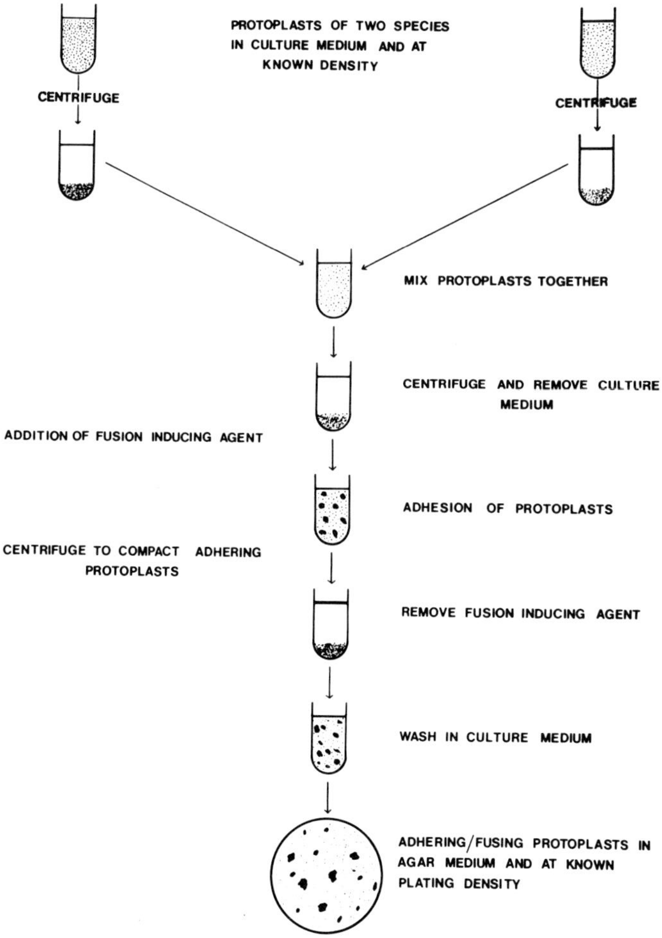

Fig. 3.2. Fusion of protoplasts.

ization has been successfully utilized for interspecific hybridization. This technique has more limited use than embryo culture because eggs of higher plants are difficult to excise and achieving *in vitro* fertilization is very difficult because fertilization is a complicated phenomenon (for a fuller discussion see Rangaswamy, 1977).

Ever since the meeting at Bellagio, Italy, in 1969 on futuristic techniques for plant improvement, it has been evident that the fusion of somatic cells from which cell walls had been removed (protoplasts) (Fig. 3.1) would provide an alternative method to sexual hybridization for combining the germplasm of two genotypes within a common nucleus. Achievement of somatic hybridization simply depended on the development of the necessary technology to cause fusion between somatic plant cells (Fig. 3.2). It also necessitated the development of methods to select such somatic hybrids from the parental protoplasts and the refinement of the procedures at the *in vitro* culture level (Fig. 3.3) for the regeneration of whole hybrid plants from selected somatic hybrid cells (Fig. 3.4). Since 1969, the basic methodology for such somatic hybridization has been perfected, and now there is available an alternative method to sexual plant hybridization.

Before discussing the present and future applications of tissue culture for crop improvement through hybridization, it is necessary to outline the differences between sexual and somatic hybridization. No reduction division is involved in somatic hybridization, and the

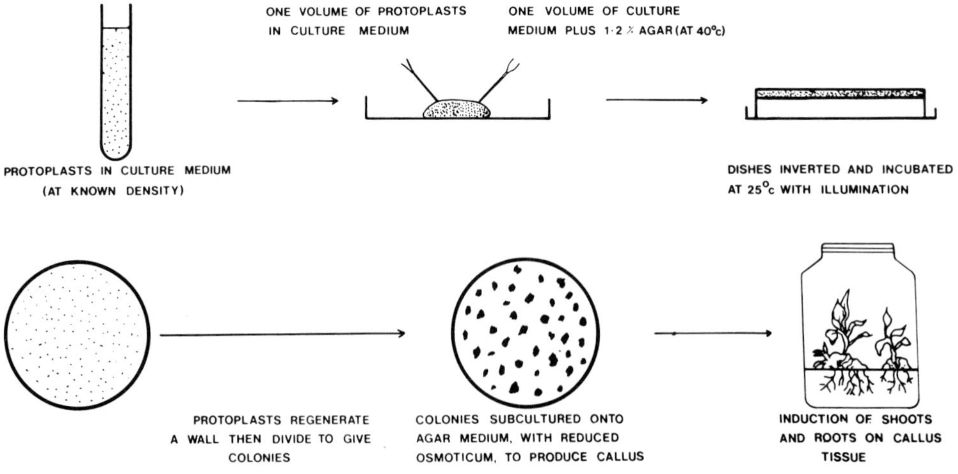

Fig. 3.3. Culture of protoplasts.

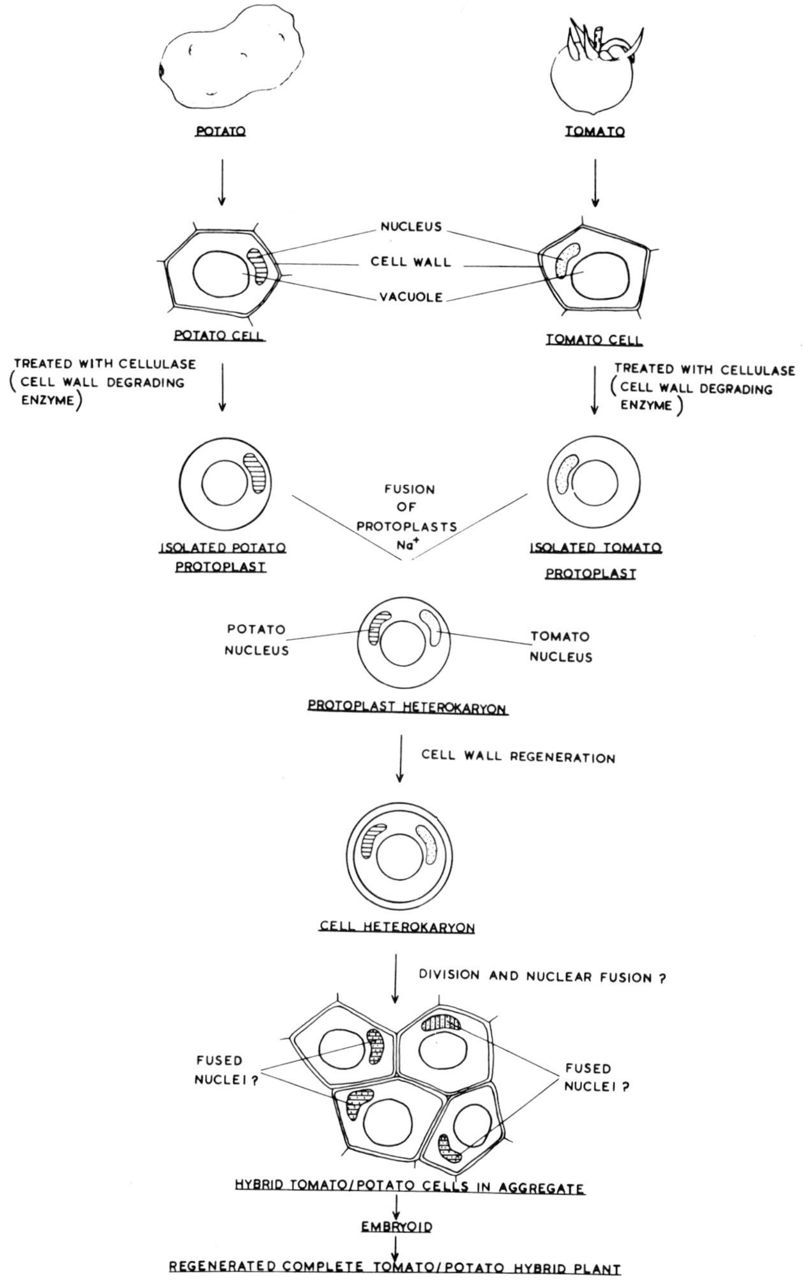

Fig. 3.4. Suggested sequence for somatic hybridization of plants: A possible tomato-potato hybrid.

fusion of somatic diploid protoplasts results in the production of allotetraploid plants. This contrasts with the sexual hybridization of the corresponding species which would produce diploid plants. An example where tetraploids from somatic fusion would be useful is the *Lolium-Festuca* hybrid. In this cross the diploid, even if obtained sexually, often lacks fertility, and chromosome doubling by colchicine treatment is frequently impossible. Thus tetraploid production somatically between these species might produce fertile hybrids.

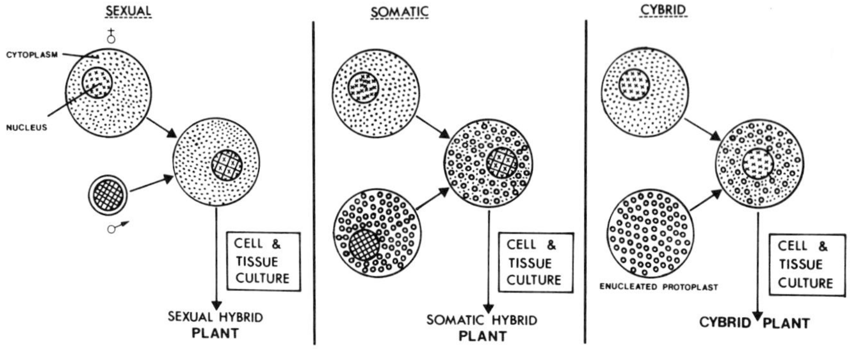

Fig. 3.5. Plant hybridizations.

In sexual hybridization, the cytoplasmic contribution from the male parent usually is very slight, and normally maternal cytoplasmic inheritance occurs. With the fusion of protoplasts from somatic cells, there is a more or less equal cytoplasmic contribution from the two parents (Fig. 3.5). As of yet, it is not known whether the two cytoplasms are stable together. There appears to be a tendency for the chloroplasts from one of the species to be eliminated, but further biochemical analysis, utilizing chloroplast and mitochondrial DNA restriction nuclease profiles is required before definite conclusions can be reached on this point. Previous methods of analysis have utilized Fraction 1 (ribulose diphosphate carboxylase-oxygenase) polypeptide electrophoresis patterns. However, chloroplast DNA may not be expressed in these novel cytoplasmic and nuclear situations. Moreover, species within the same genus (and sometimes in related genera) may have identical profiles of Fraction 1 polypetides (Gatenby and Cocking, 1978). What is of importance for plant breeding is that the cytoplasmic mix obtained from protoplast fusions is novel. Already, somatic hybridization has been used to transfer cytoplasmically based male sterility from one genotype to another (Izhar and Power, 1979). It is important to know whether such mixed cytoplasms will influence genetic variability in somatic crosses—such cytoplasmic influences are important in crop yield of *Triticales* (Hsam and Larter, 1974). The plant genetic manipulations described herein could be extended to permit the fusion of enucleated plant protoplasts with protoplasts proper, which would give cybrid plants that contained a mixed cytoplasm, and nuclear genes from only one species (Fig. 3.5). Careful comparison of the breeding behavior of cybrids with normal species would be of fundamental concern in plant breeding.

Another potentially useful feature of somatic hybridization is that it would enable plant hybridization without segregation associated with meiosis. Absence of segregation could be desirable in the breeding of tetrapolid potatoes from elite selected dihaploid strains.

Perhaps somatic hybridization will permit wider crosses to be made than is possible by sexual hybridization. Increasingly, plant breeders are utilizing *in vitro* culture methods, such as embryo culturing, to extend the range of sexual hybridizations possible. Already, it has been demonstrated that certain incompatibilities that occur with sexual hybridization, namely, those that are prezygotic (Power et al., 1979) and those due to embryo-endosperm incompatibilities (Schieder, 1978), can be bypassed by somatic hybridization (Table 3.1). It is not known whether most postzygotic incompatiblities that prevent sexual hybridization among species and genera also can be overcome by somatic hybridizations. Recently Cocking (1979) has suggested that the extent to which sexual hybridization could be used, with various *in vitro* refinements, to obtain interspecific hybrids has not been adequately explored. In such instances, success in producing somatic hybrids does not mean that sexual hybridization is impossible. Indeed, the present working hypothesis is that if diploidization is possible through the fusion of protoplasts to yield somatic hybrid plants, then (until it is proven otherwise) diploidization also should be possible by using sexual hybridization plus certain *in vitro* techniques. It is likely that for those species where there is postzygotic incompatibility, which cannot be circumvented by the application of *in vitro* culture techniques, protoplast fusions, in many instances, will produce somatic-cell hybrids which may then haploidize (see Fig. 3.6). This will result, provided plant regeneration capability is adequate, in the formation of a cybrid (Fig. 3.5), or in the formation of plants that closely resemble one parent species, but with some genetic features from the other. The current major challenge is to obtain phenotypic, chromosomal, and

Table 3.1. Methods for selection of somatic hybrids.

(1) Use of selective media for the selective growth of somatic hybrids.

Protoplasts	Fusion Procedure	Medium	Somatic Hybrid	Reference
Leaf	Polyethylene glycol (PEG)	No phytohormones	*N. glauca* ⊗ *N. langsdorffii*	Smith et al., 1976
Leaf	PEG	No phytohormones	*N. glauca* ⊗ *N. langsdorffii*	Chupeau et al., 1978
Leaf	PEG	Plus 2,4-D	*P. hybrida* ⊗ *P. parodii*	Power et al., 1977

(2) Use of light sensitive, chlorophyll deficient mutants for complementation selection.

Protoplasts	Fusion Procedure	Somatic Hybrid	Reference
Leaf	Ca^{++}/ high pH	*N. tabacum* ⊗ *N. tabacum*	Melchers & Labib, 1974
Leaf	Ca^{++}/ high pH	*N. tabacum* ⊗ *N. sylvestris*	Melchers, 1977
Leaf	PEG	*N. tabacum* ⊗ *N. tabacum* (chlorophyll deficient mutants only)	Gleba et al., 1975

Table 3.1. continued

(3) Use of complementation selection, coupled with differential media growth, by fusing wild type protoplasts with albino protoplasts.

Protoplasts	Fusion Procedure	Somatic Hybrid	Reference
Leaf (wild type) Cultured cell (albino)	PEG	*P. parodii* ⊗ *P. hybrida*	Cocking et al., 1977
Cultured cell (wild type and albino)		*D. carota* ⊗ *D. capillifolius*	Dudits et al., 1977
Leaf (wild type) Leaf (albino)	PEG/Ca^{++}/high pH	*Datura innoxia* ⊗ *D. discolor* and ⊗ *D. stramonium*	Schieder, 1978
Leaf (wild type) Cultured cell (albino)	Ca^{++}/high pH	*P. parodii* ⊗ *P. inflata*	Power et al., 1979

Table 3.1. continued

(4) Use of biochemical mutants for complementation selection of somatic hybrids.

Fusion	Selection	Somatic Hybrid	Reference
PEG	Nitrate reductase minus mutants	*N. tabacum* ⊗ *N. tabacum*	Glimelius et al., 1978
PFG	Kanamycin resistance mutant	*N. sylvestris* ⊗ *N. knightiana*	Maliga et al., 1977
PEG	Actinomycin D differential resistance (naturally occurring)	*P. parodii* ⊗ *P. hybrida*	Power et al., 1976

(5) Use of two nonallelic albino mutants for complementation selection of somatic hybrids.

Fusion	Selection	Somatic Hybrid	Reference
PEG/Ca^{++} high pH	Albino complementation to green	*Datura innoxia* ⊗ *D. innoxia*	Schieder, 1976
(Albino leaf protoplasts) Ca^{++} high pH	Albino complementation to green	*P. parodii* ⊗ *P. hybrida*	Power & Cocking, 1978

Table 3.1 continued

(6) Heterokaryon selection by visual means, mechanical isolation (single cell culture), and karyotypic identification of somatic cells.

Fusion	Selection	Somatic Cell Hybrids	Reference
PEG/Ca^{++}	Single heterokaryons isolated and cultured in microdroplets	*Arabidopsis thaliana* ⊗ *Brassica campestris* (chromosomally stable)	Gleba & Hoffman, 1978
PEG	Single heterokaryons	Rapid elimination of tobacco chromosomes in *Glycine max*/ *Nicotiana glauca*	Kao, 1977

(7) Selection of cybrids following fusion of protoplasts.

Hybrid	Selection	Reference
N. tabacum ⊗ *N. tabacum*	Plastome chlorophyll deficiency markers with genome markers	Gleba et al., 1975
N. tabacum ⊗ *N. debneyi*[a]	Plastome chlorophyll deficiency markers with genome markers	Gleba, 1978
N. tabacum ⊗ *N. debneyi*[a]	Flower morphology Leaf shape	Belliard et al., 1977
P. hybrida ⊗ *P. axillaris*[a]	Flower morphology	Izhar & Power, 1979

[a]transference of male sterility.

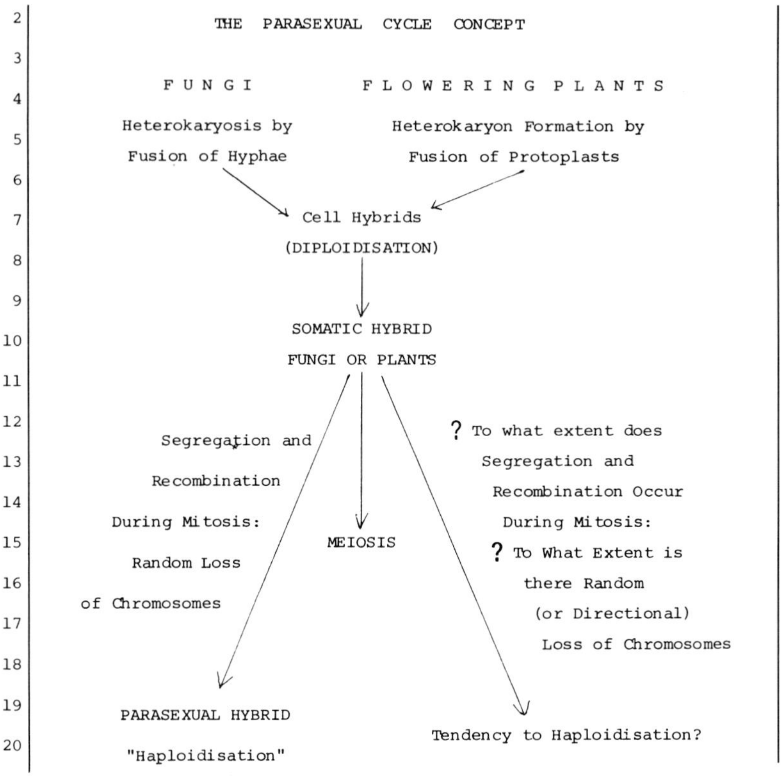

Fig. 3.6. The parasexual cycle concept.

DNA markers which will enable such parasexual-type hybrids to be identified and fully characterized.

In a few instances involving interspecific sexual crosses, x-ray treatment has effected gene substitution from alien chromosomes. In crosses between widely differing species, the breeder usually wishes to transfer only a single gene, such as a gene for disease resistance, from a wild to a cultivated species. In this exchange, it is important that deleterious genes should not be transferred. Sears (1956) utilized this approach for the transfer of leaf-rust resistance from *Aegilops umbellulata* to wheat (*Triticum aestivum*). Plant breeders are attracted to somatic cell fusions to provide experimental material for irradiation (including laser beam studies) and chemical treatment to obtain gene transfers between sexually incompatible interspecies heterokaryons. Generally, of course, yielding ability of crops is influenced by the cumulative effect of many genes, so such procedures may have limited applicability. However, somatic-cell hybridization and various *in vitro* techniques that can be appended to it could result in increased genetic variability for our crop plants.

Another approach involves the use of plant tissue cultures, particularly protoplasts, for gene transfer using plasmids as vectors. Recent studies suggest that protoplasts are ideal materials for plasmid interaction investigations, especially with *Agrobacterium* plasmids which may prove useful as a vector for the transfer of genes into plants. Protoplasts can be transformed by isolated *Agrobacterium* plasmids (Davey et al., 1979).

Clearly, cell biologists and plant breeders are still at a very early stage in the application of tissue culture technology to crop improvement. For sure though, plant breeders must be realistic in their use of this new technology as an addition to, and not a substitute for, conventional breeding methodology.

Discussion

K. L. GILES
C. E. GREEN
T. B. RICE

1. K. L. GILES. Many plant breeders seem to have an image of the use of tissue culture in plant breeding as a procedure whereby one has a small flask containing hundreds of thousands, or millions, of single cells into which is introduced an antibiotic, a selective agent, or a certain stress parameter in the environment to select out cells that when regenerated into whole plants possess a desired specific phenotype for use in an overall breeding program. Unfortunately, the idea that cells from higher plants are part-time microorganisms, which can be treated at the single cell level much like bacteria, is probably incorrect. Development and differentiation in multicellular organisms is based on a differential array and expression of particular gene sets, that is, genes that are switched on and off during different periods of the development of the plant. Selecting for certain genotypic characteristics at the cellular level is no guarantee that the selected genes will be expressed in fully differentiated and developed organisms. In some cases this may happen, but in general terms there is no guarantee that it will. Furthermore, plant breeders most often are interested in traits that are not expressed at the cellular level, that is, time to maturity or grain yield which cannot be selected at the cellular level. This imposes severe restriction on the use of single-cell cultures in plant breeding.

A point that needs additional emphasis is that many plant species have not yet had whole plants regenerated from callus or single cells. For example, with neither maize nor the forage grasses has it been possible to regenerate shoot and root tissues from single cells. These problems may be solvable. Within a year it may be possible to regenerate whole plants from monocotyledonous tissue much as can be done today with solanaceous tissue.

These comments are simply notes of caution. The real problem centers around the level of organization of the tissue that is used for selection. Single cells are the most difficult stage at which to practice selection, because even if the selection is successful, the cell must regenerate into a whole plant which makes use of its whole genetic information and not necessarily the gene selected for. Starting at a higher level of organization, such as embryo culture, may hold more promise. Possibly these considerations also have some inferences with respect to protoplast culture.

2. T. B. RICE. The application of plant tissue culture techniques to plant improvement is limited by the state of regeneration methods in field crops. When regeneration has been achieved, morphogenesis usually is associated with relatively complex tissue or organ cultures. In general, regenerating cultures have two important characteristics: (1) they produce large numbers of plantlets, and (2) they maintain the competence to regenerate plants over extended periods of subculture. Regeneration from isolated cells or protoplasts is not yet possible for most field crops. Consequently, in the near term, the important focus should be whether and how *in vitro* propagation methods themselves might be utilized to improve or extend the capability of current breeding practices.

The following is proposed to utilize *in vitro* propagation methods to extend the range of sexually mediated wide hybridization as one example of an approach that exploits current regeneration capabilities. Many examples exist of interspecific and intergeneric hybrids obtained with considerable difficulty that may survive to maturity but prove to be sterile. In this case it may be useful to employ *in vitro* vegetative propagation to save and propagate the hybrid genotype, and to promote genetic exchange which might lead to a fertile variant.

The steps in this example would be:

1. Sexual hybridization would be effected.

2. A stable regenerating culture would be initiated. This step depends upon having the hybrid develop to a stage where a tissue source can be excised. For cereals the immature embryo is the best tissue for generating cultures.

3. Use a mutagen to cause chromosomal alterations or changes in ploidy level. Colchicine is a very effective agent to use in culture, but radiation treatment could be used, or with time, spontaneous changes in ploidy and chromosome numbers likely would

occur. With directional chromosome loss, either spontaneously or induced, it might be possible to achieve introgression *in vitro*. This has been observed in hybrids among mammalian lines.

4. A large population of plants would be regenerated to permit sampling of the variability in the culture.

5. Evaluation for fertility and expression of parental traits would be done at the whole plant level. Note that no *in vitro* selection is proposed. The culture period was simply to allow a prolonged opportunity for genetic exchange.

The approach described here is currently being utilized in the barley program at Michigan State to transfer specific traits from *Hordeum jubatum* into commercial cultivars.

3. C. E. GREEN. If any tissue or cell culture technique is to have a bearing on plant breeding, there first must be efficient, productive, and reliable schemes for plant regeneration. The regenerated plants must be fertile and movable into conventional breeding programs.

For cereals, as a group, considerable progress has been made since 1975 in developing schemes by which plants can be regenerated from tissues. Cereal species for which good plant regeneration schemes have been developed are maize, oats, rice, and sorghum. For any one of these species, it is possible to generate thousands of plants from tissue culture.

With several cereals, it has been possible to initiate cultures that have a stable and high level capability for the regeneration of plants. Usually, these cultures are initiated from immature embryos of controlled crosses made on a field or greenhouse grown plant. The immature embryos, when placed on an appropriate culture medium and incubated, will develop into the desired culture. Histologically these cultures are complex as indicated by the presence of several tissue types. The appearance of chlorophyll, shoot apices, and small leaves signals early plant formation. When the shoots are apparent, the culture should be transferred onto a regeneration medium for further growth. Ultimately, the plants are moved into soil and grown via conventional procedures. Sometimes a culture transferred to a regeneration medium will not produce plants. This may be due to physiological or genetical disturbances in the culture. With these tissue culture procedures it is possible, however, to generate substantial numbers of plants, to raise them to maturity, and to obtain crosses and progeny for a conventional program. It is relatively easy to carry out such plant regeneration in maize.

A similar system for regenerating plants from cultures is available for sorghum by which large numbers of plants can be obtained. As with maize, plants derived from sorghum come from complex tissue cultures.

Regeneration of oat plants from tissue culture has been successful for several genotypes. In some cultures of oats, considerable variation in chromosome makeup occurs. For example, in one sporocyte from a regenerated oat plant a heteromorphic chromosome pair was found where a portion of one chromosome was missing. This chromosome instability can be utilized for the production and isolation of monosomic plants.

Plant regeneration schemes have been developed for some forage grasses. This technique may have particular utility for breeding forage grasses because oftentimes these species have low seed fertility and the breeder must depend on vegetative propagation for increasing plant populations. With such species, plant regeneration from tissue cultures would provide a scheme for producing large plant populations. And, of course, this system might reveal new genetic variability (such as described for oats) among regenerated plants.

In summary, it is not yet possible to regenerate plants from single cells of grass or cereal species, but it is possible to reproducibly regenerate plants from tissue cultures of an increasing number of these species.

4. K. RUSSELL. Is the inability to regenerate whole plants or tissue from protoplast cells of cereals related to the fact that protoplasts may not carry the total genome?

5. E. C. COCKING. Indeed, maize stem protoplasts will divide and produce calluses, so protoplasts from cultures of cereals are capable of division. The chief difficulty has occurred in trying to culture mesophyll protoplasts of cereals, but these are highly differentiated types of protoplasts, so they may not be representative of what can be done with protoplast cultures. So the system will work for cereal protoplasts if suitably chosen from certain parts of the plant.

With respect to the totipotency of the cells of cereals, the entire genetic complement must be present in microspores because such cells have been used to regenerate plants of wheat, maize, rice, and the like. The evidence for totipotency is that fields of these species are being grown with genotypes obtained from microspores. Now, perhaps very specialized cells, such as a mesophyll protoplast, even though totipotent, are in a state that will not permit division and regeneration.

6. T. B. RICE. It is not surprising that attempts to culture protoplasts from cereal leaves have been unsuccessful since attempts simply to initiate cultures from the leaves of most grasses have not been successful. This is probably a developmental problem due to the differentiated characteristics of this specific tissue.

7. J. DANIELS. Tissue culture techniques are being researched for culturing plantlets of Douglas fir and loblolly pine using cotyledonous tissues. The interest in this technique is to mass produce selected genotypes as a possible substitute for production of seedlings from seed, which often is time consuming and unpredictable. Selections in our seedling test nurseries are now made at eight years of age; and with tissue culture, selections could possibly be made even earlier. With these long-lived species, however, there is concern that subtle genetic changes might be induced, *in vitro*, which might not be manifested or detected early on. Will the likelihood of such occurrences be so great as to seriously limit the usefulness of tissue culture techniques for creating large numbers of trees of a selected genotype?

8. C. E. GREEN. Dr. Ono from Japan has initiated a series of cultures from a homozygous line of rice, and after maintaining these cultures for a few months, he regenerated approximately a thousand plants. These were evaluated directly and via their progenies. In subsequent generations, he found that a whole series of monogenically controlled chlorophyll traits resulted from spontaneous variation that accumulated in the culture period. We also found lines that varied substantially from the parental type in maturity, plant height, panicle length, and panicles per plant. So yes, spontaneous variation is quite prevalent in tissue cultures.

9. K. L. GILES. *Pinus radiata* and Douglas fir are maintained for a minimum period in the callus state before regeneration of the plants is begun. For *Pinus radiata*, the bud initiation comes from individual epidermal cells of the cotyledon. Because the callus phase is the stage where most chromosomal aberrations occur, the fact that this phase is minimal for these two species minimizes the chance of aberrations.

10. D. ALVEY. Are failures to regenerate plants from some maize tissue cultures due to genotype or another factor? And if they were associated with the genotype, what percent of maize genotypes give regeneration?

11. C. E. GREEN. At Minnesota, we have used immature embryos to establish tissue cultures from 30 genotypes and cultivars of maize. The desired cultures were initiated from 70 percent of these genotypes. The remaining 30 percent simply did not respond under the experimental conditions of this study.

Certainly, there are factors other than genotype that determine whether a tissue culture will generate new plants. For example, a culture may sector so that a portion is nonregenerating and another sector will continue to regenerate plants. Additionally, within the culture sector where regeneration can occur a certain fraction of the would-be plants abort. The proportion of nonregenerating areas to regenerating sectors or cultures depends on genotype, age of the culture, and the experience of the researcher.

12. D. RASMUSSON. Experience suggests that the genetic system of a plant is very finely tuned and that the system is easily disrupted by alien genes. So maybe research involving wide crosses should be minimized with a concomitant increase in research on perfecting the genetic system or devising better schemes for selection.

13. E. C. COCKING. Probably it would be unwise for science to establish what areas of plant breeding research should be emphasized. For instance, the limits of embryo culture in relation to interspecies hybridizations are unknown; so at this stage, the tissue culture work needs to proceed on as broad a front as possible.

14. K. RAWAL. There is much concern among plant breeders about the dangerously narrow genetic base of most of our field crops. Now, maybe in a few years, crop cultivars developed via these new innovative techniques will be used in agricultural production. If these cultivars are derived from plants which in turn have been derived from single-cell cultures, won't the genetic base be narrowed even further with the result that our crop production will become even more vulnerable to epidemics of pests?

15. K. L. GILES. As has been said before, there is a tendency for callus cultures to generate genetic variation with the result that regenerated plants from one callus may be genetically variable. If this is true, maybe tissue culture techniques applied to plant breeding will broaden the genetic base of a crop used in agricultural production rather than narrowing it.

16. K. RAWAL. The concern is that selection would be practiced within a genotype that responds well to the medium and techniques of tissue culture; so the initial genetic base for developing cultivars would be very narrow at the starting point, and onto this, very stringent selection would be superimposed. Of necessity, this would narrow the germplasm base of the crop.

17. K. L. GILES. Of course, you are assuming that the tissue culture technique would be the only breeding scheme used and that a single genotype would be employed. It is likely that breeders would have sufficient wisdom to use additional breeding procedures and more than a single genotype.

18. C. E. GREEN. From a philosophical point of view, tissue and cell culture techniques are not apt to supplant conventional plant breeding procedures. They will be supplements to current procedures. And their use probably will require a team of researchers that includes an individual who is competent in tissue culture and another individual who is competent in plant breeding.

It is possible that by employing a tissue culture system, a breeder could utilize only a portion of the total array of genotypes in his or her breeding nursery. From that point of view, the germplasm base represented in future cultivars might be narrowed. On the other hand, one potential application for tissue culture systems could be directed to improving elite genotypes via selection among the genetic variants that occur spontaneously in calluses.

19. M. J. GOODE. Generally, biologists accept the concept that all cells of a given plant are identical genotypically. They may react differently phenotypically according to the organ within which they occur or the function they perform. But here, it is said that regenerated plants from one callus tissue may be genetically variable. Does this result because spontaneous mutations are frequent in callus tissue, or are the millions of cells in the tissue used to establish the callus genetically variable before the callus is formed? Maybe the concept that all cells of a plant are genetically identical is in error.

20. E. C. COCKING. Of course, this is a fundamental consideration in tissue culture, and it is an area that is relatively unexplored. With mature differentiated plant tissue, cells will change from the differentiated into the division stage at different times. And, certain cells will mask other cells. It is an illusion to expect all cells of a tissue

in a medium to respond uniformly. One has greater uniformity if the cultures are initiated from a single cell. This opportunity for cells in the system to respond differentially brings all sorts of problems.

21. K. L. GILES. Development itself requires a differential pattern of gene expression; so when starting with differentiated tissue, the genes within individual cells of different tissues are at different states of expression, and on that basis, it is not surprising that once the cells are put into culture they react differently.

22. T. B. RICE. Several factors may influence the frequency of genetic variants observed in regenerated plants. First, many vegetatively propagated species are genetic chimeras. Different preexisting genotypes can be separated and cloned through tissue culture. Second, the results may be a function of time in the culture. There is a strong selection pressure on the cells when a culture is initiated, so spontaneous variants due to subtle or major genetic changes may have a growth advantage *in vitro*.

23. C. E. GREEN. There is potential for somatic mutations to occur in plants as they are growing. The plant is exposed to agents in the air, to radiation, and the like, and it is well documented that somatic mutations do occur. So, upon regenerating plants from individual cells in a culture, there is opportunity to recover variability that resulted from somatic mutations. Such variants may well be heritably stable.

24. J. DANIELS. Perhaps the tissue used as seed stock may have an impact on the growth that one gets. With Douglas fir, propagation via rooted cuttings produces plants that grow like a branch and not like a tree. When cotyledons were used to establish tissue cultures, still a majority of plantlets produced were plagiotropic in growth. Has this possible influence of the seed stock type (namely, "C-effect") been observed in other plants, and if so, how serious are its effects?

25. K. L. GILES. It is surprising that the Douglas fir species retains its plagiotropic growth because in my experience with *Populus* species, the normal products are true juvenile plants.

26. E. C. COCKING. A case that is parallel to the one with Douglas fir is one in North India, where researchers have propagated

Cinchona plants via cuttings; all regenerated plants formed adventitious roots rather than strong tap roots, which are normal for this species. Probably, if one could develop a tissue culture system to produce embryoids, some of these problems would disappear, but the development of embryoids is more difficult to achieve. Perhaps a case like this one will require fundamental study at the single-cell level. If the system can be taken to the single-cell level, it is possible to direct the regeneration via either a callus state or an embryoid stage by altering the environment of the system.

27. M. A. DO VALLE RIBEIRO. A colleague of mine has produced calli from aborted triploid embryos obtained in crosses between diploid perennial ryegrass (*Lolium perenne*) and tetraploid Italian ryegrass (*L. multiflorum*). Plants were regenerated from shoot primordia cultured in a medium in which colchicine was incorporated. However, the objective of this piece of research was not achieved as chromosome duplication did not occur. What special techniques would make duplication more sure when colchicine is incorporated into the medium?

28. C. E. GREEN. Probably, in most cases where chromosome doubling in tissue culture is sought, it will occur spontaneously. Reduplication of chromosomes has been observed repeatedly in tissue culture, particularly in haploid cultures.

29. E. C. COCKING. At the Welsh Plant Breeding Station, researchers have undertaken a program with *Lolium* and *Festuca* to fuse diploid protoplasts which would give tetraploids immediately, but to date the culture methods are not resolved.

30. R. E. FREEMAN. Anther and pollen culture generally is difficult to conduct except in the family *Solanaceae*, namely in genera *Nicotiana* and *Datura*. Is the difficulty with establishing anther or pollen cultures with other plants related to stage of microspore development?

31. E. C. COCKING. Yes, having the correct stage of microspore development must be an important element in the establishment of anther or pollen cultures. It requires a major shift in development for a microspore to go from normal development into mitotic division. A small percentage of the population of cells can be redirected by manipulation of the media empirically. If the system can be

adequately controlled and tuned so that the whole of the population is at the same stage, then success might be obtained in more species. Right now the state of the art is a trial and error and a numbers game.

32. B. C. MUSTAIN. It was suggested that selection for resistance to osmotic stress, production of certain metabolites, and the like, could be done on callus tissue. And at Minnesota, there is some indication that by using pathotoxins from *Helminthosporium maydis* race T in the media of callus cultures from maize tissue susceptible to this disease, it was possible to obtain regenerated maize plants that are resistant to race T.

To what extent do you foresee use of callus for *in vitro* selection, vs. use of suspension cultures, which can be more difficult to handle? And, in the case of the Minnesota work, was the *Helminthosporium* resistance a new source of resistance, or was the selection simply against T-cytoplasm mitochondria and in favor of normal mitochondria?

33. C. E. GREEN. At Minnesota, the toxin from race T *H. maydis* was applied to Texas-cytoplasm tissue cultures, and cell lines resistant to the toxin were found. Regenerated plants from these resistant cell lines were crossed to susceptible maize plants to check the heritability of the resistance. It was maternally transmitted and it has been transferred through four or five sexual cycles without breaking down. And finally, the resistance of the regenerated plants and their progeny was examined further by infecting plants with spores of the fungus. Plants carrying the selected trait were resistant to the disease-causing organism. Another example of selecting maize mutants at the tissue-culture level is the success in selecting heritable resistance traits that affect amino acid biosynthesis. There may be more opportunities to select at this level than are appreciated currently.

34. E. C. COCKING. Selections can be carried out at various levels of tissue organization in culture. At extreme levels, one might select at the single-cell level, as selection among protoplasts or at varying stages of aggregation, or as occurs in suspension cultures, or with callus. However, the chances of success in selection probably decrease as the complexity of the culture increases. But that is not to say that it is impossible to carry out sensible selection with aggregates of cells.

35. B. C. MUSTAIN. Were blight resistant maize plants still male sterile?

36. C. E. GREEN. No, the blight resistant maize plants also were male fertile. Of course, the question now is, Are the mitochondria selected and expressed in these regenerated cell lines and plants N-type or altered T-type? The pattern of the mitochondrial DNA from progeny from the resistant cell lines is similar to the standard DNA from T-mitochondria; so the implication is that the mitochondria in the resistant cell lines were still T- and not N-type.

There was a change in the pattern of the mitochondrial DNA from regenerated plants. So selection for resistance to *H. maydis* race T toxin in tissue culture identified small alterations in the T-mitochondrial DNA.

37. D. N. DUVICK. In Israel, *Petunia* species were crossed somatically, giving rise to a new type of cytoplasmic sterility. Is the cytoplasm of this somatic hybrid a mixture of cytoplasms from the two species, or is it predominantly from one species?

38. E. C. COCKING. That somatic hybrid has not been analyzed for cytoplasm composition, but a somewhat similar case of male sterility transfer in tobacco has been investigated. The workers were interested in whether genomes from both species were present. They found that only one of the chloroplast genomes was maintained in the somatic hybrid. However, this raises the question as to the basis for the male sterility that is cytoplasmic.

39. T. L. TEW. One use of tissue culture may be as a method for conserving germplasm of asexually propagated species over long periods of time. Considerable genetic variability has been observed among subclones derived from sugarcane cultivars as a result of chromosomal losses or gains. This type of variability has been enough that one would question whether tissue culture techniques should be used for conserving sugarcane germplasm because the conservation of its genetic integrity could not be guaranteed. Now, considering the level of genetic variation observed in tissue cultures of other asexually propagated species, can tissue culture really be considered seriously as a method for conserving or storing germplasm?

40. C. E. GREEN. It is my impression that sugarcane plants may

exhibit more genetic instability than most other agricultural species do. Certainly, if a species has a capability for high frequency of genome modification, and added to this is the potential for additional change that occurs in a tissue culture environment, probably tissue culture would not be a good method for conserving germplasm of that species. Generally, however, conservation of germplasm is done via meristem preservation and not via callus culture. Perhaps the high degree of organization that meristem primordial tissues have will permit greater genetic stability. Experience suggests that the more highly differentiated a tissue culture is, the greater is its genome stability. If this is true in general, one would expect preserved meristems to be reasonably stable.

41. T. L. LUND. There has been much furor in the press about the safety aspects of genetic engineering. Are there any legal restrictions or safety problems with the techniques that have been discussed in this session?

42. K. L. GILES. There are no legal implications with respect to the techniques discussed in this session. The National Institute of Health regulations refer solely to recombinant DNA. Probably there are no health and safety problems to the persons handling tissue cultures from higher plants. Care certainly is needed in the handling of parasitic fungi or other pathogens that may be used in selection experiments with tissue cultures. However, this applies to the use of pathogens in a conventional breeding program also.

43. E. C. COCKING. When bacteria genes for drug resistance and heavy metal resistances have been inserted into plants via the plasmid technique and have been expressed, indeed there will be a need to meet particular containment requirements. Of course, with higher plant systems, the problem of containment is much easier than with bacteria.

44. D. HILLERISLAMBERS. Have there been any experiments to relate the magnitude of differences in field resistance to chemical stress among different crop species to the magnitude of differences that show in their *in vitro* cell cultures?

45. C. E. GREEN. Not to my knowledge.

REFERENCES

Ammirato, P. V. 1980. Somatic embryogenesis and plantlet development in suspension cultures of *Dioscorea floribunda.* Ann. Bot. (in press).

Belliard, G., G. Pelletier, and M. Ferault. 1977. Interspecific hybridisation in plant breeding, pp. 237-42. *In* Sanchez-Monge, E., and F. Garcia Olmedo (eds.), Proc. of 8th Congr. of EUCARPIA, May, 1977, Madrid.

Bhojwani, S. S. 1976. Reproduction in higher plants. New Delhi Press.

Bhojwani, S. S., P. K. Evans, and E. C. Cocking. 1977. Protoplast technology in relation to crop plants: Progress and problems. Euphytica 26:343-60.

Boulter, D., and O. M. Crocomo. 1977. Plant cell culture implications: Legumes, pp. 615-31. *In* Sharp, W. R., P. O. Larsen, E. F. Paddock, and V. Raghavan (eds.), Plant cell and tissue culture: Principles and applications. Ohio State Univ. Biosciences Colloqu. No. 4.

Bourgin, J. P. 1978. Valine resistant plants from *in vitro* selected tobacco cells. Mole. Gen. Genet. 161(5):225-30.

Chupeau, Y., C. Missonier, M. C. Hommel, and J. Goujaud. 1978. Somatic hybrids of plants by fusion of protoplasts. Mole. Gen. Genet. 105(3): 239-45.

Cocking, E. C. 1978. Selection and somatic hybridisation. IAPTC Conf. August 20-25, Calgary.

Cocking, E. C. 1979. Parasexual reproduction in flowering plants, pp. 665-71. *In* Symp. on Reprod. in Flowering Plants. Christchurch, N.Z.

Cocking, E. C., D. George, J. J. Price-Jones, and J. B. Power. 1977. Selection procedures for the production of inter-species somatic hybrids of *Petunia hybrida* and *Petunia parodii.* II. Albino complementation selection. Plant Sci. Lett. 10:7-12.

Cooper, K. V., J. E. Dale, A. F. Dyer, R. L. Lyne, and J. T. Walker. 1978. Hybrid plants from the barley x rye cross. Plant Sci. Lett. 12:293-98.

Davey, M. R., and E. Thomas. 1975. From single cells to plants. Wykham Publ., London.

Davey, M. R., E. C. Cocking, J. Freeman, N. Pearce, I. Tudor, J. Schell, and M. van Montagu. 1979. Transformation of *Petunia* protoplasts by isolated *Agrobacterium* plasmids. Nature (submitted).

Dudits, D., Gy. Hadlaczky, E. Levi, O. Fejer, Za. Haydu, and G. Lazar. 1977. Somatic hybridization of *Daucus carota* and *D. capillifolius* by protoplast fusion. Theor. Appl. Genet. 51: 127-32.

Gatenby, A. A., and E. C. Cocking. 1978. The polypeptide composition of the sub-units of Fraction 1 protein in the genus *Lycopersicon.* Plant Sci. Lett. 13:171-76.

Gegenbach, B. G., C. E. Green, and C. M. Donovan. 1977. Inheritance of selected pathotoxin resistance in maize plants regenerated from cell cultures. Proc. Natl. Acad. Sci. USA 74:5113-17.

Gleba, Y. Y. 1978. Nonchromosomal inheritance in higher plants as studied by somatic cell hybridization, pp. 775-88. *In* Sharp, W. R., P. O. Larsen, E. F. Paddock, and V. Raghavan (eds.), Plant cell and tissue culture: Principles and applications. Ohio State Univ. Biosciences Colloqu. No. 4.

Gleba, Y. Y., R. G. Butenko, and K. M. Sytnik. 1975. Fusion of protoplasts and parasexual hybridisation in *Nicotiana tabacum* L. Dok. Acad. Nauk SSR, 221:1196-98.

Gleba, Y. Y., and F. Hoffman. 1978. Hybrid cell lines of *Arabidopsis thaliana* + *Brassica campestris:* No evidence for specific chromosome elimination. Mole. Gen. Genet. 165:257-64.

Glimelius, K., T. Eriksson, R. Grafe, and A. J. Muller. 1978. Somatic hybridisation of nitrate reductase-deficient mutants of *Nicotiana tabacum* by protoplast fusion. Physiol. Plant 44:273-77.

Green, C. E., and R. L. Phillips. 1974. Potential selection system for mutants with increased lysine, threonine and methionine in cereal crops. Crop Sci. 14:827-30.

Guha, S., and S. C. Maheshwari. 1964. *In vitro* production of embryos from anthers of *Datura.* Nature 204:497.

Hsam, S. L. K., and E. N. Larter. 1974. Influence of source of wheat cytoplasm on the nature of proteins in hexaploid triticale. Can. J. Genet. Cytol. 16: 529-37.

Izhar, S., and J. B. Power. 1979. Somatic hybridisation in *Petunia*: A male sterile cytoplasmic hybrid. Plant Sci. Lett. 14:49-55.

Kao, K. N. 1977. Chromosomal behavior in somatic hybrids of soybean: *Nicotiana.* Mole. Gen. Genet. 150:225-30.

Kimber, G., and R. Riley. 1963. Haploid angiosperms. Bot. Rev. 29:480-531.

King, P. J., I. Potrykus, and E. Thomas. 1978. *In vitro* genetics of cereals: Problems and perspectives. Physiol. Veg. 16(3):381-99.

Maliga, P., G. Lazar, F. Joo, A. H.-Nagy, and L. Menczel. 1977. Restoration of morphogenic potential in *Nicotiana* by somatic hybridisation. Mole. Gen. Genet. 157:291-96.

Melchers, G. 1976. Microbial techniques in somatic hybridization by fusion of protoplasts, pp. 207-15. *In* Brindley, B. R., and K. R. Porter (eds.), International cell biology, 1976-77: Papers presented at the 1st international congress on cell biology, Boston, Massachusetts. Rockefeller Univ. Press, 1977.

Melchers, G., and G. Labib. 1974. Somatic hybridisation of plants by fusion of protoplasts. 1. Selection of light resistant hybrids of haploid light sensitive varieties of tobacco. Mole. Gen. Genet. 135:277-94.

Nabors, M. W. 1976. Using spontaneously occurring and induced mutations to obtain agriculturally useful plants. Bioscience 26:761-67.

Nitzsche, W., and G. Wenzel. 1977. Haploids in plant breeding. Adv. in Plant Breed. (suppl. J. Plant Breed.). Vol. 8.

Poehlman, J. M. 1959. Breeding field crops. Holt, Rinehart & Winston.

Power, J. B., S. F. Berry, J. V. Chapman, and E. C. Cocking. 1979. Somatic hybrids between unilateral cross-incompatible *Petunia* species. Theor. Appl. Genet. 54:97-99.

Power, J. B., and E. C. Cocking. 1978. Unpublished results.

Power, J. B., S. F. Berry, E. M. Frearson, and E. C. Cocking. 1977. Selection procedures for the production of inter-species somatic hybrids of *Petunia hybrida* and *Petunia parodii.* 1. Nutrient media and drug sensitivity complementation selection. Plant Sci. Lett. 10:1-6.

Power, J. B., E. M. Frearson, C. Hayward, D. George, P. K. Evans, S. F. Berry, and E. C. Cocking. 1976. Somatic hybridisation of *Petunia hybrida* and *P. parodii.* Nature 263:500-502.

Rangaswamy, N. S. 1977. Applications of *in vitro* pollination and *in vitro* fertilization, pp. 412-25. *In* Reinert, J. and Y. P. S. Bajaj (eds.), Applied and fundamental aspects of plant cell, tissue, and organ culture. Springer-Verlag, Berlin.

Schieder, O. 1978. Somatic hybrids of *Datura innoxia* Mill + *Datura discolor* Bennh and of *Datura innoxia* Mill + *Datura stramonium* L. var. *tatula*. 1. Selection and characterization. Mole. Gen. Genet. 162:113-19.

Schieder,O. 1977. Hybridisation experiments with protoplasts from chlorophyll-deficient mutants of some Solanaceous species. Planta 137:253-57.

Scowcroft, W. R. 1977. Somatic cell genetics and plant improvement. Adv. Agron. 29:39-81.

Sears, E. R. 1956. The transfer of leaf-rust resistance from *Aegilops umbellulata* to wheat, pp. 1-22. *In* Genetics in plant breeding. Brookhaven Symp. in Biol. No. 9. Brookhaven Natl. Lab., Upton, N.Y.

Smith, H. H., K. N. Kao, and N. C. Combatti. 1976. Interspecific hybridization by protoplast fusion in *Nicotiana:* Confirmation and extension. J. Hered. 67:123-28.

Widholm, J. M. 1976. Selection and characterization of cultured carrot and tobacco cells resistant to lysine, methionine, and proline analogs. Can. J. Bot. 54:1523-29.

CHAPTER 4

Chromosomal and Cytoplasmic Manipulations

S. J. PELOQUIN

THIS CHAPTER cannot possibly include all aspects of the use of cytogenetics in plant breeding, so topics covered are intentionally selective and arbitrary.

Cytogenetics, as many other areas of research, has in the past, and will continue in the future, helped to solve certain problems in breeding particular species. This can occur either indirectly through increased knowledge of the genetic architecture of a species or by the direct application of cytogenetic techniques to plant improvement.

The topics to be discussed in relating cytogenetics to plant breeding are: (1) haploids, (2) meiotic modifications, (3) gametophytic modifications, (4) somatic instability, and (5) endosperms.

It will be apparent that in many situations the results from one area are closely associated with progress in another topic area. For example, modifications of meiosis which result in the production of 2n spores will necessarily modify the chromosome number of the gametophyte. These topics include the cytogenetic aspects of all major phases of the life cycle of an angiosperm—an intention which is desirable. Emphasis is placed on haploids and meiotic modifications, because they are particularly important and pertinent in applying chromosome and cytoplasmic manipulations to plant improvement. The lack of emphasis on chromosome aberrations is not an indication that they are unimportant; rather, it is a recognition of the excellent and thorough treatment of these topics given by Burnham (1966) in Plant Breeding I.

Professor of Genetics and Horticulture, Department of Horticulture, University of Wisconsin, Madison.

HAPLOIDS

A haploid is defined as a sporophyte with the gametophytic chromosome number. The First International Symposium on haploids in higher plants provided a thorough discussion of all aspects of haploid research including their utilization in plant breeding. Several general features emerged from the proceedings (Kasha, 1974):

1. The potential of haploids in breeding and genetic research can be explored only if large numbers of haploids from many diverse parents can be obtained efficiently.

2. No one method of producing haploids is effective with many species. In fact, in economically important species where haploids have been obtained in reasonable numbers, a different procedure has been used for each species. For example, in maize (*Zea mays*, 2n = 2x = 20) they occur spontaneously (Chase, 1949), in potato (*Solanum tuberosum*, 2n = 4x = 48) following interploid crosses (Hougas et al., 1958), in barley (*Hordeum vulgare*, 2n = 2x = 14) through chromosome elimination following interspecific crosses (Kasha and Kao, 1970), in pepper (*Capsicum frutescens*, 2n = 2x = 24) as a member of a set of twins (Morgan and Rappleye, 1950), in cotton (*Gossypium barbadense*, 2n = 4x = 52) by semigamy (Turcotte and Feaster, 1967), in tobacco (*Nicotiana tabacum*, 2n = 4x = 48) by anther culture (Nitsch and Nitsch, 1969; Sunderland and Wicks, 1969; Tanaka and Nakata, 1969), and in wheat (*Triticum aestivum*, 2n = 6x = 42) through the use of alien cytoplasm (Tsunewaki et al., 1968).

3. The frequency of haploids can be increased dramatically in some species by use of special genetic stocks. In both maize (Chase, 1969) and potato (Hougas et al., 1964) the "pollinator" (source of the pollen) has a great influence on haploid frequency. Superior pollinators increase haploid frequency 5-20 fold over a wide variety of seed parents. Turcotte and Feaster (1969) found that doubled haploid 57-4 of Pima cotton is true breeding for semigamy and produces a very high frequency of haploids. Through the use of semigametic lines with the virescent 7 marker, androgenetic haploids from selected stocks can be obtained with ease. Kermicle (1969) observed an increase in haploid frequency was associated with a mutation, indeterminate gametophyte (*ig*), in maize. With the homozygous *ig* plant as female, he obtained more than 2 percent androgenetic haploids, a phenomenal increase of more than 1,000 fold compared to the average frequency of androgenesis in maize.

4. Haploids have been of significant indirect value to plant

breeding. Several examples will illustrate these contributions:

a) One of the best examples is in wheat, where haploids were the major source of the trisomics and monosomics (Sears, 1954). From these, Sears developed the complete series of nullisomics, monosomics, trisomics, and tetrasomics. The aneuploids have provided the basic materials for extensive studies related to the breeding, genetics, cytogenetics, and evolution of wheat.

b) A haploid of wheat deficient for chromosome 5B provided the evidence (along with the results obtained by Sears and Okamoto (1958) with aneuploids among hybrids between *T. aestivum* and *T. monococcum*) that homoeologous pairing was prevented by the activity of gene(s) on the 5B chromosome (Riley and Chapman, 1958). This concept that particular genes control the pairing specificity among related chromosomes revolutionized cytogenetic thinking about pairing in polyploids.

c) In potato (Peloquin, et al., 1966) and alfalfa (*Medicago sativa*) (Bingham, 1971), haploids have provided convincing evidence for the polysomic nature of these tetraploid crops.

5. The progress in using haploids more directly in developing new cultivars has been modest. The goals and approaches of employing haploids in cultivar breeding can be quite different depending on whether a species is a diploid, a disomic polyploid, or a polysomic polyploid.

Types of Polyploids

Polyploids are classified as allopolyploid and autopolyploid, but probably disomic and polysomic are better terms to use. The latter terms give a more precise conceptual picture of the nature of polyploids from the viewpoint of both cytogenetics and breeding (MacKey, 1970). Disomic polyploids, as exemplified by wheat, have regular bivalent pairing only between strictly homologous chromosomes, disomic genetics, homozygosity at each locus; and most important, the chance for heterozygosity, and its consequent heterosis, from interactions between homoeologous loci. In contrast, a polysomic polyploid, such as the tetraploid potato, can have meiotic pairing between any two or more of the four homologous chromosomes, tetrasomic genetics, and again most important, the opportunity for up to four alleles per locus is available for intra- and interlocus interactions in optimizing heterotic combinations. Haploids have been of particular value in establishing the importance of maximum heterozygosity in breeding both alfalfa (Dunbier and Bingham, 1975) and potato (Mendiburu et al., 1974).

Haploids and Cultivar Breeding

Some specific examples where progress has been made using haploids in breeding should be considered. Doubled haploids of maize have been used successfully in commercial hybrids (Chase, 1974), but it is debatable whether the haploid method has enough advantages in identifying exceptional genotypes to outweigh the extra effort needed in obtaining doubled haploids. A valuable tool in maize is the use of the gene *ig* for the production of androgenetic haploids in high frequencies (Kermicle, 1969). Haploid androgenesis, which involves the substitution of a sperm nucleus into maternal cytoplasm, is an efficient system for cytoplasmic conversion of valuable lines. With this system, one androgenetic individual in ten was diploid. These androgenetic diploids arose through reduction to the haploid level and subsequent doubling (Kermicle, 1974). Changing the cytoplasm of a line also can be accomplished easily in cotton with semigamy (Turcotte and Feaster, 1974) and in barley with chromosome elimination (Kasha and Kao, 1970). This system has been used in barley to determine the effects of *Hordeum bulbosum* cytoplasm on *H. vulgare* (Johns and Harvey, 1974). Plants with *bulbosum* cytoplasm differed from their normal counterparts in morphological, physiological, and chemical characteristics, and were significantly lower yielding.

Barley haploids are being used intensively to develop new cultivars (Kasha, pers. commun.). The time saved to obtain homozygous lines, the ease of selecting desirable characteristics, plus the apparent increased frequency of superior genotypes obtained in doubled haploids from F_1 hybrids provide the basis for exploiting the haploid method in breeding new cultivars. In spring, oilseed rape (*Brassica napus*), a doubled haploid of the cultivar 'Oro,' exceeds Oro in oil yield and has been released as the cultivar 'Maris Haplona' (Thompson, 1972). Doubled haploids from anther culture have also been used to improve the disease resistance of the leading Japanese flue-cured cultivar 'MC1610' of tobacco (*Nicotiana tabacum*) (Nakamura et al., 1974).

The haploid method has special application with dioecious plants for the production of inbreds and hybrid cultivars. In asparagus (*Asparagus officinalis*) where male plants outyield females, haploids have been obtained from polyembryonic seeds and anther culture. Haploids from anther culture, following chromosome doubling, yield both female (xx) homozygous diploids and supermale (yy) homozygous diploids. Crosses between yy male plants and any female produce male progenies entirely. The uniformity and high yield of several F_1 hybrids favors their use in commercial production (Therenin, 1974).

Haploids from polysomic polyploids have unique and exciting potential in breeding, genetic, and evolutionary research. The potential in breeding is particularly enhanced if the polyploid has wild and cultivated relatives at lower ploidy levels as occurs in potatoes, alfalfa, strawberries (*Fragaria virginiana*), coffee (*Coffee arabica*), blueberries (*Vaccinium* spp.), sweet potato (*Ipomoea batatas*), yams (*Dioscorea sativa*); as well as in many tropical forage grasses, tropical and subtropical trees and shrubs, ornamentals and other economically important plants. Haploids (2n = 2x = 24) of the commercial potato (2n = 4x = 48) have been useful in investigations related to genetics, cytogenetics, evolution, and disease resistance (Peloquin et al., 1966), but it was not entirely obvious how they could be used in practical breeding. Recent results (Mendiburu et al., 1974) indicate that they can be used most effectively in breeding with 2n gametes, so further discussion of this topic is included in the section on meiotic modifications.

MEIOTIC MODIFICATIONS

The modifications discussed will be mainly those affecting either chromosome pairing and crossing over or the production of 2n gametes.

Genetic Control of Pairing Specificity

More than 20 years ago Sears and Okamoto (1958) and Riley and Chapman (1958), with the use of aneuploids and haploids respectively, made the fundamental discovery that a gene (or genes) on chromosome 5B of wheat suppresses pairing between homoeologous chromosomes. When this gene, *Ph*, is missing, as in nullisomic 5B plants, homoeologous chromosomes from the A, B, and D genomes can pair with each other and with the homoeologous chromosomes from related species.

This discovery had obvious applications in germplasm transfer from wild species to cultivated wheat. Many wild species can be crossed successfully with wheat, but transfer of desirable characters from the related species to wheat had not been very successful. The difficulty is due to lack of pairing between wheat chromosomes and those from the related species. The discovery that lack of pairing between homoeologous chromosomes depends on pairing suppressors present in wheat chromosomes and that certain diploid relatives of wheat (*Aegilops speltoides*) have genes that are dominant to *Ph* presents opportunities to transfer genes from alien to wheat chromosomes. Either delete the 5B chromosome or cross to *Ae. speltoides* to provide the opportunity for exchange between homoeologous chromosomes. Both methods have been used successfully (Sears, 1972; Riley et al., 1968), but both have disadvantages. The most desirable situation is a simple deletion of the *Ph* locus. The potential advantages of such a deletion were recognized more than ten years ago, and recently this type of mutant has been obtained. Wall et al. (1971) obtained a mutant, *ph*, that permitted homoeologous chromosome pairing. The level of pairing in this mutant, however, is not as high as in nullisomic 5B.

A new induced mutant has been obtained recently by Sears (1977), which appears to be a deletion of the gene (genes) on chromosome 5B that suppress homoeologous pairing. The mutants have several advantages over nullisomy for chromosome 5B: (1) reduced aneuploidy in the progeny, (2) higher fertility than nulli-5B, tri-5D, and (3) recombinants between chromosome 5B and its homoeologues can be obtained. Thus, a fundamental finding on genetic control of pairing specificity has led to a valuable practical tool to obtain recombination with the genomes of a disomic polyploid and to transfer germplasm to cultivated wheat from related species. This type of mutant would be equally valuable in other disomic polyploids.

Effects of Heterochromatin

Heterochromatin may be responsible for a variety of meiotic alterations, but only speculation is currently possible on how it brings about these effects. However, use of heterochromatin can be made for increasing crossing over, changing the position of crossing over, reducing pairing between homoeologous chromosomes, and increasing pairing between homologous chromosomes.

There are several reports relating heterochromatin (B chromosomes, telomeres, and knobs) to modifications in chromosome behavior. A well-known genetic effect of B chromosomes in maize is nondisjunction of the B centromere at the second microspore division (Roman, 1947). This behavior combined with A-B translocations in maize has been used effectively to attack a variety of genetic problems (Carlson, 1978). Modest increases in crossing over in particular regions of the chromosomes were found in the presence of B chromosomes in maize (Hanson, 1969; Nel, 1973). Rhoades (1968) determined the recombination in maize plants homozygous for a transposition (*Tp* 9—a piece of chromosome 3 was inserted into the short arm of chromosome 9) and heterozygous for several markers including *C* and *Wx* flanking the transposition. The recombination between *C* and *Wx* was more than doubled when B chromosomes were present. This increase was accompanied by a reduction in the adjacent *Yg* - *C* region. Thus, it appears that B chromosomes can change the position of crossing over.

A striking change in chromosome pairing controlled by B chromosomes occurs in *Lolium* species hybrids (Evans and Macefield, 1972). Diploid hybrids (2n = 2x = 14) between *L. temulentum* and *L. perenne* have a frequency of 6.8 bivalents per sporocyte. But, in the presence of a pair of B chromosomes from *L. perenne*, the bivalent frequency was significantly reduced to 3.6, and some cells had only univalents. In induced tetraploids of hybrids without B chromosomes there were several multivalents per cell compared to primarily bivalents when B chromosomes were present. The B chromosomes, therefore, suppress pairing and chiasma formation between homoeologous chromosomes. Thus, they could be used to obtain meiotic stability in synthetic disomic polyploids. A similar effect of B chromosomes on homoeologous pairing has been detected in wheat (Dover and Riley, 1972). The presence of B chromosomes from *Aegilops mutica* or *Ae. speltoides* prevents pairing of homoeologous chromosomes in *T. aestivum* x *Ae. speltoides* hybrids that lack chromosome 5B; the presence of B chromosomes compensated for the absence of 5B in preventing homoeologous pairing. Telomeric heterochromatin of rye (*Secale cereale*) may affect homologous chromosome pairing in hexaploid triticale. Merker (1976) demonstrated, with isogenic lines differing only in the presence or absence of telomeric heterochromatin of a rye chromosome (probably 7), that in the absence of this heterochromatin the number of univalents at Met I was reduced by one-third. The presence or absence of telomeric heterochromatin of the short arm of chromosome 6 of rye, investigated by Roupakias and Kaltsikes (1977), had no significant effect on the number of univalents in plants. However, the overall chromosome pairing was significantly higher in plants without this heterochromatin as measured by reduction in open and increase in closed bivalents.

Seemingly, elimination of the large blocks of telomeric heterochromatin of rye chromosomes might increase the meiotic stability of triticale, because this heterochromatin may be responsible for some pairing failures in triticale.

2n Gametes

Meiotic modification that results in the systematic production of 2n gametes (gametes with the sporophytic chromosome number) have been found in *Datura* (Satina and Blakeslee, 1935), maize (Rhoades and Dempsey, 1966), and diploid potatoes (Mok and Peloquin, 1975a).

Unilateral Sexual Polyploidization

The utilization of 2n gametes and haploids are essential features of a new cytogenetic approach to breeding (Mendiburu et al., 1974). Haploids (2n = 2x = 24) of the common potato, *Solanum tuberosum* Group Tuberosum (2n = 4x = 48) have been hybridized with many related 24-chromosome, tuber-bearing species. Many of the F_1 hybrids obtained from crossing Tuberosum haploids and Phureja (cultivated 24-chromosome group from Colombia and Peru) were very vigorous and high yielding. When these 24-chromosome F_1 hybrids were crossed to 48-chromosome cultivars the progeny were almost all tetraploids; a process called unilateral sexual polyploidization (Mendiburu and Peloquin, 1976).

Many tetraploid families produced from tetraploid x diploid crosses where the diploid parent was a Phureja-haploid Tuberosum hybrid were very vigorous and manifested striking heterotic responses in terms of tuber yield (Mendiburu and Peloquin, 1971, 1977a; Mok and Peloquin, 1975b). The mean yield of several families significantly exceeded the yield of the parents. These families were also much more uniform than one would expect from crosses between parents that were highly heterozygous.

The basis of the very high yields in the progeny from particular tetraploid x diploid crosses resides in the method of 2n pollen formation in the diploid parent (Mok and Peloquin, 1975a). Cytological examination of microsporocytes revealed that in many cells the Met II spindles were parallel, rather than oriented as they are normally where the poles of the two spindles define a tetrahedron. The result of parallel spindles, following one cleavage furrow at Telo II, is two 2n microspores instead of the normal four n spores (Fig. 4.1.).

The genetic significance of parallel spindles is that it is es-

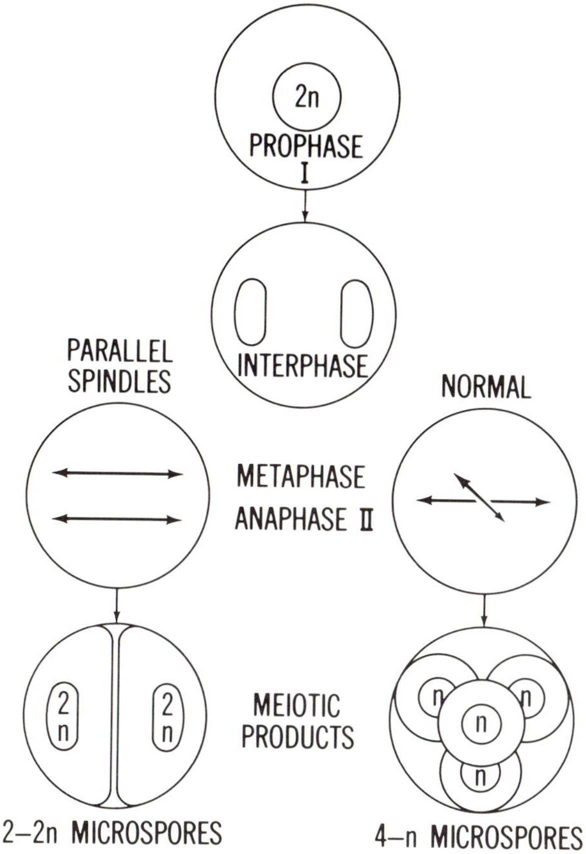

Fig. 4.1. A schematic representation comparing meiosis in cells with normal and parallel orientation of second division spindles.

sentially a first division restitution (FDR) mechanism. Therefore, all heterozygous loci between the centromere and the first crossover on all chromosomes are also heterozygous in the gamete. Approximately 75-80 percent of the heterozygosity is transmitted from parent to offspring with FDR, and a large fraction of the epistatic interactions are maintained also (Fig. 4.2).

FDR 2n pollen is a new method of transferring intra- and interlocus interactions from parent to offspring intact. It is also obviously an effective and efficient method of germplasm transfer. It is interesting that such a small cytological change (parallel spindles) provides the basis of a powerful breeding method. The meiotic mutant, parallel spindles (*ps*), is inherited as a simple

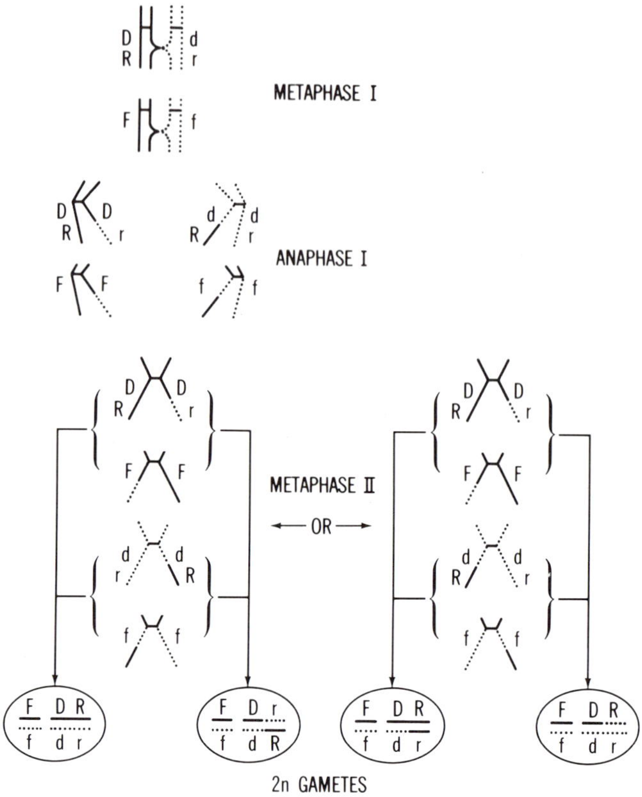

Fig. 4.2. Diagramatic representation of the genetic consequences of FDR as a result of parallel orientation of second division spindles. The genotypes of the meiotic products will be determined by the presence or absence of an effective crossover between the locus and the centromere. All loci between the centromere and the first exchange that are heterozygous in the parent will remain heterozygous in the gamete (D and F). Heterozygous loci between the first and second exchange will have an even chance of remaining heterozygous in the gamete (R). The loci from the centromere to the first exchange, on all pairs of homologous chromosomes, will display "horizontal linkage," such as F and D, so that all interactions involving these loci will be transmitted intact.

recessive (Mok and Peloquin, 1975c).

Heterotic responses, similar to those in tuber yield, have been obtained for the male gametophyte (Simon and Peloquin, 1976), and for both vegetative and seedling vigor. Vegetative vigor is used to evaluate potato vines as silage for ruminant animals. Selected clones have both large vine and tuber yields (Parfitt and Peloquin, 1977). The seedling vigor plus the relative uniformity of the progeny make FDR derived hybrids excellent material for investigating the use of botanical seed in potato production.

Two other meiotic mutants, premature cytokinesis one and two (pc_1 and pc_2) that form 2n pollen occur in diploid potatoes (Mok and Peloquin, 1975a). Premature cytokinesis follows the first meiotic division, no second division occurs, and two 2n microspores are formed. The 2n pollen formed by premature cytokinesis is genetically equivalent to second division restitution (SDR). In contrast to FDR, all heterozygous loci from the centromere to the first crossover will be homozygous in the gametes. Tuber yields obtained from tetraploid x diploid (FDR) hybrids were 40-60 percent larger than from similar tetraploid x diploid (SDR) hybrids (Mok and Peloquin, 1975b). This indicates the importance of intra and interlocus interactions because with SDR only 40-45 percent of the heterozygosity of the parent is transmitted to the progeny.

Bilateral Sexual Polyploidization

Tetraploids also can be obtained in potatoes from matings between diploids that produce 2n pollen and 2n eggs (bilateral sexual polyploidization). Intercrossing diploid hybrids (Phureja x haploid Tuberosum) produces families containing both diploid and tetraploid individuals. The tetraploid progeny are more vigorous and significantly outyield, by 30-40 percent, their diploid "full sibs." Several tetraploids from diploid-diploid crosses yield as much as standard cultivars (Mendiburu and Peloquin, 1977b). In contrast, tetraploids obtained from diploids via somatic doubling with colchicine have the same or less yield than their undoubled counterparts. These results have important breeding and evolutionary significance.

The synthesis of tetraploids from diploid-diploid matings provides a breeding method with great potential (Mendiburu et al., 1974). Superior unrelated diploid hybrids are developed from crosses between unrelated diploids whose chromosome sets have been upgraded by intensive breeding and selection before synthesis of the hybrids. The diploid hybrids are selected for adaptation, yield, other desired characteristics, and 2n gamete formation. Tetraploids with maximum intra and interlocus interactions can be obtained from intermating the diploid hybrids.

Near maximum benefits would be expected if both diploid

hybrids produced 2n gametes by FDR, and clones that produce 2n pollen by FDR have been found (*ps* mutant). However, it is difficult to identify the mode of 2n egg formation cytologically, so 2n eggs must be evaluated on the basis of progeny performance or genetics. Recently, a synaptic mutant in diploid potatoes which affects only megasporogenesis has been found (Iwanaga and Peloquin, 1979). A few FDR 2n eggs per ovary appear to be a class of meiotic product from the synaptic difficulty.

Note that the optimum genetic situation for maximum heterozygosity would be for the 2n gametes to have the same genotypes as their hybrid diploid parents. Fusion of these gametes would be the nuclear equivalent of somatic cell hybridization of the two diploid hybrids. Meiotic modifications (mutants) in which 2n gametes were formed through a "mitosislike" process (namely, all univalents, no crossing over, either all univalents divide in the first division and no second division follows or all univalents divide in the second division following an incomplete first division) would be ideal. Equally desirable would be the formation of 2n eggs by apospory, as occurs in some apomicts. The advantages of 2n gametes for breeding would appear to justify intensive searches for mutants which modify meiosis so that such gametes are produced.

Sexual Polyploidization vs. Somatic Doubling

The principles and procedures underlying the breeding methods with unilateral and bilateral sexual polyploidization are adaptable to the improvement of other polyploid crops. They also provide a more realistic method for producing tetraploid cultivars of normally diploid crops, particularly in contrast to somatic doubling. The advantages of unilateral sexual polyploidization over somatic doubling with colchicine in developing tetraploid from diploid cultivars have been clearly demonstrated by Skiebe (1958). He made excellent breeding progress in obtaining tetraploid from diploid x tetraploid crosses even though he did not know the mode of 2n egg formation.

Sexual polyploidization via 2n gametes is a distinctly separable process from somatic cell doubling. It has the advantages of providing heterosis, variability, minimal inbreeding, high fertility, and maximum heterozygosity which leads to new intra and interlocus interactions. These differences are of utmost importance in breeding and in evolution.

Odd Ploidy Levels

Meiotic modifications in uneven ploidy individuals (such as

triploids, pentaploids, and the like) lead to unusual results. The unique capacity of FDR by parallel spindles to overcome numerical nondisjunction provides triploids with a mechanism to significantly increase male fertility (Mok et al., 1975). A similar result was detected in monoploid peaches (*Prunus persica*) (Hesse, 1971). Maybe these findings are of breeding and evolutionary importance, because parallel spindles endow plants with uneven ploidy levels with the ability to produce balanced gametes in a systematic fashion; so triploids, for example, do not need to be "evolutionary dead ends." Another mechanism which increases the fertility of triploids has been described recently by Maan and Sasakuma (1977). They found that F_1 hybrids (2n = 3x = 21) of *Aegilops heldreichii* (2n = 2x = 14) and *Triticum durum* (2n = 4x = 28) had a high frequency of 2n male and female gametes and amphidiploid progeny. In about one-third of the microsporocytes the 21 univalents formed a Met I plate, and divided equationally at Ana I. Telo I was followed by cytokinesis, no second division occurred, and 21 chromosome gametes were formed.

GAMETOPHYTIC MODIFICATIONS

Several mutations have been found recently that significantly modify the postmeiotic events in the formation of the female gametophyte. These are not only of interest cytogenetically, but also for their direct or indirect value in plant breeding. Three examples, two in maize and one in soybeans (*Glycine max*) will be explained. The indeterminate gametophyte gene (*ig*) in maize has pleiotropic effects on seed development when homozygous *igig* plants are used as females (Kermicle, 1971). These effects include defective seeds due to increased ploidy level in the endosperm, polyembryony, heterofertilization, high frequencies of maternal haploids, and extremely high frequencies of paternal haploids with both types of haploids occurring nonpreferentially among seed with either single or multiple embryos. The pleiotropic action of *ig* is not, however, based on its effect on seed development, but instead on the female gametophyte. Normally the female gametophyte contains one egg and two polar nuclei in the central cell, but in *igig* plants proliferation of both eggs and polar nuclei occurs during development. Continued cell and nuclear division in the female gametophyte before fertilization accounts for the wide range of effects of the *ig* mutant on seed development. The value of *ig* in producing high frequencies of androgenetic haploids for nuclear-cytoplasmic substitution was discussed in the section on haploids. Its use in elucidating the nature of normal endosperm development is treated in the section on endosperm.

Another intriguing mutation in maize is the *r-X1* deficiency, an X-ray induced deficiency that includes the *R* locus on chromosome 10. Satyanarayana (unpublished) discovered that plants heterozygous for *r-X1*, when used as females, produced monosomic progeny. Further work with this deficiency indicated that when gametes carrying *r-X1* deficiency are fertilized by haploid pollen, approximately 11 percent monosomic and 11 percent trisomic plants occur (Plewa and Weber, 1973). Large numbers of monosomics have been obtained with this method, and they include monosomics for all chromosomes except number 5.

This is the first time most, if not all, of the monosomics have been produced in a diploid angiosperm. Ordinarily, monosomics are not transmitted through the gametophytes, because the n - 1 (nullisomic) condition is usually lethal to the gametophyte. The *r-X1* system induces a high rate of chromosomal nondisjunction during the mitotic divisions when the female gametophyte is formed from a megaspore. Because the developing gametophyte is either two, four, or eight nucleate, the n - 1 chromosome condition in one or more nuclei is compensated for genetically by the presence of one or more nuclei with n or n + 1 chromosomes. Thus, eggs with the n - 1 chromosome number are functional.

The monosomics have been used by Plewa and Weber (1973, 1975) to detect genes that condition lipid content and the composition of fatty acids in embryos by comparing plants monosomic for a specific chromosome with their diploid sibs. Detected differences are due to dosage effects of a gene(s) on that chromosome. Weber (1978) also had used monosomics for analysis of nucleolar formation in microspores. Microsporogenesis in a monosomic plant results in two of the four cells of a microspore quartet being nullisomic. Plants monosomic for chromosome 6 never contain a nucleolus in more than two of the four cells. In contrast, all microspores of a quartet regularly contain a nucleolus in plants monosomic for chromosome 1, 2, 4, 7, 8, 9, 10. Since only microspores nullisomic for chromosome 6 lack nucleoli, no factors on the other chromosomes are necessary for nucleolus formation in microspores. The availability of monosomic plants in maize provides the opportunity for developing and using chromosome substitution lines similar to those employed so successfully in wheat (Sears, 1969). Substitution lines can be used directly in breeding to transfer desirable genes from one line to another. Further, the contributions of different chromosomes to specific traits can be determined, and biometrical analyses appear possible (Law, 1966).

The male-sterile character in soybeans, controlled by a recessive gene ms_1, is an interesting mutant (Brim and Young, 1971). The male sterility is due to failure of cytokinesis following microsporogenesis, resulting in four nuclei in each microspore (Albertson and Palmer, 1979). More pollen mother cells occur in the anthers of male-sterile plants than in those of fertile sibs. Other characteristics associated with male-sterile plants include reduced female fertility, polyembryonic seeds, and haploids and polyploids in the progeny. Haploids, triploids, and tetraploids were found among polyembryonic seeds; and haploids, triploids, tetraploids, pentaploids, and hexaploids among monoembryonic seeds (Beversdorf and Bingham, 1977). Cytological analysis of the female gametophytes of male-sterile and related normal soybean plants indicated the presence of extra cells and nuclei in the female gametophytes in almost 60 percent of the ovules of male-sterile plants; up to four egg cells and as many as nine extra nuclei were observed in the central cell.

The presence of the extra cells and nuclei provides an explanation for the occurrence of haploids and polyploids in both monoembryonic and polyembryonic seeds (Cutter and Bingham, 1977). The action of the ms_1 gene on the female gametophyte appears to be similar to that of the indeterminate gametophyte mutant on the female gametophyte of maize—the eight nuclei of the female gametophyte continue to divide rather than differentiate into a seven-celled, mature female gametophyte. Also note that ms_1, like *ig*, provides for ploidy alterations in the embryo and endosperm that permit recovery of plant types not obtained with usual procedures. For example, triploids have not been obtained from diploid x tetraploid or tetraploid x diploid crosses in soybeans. This triploid block appears to be associated with endosperm failure in the tetraploid and pentaploid endosperms in the progeny of interploidy crosses. Triploids, however, occur regularly in the progeny of ms_1 plants. This circumvention of the triploid block can occur when a 2n egg occurs in the same female gametophyte as two n polar nuclei—double fertilization would give a normal triploid endosperm and an abnormal triploid embryo.

These three examples illustrate how cytogenetic modifications of the female gametophyte can generate unusual plant material for use in plant breeding. They were, however, essentially byproducts of research with other primary objectives. Concerted efforts designed to obtain similar and other variants in the gametophyte should be rewarding.

SOMATIC INSTABILITY

Somatic instability probably has some value in plant breeding, but with a few exceptions, it has been neglected. Herein, somatic instability will be used in a generic sense to include such apparently diverse phenomena as intraplant variations in chromosome number, somatic sectoring, "complement fractionation," reductional groups, somatic reduction, and chromosome substitution. Probably the common denominator of these phenomena resides in spindle abnormalities. Further, these abnormalities occur in a very low frequency in most plants, but this frequency can be greatly increased either by the effect of a particular genotype or by temperature and chemical treatments.

The most relevant and interesting example of somatic instability to plant breeding occurs in sorghum (*Sorghum bicolor*). True-breeding mutants were obtained from sorghum line Experimental 3, following colchicine treatment of seedlings (Franzke and Ross, 1952). These mutants differed dramatically from the parental line. As a direct contribution to sorghum breeding, one mutant, a dwarf, white-seeded sorgo-type, was released as the cultivar 'Winner.'

Several results support the hypothesis that diploid, complex, true-breeding mutants arise through chromosome substitution during reduction of colchicine-induced tetraploid cells back to diploid cells (Franzke and Sanders, 1964; Sanders and Franzke, 1969). These results include the following: (1) mutants differ from the parental line in many characters, (2) repeated occurrence of the same mutant type from a true-breeding parental line, (3) true-breeding mutants can be classified into distinct classes, (4) recovery of parental-like types from each of three colchicine-treated F_1 hybrid seedlings, and (5) similar diploid true-breeding mutants were obtained from diploid and tetraploid cytotypes of sorghum lines.

The concept of chromosome substitution is based on several assumptions for which there is support in the literature (Sanders and Franzke, 1976). These assumptions are that (1) sorghum is a disomic polyploid in which homoeologs can substitute for each other genetically, (2) bivalent pairing is genetically controlled, and (3) somatic reduction occurs. Sorghum has provided better evidence for somatic reduction than has any other higher plant. Plants heterozygous for two independent reciprocal translocations were treated with colchicine to induce true-breeding mutants. All true-breeding mutants obtained were homozygous for the nontranslocated chromosome in the four pairs of chromosomes involved in the translocations (Simantel and Ross, 1963). This could occur only if reduction and doubling or doubling and reduction of the chromosome number occurred; thus, it provides convincing evidence for the occurrence of somatic reduction in sorghum.

Somatic instability has been explored in grasses in relation to its breeding and evolutionary significance. Intraplant morphological variations were found in several grass species, interploid hybrids, interspecific hybrids, and an intergeneric hybrid (Nielsen, 1968). The materials included timothy (*Phleum pratense*), bromegrass (*Bromus inermis*), sudangrass (*Sorghum vulgare* var. *sudenense*), and sorghum-sudangrass hybrids. Differences in plants propagated asexually from the same clone were striking, and both sorghum and sudangrass types were isolated from asexual propagation of somatically unstable sorghum-sudangrass hybrids (Nielsen et al., 1969).

The results in grasses also can be explained by chromosome substitution as hypothesized in the true-breeding sorghum mutants. The important difference is that they occurred without treatment in the grasses. Spontaneous somatic instability has occurred in *Rubus* (Thompson, 1962; and Britton and Hull, 1957) and *Hymenocallis* (Snoad, 1955). Likely, many of the so-called "bud sports" (some of which have been very valuable) that have occurred in cultivars of many species arose via chromosome substitution. They often differ from the asexual parent in several to many characters suggesting chromosome rather than gene changes. Certainly, chromosome substitution should be explored as a breeding tool with particular species. Obviously, polyploid crops are the most likely candidates, because one could obtain homologous chromosome substitutions in polysomic polyploids and homoeologous substitutions in disomic polyploids. Also, it would provide an alternative breeding scheme for improving cultivars of species where sexual alteration of the genotype is either undesirable or impossible. Further, the success in obtaining desirable variants in plants regenerated from tissue culture of a clone (where no variation has been induced) may be dependent primarily on chromosome substitutions.

ENDOSPERM

The endosperm is an important and unique tissue. It is important in that (1) it is a major food and feed source, and (2) in

99 percent of the angiosperms it is necessary for normal seed development. It is unique in that (1) it occurs only in angiosperms, and (2) although it, like the embryo, is a product of fertilization, it does not provide for genetic continuity over generations. The cytogenetics of the endosperm is significant to plant breeding in interploid crosses, interspecific crosses, and haploid extraction that involves seed formation.

Little is known about endosperm genetics in relation to seed development and failure, even for maize, where considerable research has been done on the endosperm (Walden, 1978). Two areas of endosperm research that are relevant to plant breeding, (1) the basis of endosperm failure in interploid and interspecific crosses, and (2) the possible genetic basis of endosperm heterosis, will be discussed herein.

Many hypotheses have been proposed to explain the basis for normal seed development, and thus, the modifications that result in seed failures. Early hypotheses suggested that a particular balance of chromosome sets between the major plant parts of the developing seed (maternal tissue:endosperm:embryo) was necessary for normal seed development. Muntzing (1933) suggested that a 2:3:2 ratio of the ploidy levels of the maternal tissue:endosperm:embryo was a requisite for normal seed development. Others indicated either that the 3:2 ratio of chromosome sets in the endosperm and embryo was involved (Watkins, 1932; Boyes and Thompson, 1937) or that the important relation was a 2:3 ratio of maternal tissue to endosperm (Valentine, 1956). Various modifications of these schemes have been suggested to account for failure of seeds to develop.

Recent evidence from haploid extraction in potatoes (von Wagenheim et al., 1960) and from parthenogenesis and meiotic restitution in maize (Chase, 1964) indicates that normal seed development can occur when the standard ploidy balance between the maternal, endosperm, and embryo components of the seed has been significantly altered. For example, following tetraploid x diploid crosses in potatoes, normal hexaploid endosperms are present in seeds with either tetraploid, diploid, or no embryo. The results from potatoes and maize suggest that normal seed development and related seed failure problems are intrinsic to the endosperm. Consequently, Nishiyama and Inomata (1966) have hypothesized that the important factor for normal seed development is to have a 2:1 ratio of maternal to paternal genomes in the endosperm.

Lin (1975) provided convincing evidence to support this hypothesis with his elegant cytogenetic research with maize endosperm. He could generate endosperms with varying ploidy levels by using as females diploid plants homozygous for *ig* in crosses with normal diploid and tetraploid males. By correlating various endosperm ploidy levels with mature seed classes, he found that the normal seeds obtained following diploid (*ig*) x tetraploid crosses possessed hexaploid endosperms. Lin (1975) also used cytological markers to determine the parental origin of the sets of chromosomes in the endosperm. He found that hexaploid endosperms with a ratio of 4:2 chromosome sets of maternal and paternal origin, respectively, were normal, and those with 5:1 female to male sets were abortive. The 2:1 genome ratio hypothesis appears applicable to intraspecific, interploid crosses in most angiosperms.

This hypothesis, however, does not accommodate some results obtained with interspecific crosses in the tuber-bearing *Solanums* and in *Avena*. In both groups the ploidy level of the parents does not accurately reflect crossability (that is, some tetraploids behave as diploids in *Solanum* and some diploids as tetraploids in *Avena*); thus abnormal endosperm development can occur with a 2:1 balance of maternal to paternal genomes.

Hypotheses proposed to encompass these exceptions are both based on the premise that there are qualitative factors and not chromosome sets, which must be in a 2 maternal:1 paternal balance for normal endosperm development. These hypothetical qualitative factors have been labelled endosperm balance numbers (EBN) in *Solanum* (Johnston et al., 1980) and activation index (AI) in *Avena* (Nishiyama and Yabuna, 1978). Normal endosperm development occurs in the tuber-bearing *Solanums* when the EBN from female and male parents are 2:1. Deviations from this ratio result in abnormal endosperm development. The AI in *Avena* is expressed as the ratio of the activating value (AV) of a male nucleus to the response value (RV) of two polar nuclei. Normal endosperm results when AI = AV/2RV = 0.5. If AI deviates from 0.5, development of the endosperm is often abnormal, arrested, or aborted. In practice, EBN or AI values are arbitrarily assigned to species based on their crossing behavior, and these values are tested by further crossing experiments. These qualitative factors, although probably an oversimplification, have proven valuable in predicting the success or failure of interspecific crosses, the expected chromosome numbers of the offspring, and the functioning of 2n gametes (Johnston and Hanneman, 1980).

Thus, the explanation for endosperm development or failure has progressed from a general concept of a required balance of genomes in embryo, endosperm, and maternal tissue, to a proper ratio of maternal to paternal genomes, to the suggestion for a balance between qualitative genetic factors. Probably, future research will provide a thorough understanding of the role of the endosperm in seed development and failure.

A broader problem concerns the significance, need, and value of a second fertilization in angiosperms. Brink (1952) suggested that the unique significance of the endosperm being a product of fertilization resides in the heterosis that could result. Until recently, it was difficult to envision endosperm heterosis in a self-pollinating species where both polar nuclei and the male gamete had identical genotypes. However, the imprinting of gene action in maize endosperm done by Kermicle (1978) clearly demonstrated that the particular expression of a gene in the endosperm can be dependent on whether it came from the female or the male parent. He uses the term *epiallelic* to indicate the relation between maternally and paternally imprinted forms of a gene and *epilleles* for the individual gene forms. Thus, it is possible that different genes or gene combinations are derepressed in the polar nuclei and the sperm nucleus. Similar results were obtained by Lin (1975). Gill and Waines (1978), who worked with interspecific hybrids of wheat, also found a factor which affected endosperm development only when transmitted paternally. The imprinting concept provides a plausible explanation for heterosis in the endosperm in general, and a possible basis for hybridity of the endosperm in self-pollinated species—a phenomenon Kermicle (1978) labels "epihybridity".

It is important to recognize and emphasize that the knowledge about the significance of either double fertilization or the triploid (or higher ploid) nature of the endosperm in more than 90 percent of the angiosperms is meager. There is considerable urgency to increase this knowledge base, because it likely will be of considerable benefit to plant breeding.

CONCLUDING COMMENTS

Many contributions from cytogenetics have been and can be used in plant improvement. But, it is time to assess the critical gaps in our knowledge about the cytogenetics of crop plants, so that cytogenetic work can be done in the areas that will most effectively increase the application of this research discipline to plant breeding. Actually, the critical factors are reasonably easy to define:

1. More cytogeneticists are needed. Cytogenetics research is considered old-fashioned, but it is being rediscovered. Fortu-

nately, a few bright young students have recognized the fact that there are many important fundamental and applied problems in chromosomal and cytoplasmic manipulations that need solution and also are exciting.

2. More money is needed to support cytogenetic research and to educate students to be cytogeneticists.

3. Cytogeneticists need to work on problems whose solutions will be of indirect value to plant breeding. Many problems involving meiotic chromosome pairing, crossing over, and the integration and organization of the total genotype have cytogenetic approaches.

4. Most important, cytogeneticists need to work on economically important plants. Significant progress will be made in relating cytogenetics to plant improvement, when and only when, cytogeneticists work directly, cooperatively, and continuously with plant breeders. There must be mutual understanding of the objectives, approaches, goals, and limitations of the two disciplines.

Finally, and most important is the fact that all progress and new discoveries in cytogenetics, as well as in cell culture and molecular biology, will not increase the world's food, feed, and fiber supply, unless there are continuous, strong, and adequately supported cultivar breeding programs. These must exist to take advantage of progress in other disciplines and to convert this progress into better cultivars.

Discussion

E. C. BASHAW
J. JAMES
J. R. LAUGHNAN
R. L. PHILLIPS

1. J. JAMES. Sometimes, a desired trait cannot be found within a given crop or species, but it may be present in a distant or unrelated species. The manipulations necessary to bring such alien germplasm into plant improvement are complex and delicate. Before a breeder can begin any sort of conventional breeding program with the trait from the alien species, the domestic and alien genomes must be brought together into the same cell. An example is provided by a program carried out at the International Center for the Improvement of Maize and Wheat (CIMMYT) to utilize chromosomal and cytoplasmic variation from alien genera for maize improvement.

At CIMMYT, the alien genera used for the intergeneric hybridization program are *Tripsacum* and sorghum. Since 1975, over 32,000 maize and *Tripsacum* crosses have been made, and 36 hybrids have been obtained. Twenty-five of these have had the expected number of chromosomes per cell, namely, 10 from maize and 18 or 36 from *Tripsacum*, depending on which ploidy level of *Tripsacum* was used. The hybrid plants are more like *Tripsacum* than maize in morphology, and they are perennial. They take approximately two years to flower. Eleven hybrids were not classical, in that they possessed a variable number of chromosomes in cells from the same root tip. Most cells contained 20 maize and various numbers of alien chromosomes. Probably these hybrids were produced by the fertilization of an unreduced maize gamete with subsequent progressive elimination of the alien chromosomes. At flowering, less than 1 percent of these root tip cells had any *Tripsacum* chromosomes. In 3 of the 11 hybrids, however, cells had 20 chromosomes with up to 4 substituted from *Tripsacum*. None of these plants produced viable pollen, but some set seed after backcrossing with maize. Often embryo culture was necessary to obtain backcross progeny because of seed breakdown. Some of these backcross progeny possessed alien chromosomes. The alien chromosomes were transmitted through viable female gametes. With backcross hybrids that contain from one to a few alien chromosomes, the incorporation of desirable *Tripsacum* characteristics into maize should be much easier to accomplish than when the classical F_1 must be used.

Also since 1975, about 30,000 pollinations have been made between maize and sorghum. From these, 25 intergeneric hybrids were obtained, and all were nonclassical hybrids with 20 maize chromosomes and up to 10 sorghum chromosomes. Probably, these maize-sorghum hybrids were produced from an unreduced maize egg cell, and a haploid sorghum pollen cell with subsequent chromosome elimination. All were maizelike and flowered within ca. 90 days after germination.

Eleven maize-sorghum hybrids produced backcross seeds upon pollination with maize. Subsequent generations have been produced by intercrossing among backcross progenies. Alien chromosomes behave erratically in progeny but some are included in viable gametes in subsequent generations. Even after intercrossing the backcross progenies for six generations, the derivatives still retained abnormal cytology and/or morphology. For example, a backcross derivative that possesses one sorghum chromosome grew to about 30 cm tall and it was very maizelike. Another plant, completely maizelike in cytology, grew to about 12 cm tall and consisted of three ears and no tassle on a 3 cm long stem. Another typical abnormality was to have male and female sectors within the same inflorescence.

Of course, the success of any plant improvement project depends on the application of selection pressure in populations of plants. Obviously, the derivatives of maize x *Tripsacum* and maize x sorghum crosses are weak, and they have reduced fertility, so strong selection pressure cannot be exerted on them for agronomic traits and performance. Therefore, selection on cytological factors is necessary to attempt to identify plants which retain alien chromosomes and are more vigorous. A better understanding of the cytological behavior and control of alien germplasm is needed to show how to accomplish selective retention of desirable alien germplasm in a maize genome and to elucidate the abnormal morphology of the plants.

2. E. C. BASHAW. Apomictic reproduction in plants presents

some serious problems in chromosomal manipulation. But when hybridization is possible, there are unique opportunities for use of the new genotype. The presence of rare sexual plants in obligate apomictic species and various degrees of sexuality in facultative apomicts allow for hybridization and manipulation. In several species, obligate apomictic plants have been used successfully as male parents in both intra and interspecific crosses with sexual plants. The apomictic barrier has been broken in some of these crosses with the creation of useful new genotypes and information about the inheritance of obligate apomixis. Obligate apomixis is simply inherited in some species with the result that true breeding apomictic hybrids can be recovered readily for use as new cultivars.

In Texas, the entire breeding program with bufflegrass is based on the hybridization of numerous apomictic plants as males with a single sexual plant and the recovery and evaluation of true breeding apomictic F_1 hybrids. Because apomixis bypasses the sexual mechanism of reproduction, recovery of obligate apomicts provides a means for seed propagation of aneuploids, polyploids, structural hybrids, and other chromosomal aberrants that might otherwise be sterile. A classic example is dallisgrass. It is a relatively fertile, naturally occurring, tri-hybrid pentaploid with 20 bivalents and 10 univalents at metaphase. This species occupies an important place in cultivated pastures, but if it was not an obligate apomict, such an aberrant plant could not have survived. Embryo culture might permit recovery of even wider hybrids, and perhaps tissue culture techniques might make it possible to recover apomictic progenies from especially wide crosses.

Fertilization of the unreduced egg of the apomictic plant has great potential for incorporating alien genomes and chromosomes and for increasing ploidy level. Crosses of this type have been accomplished in some polyploid species in which hybridization with sexual females was extremely difficult or impossible. Apparently, the apomictic egg is so well buffered that the addition of an extra genome or chromosomes has little or no detrimental effect on plant metabolism and growth. Because many hybrids from fertilization of an unreduced egg are apomictic and fertile, they may provide excellent material to investigate potential use of the products resulting from somatic instability.

The potential for cytoplasmic manipulation in an obligate apomict is unknown.

3. J. L. LAUGHNAN. In spite of the widespread use of cytoplasmic

male sterility in the production of plant hybrids, there is still poor understanding of the molecular basis for this phenomenon. Sources of genetic variability do exist for cytoplasmic-sterility systems, and recent evidence indicates that this variability is due to altered mitochondrial DNA. By using electrophoresis on enzyme digests of mitochondrial DNA, it is possible to show that the digests from normal maize and strains of IP, S, C, and T sterile cytoplasms possess different molecular classes.

With undigested, mitochondrial DNA, the bandings in the electrophoretic patterns of C, T, and normal bytoplasm are similar, but that of the S male-sterile mitochondrial DNA is unique. Work by Levings and Pring has shown that many sources of S cytoplasm from different genetic backgrounds share this unique pattern.

Modified forms of mitochondrial DNA represent mutations. Mutations have occurred in male-sterile systems, whereby the sterile form reverted to the fertile form in pearl millet, the broad bean, and in maize where the reversion was from T to N cytoplasm. The reversion mutants from cytoplasmic male sterile to male fertile could occur due to DNA changes in the nucleus or mitochondrion. Both types have occurred in relatively large numbers in maize cultures in Illinois, and there may be operative a fertility episome which can integrate into either the cytoplasmic organelle or into a site on one of the maize chromosomes. Seven of our cytoplasmic revertants were analyzed by Levings and Pring, and in all seven cases the bands peculiar to S cytoplasm have disappeared indicating some correlation between the behavior of these plasmids and the reversion to fertility.

4. R. L. PHILLIPS. For most higher plants, less than half the DNA in the nucleus represents the unique type that comprises the individual genes that code for a particular enzyme or protein. More than half of the nuclear DNA is repetitive in nature. In the repetitive class, certain sequences repeat between 10 and 1,000 times, other sequences repeat between 1,000 and 10,000 times, and some even repeat from 10,000 to 1,000,000 times. In soybeans, 40 percent of the DNA is unique, 20 percent is represented by sequences repeated 19 times, 35 percent by sequences repeated 2800 times, and about 5 percent by sequences repeated 290,000 times (Walbot and Goldberg, pers. commun.). Most of the DNA manipulated by plant breeders probably is not of the single-copy type. Most likely, breeders are manipulating regulatory DNA. Regulatory DNA signals structural genes relative to the timing, level, and frequency of their expression in various tissues.

Multiple copies of particular genes have a potential for unequal crossing over, and a familiar case is the unequal crossing over associated with bar-eye in *Drosophila.* Ribosomal RNA genes are repeated many times in plants and animals. With unequal crossing over, one can go from a few copies of a DNA sequence to many, or from few to fewer. Multiple-copy genes, therefore, have a potential for variability under any breeding scheme if that variation depends upon number of copies of a gene, and fixation of a genotype may never be realized. This may have important implications in long-term selection experiments where variability is never exhausted.

Another aspect of molecular cytogenetics has to do with the chromosome constitution of maize endosperm cells. Normally, they are considered triploid, but measurements of nuclear size and DNA quantity suggest that the older endosperm cells may be highly polytene. Nuclear diameters at two days postpollination were from 5 to 15 microns, whereas at 29 days postpollination, the diameters ranged from 10 to 70 microns. This increase in diameter of the nucleus parallels increased DNA contents in endosperm cells; no mitotic figures were seen after 14 days postpollination. Chromosomes of older endosperm cells are not condensed into distinct individual chromosomes, but fibrous chromatin is observed. A single chromosome does not appear to be of constant dimensions throughout. These observations may be related to changes in the number of ribosomal RNA genes during endosperm development. The amount of DNA that codes for ribosomal RNA begins to increase on the 12th day postpollination. By about the 20th day, the proportion comes back to near normal. Seemingly, at the 12th day the ribosomal DNA is disproportionately replicated, and by the 20th day, the remainder of the genome or part of it has replicated to substantially reduce the proportion of ribosomal DNA. The absolute number of genes would still remain high. Interestingly, the timing of increase in ribosomal RNA genes parallels the increase of protein synthesis in the corn endosperm. Perhaps the corn endosperm behaves like a *Drosophila* salivary gland cell. In *Drosophila,* the heterochromatin apparently does not replicate, but much of the euchromatin replicates about ten times, so a single chromosome may have 1024 strands. The corn endosperm cell may replicate certain genes more than others during development and thereby regulate the level of expression of these genes. An example of multiply-replicated genes active during seed filling may be those for zein. Zein protein comprises more than 50 percent of the protein in the corn endosperm.

In conclusion, these are but a few speculations on where information from molecular cytogenetics may expand our understanding of fundamental developmental processes in plants that could lead to new and beneficial manipulation.

5. H. KIDD. What is the mechanism that causes the loss of *Tripsacum* and sorghum chromosomes from the maize x *Tripsacum* and maize x sorghum hybrids?

6. J. JAMES. This requires more study, but probably the sorghum chromosomes do not align properly on the metaphase plate; thus they are lost during cell division.

7. B. ROBERTSON. It was said that the use of unreduced gametes would be preferable to using somatic doubling by colchicine for the production of synthetic polyploids. Are there any general methods for increasing the production of unreduced gametes by plants?

8. S. J. PELOQUIN. No general method exists. The 2n pollen is larger than normal pollen, so anytime that a plant produces abnormally large pollen grains, it is probably producing 2n pollen. Then, by studying meiosis in that plant, one can determine how the 2n pollen is formed. One mechanism for diploid pollen production is by FDR. Another mechanism occurs through premature cytokinesis where cleavage occurs after the first meiotic division, and there is no second division. This latter mechanism is not as valuable as FDR because it only transmits about 40 percent of the heterozygosity to the offspring, and 2n pollen from it produces offspring with much less yield than the offspring from FDR 2n gametes. To find 2n eggs is more difficult. Usually, a tetraploid is crossed onto a diploid. In many plants, such as the potato, the only seeds that survive are mainly tetraploids. Thus, the number of seeds obtained following a diploid-tetraploid cross is a measure of egg production in the diploid.

9. J. MILLER. It has been found that supernumerary chromosomes can increase crossing over in some chromosome regions. Therefore, would it be beneficial for maize breeders to incorporate B chromosomes into breeding populations in an effort to increase genetic recombination and possibly genetic variability?

10. S. J. PELOQUIN. Yes, it probably would. With most plant

chromosomes only one exchange occurs per chromosome arm, and this exchange probably occurs near the middle of the arm. One way to change the position of crossing-over might be to use supernumerary chromosomes. Such a juxtaposition of crossover site has been demonstrated in some animals, where the supernumerary chromosome changes the position of the chiasma. If genes were located near the centromere, one might be able, by changing the position of crossing-over, to release them into new linkage relationships by increasing the frequency of crossing-over; or two exchanges per arm might occur, and one of them would likely be near the centromere.

11. R. L. PHILLIPS. There are differences in recombination between male and female inflorescences in maize, and usually, the male is higher primarily in the regions near to the centromere. So if one is making a backcross, using the F_1 as male, it might give a higher frequency of recombination for specific chromosome segments.

12. L. W. KANNENBERG. Dr. Phillips suggested that unequal crossing-over within highly repeated DNA sequences may provide the genetic variability that permits continued progress in the long-term selection experiment for quantitative traits. If this is a general phenomenon in higher plants, maybe there need be no concern about the loss or erosion of genetic variability for quantitative traits. In other words, one could select within pure lines and expect to make progress from selection because unequal crossing-over or even, perhaps, mutations occurring at the normal rate in these highly repeated DNA sequences, is continually renewing the pool of genetic variability.

13. R. L. PHILLIPS. Probably, it is desirable to create genetic variability to the greatest extent and with any method possible. Certainly, if unequal crossing-over in repeated types of genes does release variability, then variability in inbred lines would occur.

In a somewhat crude way, this idea has been tested by using the low- and high-protein maize lines developed in the long-term selection experiments at the University of Illinois. Selection has been practiced for high- and low-grain protein percentage annually since 1896. The number of ribosomal RNA genes (multiple-copy genes) was 8,000 per diploid nucleus in the high-protein line and 13,000 in the low-protein line. This is the opposite of what one might expect. The reverse low-protein line had 11,000 ribosomal RNA genes per diploid nucleus and the reverse high-protein line had 23,100. Next the high- and low-protein lines were crossed and backcrossed to high protein followed by self-pollination. In these progenies, correlation between number of ribosomal RNA genes and seedling protein content was -0.86. The cross of high-protein x reverse high-protein gave progeny that showed no relationship. But, in two different years, the negative relationship was found in the high x low study. So it is possible that changes in the number of copies of repeated genes could account for some of the genetic variability in these long-term selection experiments.

14. D. DUVICK. What particular traits might be sought in the maizelike plants that arise from the maize x *Tripsacum* or the maize x sorghum crosses?

15. J. JAMES. Maize is a highly variable crop plant, so there is no need to seek genetic variability in general. However, maize could be a more environmentally stable plant. Drought and water logging tolerance from sorghum and resistance to diseases from *Tripsacum* are examples of traits that could be used in maize breeding.

16. D. MIES. Would the production of tetraploid potatoes by protoplast fusion of 2n cells give similar results to using the FDR sexual route to obtain tetraploids?

17. S. J. PELOQUIN. If one had a large amount of nonadditive genetic variance, the use of somatic cell fusion should be a good technique to use. As second choice, it would be desirable to have FDR for both egg and pollen cells which would transmit approximately 80 percent of the parent genotype to the offspring. It would permit the combining of nonadditive genetic variance available within diploid hybrids into a single tetraploid. Heterozygosity would be maximized, and all of the epistatic interaction among genes would be combined much like occurs in single crosses of maize. Of course, one big difference between FDR and somatic cell fusion is that with the latter, hybrid cytoplasms would be obtained.

18. R. L. PHILLIPS. Somatic hybrids of potato and tomato did not lead to hybrid or mixed cytoplasms. The hybrid had the cytoplasm of only one species. So the expectation of mixing cytoplasms via somatic cell hybridization should be viewed with caution.

19. T. E. MORELOCK. A number of species have cytoplasmic male

sterility and its manifestations are different, so is it logical to assume that male sterility per se is due to one cause or different causes?

20. J. R. LAUGHNAN. There are indications that the causes may be generic. For example, with T and C cytoplasms in maize, the banding patterns are similar to normal cytoplasm, whereas the S type produces a different pattern. In several species, including wheat where the *Triticum aestivum* genome is placed in *T. timopheevi* cytoplasm, the digest shows a central band difference. But it is too soon to say whether all cytoplasmic cases have a common mechanism for causing male sterility.

21. D. P. WEST. What is the mechanism by which fertility is restored in the various maize cytoplasms?

22. J. R. LAUGHNAN. Two problems are involved.

1) Past experience has shown that T cytoplasm in maize is extremely stable in producing sterility in the field. Now, whether a modified environment such as the one of tissue culture may cause reversions is unknown.

2) At the molecular level, it appeared early on that some kind of foreign DNA, perhaps a varietal DNA which is inhibitory somewhere in the pathway of male fertility, acted to produce male sterility. By ridding the maize plant of this inhibiting DNA, fertility was restored, and this is what early electrophoretic work on undigested mitochondrial DNA showed to be true. But later hybridization studies with labelled plasmids suggest that rather than a presence or absence, it is an integration vs. excision phenomenon. In other words, the mutations that occur at the nuclear level are insertions and perhaps exsertions.

23. S. N. ACHARYA. What technique was used to discover that the ribosomal RNA in low- and high-protein maize lines was governed by 13,000 or 8,000 genes?

24. R. L. PHILLIPS. These estimates were based on DNA ribosomal and RNA hybridization techniques. It is done in solution on filters instead of by *in situ* hybridization on chromosomes. Ribosomal RNA is labelled and hybridized with a known amount of denatured DNA. An estimate of the percent of the DNA that hybridizes the ribosomal RNA and knowledge of the DNA content per nucleus and ribosomal RNA size permits one to calculate gene numbers from these data.

REFERENCES

Albertson, M. C., and R. G. Palmer. 1979. A comparative light-and-electron microscope study of microsporogenesis in male sterile (ms_1) and male-fertile soybeans (*Glycine max* (L.) Merr.). Am. J. Bot. 66(3):253-65.

Beversdorf, W. M., and E. T. Bingham. 1977. Male-sterility as a source of haploids and polyploids of *Glycine max*. Can. J. Genet. Cytol. 19:283-87.

Bingham, E. T. 1971. Isolation of haploids of cultivated alfalfa. Crop Sci. 11:433-35.

Boyes, J. W., and W. P. Thompson. 1937. The development of the endosperm and embryo in reciprocal interspecific crosses in cereals. J. Genet. 34:203-27.

Brim, C. A., and M. F. Young. 1971. Inheritance of a male-sterility character in soybeans. Crop Sci. 11:564-66.

Brink, R. A. 1952. Inbreeding and crossbreeding in seed development, pp. 81-99. *In* Gowen, J. W. (ed.), Heterosis. Iowa State Univ. Press, Ames, Ia.

Britton, D. M., and J. W. Hull. 1957. Mitotic instability in *Rubus*. J. Hered. 48:11-20.

Burnham, C. R. 1966. Cytogenetics in plant improvement, pp. 139-88. *In* Frey, K. J. (ed.), Plant breeding. Iowa State Univ. Press, Ames, Ia.

Carlson, W. R. 1978. The B chromosome of corn. Annu. Rev. Genet. 16:5-23.

Chase, S. S. 1949. Monoploid frequencies in a commercial double cross hybrid maize, and its component single cross hybrids and inbred lines. Genetics 34:328-32.

Chase, S. S. 1964. Monoploids and diploids of maize: A comparison of genotypic equivalents. Am. J. Bot. 51:928-32.

Chase, S. S. 1969. Monoploids and monoploid derivatives of maize (*Zea mays* L.). Bot. Rev. 35:117-76.

Chase, S. S. 1974. Utilization of haploids in plant breeding: Breeding diploid species, pp. 211-20. *In* Kasha, K. J. (ed.), Haploids in higher plants. Univ. of Guelph, Guelph, Can.

Cutter, G. L., and E. T. Bingham. 1977. Effect of soybean male-sterile gene ms_1 on organization and function of the female gametophyte. Crop Sci. 17:760-64.

Dover, G. A., and R. Riley. 1972. Prevention of pairing of homoeologous meiotic chromosomes of wheat by an activity of supernumerary chromosomes of *Aegilops*. Nature 240:159-61.

Dunbier, M. W., and E. T. Bingham. 1975. Maximum heterozygosity in alfalfa: Results using haploid-derived autotetraploids. Crop Sci. 15:527-31.

Evans, G. M., and A. J. Macefield. 1972. Suppression of homoeologous pairing by B chromosomes in a *Lolium* species hybrid. Nature 236:110-11.

Franzke, C. J., and J. G. Ross. 1952. Colchicine induced variants in sorghum. J. Hered. 43:107-15.

Franzke, C. J., and M. E. Sanders. 1964. Classes of true-breeding diploid mutants obtained after colchicine treatment of sorghum line, experimental 3. Bot. Gaz. 125:170-78.

Gill, B. S., and J. G. Waines. 1978. Paternal regulation of seed development in wheat hybrids. Theor. Appl. Genet. 51:265-70.

Hanson, G. B. 1969. B-chromosome stimulated crossing over in maize. Genetics 63:601-09.

Hesse, C. O. 1971. Monoploid peaches, *Prunus persica* Batsch: Description and meiotic analysis. J. Am. Soc. Hort. Sci. 96:326-30.

Hougas, R. W., S. J. Peloquin, and A. C. Gabert. 1964. Effect of seed parent and pollinator on frequency of haploids in *Solanum tuberosum*. Crop Sci. 4:593-95.

Hougas, R. W., S. J. Peloquin, and R. W. Ross. 1958. Haploids of the common potato. J. Hered. 47:103-7.

Iwanaga, M., and S. J. Peloquin. 1979. Synaptic mutant affecting only megasporogenesis in potatoes. J. Hered. 70:385-89.

Johns, W. A., and B. L. Harvey. 1974. The effects of *Hordeum bulbosum* L. cytoplasm on *H. vulgare* L., pp. 276-77. *In* Kasha, K. J. (ed.), Haploids in higher plants. Univ. of Guelph, Guelph, Can.

Johnston, S. A., and R. E. Hanneman. 1980. Support of the endosperm balance number hypothesis utilizing some tuber-bearing *Solanum* species. Am. Potato J. 37:7-14.

Johnston, S. A., T. P. M. den Nijs, S. J. Peloquin, and R. E. Hanneman. 1980. The significance of genic balance to endosperm development in interspecific crosses. Theor. Appl. Genet. 57: 5-9.

Kasha, K. J., (ed.) 1974. Haploids in higher plants. Univ. of Guelph, Guelph, Can.

Kasha, K. J., and K. N. Kao. 1970. High frequency haploid production in barley (*Hordeum vulgare* L.). Nature 225:874-76.

Kermicle, J. L. 1969. Androgenesis conditioned by a mutation in maize. Science 166:1422-24.

Kermicle, J. L. 1971. Pleiotropic effects on seed development of the indeterminate gametophyte gene in maize. Am. J. Bot. 58:1-7.

Kermicle, J. L. 1974. Origin of androgenetic haploids and diploids induced by the indeterminate gametophyte (*ig*) mutation in maize, p. 137. *In* Kasha, K. J. (ed.), Haploids in higher plants. Univ. of Guelph, Guelph, Can.

Kermicle, J. L. 1978. Imprinting of gene action in maize endosperm, pp. 357-71. *In* Walden, D. B. (ed.), Maize breeding and genetics. John Wiley and Sons, New York.

Law, C. N. 1966. Biometrical analysis using chromosome substitutions within a species, pp. 59-85. *In* Riley, R., and K. R. Lewis (eds.), Chromosome manipulations and plant genetics. Oliver and Boyd, Edinburgh.

Lin, B. Y. 1975. Parental effect on gene expression in maize endosperm development, Ph. D. thesis. Univ. of Wis., Madison, Wis.

Maan, S. S., and T. Sasakuma. 1977. Fertility of amphihaploids in Triticinae. J. Hered. 94:87-94.

MacKey, J. 1970. Significance of mating systems for chromosomes and gametes in polyploids. Hereditas 66:165-76.

Mendiburu, A. O., and S. J. Peloquin. 1971. High yielding tetraploids from 4x - 2x and 2x - 2x matings. Am. Potato J. 48:300-301.

Mendiburu, A. O., and S. J. Peloquin. 1976. Sexual polyploidization and depolyploidization: Some terminology and definitions. Theor. Appl. Genet. 48:137-43.

Mendiburu, A. O., and S. J. Peloquin. 1977a. The significance of 2n gametes in potato breeding. Theor. Appl. Genet. 49:53-61.

Mendiburu, A. O., and S. J. Peloquin. 1977b. Bilateral sexual polyploidization in potatoes. Euphytica 26:573-83.

Mendiburu, A. O., S. J. Peloquin, and D. W. S. Mok. 1974. Potato breeding with haploids and 2n gametes, pp. 249-58, *In* Kasha, K. J. (ed.), Haploids in higher plants. Univ. of Guelph, Guelph, Can.

Merker, A. 1976. The cytogenetic effect of heterochromatin in hexaploid triticale. Hereditas 83:215-22.

Mok, D. W. S., and S. J. Peloquin. 1975a. Three mechanisms of 2n pollen formation in diploid potatoes. Can. J. Genet. Cytol. 17:217-25.

Mok, D. W. S., and S. J. Peloquin. 1975b. Breeding value of 2n pollen (diplandroids) in tetraploid x diploid crosses in potatoes. Theor. Appl. Genet. 46:307-14.

Mok, D. W. S., and S. J. Peloquin. 1975c. The inheritance of three mechanisms

of diplandroid (2n pollen) formation in diploid potatoes. Heredity 35: 295-302.
Mok, D. W. S., S. J. Peloquin, and T. R. Tarn. 1975. Cytology of potato triploids producing 2n pollen. Am. Potato J. 52:171-74.
Morgan, D. T., and R. D. Rappleye. 1950. Twin and triplet pepper seedlings. A study of polyembryony in *Capsicum frutescens.* J. Hered. 41:91-95.
Muntzing, A. 1933. Hybrid incompatibility and the origin of polyploidy. Hereditas 18:3-56.
Nakamura, A., T. Yamada, N. Kadotani, and R. Itagaki. 1974. Improvement of flue-cured tobacco variety MC1610 by means of haploid breeding method and some problems of this method, pp. 277-78. *In* Kasha, K. J. (ed.), Haploids in higher plants. Univ. of Guelph, Guelph, Can.
Nel, P. M. 1973. The modification of crossing over in maize by extraneous chromosomal elements. Theor. Appl. Genet. 43:196-202.
Nielsen, E. L. 1968. Intraplant morphological variation in grasses. Am. J. Bot. 55:116-22.
Nielsen, E. L., J. Franckowiak, and P. N. Drolsom. 1969. Characteristics of inbred progenies from morphologically unstable *Sorghum* plants. Euphytica 18:227-36.
Nishiyama, I., and N. Inomata. 1966. Embryological studies on cross-incompatibility between 2x and 4x in *Brassica.* Jap. J. Genet. 41:27-42.
Nishiyama, I., and T. Yabuna. 1978. Causal relationships between the polar nuclei in double fertilization and interspecific cross-incompatibility in *Avena.* Cytologia 43:453-66.
Nitsch, J. P., and C. Nitsch. 1969. Haploid plants from pollen grains. Science 163:85-87.
Parfitt, D. E., and S. J. Peloquin. 1977. Variation of vine and tuber yield as a function of harvest date and cultivar. Am. Potato J. 54:411-17.
Peloquin, S. J., R. W. Hougas, and A. C. Gabert. 1966. Haploidy as a new approach to the cytogenetics and breeding of *Solanum tuberosum,* pp. 21-28. *In* Riley, R., and K. R. Lewis (eds.), Chromosome manipulations and plant genetics. Oliver and Boyd, Edinburgh.
Plewa, M. J., and D. F. Weber. 1973. The use of monosomics to detect genes conditioning lipid content in *Zea mays* L. embryos. Can. J. Genet. Cytol. 15:313-20.
Plewa, M. J., and D. F. Weber. 1975. Monosomic analysis of fatty acid composition in embryo lipids of *Zea mays* L. Genetics 81:277-86.
Rhoades, M. M. 1968. Studies on the cytological basis of crossing over, pp. 229-41. *In* Peacock, W. J., and R. D. Brock (eds.), Replication and recombination of the genetic materials. Aust. Acad. Sci., Canberra, Aust.
Rhoades, M. M., and E. Dempsey. 1966. Induction of chromosome doubling at meiosis by the elongate gene in maize. Genetics 54: 505-22.
Riley, R., and V. Chapman. 1958. Genetic control of cytologically diploid behavior of hexaploid wheat. Nature 182:713-15.
Riley, R., V. Chapman, and R. Johnson. 1968. The incorporation of alien disease resistance in wheat by genetic interference with the regulation of meiotic chromosome synapsis. Genet. Res. 12:199-219.
Roman, H. 1947. Mitotic nondisjunction in the case of interchanges involving the B-type chromosome in maize. Genetics 32:391-409.
Roupakias, D. G., and P. J. Kaltsikes. 1977. The effect of telomeric heterochromatin on chromosome pairing of hexaploid triticale. Can. J. Genet. Cytol. 19:543-48.
Sanders, M. E., and C. J. Franzke. 1969. Colchicine-induced complex diploid mutants from tetraploid seedlings of four sorghum lines. J. Hered. 60: 137-48.
Sanders, M. E., and C. J. Franzke. 1976. Effect of temperature on origin of

colchicine-induced complex mutants in sorghum. J. Hered. 67:19-29.
Satina, S., and A. F. Blakeslee. 1935. Cytological effects of a gene in *Datura* which causes dyad formation in sporogenesis. Bot. Gaz. 96:521-32.
Sears, E. R. 1954. The aneuploids of common wheat. Mo. Agr. Exp. Stn. Res. Bull. 572:1-58.
Sears, E. R. 1969. Wheat cytogenetics. Annu. Rev. Genet. 3:451-68.
Sears, E. R. 1972. Chromosome engineering in wheat. Stadler Genet. Symp. 4:23-38.
Sears, E. R. 1977. An induced mutant with homoeologous pairing in common wheat. Can. J. Genet. Cytol. 19:585-93.
Sears, E. R., and M. Okamoto. 1958. Intergenomic chromosome relationships in hexaploid wheat. Proc. X Int. Congr. Genet., Montreal, Can. Univ. of Toronto Press. 2:258-59.
Simantel, G. M., and J. G. Ross. 1963. Colchicine-induced somatic chromosome reduction in *Sorghum.* III. Induction of plants homozygous for structural chromosome markers in four pairs. J. Hered. 54:277-84.
Simon, P. W., and S. J. Peloquin. 1976. Pollen vigor as a function of mode of 2n gamete formation in potatoes. J. Hered. 67:204-8.
Skiebe, K. 1958. Die Bedeutung von unreduzierten Gameten fur die Polyploidiezuchtung bei der Fliederprimel (*Primula malacoides* Franchet). Zuchter 28:353-59.
Snoad, B. 1955. Somatic instability of chromosome number in *Hymenocallis calathium.* Heredity 9:129-34.
Sunderland, N., and F. M. Wicks. 1969. Cultivation of haploid plants from tobacco pollen. Nature 224:1227-29.
Tanaka, M., and K. Nakata. 1969. Tobacco plants obtained by anther culture and experiments to get diploid seeds from haploids. Jap. J. Genet. 44: 47-54.
Therenin, L. 1974. Haploids in asparagus breeding, p. 279. *In* Kasha, K. J. (ed.), Haploids in higher plants. Univ. of Guelph, Guelph, Can.
Thompson, K. F. 1972. Oilseed rape. Rep. Plant Breed. Inst., Cambridge, pp. 94-96.
Thompson, M. M. 1962. Cytogenetics of *Rubus.* III. Meiotic instability in some higher polyploids. Am. J. Bot. 49:575-82.
Tsunewaki, K., K. Noda, and T. Fujisawa. 1968. Haploid and twin formation in a wheat strain Salmon with alien cytoplasms. Cytologia 33:526-38.
Turcotte, E. L., and C. V. Feaster. 1967. Semigamy in Pima cotton. J. Hered. 58:54-57.
Turcotte, E. L., and C. V. Feaster. 1969. Semigametic production of haploids in Pima cotton. Crop Sci. 9:653-55.
Turcotte, E. L., and C. V. Feaster. 1974. Semigametic production of cotton haploids, pp. 53-64. *In* Kasha, K. J. (ed.), Haploids in higher plants. Univ. of Guelph, Guelph, Can.
Valentine, D. H. 1956. Studies in British primulas. V. The inheritance of seed compatibility. New Phytol. 55:305-18.
von Wangenheim, K. H., S. J. Peloquin, and R. W. Hougas. 1960. Embryological investigations on the formation of haploids in the potato (*Solanum tuberosum*). Z. Vererbgslhre. 91:391-99.
Walden, D. B., (ed.), 1978. Maize breeding and genetics. John Wiley and Sons, New York, pp. 353-405.
Wall, A. M., R. Riley, and V. Chapman. 1971. Wheat mutants permitting homoeologous meiotic chromosome pairing. Genet. Res. 18:311-28.
Watkins, A. E. 1932. Hybrid sterility and incompatibility. J. Gent. 25:125-62.
Weber, D. F. 1978. Nullisomic analysis of nucleolar formation in *Zea mays.* Can. J. Genet. Cytol. 20:97-100.

CHAPTER 5

Breeding Plants for Stress Environments

C. F. LEWIS
M. N. CHRISTIANSEN

ANY TIME farmers get together, their first topic of conversation is how their crops are growing. Invariably, they lament some unfavorable weather or soil condition that minimizes to some extent their prospects for a good crop. Scientists who conduct field experiments are not different from farmers in this regard. An experiment station annual report once began, "As usual, we had abnormal weather this year . . . ."

In addition to variable weather, farmers must contend with mineral toxicities or deficiencies, salinity, acidity, alkalinity, and air pollution. Insects, diseases, nematodes, and other creatures attack plants, but stresses imposed by organisms or their interaction with environment are beyond the scope of this paper. As Lewis (1976) and Epstein (1976) pointed out, plant growth and development is a result of the interplay between the genetically governed potential of the plant and the plant environment in which it grows. This may be symbolized by the equation:

$$p = ge$$

where p = the phenotype (what actually develops)
g = the genotype (the genetic potential)
e = the environment (for this discussion all soil, water, air, light, and other nonliving factors influencing plant growth and development)

C. F. Lewis, Staff Scientist for Plant Genetics and Breeding, National Program Staff, Agricultural Research, Science and Education Administration, USDA, and M. N. Christiansen, Supervisor and Plant Physiologist, Plant Physiology Institute, Agricultural Research, Science and Education Administration, USDA, Beltsville, Maryland.

Because plants are seldom grown for the total phenotype but rather for special plant parts such as grains, fruits, tubers, or fiber, the equation should be modified to:

$$yp = ge$$

where yp = the yield and quality of the product for which the plants were grown

Yield and quality of crops may be improved by manipulating the environment to suit the crops or by genotypically fitting the crops to better cope with the environment. Dilley et al. (1975) listed three major research imperatives:

> I. Manipulate crops or their environments in ways which avoid or reduce stress injury and increase productivity; II. Exploit the genetic potential by developing new cultivars of crops resistant to environmental stresses; III. Elucidate the basic principles of stress injury and resistance in plants, and evaluate the scope and nature of stress damage to crops.

Fitting crops to environments is as old as plant culture in agriculture. Klages (1947) described the social, physiological, and ecological factors that have shaped the geographic distribution of crop plants. Plants are grown where they are grown for good reasons. Each species has its limits of adaptability, and if environmental stress becomes severe enough, it is better to turn to species more suited to the environment than to attempt to grow unadapted ones. For example, cotton is not a good crop for Iowa, but corn does beautifully there. Certain land slopes, soil conditions, and other environmental restrictions determine the cropping system and strongly influence plant breeding objectives. Tree crops and perennial forage crops can be grown on steeper slopes and more marginal land than can annual row crops. In general, the larger the management inputs, the greater the need for assurance that environmental stresses do not reduce yield markedly. Row crops on prime land demand breeding for high yield. With permanent forage crops, emphasis is placed on survival and stability of yield rather than maximum production capacity. Finally, environments may become so unsuitable for crop production that agriculture is not practiced at all.

The earth's surface is covered about 70 percent by salt water and 30 percent by land. Carter (1975) has estimated that 1,461 million hectares were available for agricultural use in 1970, so over half the earth's land surface is not suitable for economic agricultural production, primarily due to temperature and water limitations and to terrain slope. Moreover, land that is used for agricultural production presents many environmental stress problems. Millions of hectares require irrigation, and more millions have mineral stresses of various kinds.

Several studies of the world food and nutrition situation have been sponsored by National Research Council (1975, 1977), Agricultural Research Policy Advisory Committee (1975), American Society of Agronomy (1965), Iowa State University, United Nations, and others. All reach the general conclusion that world population will increase from four to seven or eight billion in the next two or three decades. Mayer (1975) concluded, "This means that we have to find in the next 25 years food for as many people again as we have been able to develop in the whole history of man 'til now." If this task is to be accomplished, increased production must come not only from land already cultivated but also from new land brought into production. Losses caused by environmental stresses must be reduced, and crops that can be grown on marginal land must be developed.

In this context of the necessity for producing more crops on marginal land, plant breeders must turn their collective attention to producing new cultivars better suited to stress environments. Modifying or controlling the environment is another option, and genetic-environmental interaction is optimized through research on both breeding and cultural practices.

Before citing examples of breeding crops for specific environmental stresses, it is instructive to discuss some general principles of plant breeding in order to put breeding for stress environments into context.

Experience and textbooks on plant breeding teach that plant breeders do two things: they develop strains of plants with genetic constitutions different from other strains, and they evaluate the performance of these strains relative to that of currently available cultivars or other candidate strains. All plant breeding methods involve selecting plants from genetically heterogenous populations. It does not matter whether the procedure is simple, as with pedigree selection, or complex, as with recombinant DNA technique. The end product is a strain with a specific set of DNA molecules that in composite gives the genotype. Theoretically, it is possible to propagate a single genotype vegetatively so that the resulting population is composed of genetically identical individuals, but in sexually reproducing crops, this situation probably is never achieved. Cultivars almost always contain a certain amount of intracultivar variability.

After several generations of selection and preliminary testing, breeders develop a number of populations of plants that have enough promise to justify extensive evaluation as potential new cultivars. The sciences of agronomy, horticulture, plant breeding, and statistics have developed a relatively dependable way of evaluating the performance of populations of plants. If this evaluation is properly done over enough sites and seasons in the territory that the cultivar is expected to serve, generally enough environments will be sampled to allow statistical inference about the cultivar's adaptation.

Sometimes after much testing, the decision to release a cultivar is not an obvious one because it may excel in some traits and be deficient in others. Generally, a successful cultivar is one with a total balance of traits that makes it more profitable for growers than any other one they might choose. This is why breeders are wary about emphasizing one trait to the exclusion of others. It is distressing to have a susceptible diseased check yield more than the resistant strains. For example, several years ago, cotton (*Gossypium hirsutum*) cultivars resistant to Verticillium wilt ended the season with a marvelous canopy of green leaves but with few bolls of cotton; the susceptible check was nearly defoliated but it set a fairly good crop, by now, wilt resistance and high productivity have been combined and the new cultivars have less loss from the disease.

Remember that it is easy to elevate a given trait to number one priority, but other traits need attention also, such as pest resistance, nutrition, taste, appearance, shipping or processing quality, early maturity, uniform height, and especially high yield—the Holy Grail of plant breeders. Plant breeders must be careful that too much is not promised too soon, or their credibility may be jeopardized. Also, breeding and management should be at levels that will optimize investment and yields for maximum economic return. In the free enterprise system maximum yields may not be desired, but rather the combination of yields and cost of production that will give the greatest profit per unit of production, land area, or hours worked.

Finally, plant breeding takes time. The time from initial cross to release of cultivar can take 15 years and even longer if interspecific hybrids are involved. Because the process is begun each year, finished cultivars, elite germplasm, and parental lines are released each year.

INDIRECT BREEDING FOR STRESS ENVIRONMENTS

Regular Field Performance Trials

The first major category of breeding for stress environments is the indirect method, in which the material, although not being tested directly for environmental stresses, is exposed to such stresses in regular field performance trials.

If stress environments are a problem within the territory a cultivar is supposed to serve, the testing program should screen for those stresses; therefore, strains with the ability to perform well will excel under such conditions. In selecting for high yield and quality, the breeder automatically will choose those strains with stress-resistant characteristics. As such, breeders generally do not make any measurement or select directly for stress tolerance.

For example, Harvey (1977) analyzed the yield of open-pollinated maize (*Zea mays*) cultivars grown in seven states during the 1928 to 1936 prehybrid era and of maize hybrids grown in those same seven states from 1972 to 1976. The years chosen included those with both good and poor yields. The open-pollinated cultivars averaged 1.5 tons/ha (24.2 bu/A) in good years and 0.5 tons/ha (8.8 bu/A) in poor years. The hybrids yielded 5.7 tons/ha (91.3 bu/A) in good years and 4.0 tons/ha (64.0 bu/A) in poor years. Note especially that hybrids yielded much more in good years as well as in poor ones and the drop in yield was 30 percent for hybrids and 63 percent for open-pollinated cultivars.

Russell (1974) reported experiments that compared the performance of one open-pollinated cultivar and 24 maize hybrids developed between 1930 and 1970. Population densities of about 30, 45, and 60 thousand plants per ha were used, and the tests were conducted in 11 environments. The two best new hybrids (from the 1970 era) yielded 40 percent more than the two best old hybrids (from the 1930 era) and 55 percent more than the open-pollinated cultivar. New hybrids had maximum yields at high plant densities, indicating a tolerance to crowding, and they showed a more stable performance over variable environments. The new hybrids were especially superior in low-yield environments because they were less sensitive to drought stress.

Jensen (1978) reported that wheat (*Triticum aestivum*) yields doubled in New York between 1935 and 1975. Half the gain was due to genetic advances and half to technological and cultural factors, and part of the genetic gain came from shorter-strawed wheats with resistance to lodging.

Air pollution tolerance illustrates location effect on breeding progress. Menser (1974) found that five sugar beet (*Beta vulgaris*) cultivars exposed to controlled ozone rates showed various degrees of leaf injury. The two California cultivars that were more ozone-tolerant than the others were selected and tested in the presence of air pollution.

Heggestad et al. (1977), by using greenhouse tests, found 'Acala SJ-1' cotton, bred in the San Joaquin Valley of California, was the most resistant of eight cultivars tested for their reaction to oxidant air pollutants. 'Paymaster 202' and 'Gregg 45', bred on the High Plains of Texas, were more sensitive. This variation among cultivars probably was caused by atmospheric differences that caused a selection pressure in the San Joaquin Valley, which has high oxidant levels, and no selection pressure in the Texas High Plains, which has low levels.

Lafever et al. (1977) reported that wheat and barley (*Hordeum vulgare*) cultivars were known to differ widely in their tolerance to highly acid soils containing high levels of free aluminum (Al). From detailed tests, Foy et al. (1965, 1974) concluded that many cultivars bred in the eastern USA appeared to have been selected unconsciously for Al tolerance, whereas cultivars tended to lack Al tolerance if bred on Indiana soils where no Al stress problems occur. Brown and Jones (1977) discussed ways to fit soybeans (*Glycine max*), cotton, and sorghum (*Sorghum bicolor*) plants nutritionally to soils. Cultivars of these three crops differed in tolerance to iron (Fe), zinc (Zn), and copper (Cu) stress and to manganese (Mn) and Al toxicities. They advised that care should be taken in choosing cultivars for planting so that stress losses would be minimized.

These examples show that cultivars bred without direct selection pressure for stress resistance nevertheless have considerable variability in resistance to stress. This is probably due to the indirect selection pressure imposed by performance trials conducted in the area of growth where the stress conditions exist. If breeders place test locations on soil that is atypical of the region in which the crop is normally grown, they may release cultivars that are susceptible to some stress in the environment. For example, Brown (1978) reported "de Mooy observed that when 'Hawkeye' soybeans (Fe efficient) were replaced by new soybean cultivars (Fe inefficient) in central and north central Iowa (calcareous soil), Fe chlorosis developed in the new soybean strains. These plants were probably not tested for Fe efficiency before they were released in the field."

DIRECT BREEDING FOR STRESS ENVIRONMENTS

Deliberate Choice of Fields with Stress Environments

The second category of breeding for stress environments involves deliberately choosing testing sites that represent stress conditions reliably and uniformly. Temperature and moisture are notorious for their unpredictable variability from location to location and year to year. Soil problems generally are not variable from year to year but may vary greatly from location to location. This makes it difficult for the breeder to get an adequate sample of environments within the scope of his testing program. Therefore, breeders select fields that have a stress factor at a level that will discriminate between resistant and susceptible genotypes. There is no value to a test if it is located in an environment that is stress-free or if the stress is so severe that nothing survives. Pathologists and nematologists use this technique by screening plants in fields known to be infested with pathogens and nematodes, respectively.

This technique has been used in a wheat breeding program in Brazil, where da Silva (1976) chose breeding sites which had soil acidity and Al toxicity so severe that selection pressure was absolute. Susceptible strains died, and Al-tolerant cultivars made good yields without liming. These tolerant wheat cultivars were grown in areas where wheat production would be impossible without tolerance to Al.

Meiners and Heggestad (1980) used fields at Beltsville and Salisbury, Maryland, to screen snap bean (*Phaseolus vulgare*) cultivars and advance breeding lines for resistance to oxidant (ozone) air pollution. These locations have a natural high incidence of ozone, and they can be used effectively to classify resistance levels of plants to ozone. So many cultivars were resistant that a special breeding program for resistance to air pollutants for snap beans was not necessary; however, these researchers recommended that all breeding lines be exposed to ambient pollutants before being released. They exposed snap bean seedlings to controlled levels of ozone in greenhouses, but seedling response had little value as an indicator of the mature plant response in the field. Snap beans may vary in degree of susceptibility to ozone according to stage of development.

According to Menser and Hodges (1972), the cigar wrapper industry in the Connecticut River Valley lost millions of dollars because of weather fleck, a disorder caused by air pollution. Breeding of fleck-resistant cultivars saved this industry from almost certain economic disaster. Povilatis (1966) selected tolerant plants from within 'Delcrest' cultivar of tobacco (*Nicotiana tabacum*) to develop the more fleck-tolerant cultivar 'Delcrest 66.' Additional progress has been made by crossbreeding and selection.

Blum (1977) showed that sorghum hybrids are superior to

open-pollinated cultivars in both optimum and suboptimum growing conditions. Hybrids not only yielded more in good environments, but they also had a greater stability in yield performance over variable environments. Blum (1977) and Sullivan and Blum (1970) present a thorough discussion of the philosophy of breeding for stress resistance and the need to be precise about heat and drought resistance, yield potential, and specific genes for stress resistance. In short, better knowledge about the mechanism for resistance and genetics of the response should enhance breeding progress for stress resistance.

Breeding Under Precise, Laboratory-Controlled Conditions

The third category of breeding for stress environments involves providing precisely controlled test conditions. Field environments, even those specifically chosen for the purpose, are highly variable. To bring more precision into the selection process, particularly in the early screening and selection phases, solution culture tanks or potting mixtures have been developed with a range in salt, pH, or mineral content levels. Controlled environment chambers may be employed to provide closely regulated temperature, moisture, and light regimes.

For example, Zimmerman and Buck (1977) found that success of selection in the field for cold tolerance in safflower (*Carthamus tinctorius*) seedlings was sporadic because the occurrence of freezing temperatures was unpredictable. By using a chamber in which a specified cold temperature could be maintained, the precision of selection was so good that they were able to introduce gene(s) for seedling cold tolerance from a wild species into a segregating population derived from a four-way cross. With controlled hardening and controlled cold stress, the breeding of safflower cultivars with even greater cold tolerance is feasible.

Foy et al. (1977) found that weeping lovegrass strains differed in their susceptibilities to severe Fe-deficiency chlorosis in field tests in Oklahoma. By using a greenhouse test with potting soil mixtures carefully prepared to detect Fe-deficiency chlorosis in crops, they were able to show that Fe chlorosis was related to Fe metabolism and not Fe uptake. Several experimental strains were shown to be more resistant to Fe-deficiency chlorosis than current cultivars with this greenhouse technique.

Dexter (1956) has said that ordinary field testing is the most common method of evaluating cultivars of crop plants for cold hardiness. However, the presence of variable growing conditions, diseases, insects, and other uncontrolled factors causes highly variable results that are difficult to interpret. The hardiness ratings may not reflect a tolerance to low temperatures, but rather, a false reading compounded from reactions to numerous factors. Also, winterhardiness is difficult to measure and express on a precise scale. According to Dexter (1956), the use of refrigeration to simulate field freezing conditions is a virtual necessity in winterhardiness selection studies. Two fundamental problems with artificial freezing are: (1) what the proper temperature should be for discriminating winterhardiness, and (2) what scale should be used to rate degree of hardiness. Degrees of winterhardiness measured in the field are remarkably well correlated with tolerance to low temperatures in freezing tests. Freezing tolerance is a multiphase situation that involves membrane changes and dehydration during hardening. The level of tolerance may involve rapidity of hardening response to initial cold and the minimum temperature level attainable.

Howell et al. (1971) reported that 'Team' alfalfa (*Medicago sativa*) was developed with a recurrent selection procedure based on greenhouse-grown plants that were relatively free of injury from naturally occurring levels of ozone in greenhouses at Beltsville, Maryland; and in subsequent tests in control chambers, 'Team' had less ozone damage than cultivars developed elsewhere. Menser and Hodges (1969) reported that tolerance to ozone can be evaluated in control chambers as well as in fields where the stress occurs. Results from field and greenhouse tests were not perfectly correlated; however, the greenhouse test does serve as a way for the breeder to find resistant parental material to make early generation selections and to evaluate cultivars.

Lafever et al. (1977) found that root lengths of wheat strains grown in nutrient solutions containing Al in the greenhouse were correlated with grain yields when these same strains were grown in fields with high Al soils. This technique would be useful for screening segregating populations.

Epstein (1976) discussed the problem of salinity in agricultural soils and pointed out that the problem was solved primarily by manipulating the environment through reclamation and drainage projects, but he made a plea that an easier solution would be to breed crop plants better able to grow normally under saline conditions. He screened barley lines in the greenhouse in solution culture tanks with salinity levels as high as that of seawater. Surviving strains were field tested on ocean beach areas with undiluted seawater being used for irrigating. All lines survived to maturity, and the three best ones yielded about 20 percent as much as barley yields in nonsaline soils. Work is underway to combine the high salt tolerance of a wild tomato (*Lycopersicon* spp.) with the acceptable horticultural properties of susceptible cultivars (*Lycopersicon esculentum*). Little is known about the mechanisms that fit salt-tolerant plants to conditions so saline that salt-sensitive plants are killed. Genetic, biochemical, and physiological studies are needed on near isolines which have sharp contrasts in salt tolerance. Mass and Nieman (1976) presented a thorough review of the physiology of plant tolerance to salinity.

Dewey (1962) observed wide differences in salt tolerance among five strains of crested wheatgrass (*Agropyron cristatum*) in artificially salinized field basins. Moreover, selection for tolerance in artificially salinized basins increased the salt-tolerance level of selected progeny and appeared to be an effective way to breed more salt-tolerant strains.

Tanimoto and Nickell (1965) tested cuttings of sugarcane (*Saccharum officinarum*) in pots of soil treated with various levels of salt, and wide differences in salt tolerance were found among strains; moreover, the reaction to salt was highly correlated with field drought resistance. Salt shock treatment is now a standard practice in Hawaii for selecting drought-tolerant sugarcane clones (Nickell, 1977).

Breeding for Fundamental Causes of Stress Resistance

The fourth category in breeding for stress environments involves breeding for the fundamental causes of stress resistance rather than measuring the phenotypes of plants. Generally, breeders have no knowledge about the chain of biochemical and physiological events between genes and phenotype, but if some key step in the chain is known to be highly correlated with stress resistance, the breeder can select directly for it rather than for the final phenotype. For example, in cotton, high gossypol levels in the fruit reduce damage by bollworms; so instead of measuring bollworm damage, the breeder may breed directly for gossypol level.

According to Sullivan (1972), most of the progress in breeding for drought resistance in crop plants has been achieved by using empirical methods. In practice a drought resistant cultivar is defined as the one making the greatest yield with limited soil moisture, but better progress likely would be made if the internal water status and physiological responses of the plant to drought were known. Sullivan (1972) discussed methods for measuring tolerance and avoidance of heat and drought stresses, and he suggested that breeders should select for intermediate levels of tolerance and avoidance as the best compromise.

Sullivan and Ross (1977) discussed the relationship of heat and drought tolerance in grain sorghum, and they pointed out the need to measure the two responses separately. They described a leaf-disc test for evaluating heat and desiccation tolerance in sorghum via measurements of photosynthetic rate which is sensitive to drought and heat stress. A simple portable photosynthesis chamber made it possible to measure photosynthesis of many plants within a short time. Such an apparatus makes it feasible to use photosynthesis as a criterion for selection for drought and heat resistance. Sullivan and Ross (1977) also employed hydroponics tanks to evaluate root growth as a screening procedure for seedling drought resistance.

Foy et al. (1978) have written a comprehensive review of the physiology of metal toxicity in plants. They suggest the need for a multidisciplinary approach to identify the specific mechanisms that make plants resistant to various metals. Such cooperation will lead to an understanding of the mechanisms of tolerance and their genetic control, which in turn, will lead to greater progress in breeding for resistance and rapid, economical, and reliable artificial testing techniques.

O'Toole and Chang (1977) present a thorough review of research on drought tolerance in cereals, with rice (*Oryza sativa*) as a case study. They reported progress in breeding rice for drought tolerance at the International Rice Research Institute. Progress to date has been based on empirical methods; however, they expect greater progress will be forthcoming when physiological and genetic research allow breeding for specific genes responsible for the underlying causes of drought resistance.

The growth habit of a crop may account for tolerance to stress. Quisenberry and Roark (1976) found that a cotton cultivar with relatively indeterminate growth and fruiting is better adapted to an environment with seasonally limited soil moisture than one with determinate growth. Indeterminate growth gave the plants the flexibility to fruit throughout the growing season whenever sufficient moisture was available, whereas a determinate cultivar would not likely begin fruiting again should moisture become available late in the growing season.

Specific stress control mechanisms can also be a selection goal. Roark and Quisenberry (1978) described a laboratory method for measuring transpiration of detached cotton leaves. Ultimately, it was learned that stomatal behavior was an effective selection criterion in breeding cotton for drought tolerance.

Engle and Gabelman (1966) demonstrated that resistance to

ozone in onion (*Allium cepa*) was determined by hypersensitivity of guard cells. Guard cells of resistant plants collapsed more readily than those of susceptible ones, which decreased the ingress of ozone.

It should be pointed out that field testing is required regardless of how early selections and preliminary evaluations are made. Field test validation of laboratory procedures is the surest way of predicting the experience farmers will have with a cultivar.

Genetic Engineering

The final category in breeding for stress environments involves genetic engineering as defined by Rieger et al. (1976). That is, genetic engineering is "genetic manipulation (by-passing the sexual cycle) by which an individual having a new combination of inherited properties is established. Genetic engineering presently follows two major approaches: (1) The cellular approach involving the *in vitro* culturing of (haploid) cells and the hybridization of somatic cells; (2) The molecular approach involves the direct manipulation of DNA."

Direct manipulation involves recombinant DNA molecules which has been defined by HEW (1978) as either "(i) molecules which are constructed outside living cells by joining natural or synthetic DNA segments to DNA molecules that can replicate in a living cell, or (ii) DNA molecules that result from the replication of those described in (i) above."

These genetic engineering techniques possibly would be useful for combining germplasms across species and genera which are presently insurmountable by sexual means. However, the technique most likely to be useful in breeding plants for stress environment would be cell culture. For certain stresses, such as salinity, it would seem possible to subject millions of cells in a flask to an elevated level of saline solution with the result that only resistant or tolerant cells would survive. The surviving cells would be treated to regenerate plants that were saline tolerant.

The genetic manipulation by recombinant DNA techniques or by selecting at the cellular level has not resulted in any stress-resistant cultivars of plants or cultivars of any kind for that matter.

After an extensive literature review of plant tissue culture, Vasil (1978) concluded that "many of us have been over-enthusiastic or even uncritical in forecasting the future uses of plant tissue culture, and have generally failed to adequately realize the variety and severity of problems that must be resolved before today's dreams can be translated into tomorrow's realities." This note of caution was not meant to discourage research on plant tissue culture, "but rather to stimulate some serious thinking and self-examination. . . ."

At this time any plant genetically modified by these techniques could not be field tested or used by farmers, because a HEW sponsored law prohibits "deliberate release into the environment of any organism containing recombinant DNA." A mechanism exists for obtaining exemptions to this prohibition, but to date, there has been no occasion for making an appeal for an exemption. Vasil (1978) says that "uptake of DNA, chloroplasts, and bacteria has been suitably demonstrated, but there is no clear evidence that these remain biologically functional or are fully stabilized in the host cells or protoplasts." Day (1977), offers a well-balanced approach of cautious optimism that the techniques of cell culture systems, protoplast fusion, and recombinant DNA can eventually contribute to increased crop yields.

CONCLUSIONS

The world demand for food and fiber means that crops must be grown on marginal land that is likely to have stress problems. Land now in cultivation experiences stresses from widely fluctuating temperature and moisture conditions, but marginal lands present chronic problems with salinity, acidity, alkalinity, air pollution, and mineral deficiency and toxicity.

Breeding for stress resistance and good agronomic procedures are two approaches to increase agricultural production and stabilize productivity. Germplasm offers useful heritable variation for breeding for stress resistance, and this variability has only begun to be exploited. Considerable success has already been achieved in breeding for stress environments, largely by the indirect method.

Five categories of breeding approaches, ranging from indirect selection through field performance testing to genetic engineering, were discussed. Selected examples of progress in each category suggest that much more can be accomplished if breeders and agronomists cooperate to codify stress environments and breed plants more tolerant of stress conditions.

It seems clear that scientists are becoming more deliberate in breeding for stress resistance; they are looking for ways to measure resistance and seeking the underlying mechanisms for resistance.

True, plants perform better in optimum than in suboptimum environments, but plant genotypes can be developed that perform better in stress conditions than cultivars now available; and cultivars can be developed that make agricultural production practical on marginal lands not now in cultivation.

Discussion

M. N. CHRISTIANSEN
S. D. JENSEN
M. C. SHANNON

1. M. N. CHRISTIANSEN. Because the first defense of plants against stress is genetic resistance, the work in stress physiology has a primary goal of providing breeders with quick, efficient selection tools to identify resistant germplasm. The terminology used by physiologists may differ slightly from that of plant breeders, but our common goals are to improve the efficiency and productivity of crop plants under extreme environments. Scientists who work on research in environmental stress seek to define what types of environmental regimes cause stress, the nature of plant injury from stress, and how the stress factors can be ameliorated. And, the use of genetics is the first approach for amelioration. Cultural techniques, chemical therapy, and the like, are other approaches. And, the ultimate goal is to grow a productive crop where no crop grows today. To date, plant breeders have done a remarkable job in breeding crops for stress environments, but given better screening tools, the rate of progress will be accelerated markedly.

If pests, that is, diseases and insects, as factors that limit crop production are assigned to pathologists and entomologists, the remaining major environmental factors that limit crop productivity and delimit cropping areas are temperature, water, and soil. These factors determine where crops are grown and how much they produce. My remarks will be confined to mineral stress and temperature.

In past eras of low cost fertilizer and low cost land, the idea of fitting crop plants to problem soils attracted little interest. With the great changes in the world energy situation, however, ideas and viewpoints may need to be altered. Researchers currently are discussing fertilizer efficiency and genotype tolerance to extremes in soil chemical factors. A good example of the use of a genetic solution to a soil chemical problem is the breeding of soybeans for resistance to iron (Fe) chlorosis. Certain Corn Belt soils, such as the Harpster silt loam, have a pH = 7.2 or higher; and in alkaline soils, much of the Fe is unavailable to plant uptake. Most plants take up Fe in a reduced form; so the oxidized form, which occurs in high pH soils, is unavailable to the plant. Certain genotypes of soybeans, when grown on such soils, will turn yellow, have stunted growth, and even die. There are over three million ha of alkaline Clarion-Webster soil type in Iowa on which this problem occurs if the proper genotype is not utilized. Efficient genotypes, on the other hand, remain healthy when grown on alkaline soils. The yellowing and stunted growth can be solved by spraying an Fe solution onto the plants, but this is temporary and expensive. So the primary solution to the problem is genetic. The mechanism by which a cultivar is Fe efficient is fairly simple. It involves the exudation of substances from the roots that reduce the Fe in the rhizosphere to make it available to the plant.

Another interesting case of the genetic solution to a mineral stress is the production of wheat cultivars that are tolerant to high aluminum (Al) availability on acid soils. The cultivar 'Thorne' developed in Ohio is tolerant to the Al toxic soils with pH of 4.0 to 4.5. On the other hand, 'Arthur,' which was bred in Indiana and is a widely planted cultivar, cannot grow on such soils. Of course, liming the soil could alter the situation, but liming only corrects the pH in the upper 15 to 20 cm of soil. Wheat roots normally will penetrate 30 cm into a favorable soil, so even with liming of acid soils, Arthur does not produce well when grown on them. The breeding of the Al tolerant wheat cultivar 'BH 1146' has permitted wheat production on several million ha of high Al soils in Brazil, where wheat could not otherwise be produced. The mechanism involved in tolerance to Al entails an ability of plant roots to exude substances that raise the rhizosphere pH, which in turn precipitates Al from the soil solution. Such tolerance mechanisms are known to exist in corn, sorghum, weeping lovegrass, fescue, soybeans and many other crops.

Temperature stress, including heat, chilling, and freezing, is a complex subject. For example, temperature stress can cause fruit blossom loss, failure of germination, or failure of grain set. Each of these is a specialized case that involves the stage of plant development, species, nature of the injury, dosage of stress, interaction with water availability, and the nature of hardening. In temperature stress studies, various criteria are used to quantify temperature susceptibility. Usually readings are empirical, such as survival in the field or under some specified controlled conditions. Survival includes the impacts that temperature has on the plant, respiration, photosynthesis, and water uptake. There is a growing interest in the membranes of cells as a site of temperature impact. Temperature may affect the form, chemistry, and function of membranes. After all, membranes compartmentalize the metabolic organization of the plant, and they function as osmotic barriers in solute movement and Fe and water uptake; so it is sensible that quite dramatic alterations of membrane mediated functions should occur under different temperature regimes. Photosynthesis, respiration, ion and solute movement are just a few of the functions altered by temperature extremes.

The basic approach to working with temperature stress is to first, determine what temperature regimes cause injury at what stage in the plant development; second, to quantify the response or injury; and third, to screen for genetic variants under controlled or semicontrolled conditions.

2. M. C. SHANNON. Salt stress probably causes more damage to plants in terms of economic losses than any other crop production stress factor. Additionally, its effects are difficult to detect because they are so subtle, and its primary mode of physiological damage remains elusive. A primary reason for so much concern with breeding plants for stress environments is that most expansion of crop production in the future will have to be onto marginal land that is not now suited to agricultural use. There are several reasons for the increased interest in breeding crop plants for salt tolerance:

1. Water quality is deteriorating on a worldwide scale. Increased salinity results from the increased metropolitan and agricultural demands on water resources, and in the USA, from over-irrigation and the excessive application of fertilizer.

2. The depletion of energy resources has increased the cost of salinity management techniques such as land reclamation, drainage, and irrigation management.

3. The trends toward crop diversity cause a need to fit several crops for growing on saline soil where only one crop has been used in the past. With the future significance of these factors, the development of salt-tolerant plants can be an energy efficient and affordable technology for use in both developed and underdeveloped countries.

3. S. D. JENSEN. Probably the most difficult problem for the maize breeder who wishes to select for drought resistance is how to evaluate the trait. Drought resistance has a low heritability caused by high genotype x environmental interaction. Even when successfully selected, it is difficult to demonstrate drought resistance consistently. A second problem is that high-yielding hybrids under stress conditions may not have as favorable performance overall relative to those that yield well under more optimum conditions; thus, they tend to be overlooked in favor of hybrids that have higher mean yield but poorer performance under stress.

We have used the regression response technique to measure the stability of hybrids over a range of stress environments, and comparisons are always paired, that is, two hybrids being compared were evaluated in the same test. As a first example, the veteran commercial hybrid, 'Pioneer 3388,' which has a long history of excellent performance under conditions of drought stress in the western Corn Belt, had a b = 0.77, whereas the widely grown hybrid, 'B 73' x 'Mo. 17,' had a b = 1.17. For comparing these values, t = 8.33, (based on 168 locations) so these slopes are significantly different. The two regression lines intersect at 4.5 t/ha (74 bu/a). Therefore, one would conclude that for high productivity environments, B 73 x Mo. 17 would be the better hybrid to grow, and that for low productivity ones, Pioneer 3388 would be best. However, this is not an entirely safe conclusion. In 1976, the lines crossed at 1.5 t/ha (25 bu/a), in 1977, at 6.2 t/ha (100 bu/a), and in 1978, at 3.1 t/ha (50 bu/a). In the central Corn Belt, the regression lines cross at 6.2 t/ha (100 bu/a) and in the western Corn Belt at 4.5 t/ha (75 bu/a). In the case of these two hybrids, the slopes of the regression lines were different in each of three years, in two regions of the country and in the overall comparison, and the regression lines intersected each other between 1.5 and 6.2 t/ha (25 to 100 bu/a).

As a second example, the slopes of the regression lines were 1.03 for 'Pioneer 3780' and 0.93 for 'Pioneer 3709' (based on 381 locations). The slopes were significantly different with t = 4.01. Pioneer 3709 had a somewhat flatter slope in all three of the comparison years, but the point of intersection was 9.3 t/ha (150 bu/a) in 1976, and there was no intersection in 1977 and 1978. The regression lines for these hybrids were highly and significantly different in central, significantly different in eastern, and not different in western USA. Thus, their responses varied from year to year and region to region.

In summary, drought resistance in maize does exist. Hybrids

respond differently to drought stress, but their responses vary so much from year to year and location to location that it is difficult for the breeder to identify heritable differences.

4. S. JANA. Some components of a stress environment are predictable and some others are not. Would different breeding strategies be used to cope with predictable components?

5. M. N. CHRISTIANSEN. A salt or an acidity problem would be predictable whereas rainfall or temperature would not. Theoretically, it should be easier to select for resistance to a stable stress than to a fluctuating one. But, of course, one could resort to using artificially controlled levels of stress in the greenhouse or control chamber for discriminating resistant genotypes.

6. F. M. MARQUEZ. Plant breeders usually try to develop genotypes that are good for both yield and reaction to stress environments. For maize breeding in Mexico where combining these traits is a goal, we assume that: (1) genes for yield are different than genes for resistance to stress environment; (2) selection for yield is more efficient in a good than in a stress environment; and (3) selection for resistance to stress has to be done in a stress environment.

The approach is to select for both traits simultaneously, but in separate parts of a population. Select for yield in one part of the base population in a nonstress environment and for resistance to stress in the second part in a stress environment. After several cycles of selection, the two selected populations would be recombined, and the resultant population presumably would have the combined traits.

7. S. D. JENSEN. Maize breeders use this procedure currently. Probably, if the initial selection is for drought resistance, the materials selected will tolerate stress, but they may not yield well under optimum conditions. And, if the initial selection is under optimum conditions, the material may not produce well under stress environments. Since information is not available on the relative efficiency of the two selection schemes, simultaneous selection under both stress and optimum environments would seem to be a desirable choice. This would help insure that genes for both drought resistance and yield would be maintained.

8. P. BUSEY. The regression response technique for measuring

reaction to stress environments has been used to select turfgrasses for resistance to shade, salt, and low fertility stresses. The problems with regression analysis are: (1) gross scale differences within a broad germplasm base can cause uncertainties in data interpretation; (2) when the regression procedure is applied to data from a broad spectrum of stress environments, the response is sometimes sigmoidal; and (3) there is increased variability with really high stress levels.

Now with such problems, is it really possible to relate overall yield response over all environments with specific adaptations to stress?

9. S. D. JENSEN. In the regression analyses quoted for maize, the values for coefficients of contingency were very high, which means that the fit of the data to the regression lines was good. So there is no reason to believe that the yield data were highly variable.

10. C. F. LEWIS. In a regression analysis on cotton cultivars done several years ago, the regression lines for genotypes intersected quite complexly. However, cotton farmers were way ahead of us. In areas with good environments, they planted the cultivar that exploited good environments, and in territory where poor environments were highly variable, they planted cultivars that did well under stress.

11. K. DIESBURG. Is it true that soil variability, and the high error variance it causes, is the major factor that limits the progress in selecting for resistance to drought stress? If so, would it not make sense to add amendments to the soils to make the conditions more uniform?

12. S. D. JENSEN. Soil variability is not the most important problem in selecting for drought resistance. True, error variances increase relatively as yield levels go down. But the greater problem is the genotype x environment interaction that occurs over and above the problem of soil variability. Selection from one location to another, and selection from one year to the next, is not reliable because of genotype x environment interaction.

13. R. J. BAKER. Have definitive experiments been conducted to tell whether the response regression lines can be used to predict what environment type a given genotype will be adapted to?

14. S. D. JENSEN. If such experiments were conducted, they

would require many locations and several years due to the unstable nature of the characteristic.

15. A. F. TROYER. Are the production regressions for maize hybrids related to the popularity or customer acceptance of the hybrids?

16. S. D. JENSEN. Yes, the regression lines of hybrids can be related to farmers' preferences in specific environments. In certain areas, hybrids with low values are in demand. It is especially notable that in areas where there is irrigation or higher rainfall, hybrids with the steep slopes are in demand. If the slope is very steep, the hybrid must be vulnerable to stress environments. Specifically, the hybrid, B 73 x Mo. 17, produces very low yield in droughty environments.

17. W. J. VAN DER WALT. What are the specific plant characteristics of maize, especially those that permit the plants to go dormant, which render corn tolerant to drought?

18. S. D. JENSEN. The so-called "latente" trait is used to denote a condition whereby during a droughty period a plant tends to go dormant with respect to growth. When adequate moisture becomes available, the plant resumes growth. The trait is fairly common in sorghum, and for maize, it was first identified in some tropical collections.

Probably the latente trait can be found in most breeding populations of maize. Such genotypes when grown under severe drought stress will roll up and look as if in very serious condition; but upon application of water, they show shoot or silk emergence in two or three days, whereas others will not. It is a good and necessary characteristic to have in cultivars that will be grown in droughty conditions.

19. S. JANA. Likely, crop cultivars have several mechanisms for coping with stress to provide stability of production over a range of environments, and it is also likely that these mechanisms have different genetic controls. Wouldn't it be possible and desirable to combine these several mechanisms for stability in performance into one cultivar so as to increase its ability to cope with a range of environmental stresses?

20. S. D. JENSEN. This is a desirable goal and would lead to the

selection of genotypes with a high mean yield and a slope of b = 1. Experience suggests that it would be difficult to achieve because the highest yielding maize hybrids under stress have b values of less than 1.0 and hybrids that perform best under optimum conditions have b values greater than 1.0.

21. D. NANDA. Is there any correlation between prolificacy and stress tolerance in maize?

22. S. D. JENSEN. There does not appear to be a high correlation between prolificacy and tolerance to drought stress in maize. The idea is, of course, that prolific plants that normally produce two ears, when under drought stress will produce at least one. A tendency to prolificacy may be helpful under some conditions of drought stress, but drought resistance probably has a physiological basis that is not necessarily expressed in multiple ears. Under severe drought stress some prolific types are just as barren as the non-prolifics.

23. D. GROSS. Does maize selected for drought resistance also have good seedling vigor in cold environments?

24. S. D. JENSEN. Generally, there is negative correlation between heat plus drought tolerance and cold tolerance. So genotypes selected for good heat and drought tolerance do not germinate well or have good early vigor under cold soil conditions, and genotypes selected for cold tolerance (that is, good germination below 10°C) are susceptible to high temperature during pollinating and grain filling. The correlation between cold tolerance and yield was negative. Perhaps these associations can be broken.

25. S. N. NIGAM. When selecting genotypes for a production environment that has several factors of stress, is it better to select for each component separately or to select for all factors simultaneously?

26. C. F. LEWIS. In one study, sorghum was selected for a production environment that had both heat and drought stresses. These two factors of stress usually occur together, but the mechanisms for resistance or avoidance in the plants are different. So, under this circumstance, each should be selected for separately.

27. A. M. THRO. There is a phenomenon termed variously "developmental flexibility" or "phenotypic plasticity" which is a mechanism for adaptation to stress such as drought or shading. This characteristic is probably already present in crop cultivars, but is it of value or specific use in crop breeding?

28. M. C. SHANNON. If a population has a sufficiently broad genetic background, plasticity will show when an environmental stress is applied to that population. As an example, with lettuce, much more variation shows when stress is applied to the plants than when they are grown in nonstress environments. Reaction to salinity is not simple and may be due to several physiological mechanisms, so probably plasticity to it results from the sum of adaptive responses in a population of genotypes.

29. S. JANA. Is there much information available that shows a common genetical basis for tolerance to more than one stress?

30. M. N. CHRISTIANSEN. Yes, and as an example, hardening of a plant to drought conveys a certain amount of low temperature tolerance in the plant.

31. D. D. WEST. It has been shown that in maize good correlation exists between responses in certain laboratory or greenhouse tests and field responses to drought and heat tolerance where one uses small numbers of inbreds or hybrids. Has any population improvement for drought or heat tolerance been accomplished via recurrent selection for responses to drought in laboratory or greenhouse tests?

32. M. N. CHRISTIANSEN. Some work has been done with respect to water relationships in peppers. Correlations were run between pressure bomb readings and diffusive resistance ratings obtained in the greenhouse with water usage of peppers in the field.

33. C. F. LEWIS. Team cultivar of alfalfa was developed through a recurrent selection procedure based on greenhouse plants screened for freedom from ozone injury.

34. K. SAMMETA. Cold resistance in plants can be selected under control conditions. The procedure requires well-built chambers and

elaborate equipment, but it is possible to separate cold susceptible from cold resistant plant genotypes by using temperatures of −5°C for a few hours. Once standardized, such a screening method can be used on 100-120 plants per test.

35. M. MARTIN. One tolerance mechanism of wheat cultivars to Al toxicity is that such cultivars exude substances that precipitate the excess aluminum in the soil. What is the effect of these exuded substances on other minerals in the soil?

36. M. N. CHRISTIANSEN. For review, when pH of the soil drops below 7, Al solubility increases greatly. Fe solubility also increases. Now, the exudates from plants that are tolerant to Al do not immobilize the Al directly—rather they raise the pH to the point at which Al is no longer at sufficiently soluble levels to be damaging to the plant. It is unknown whether Fe decreases in solubility in these cases, but it is unlikely that pH levels would reach a sufficiently high level to make Fe unavailable.

37. A. A. QUINN. It is likely that breeding for salt tolerance has been done in routine breeding programs inadvertently. Over time, breeders have selected crops and genotypes within crops for irrigated, arid environments where high salt concentrations have built up, so that salt tolerance could have become a basic component of adaptation in these regions.

This seems to be the case in sugar beets. In the USA, there is a difference in adaptability of sugar beet cultivars from east to west. In the humid Red River Valley (that is, parts of western Minnesota and eastern North Dakota), many cultivars bred specifically for humid Western European environments seem well adapted. However, as these cultivars are grown farther west in the irrigated areas of the Great Plains and west of the Rockies, their root tissue takes on a higher sodium content than adapted cultivars, and there is more variability in their performance. A higher sodium concentration in the root tissue is detrimental because it causes a decrease in sugar extraction percentage. It is also possible that the higher salt concentrations are deleterious to the plant growth of these unadapted cultivars, and that their variability in performance is due to a variability in environmental salt concentrations.

This situation suggests that inadvertently, breeders of sugar beets in the western USA have selected cultivars resistant to the higher salt concentrations, the mechanism of resistance being a selective exclusion of sodium ions. Probably, this type of inadvertent selection has occurred in other crops as well, especially in those grown under irrigation in arid environments.

38. M. C. SHANNON. This falls under indirect selection, and it happens quite often in the case of salinity. As examples, Egyptian cottons are more salt tolerant than American cottons, and wheats from Mexico, India, and Pakistan, have been indirectly selected for salt tolerance because salinity problems are indigenous to those countries. Sugar beet, as a species, is highly salt tolerant, but sodium uptake causes an impurity problem in beets from the standpoint of sugar extraction.

Certainly, variability in soil salinity could cause variable production if one used susceptible sugar beet cultivars. Breeders have attempted to select for salt tolerance in the field, but because soil salinity is so variable, selection for tolerance has not been especially successful. In any one field, regions of high salt vary horizontally and with depth. So, often the root zone of a particular plant may be in very saline soils and in soils that are not so saline. To obtain a quantitative assessment of how much salinity a plant is being subjected to is very difficult.

39. T. E. DEVINE. Sources of genetic tolerance to specific environmental stresses probably exist in the germplasm collections of the world. These collections have very large numbers of samples and to assay them completely for tolerance to a specific edaphic stress, such as Al tolerance, would require a large scale effort. Information on the soil characteristics of the site from which the plant introductions(PI's) were obtained could prove a valuable guide in selecting the PI's most apt to have the desired characteristics. Often this information is lacking. I would hope that future plant collection programs would make an effort to include this information.

40. W. W. ROATH. Land has been taken out of production due to decrease in water quality, and one causal factor for the decrease in water quality is the excess use of fertilizer. Are there specific examples of where and to what extent this has occurred in the USA?

41. M. C. SHANNON. Water quality can be diminished by the excessive use of fertilizer and overirrigation. One example occurs along the lower regions of the Colorado River. Every time water is used for irrigation and returned to the aquifer the salt content

increases. Possible alternatives for improving water quality include the construction of a desalination plant, switching the farmers and ranchers to minimum irrigation practices, or terminating agriculture in those areas. Some factions believe that taking the land out of production is the fastest and most economical way to solve the problem.

Similar water-quality problems exist in the Central Valley of California. Here, separate canals have been proposed as a means of eliminating used agricultural waters of high nitrate and salt contents.

REFERENCES

Blum, A. 1977. The genetic improvement of drought resistance in crop plants: A case for sorghum. Contrib. from the Agric. Res. Organ., Volcani Cent. Bet Dagan, Israel. Series 136-E.

Brown, J. C. 1978. Physiology of plant tolerance to alkaline soils, pp. 257-76. *In* J. A. Jung (ed.), Crop tolerance to suboptimal land conditions. Special Pub. 32 Am. Soc. Agron.

Brown, J. C., and W. E. Jones. 1977. Fitting plants nutritionally to soils. I. Soybeans. Agron. J. 69:399-404.

Brown, J. C., and W. E. Jones. 1977. Fitting plants nutritionally to soils. II. Cotton. Agron. J. 69:405-9.

Brown, J. C., and W. E. Jones. 1977. Fitting plants nutritionally to soils. III. Sorghum. Agron. J. 69:410-14.

Carter, H. O. 1975. Prospectives on prime land. USDA. Background papers for seminar on retention of prime lands. Sponsored by USDA Comm. on Land Use, July 16-17, 1975.

da Silva, A. R. 1976. Application of the genetic approach to wheat culture in Brazil. pp. 223-31. *In* Wright, M. J. (ed.), Proceedings of workshop on plant adaption to mineral stress in problem soils, Beltsville, Md., Nov. 22-23, 1976.

Day, P. R. 1977. Plant genetics: increasing crop yields. Science 197:1334-9.

Dewey, D. R. 1962. Breeding crested wheatgrass for salt tolerance. Crop Sci. 2:403-7.

Dexter, S. T. 1956. The evaluation of crop plants for winterhardiness, pp. 203-39. *In* Norman, A. G. (ed.), Advances in agronomy VIII. Academic Press, New York.

Dilley, D. R., H. E. Heggestad, W. L. Powers, and C. J. Weiser. 1975. Environmental Stress, pp. 319-20. *In* Brown, A. W. A., T. C. Byerly, M. Gibbs, and A. San Peitro (eds.), Crop productivity—research imperatives. Mich. Agric. Exp. Stn. Publ.

Engle, R. L., and W. H. Gabelman. 1966. Inheritance and mechanism for resistance to ozone damage in onion. Proc. Am. Soc. Hort. Sci. 89: 423-30.

Epstein, E. 1976. Genetic potentials for solving problems of soil mineral stress: Adaptation of crops to salinity, pp. 73-82. *In* Wright, M. J. (ed.), Proceedings of workshop on plant adaptation to mineral stress in problem soils, Beltsville, Md., Nov. 22-23, 1976.

FAO. 1974. Assessment of the world food situation, present and future. FAO, Rome.

Foy, C. D., W. H. Armiger, L. W. Briggle, and D. A. Reid. 1965. Differential aluminum tolerance of wheat and barley varieties to acid soils. Agron. J. 57:413-17.

Foy, C. D., R. L. Chaney, and M. C. White. 1978. The physiology of metal toxicity in plants. Annu. Rev. Plant Physiol. 29:511-66.

Foy, C. D., H. N. Lafever, J. W. Schwartz, and A. L. Fleming. 1974. Aluminum tolerance of wheat cultivars related to region of origin. Agron. J. 66: 751-58.

Foy, C. D., P. W. Voight, and J. W. Schwartz. 1977. Differential susceptibilities of weeping lovegrass strains to an iron-related chlorosis on calcarious soils. Agron. J. 69:491-96.

Harvey, P. H. 1977. ARS National Research Program, NEP No. 20040 breeding and production: Corn, sorghum and grain millets. American Seed Trade Assoc. 1977 Yearb. and Proc. of 94th Annu. Conv., Louisville, Ky., June 26-30, 1977, pp. 178-82.

Heggested, H. E., M. N. Christiansen, W. L. Craig, and W. H. Heatley. 1977. Effects of oxidant air pollutants on cotton in greenhouses at Beltsville, Md. Presented at Cottrell Centen. Symp. Air pollution and its impact on agriculture, Turlock, Calif., California State College, Jan. 13-14, 1977.

Health, Educ. Welfare, U.S. 1978. NIH guidelines for research involving recombinant DNA molecules. Federal Register 43 (no. 247, Dec. 22, 1978): 601108.

Howell, R. K., T. E. Devine, and C. H. Hanson. 1971. Resistance of selected alfalfa strains to ozone. Crop Sci. 11:114-15.

Jensen, N. F. 1978. Limits to growth in world food production. Science 201: 317-20.

Klages, K. H. W. 1947. Ecological crop geograph. Macmillan Co., New York.

Lafever, H. N., L. G. Campbell, and C. D. Foy. 1977. Differential response of wheat cultivars to Al. Agron. J. 69:563-68.

Lewis, C. F. 1976. Genetic potential for solving problems of soil mineral stress: Overview and evaluation. pp. 107-9. *In* Wright, M. H. (ed.), Proceedings of workshop on plant adaptation to mineral stress in problem soils, Beltsville, Md.; Nov. 22-23, 1976.

Maas, E. V., and R. H. Nieman. 1976. Physiology of plant tolerance to salinity. *In* Jung, G. A. (ed.), Crop tolerance to suboptimum land conditions. Proc. Symp. Houston, Tex., Nov. 28-Dec. 3, 1976. ASA Spec. Publ. 32:277-99.

Mayer, J. 1975. Agricultural production and world nutrition, pp. 97-108. *In* Brown, A. W. A., T. C. Byerly, M. Gibbs, and A. San Pietro (eds.), Crop productivity—research imperatives. Mich. Agric. Exp. Stn. Publ.

Meiners, J. P., and H. E. Heggestad. Evaluation of snap bean cultivars for resistance to ambient oxidants in field plots and to ozone in chambers. Plant Dis. Rep. (in press).

Menser, H. A. 1974. Response of sugarbeet cultivars to ozone. J. Am. Soc. of Sugar Beet Technol. 18:81-86.

Menser, H. A., and G. H. Hodges. 1969. Tolerance to ozone of flue-cured tobacco cultivars in field and fumigation chamber tests. Tobacco Sci. 13: 176-79.

Menser, H. A., and G. H. Hodges. 1972. Oxidant injury to shade tobacco cultivars developed in Connecticut for weather fleck resistance. Agron. J. 64:189-92.

Nickell, L. G. 1977. Crop improvement in sugarcane: Studies using *in vitro* methods. Crop Sci. 17:717-19.

O'Toole, J. C., and T. T. Chang. 1977. Drought resistance in cereals: Rice a case study. Proc. Int. Conf. Stress Physiol. of Plants Useful in Food Prod. Boyce Thompson Inst., Yonkers, N.Y., June 28-30, 1977.

Povilatis, B., and F. H. White. 1966. Delcrest 66, a new variety of bright tobacco. Can. J. Plant Sci. 46:457-8.

Quisenberry, J. E., and B. Roark. 1976. Influence of indeterminate growth habit on yield and irrigation water-use efficiency in uplant cotton. Crop Sci. 16:762-66.

Rieger, R., A. Michaelis, and M. M. Green. 1976. Glossary of genetics and cytogenetics-classical and molecular. 4th ed. Springer-Verlag, New York.

Research to meet U.S. and world food needs. 1975. Report of a working conference sponsored by the Agric. Policy Adv. Comm. (ARPAC). 326 pp.

Roark, B., and J. E. Quisenberry. 1978. Stomatal control of transpiration. Agron. Abstr. 84.

Russell, W. A. 1974. Comparative performance for maize hybrids representing different eras of maize breeding, pp. 81-101. *In* Proc. 29th Annu. Corn and Sorghum Res. Conf., Chicago, Dec. 10-12, 1974.

Sullivan, C. Y. 1972. Mechanisms of heat and drought resistance in grain sorghum and methods of measurement, pp. 247-64. *In* Rao, N. G. P., and L. R. House (eds.), Sorghum in the seventies. Oxford and IBH, New Delhi.

Sullivan, C. Y., and A. Blum. 1970. Drought and heat resistance of sorghum and corn, pp. 55-66. *In* Proc. 25th Corn and Sorghum Res. Conf., Chicago, Dec. 8-10, 1970.

Sullivan, C. Y., and W. M. Ross. 1977. Selecting for drought and host resistance in grain sorghum. Proc. Int. Conf. Stress Physiol. in Crop Plants, Boyce Thompson Inst., Yonkers, N.Y., June 28-30, 1977.

Tanimoto, T., and L. G. Nickell. 1965. Estimation of drought resistance in sugarcane varieties. Proc. Int. Soc. Sugarcane Technol. 12:893-96.

Vasil, I. K. 1978. Plant tissue culture and crop improvement: Fact and fancy. Int. Assoc. Plant Tissue Culture Newsl. 26.

World food and nutrition study: Enhancement of food production for the United States. 1975. Rep. Board of Agric. and Renewable Resour., Comm. on Nat. Resour., NRC. 174 pp. Available from NAS, Washington, D.C.

World food and nutrition study: The potential contributions of research. 1977. Prepared by the steering committee, NRC study on world food and nutrition of the Comm. on Int. Relat. NRC. 192 pp. Available from NAS, Washington, D.C.

World population and food supplies, 1980. 1965. ASA Spec. Publ. 6. 50 pp. Available from Am. Soc. Agron., Madison, Wis.

Zimmerman, L. H., and B. B. Buck. 1977. Selection for seedling cold tolerance in safflower with modified controlled environment chambers. Crop Sci. 17:679-82.

CHAPTER 6

Development of Plant Genotypes for Multiple Cropping Systems

C. A. FRANCIS

IN THE RECENT BOOK, *To Feed the World: The Challenge and the Strategy* (Wortman and Cummings, 1978), the current food situation in developing countries is reviewed in detail, and strategies are presented for the application of technology to overcome the world's growing food deficits. Multiple cropping has been practiced over the centuries to produce basic food crops, but has been relatively unaffected by research and technology.

BACKGROUND ON MULTIPLE CROPPING

Traditional Cropping Systems

Small farm agriculture predominates in most regions of the developing world. The cropping systems are characterized by low levels of purchased inputs, intensive labor use, traditional cultivars, and relatively low yields of component crops. Multiple cropping systems provide a stability of production, diversity of diet and income sources, and a distribution of labor through the year. These characteristics are summarized in Table 6.1 with suggested implications for plant breeding. Multiple cropping systems make efficient use of land—usually the most scarce resource—through much of the potential cropping year. New technology has had little impact on these systems. It is difficult to generalize about multiple cropping

Associate Professor and Sorghum Breeder, Department of Agronomy, University of Nebraska, Lincoln. Contribution from the Nebraska Agr. Exp. Stn., Univ. of Nebraska Journal Series No. 5781.

Table 6.1. Comparison of monoculture and multiple cropping systems with implications for crop improvement. (Adapted from Altieri et al., 1978; Dickinson, 1972.)

Characteristic	Monoculture	Multiple Crop	Breeding Implication for Multiple Crop Genotypes
Net production	High (with fossil fuels)	Moderate (near monoculture)	Specific cultivars may be needed for some multiple cropping systems.
Species diversity	Low	Moderate-high	System approximates native vegetation and crop variability may be desirable.
Nutrient/light use	Poor-moderate	Moderate-high	Component crops may complement each other in light interception and rooting patterns.
Nutrient cycles	Open (leaching losses)	Closed (with perennials)	More efficient use of lower levels of applied fertilizer desirable.
Weed competition	Intense	Moderate	Crop competition suppresses some weeds; more difficult to use herbicide mixes.
Insects/diseases	Severe	Moderate	Some insect control from system; less difference in disease incidence.
Labor requirements	Seasonal	Distributed	More operations possible by hand on small areas; multiple hand harvests possible.
Diet contributions	Low	High	Nutritional quality a desirable trait for most consumed crops.
Economic stability	Low	High	Risk reduced by diversity, range of crop maturities to spread income.
Social viability	Volatile	Stable	Different client groups and levels of technology generally involved.

systems. Nevertheless, the table lists comparisons that generally describe the differences between contrasting systems in most zones where a dichotomy of farm size and use of technology exists in the tropics.

National production of many food crops has not kept pace with increased demand in most developing countries, partly due to their production in multiple cropping systems at the subsistence level, and the use of the best land for production of export crops. The net result has been a manyfold increase in food imports over the past 40 years (Wortman and Cummings, 1978). An increased awareness of the importance of multiple cropping systems by scientists in the tropics is leading to greater emphasis on research to improve components of these systems.

There was much early work in the U.S. on maize-soybean mixtures for silage (see Wiggans, 1935, for other references), but this practice apparently was never widely accepted. More recent work on double and triple cropping in temperate zones has emphasized agronomic aspects (ASA, 1976, Chaps. 3, 6, 10, 13, 16).

Reviews of multiple cropping systems have appeared through the years (Aiyer, 1949; Dalrymple, 1971; Kass, 1978; Willey, 1979a, 1979b), and recently as the result of technical symposia (Papendick et al., 1976; Jain and Bahl, 1975). The concept of improving traditional cropping systems as an alternative to complete renovation and introduction of massive technology is receiving increasing support from some research leaders in the developing world.

The potentials for increased production through the integration of appropriate technology into traditional systems have been summarized (Francis, 1979). Crop cultivars in present systems often represent years of natural evolution for survival and selection by the farmer for production. Though relatively low in yield potential compared to improved cultivars grown in monocultures with high levels of technology, these traditional cultivars generally are good competitors with weeds and other associated crop species, are relatively resistant to prevalent insect and disease pests, and possess a high level of variability. This low potential yield level may be due in part to the coevolution of pests that limit productivity in the centers of origin of basic food crops (Jennings and Cock, 1977). Moreover, there has been very limited use of improved cultivars from the experiment stations in these traditional cropping systems.

Cultivars Used in Cropping Systems

Broadly defined as the culture of more than one crop in a given area in one year, multiple cropping is further described by a range of more specific terms: associated cropping, double or triple cropping, intercropping, mixed cropping, relay cropping, and ratoon cropping. An attempt to reach agreement on the definitions of these terms was made in the symposium sponsored by ASA in 1975 (Andrews and Kassam, 1976). Some of the most frequently used terms are:

Multiple Cropping: intensification of cropping in time and space dimensions; growing two or more crops on the same field.

Sequential Cropping: growing two or more crops in sequence on the same field; crop intensification in the time dimension only.

Double cropping: growing two crops in sequence.

Triple cropping: growing three crops in sequence.

Ratoon cropping: cultivation of crop regrowth after harvest.

Intercropping: growing two or more crops simultaneously on the same field; crop intensification in both time and space dimensions.

Mixed intercropping: growing two or more crops simultaneously with no distinct row arrangement.

Row intercropping: growing two or more crops simultaneously where one or more crops are planted in rows.

Strip intercropping: growing two or more crops in different strips wide enough to permit independent cultivation, but narrow enough for crops to interact agronomically.

Relay intercropping: growing two or more crops simultaneously during part of the life cycle of each with second crop planted before harvest of first.

Related Terminology:

Sole cropping: one crop grown alone in pure stands at normal density; synonymous with solid planting.

Monoculture: repetitive growing of the same sole crop on the same land.

Rotation: repetitive cultivation of an ordered succession of crops on the same land; one cycle often takes several years to complete.

Associated cropping: general term synonymous with intercropping.

Simultaneous polyculture: synonymous with intercropping.

Cropping pattern: yearly sequence and spatial arrangement of crops, or of crops and fallow, on a given area.

Cropping system: cropping patterns used on a farm and their interactions with farm resources, other enterprises and available technology.
Mixed farming: cropping systems that involve the raising of crops, in combination with animals, and/or trees.
Cropping index: number of crops grown per annum on a given land area x 100.
Land equivalent ratio (LER): ratio of area needed under sole cropping to that of intercropping at the same management level to produce an equivalent yield.

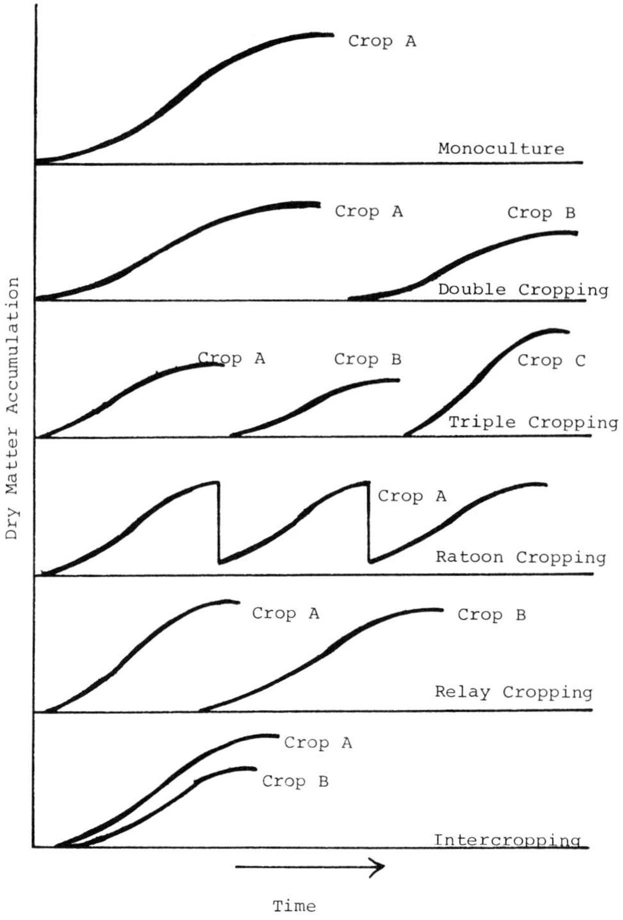

Fig. 6.1. Diagrammatic comparisons of principal multiple cropping systems.

A comparison of several common systems is diagrammed in Fig. 6.1. This summary illustrates the complexity of systems and our limited ability to research and communicate about them.

There is no clear definition of a multiple cropping system from the genetic point of view. But there is no question about the diversity of genotypes in a 15-crop mixed culture of food crops, including perennial species, in the humid tropics of West Africa. Nor is there debate about a potato-maize-bean system in Colombia, a rice-maize association in Ecuador, nor a traditional wheat-barley-oat cereal mixture in northern Europe. A multi-line cereal variety is genetically diverse, as is a maize (*Zea mays*) composite or population, or a mixture of bean types grown by small farmers in the tropics. Yet we generally would not consider these multiple crops. Figure 6.2 illustrates schematically the range of genetic diversity which exists in cropping systems, from the extremely diverse shifting cultivation and 15-crop mixtures to the extremely narrow single-cross maize hybrids. For this discussion, multiple cropping will refer to those systems that include more than one species in the field during the same year, or the same species grown in ratoon, relay, or sequential plantings. From a multiple genetic system interpretation, the world's cropping systems clearly represent a

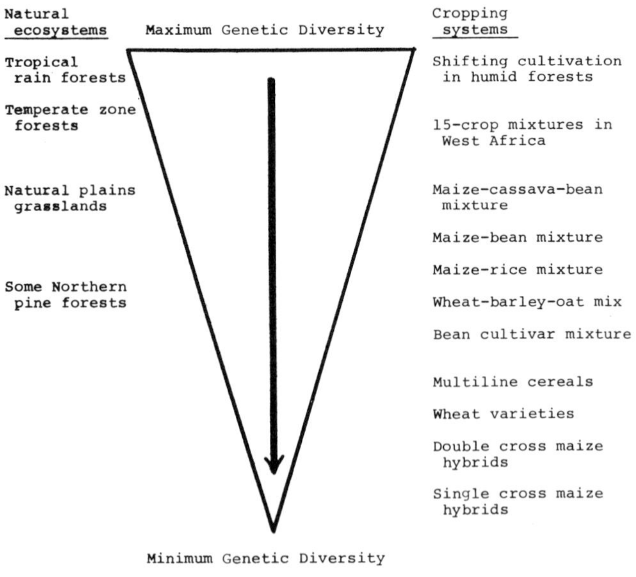

Fig. 6.2. Schematic representation of genetic diversity in cropping systems and natural ecosystems.

spectrum of genetic diversity as illustrated above. A combination of high productivity and long-term stability of production probably can be maintained by choosing an appropriate point on this spectrum in each ecological situation. This is a rational alternative to the trend of current agricultural technology which is moving rapidly to monoculture and to genetic uniformity across large areas. The dangers of genetic uniformity have been described in Adams et al. (1971), Allard and Hansche (1964), Borlaug (1959), Browning and Frey (1969), Jensen (1952), and Trenbath (1975a).

Use of improved cultivars to better exploit total available resources in a specific crop environment is central to applied plant breeding. In complex traditional cropping systems, or where the growing season is long enough to permit alternatives to monocultures, the concepts of time, space, and production per day must be considered in the design of cultivars to best use total available moisture, light, and nutrients (Bradfield, 1970). The plant breeder's challenge is to develop new cultivars appropriate to the range of cropping systems and microclimates which characterize many small farm regions. The questions of whether specific cultivars should be developed for multiple cropping systems, or for different levels of technology, have not been addressed by the majority of our crop improvement programs.

The following sections emphasize the more intensive intercropping systems that combine two or more crops in the field at the same time. This is not to minimize the importance nor the potential that double and triple cropping systems have today or will have in the future. There are problems to be solved agronomically, as well as a challenge to improve cultivars specific to new planting dates (with changes in day lengths, temperatures, and moisture levels). These problems can be solved through application of known technology, use of existing improved cultivars, and the techniques of traditional agricultural research. A concentration of the discussion on intensive intercropping is justified because only limited work has been done in genetic improvement for these systems, and new methodology may be needed to efficiently and rapidly achieve the genetic advances necessary to increase productivity.

Variables Inherent in Multiple Cropping Research

Before addressing the central theme of improved cultivars, it is useful to consider carefully the limitations of existing research results. Cultivars used to evaluate cropping systems have been of two types: traditional genotypes grown by farmers, often with limited yield potential, and improved genotypes developed for high input monoculture systems. Subjecting traditional cultivars to increased levels of fertilizer or higher densities in multiple cropping systems often meets with the same lack of yield advance that this approach achieves in monoculture. Introduction into multiple cropping systems of new and high-yielding strains developed in monoculture has met with greater success, especially when done in coordination with well designed and comprehensive agronomic trials with these same systems. With maize and climbing beans (*Phaseolus vulgaris*), the best combinations of improved cultivars, high plant densities, adequate fertility, and plant protection have given yields of 5000 kg/ha and 2000 kg/ha for maize and beans, respectively, in about 140 days in the Cauca Valley of Colombia (Francis, 1978). With genotypes developed for monoculture and without a specific program to select genotypes that are optimum for the intercrop system, these yields cannot be expected to approach the genetic potentials possible in multiple cropping.

To evaluate genotypes for multiple cropping systems, or to evaluate the contribution of any other single component, it is necessary to hold a number of other factors constant. A summary of the more important factors—genetic, cultural, and climatic/soil—is shown in Fig. 6.3 for a monoculture system. Genetic and cultural factors

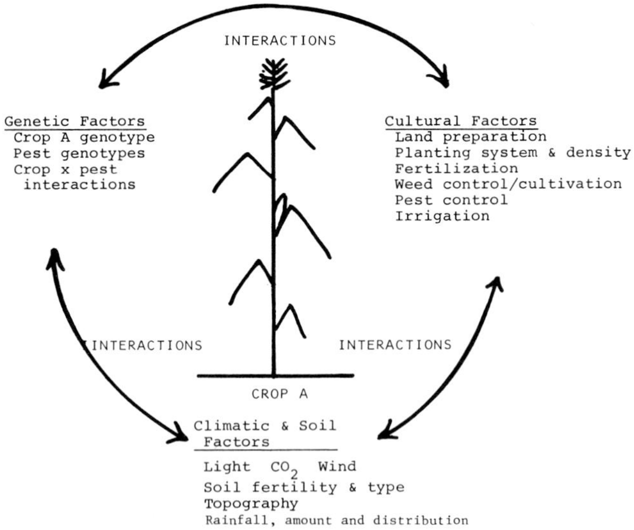

Fig. 6.3. Factors which may vary and interact in a monocrop system in one location.

generally are under control of the researcher and to some degree are controlled by the farmer. Interactions among these three groups of factors add to the complexity of research and to the uncertainty of farming, especially for the small farmer with little control over his natural environment.

Consider next a simple case of multiple cropping, with two crops planted at about the same time in the same field. Figure 6.4 illustrates additional variables which must be considered when two crops are grown in association, with the increased number of possible interactions between and among these new genetic and cultural variables and the environment. With 17 factors in the monocrop situation, there are 136 combinations of two factors which might possibly interact. With 26 factors listed in Fig. 6.3, there are 325 such combinations—and this is among the simplest of possible multiple cropping situations.

Varying one or more cultural factors in an experiment vastly increases the amount of seed, space, and other resources needed. Thus evaluation of cultivars should take place at a specified level of fertility, water control, pest and disease management, and weed control. If one component of a two-crop association is to be evaluated, the simplest procedure is to choose and maintain an appropri-

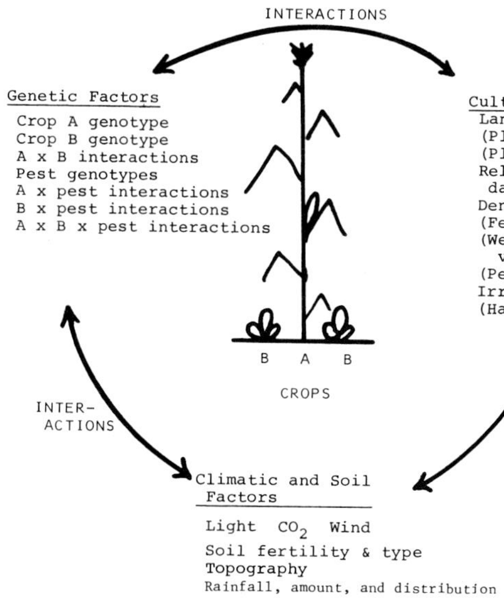

Fig. 6.4. Factors which may vary and interact in a two-crop multiple cropping system in one location; cultural factors complicated by intercropping are in parentheses.

ate genotype of the other. A more complex scheme to simultaneously improve two species may be possible. Densities, planting dates, and spatial location of each component should be held constant. As in any experimental design, uniformity of soil and topography will enhance the precision of the experiment. This series of constraints is complicated by the possibility that results and conclusions could be specific to the location, soil type, and prevailing climate in each season. The complexity of genetic improvement for multiple cropping systems is clear. Within this context and these limitations, we can consider the improvement of cultivars.

SPECIES CHOICE AND GENETIC SELECTION

Species Choice in Cropping Systems

Most research on multiple cropping systems has focused on agronomic aspects—planting dates, densities, spatial orientation, fertilization, pest control, and other appropriate cultural practices. There has been some emphasis on the selection of appropriate crops to associate under a specific set of conditions. Since this does not involve what breeders consider genetic selection, species choice is preferred to describe this type of agronomic activity.

Historical data indicated an interest in the testing of species in combinations to seek yield advantage over monoculture (Zavitz, 1927). Most studies have appeared in the past decade. Agboola and Fayemi (1971) found that cowpea (*Vigna sinensis*) and greengram (*Phaseolus aureus*) have less effect on maize yields, and were more tolerant to shade, than seven other legumes in Nigeria. Short cycle pulse crops were found to fit best into double and triple cropping sequences in India (Saxena and Yadov, 1975). Studies in Tanzania (Enyi, 1973) explored the best combinations of cereals and legumes for total food production, with sorghum/pigeon pea (*Sorghum bicolor/Cajanus cajun*) giving the highest total yields. The screening of twelve potentially useful shade tree species in India was accomplished by measuring tea (*Thea surensis*) yields as the criterion for evaluation (Hadfield, 1974). These are but a few examples of the many trials which have been conducted in many parts of the world to determine which species to choose in combination with appropriate agronomic practices. The crop species by system interactions which are obvious in these tests led to the logical question of which cultivars of each species are most appropriate for multiple cropping systems.

Cultivar Choice in Cropping Systems

Having determined which species to emphasize in a multiple cropping system, researchers often have screened or tested a range of available genotypes for their performance under some set of environmental and cultural conditions. This is a logical first step in genetic improvement for multiple cropping systems. Examples are many. A late cotton (*Gossypium hirsutum*) cultivar associated with groundnut (*Arachis hypogaea*) is preferred over an early cultivar, since late flowering produces most of the cotton after harvest of the understory crop (Rao et al., 1960). Several authors who tested pigeon pea cultivars reported that early and dwarf genotypes (Singh, 1975), nonbranching and heavy terminal bearing genotypes (Tarhalkar and Rao, 1975), and spreading plant types (Tiwari et al., 1977) were preferable under each specific system and set of conditions. This illustrates the specificity of plant type needed for contrasting intercropping systems.

Traditional cultivars of maize provided better support than improved cultivars of maize for associated climbing beans in Guatemala (ICTA, 1976). Maize of medium maturity gave best total system yields when double cropped with legumes in Florida (Guilarte et al., 1974). Crookston and colleagues (1978) followed winter rye (*Secale cereale*) with three maize hybrids and achieved highest total biomass yields per year with a maize about 14 percent later maturing than normal full season, planted at two times normal density (total dry matter 25.9 MT/ha).

Dry beans (*Phaseolus vulgaris*) commonly are planted in association with maize in Latin America. Among four cultivars tested in Brazil, the strong climber was lowest yielding in simultaneous planting and highest yielding in a relay system, compared to bush and weakly indeterminate types (Santa-Cecilia and Vieira, 1978). In contrast, research in Peru indicated higher yields from indeterminate climbers planted simultaneously with maize, and higher yields for bush types planted near harvest time of the maize (Tuzet et al., 1975). Prostrate cultivars of cowpea generally were less affected by shading of intercropped maize than erect types tested in Nigeria (Wien and Nangju, 1976). The leafy and semierect type 'VITA-4' has proven to be one of the best individual genotypes in association with maize (IITA, 1976). In another test at the International Institute for Tropical Agriculture (IITA), the strong climber 'Pole Sitao' was least reduced in yield in association, compared to potentials in monoculture.

Choice of cultivar may depend on its effect on another principal crop. In sugarcane (*Saccharum officinarum*) culture in Taiwan, sweet potato (*Ipomoea batatas*) may be intercropped during the early part of the cycle; short, dwarf-vined types of early maturity must be selected to minimize competition with the cane crop (Shia and Pao, 1964; Tang, 1968). Vegetable crops developed for these intercrop systems need to be shallow rooted (to plant with sugarcane), shade tolerant (if designed for relay systems), or relatively drought tolerant if developed to follow rice (*Oryza sativa*) at the end of the rainy season (Villareal and Lai, 1976). Thus cultivar choice depends on the relative importance of the two or more crops in the system, the potential growing season and optimum planting system (simultaneous, relay, sequential), and the genotype by system interaction of available germplasm with predominant cropping systems. Conflicting results from different studies with the same species reflect the complexity of interactions already described for these traditional systems, as well as the specificity of environmental conditions which surround each research location.

Genotype by Cropping System Interactions

Several examples of genotype by system interaction were given in a previous symposium (Francis et al., 1976). Significant interactions were described for cultivars of beans (intercrop with dwarf maize vs. intercrop with normal maize; Buestan, 1973), soybeans (*Glycine max*) (monocrop vs. intercrop with maize, sorghum, or millet; Finlay, 1974), and mungbeans (monocrop vs. intercrop with maize over three seasons; IRRI, 1973, 1974). The only significant correlation of monoculture yield with that in intercropping was reported by Baker (1975) for sorghum, though only four genotypes were included. We concluded that interaction of genotype by cropping system was an important reality in some crops and deserved study by the plant breeder.

Additional data now are available on various crop species and over a wide range of environments. Genotypes by system interaction may be evaluated by calculating the correlation of monocrop with intercrop yields. This is a rapid and uniform method of evaluating data from the literature and from annual reports (Francis et al., 1976).

Sorghum, millet (*Setaria italica*), and maize data are summarized in Table 6.2. A number of comparisons from the University of Philippines College of Agriculture-International Rice Research Institute-International Development and Research Center (UPCA-IRRI-IDRC) Program in Los Baños, Philippines (Gomez, 1976, 1977)

Table 6.2. Correlations of monocrop with intercrop yields in cereals.

Crop	n	Average Yield (kg/ha)		r_{yield}	r_{rank}	Reference
		Monocrop	Intercrop (system)			
Maize	18	4413	4220 (climbing beans)	.44	.36	Torregroza, 1978
Maize	20	5619	4681 (bush beans)	.90**	.83**	Francis et al., 1979
Maize	20	5003	5768 (bush beans)	.40	.27	Francis et al., 1979
Maize	20	5619	3479 (climbing beans)	.89**	.83**	Francis et al., 1979
Maize	20	5003	3836 (climbing beans)	.73**	.62**	Francis et al., 1979
Sorghum	4	2393	3056 (millet)	.95*	.80	Baker, 1975
Pearl Millet	40	1050	1270 (pigeon pea)	.68**	.64**	ICRISAT, 1977
Pearl Millet	40	1050	1040 (sorghum)	.61**	.61**	ICRISAT, 1977
Sweet Maize	15	584	2481 (40% shade)	.28	.29	Gomez, 1977
Green Maize	5	3200	2450 (40% shade)	.07	.38	Gomez, 1976
Popcorn	7	2730	2120 (40% shade)	−.30	−.50	Gomez, 1976
Flint Maize	38	3770	2540 (40% shade)	.37*	.35*	Gomez, 1976
Glutinous Maize	10	671	1816 (40% shade)	.12	.27	Gomez, 1977
Flint Maize	58	588	1149 (40% shade)	.35**	.35**	Gomez, 1977
Sorghum	16	2644	2706 (40% shade)	.65**	.67**	Gomez, 1977
Sorghum	16	2920	1670 (40% shade)	.43	.46	Gomez, 1976

contrasted monoculture following rice with the same series of genotypes in monoculture under 40 percent shade. Artificial shading in this ambitious tropical screening program simulates in monoculture the competition for light from an associated taller crop such as maize. Average yields in the trials range from less than one MT/ha to more than five MT/ha, and cereal yields in association are neither consistently lower nor consistently higher than monoculture. The correlations of monoculture with intercrop yields likewise are variable, generally positive, but not always significant. Though a number of the r-values are significant, this statistic must be greater than 0.7 to give a coefficient of determination (r^2 value) greater than 0.5; only in four of the comparisons does genotype explain more than half of the variation in yields across systems. Correlations for rank generally follow the yield results and may be more important than yields if a breeder intends to select a certain percentage of the tested lines without evaluating in both systems.

Though no specific data were presented, Kass (1976) indicated a positive correlation of rice yields in monoculture and association, when six cultivars were grown in three locations. Sayed Galal et al. (1974) reported strong positive correlations ($r = 0.91$, $r = 0.98$) in two consecutive seasons between intercropping tolerance of parental stocks and their topcrosses of maize. They concluded that this indicated a hereditary component to intercropping tolerance.

Data for grain legumes and sweet potato are summarized in Table 6.3. A number of correlations are significant between monoculture and intercropping. These correlations are not always consistent from one season to the next, as illustrated by lines 2 and 3 where the same 20 climbing bean cultivars were tested in two consecutive seasons with the same intercropped maize hybrid. The correlation coefficient was highly significant in one season (0.82) and nonsignificant in the next (0.41). Two consecutive seasons with the same 20 bush bean cultivars (lines 5 and 6) gave more consistent results, with significant correlation coefficients in both seasons. Mungbean correlations in lines 14 and 15 were not consistent in two seasons. Correlations consistently were positive in these comparisons. Of special interest is the unreplicated trial with 500 genotypes screened in two systems (line 8); the correlation was 0.33 between yields in monoculture and those with intercropping. Significant correlations between bean yields in monoculture and in association with maize also were reported by Clark et al. (1978) and by Chiappe and Huamani (1977).

Soybean and mungbean data from the Philippines were similar

Table 6.3. Correlations of monocrop with intercrop yields in legumes and sweet potatoes.

Crop	n	Average Yield (kg/ha) Monocrop	Intercrop (system)	r_{yield}	r_{rank}	Reference
Beans, climbing	9	1700	377 (maize)	.90**	.88**	Francis et al., 1978b
Beans, climbing	20	2024	615 (maize)	.82**	.80**	Francis et al., 1978b
Beans, climbing	20	2897	1038 (maize)	.41	.09	Francis et al., 1978b
Beans, bush	9	1318	954 (maize)	.91**	.93**	Francis et al., 1978c
Beans, bush	20	1873	1157 (maize)	.88**	.58*	Francis et al., 1978c
Beans, bush	20	2295	971 (maize)	.51*	.54*	Francis et al., 1978c
Beans, climbing	64	2212	995 (maize)	.82**	.83**	Francis (unpublished)
Beans, climbing	500	2531	1118 (maize)	.33**	. . .	Francis (unpublished)
Beans, climbing	10	2986	840 (maize H210)	.61*	.44	CIAT, 1978
Beans, climbing	10	2986	847 (maize Suwan)	.24	.52	CIAT, 1978
Beans, climbing	10	2986	649 (maize LaPosta)	.41	.37	CIAT, 1978
Soybeans	16	1019	714 (40% shade)	.53*	.50*	Catedral & Lantican, 1977
Mungbeans	20	1149	368 (40% shade)	.53*	.48*	Lantican & Catedral, 1977
Mungbeans	18	1511	558 (maize)	.13	.24	IRRI, 1973
Mungbeans	20	1170	570 (maize)	.67**	.61**	IRRI, 1974
Sweet Potato						
Group I	11	16850	2640 (40% shade)	.14	.22	Gomez, 1977
Group II	12	14140	2730 (40% shade)	.87**	.89**	Gomez, 1977
Group III	12	13600	1700 (40% shade)	.58*	.44	Gomez, 1977
Group IV	12	20100	3400 (40% shade)	.40	.52	Gomez, 1977

to those of beans. Sweet potato correlations were positive but variable in screening trials comparing normal monoculture with a 40 percent shade situation, analogous to the light environment when an understory crop is associated with maize.

A number of authors have observed the importance of genotype by system interaction, and concluded that it cannot be ignored by the plant breeder. Working in wheat (*Triticum aestivum*) and barley (*Hordeum vulgare*), Sakai (1955) found no consistent relationship between yield of a cultivar in mixture and its yield in pure culture. Over the years, the experience with mixtures of pasture species has led some researchers to the conclusion that genotypes expected to yield well as components of a mixture must be selected specifically for that objective (Dijkstra and de Vos, 1972; Fyfe and Rogers, 1965; Harper, 1967). Bean cultivars tested across environments led Hamblin (1975) to conclude that relative performance of genotypes in mixtures in one environment is not necessarily the same as relative performance in another set of conditions, since competition between genotypes interacts with the environment. This same conclusion has been reached by Lantican (1977) working with field crops in the Philippines and by Villareal in the Asian Vegetable Research and Development Center (AVRDC) in Taiwan (Villareal and Lai, 1976; Villareal, 1978) with sweet potato, tomato (*Lycopersicon esculentum*), and other horticultural crops.

When cultivars are compared among different intercropping systems, these correlations generally are high. Examples in Table 6.4 indicate significant r-values for yields of maize, climbing beans, and soybeans across several comparable cropping systems. The differences in environments between two intercropping systems generally may be less than between a monoculture and an intercropping system. The interaction of genotype by cropping system is one type of genotype (G) by environment (E) interaction. The magnitude and significance of G × E interactions may vary over years, locations, and planting dates. These also will vary among cropping systems, depending on how great the differences are among the environments where genotypes are tested, and on how the specific genotypes in a trial react to the specific environments included.

Firm conclusions on the importance of the genotype by system interaction are difficult to achieve, and possibly misleading. It is dangerous to generalize at this point. The results of any specific comparison are highly influenced by the lines that are chosen for that comparison. This important point may explain the conflicting results which we observe (Hamblin, 1979).

Table 6.4. Correlations of crop yields between two associated cropping systems.

		Average Yield (kg/ha)				
Crop	n	Association 1 (system)	Association 2 (system)	r_{yield}	r_{rank}	Reference
Maize	20	4681 (bush bean)	3479 (climbing bean)	.93**	.89**	Francis et al., 1979
Maize	20	5768 (bush bean)	3836 (climbing bean)	.68**	.58**	Francis et al., 1979
Bean, climbing	10	840 (maize H210)	847 (maize Suwan)	.67*	.60	CIAT, 1978
Bean, climbing	10	840 (maize H210)	649 (maize LaPosta)	.90**	.84**	CIAT, 1978
Bean, climbing	10	847 (maize Suwan)	649 (maize LaPosta)	.89**	.75**	CIAT, 1978
Bean, climbing	9	941 (dwarf maize)	829 (normal maize)	.26	.36	Buestan, 1973
Soybean	12	560 (maize)	650 (sorghum)	.60*	.39	Finlay, 1974
Soybean	12	560 (maize)	280 (millet)	.44	.34	Finlay, 1974
Soybean	12	650 (sorghum)	280 (millet)	.69**	.60*	Finlay, 1974

Statistical Alternatives for Genotype by System Comparisons

There are a number of statistical alternatives for evaluating the magnitude and nature of G × E interactions. The analysis of variance and partitioning of sums of squares due to the several sources of variation has been used most extensively. Though more rigorous and precise than correlations, an analysis of variance requires access to the original data by replication, which usually are not included in publications or annual reports where much of these data are found. Other estimates of G × E interaction are possible using the means of genotypes in each system.

Data are presented in Table 6.5 from three trials of climbing beans associated with maize; two trials of bush beans associated with maize; two trials of mungbeans associated with maize; and two trials of maize cultivars associated both with climbing beans and with bush beans, in which the contrasting monoculture systems for each cultivar were included for comparison. The bean and maize trials were conducted at the International Center for Tropical Agriculture (CIAT) in Colombia, and the two mungbean trials were conducted on the IRRI station in the Philippines (data from R. R. Harwood). Mean yields in each system, and average differences in yield (with their respective variances), are presented in the first seven columns.

Yield reductions ranged from a high of 69 percent in the first climbing bean trial to a low of 38 percent in the first bush bean trial, among the trials of legume species. It is interesting to note the similarities between paired trials, since these included most of the same genotypes of the test crops in two seasons. These comparisons in Table 6.5 are for climbing beans (lines 1 and 2), bush beans (lines 4 and 5), and mungbeans (lines 6 and 7). The standard deviations of the proportionate reductions in bean yield are remarkably similar in the trials with four replications each conducted at CIAT (lines 1, 2, 4, and 5); these range from 0.060 to 0.063 in the four trials. The other climbing bean (line 3) and two mungbean trials included only two replications in each cropping system, and have a range from 0.075 to 0.104 in standard deviations of the proportionate reductions.

The analyses of variance using replicated data from these same trials are summarized in the first five columns of Table 6.6. Genotype (G) by system (S) interactions were highly significant in four trials, and significant in two more. From this analysis, one may suggest that selection for specific genotypes in each system could be indicated in those systems with a highly significant G × S interaction. F-values for G × S from two seasons and the same genotypes, as indicated above, are similar.

If data were not available from replications, but number of

Table 6.5. Statistical alternatives for comparing yields in two cropping systems, I.

Test Crop	n	Yield (kg/ha)				Proportionate Yield Reduction	
		Monocrop	Associated (crop)	Difference	$S_{difference}$	Reduction	$S_{reduction}$
Beans, climbing	20	2024	615 (maize	1409	429	.69	.063
Beans, climbing	20	2897	1038 (maize)	1859	323	.64	.061
Beans, climbing	64	2212	995 (maize)	1217	633	.54	.104
Beans, bush	19	1873	1157 (maize)	716	232	.38	.063
Beans, bush	20	2295	971 (maize)	1324	225	.58	.060
Mungbeans	18	1511	558 (maize)	600	182	.63	.075
Mungbeans	20	1170	570 (maize)	600	185	.51	.100
Maize	20	5619	4681 (beans, bush)	938		.17	
			3479 (beans, climbing)	2140		.38	
Maize	20	5003	5768 (beans, bush)	−765		(.15)	
			3836 (beans, climbing)	1167		.23	

Table 6.6. Statistical alternatives for comparing yields in two cropping systems, II.

Test Crop	n	Analysis of Variance			CV	Estimated	Correlation Coefficient	
		Genotype	System	Gen. × Sys.		Gen. × Sys.	Yields	Ranks
			F values		%	*F values*	*r values*	
Beans, climbing	20	2.96**	127.68**	4.74**	21.7	9.90**	.82**	.80**
Beans, climbing	20	1.65	679.31**	5.01**	10.7	2.53**	.41	.09
Beans, climbing	64	3.26**	85.65**	4.72**	18.7	7.32**	.82**	.83**
Beans, bush	19	7.76**	143.77**	1.85	14.3	1.69	.88**	.58*
Beans, bush	20	2.19*	138.64**	1.78*	15.7	1.93*	.51*	.54*
Mungbeans	18	1.28	536.45**	2.13	11.8	1.96	.13	.24
Mungbeans	20	3.86**	57.97**	1.33	18.8	2.10*	.67**	.61**
Maize	20	42.22**	171.35**	1.55*	14.3	1.41	.90** (bush)	.83**
							.89** (climbing)	.83**
Maize	20	12.26**	48.02**	2.21**	23.8	5.55**	.40 (bush)	.27
							.73** (climbing)	.62**

replications and coefficient of variation (CV) (or lsd) were known, it would be possible to calculate an error term for testing G X S interactions. If CV's (or lsd) were not reported, it is possible to assume a value based on experience and other literature on that crop. If replication number was not known, some knowledge about the source of the data or verbal description in the text might indicate a number of replications which could be assumed to generate an error term. For example, the nine trials were analyzed using cultivar means in each system, using known replicate numbers and assuming a CV of 15 percent. Error mean squares (EMS) for the F test of G X S were calculated for each trial as follows:

$$CV = \frac{s}{\bar{x}} = 0.15 \qquad EMS = \hat{s}^2 = (0.15\ \bar{x})^2$$

These F-values are presented in the sixth column of Table 6.6. Although numerically different from the calculated F-values using the entire data set, significance and conclusions are similar. The more values that must be estimated to generate the error term, the less precise this test will be.

Finally, the simple correlations between yields and ranks for the two systems are presented in the last two columns (presented previously in Tables 6.3 and 6.4). Correlations for rank correspond closely to those for yield, and are not discussed further. Correlations for yield are variable, and do not correspond closely to the results for the calculated analysis of variance. One would expect that a large F-value (highly significant G X S interaction) would correspond to a low correlation coefficient, and vice versa. This was not the case in lines 1, 3, and 6 of the table. The apparent lack of correspondence of r-values from one season to the next with the same two systems and the same genotypes (especially line 1 vs. line 2, line 6 vs. line 7) probably is due to the differences in magnitude of the genotype differences.

Based on the results of these analyses of nine trials, the complete analysis of variance appears to be preferable. The computation of an experimental error term by using known values for C.V. and replication number, or by assuming some rational values for these parameters, appears to give more consistent results than do correlation coefficients between cultivar means in the two systesm.

Breeding Methodology

The optimum breeding systems for intercropping have yet to be designed and tested. The obvious complexity of both inter and intraspecific plant to plant interactions suggests that no single breeding procedure will be indicated for all crops and cropping systems. The best recommendation which can be drawn from the limited experience to date is to concentrate on easily identified qualitative traits in the early generations: seed color; endosperm type; disease and insect resistance; plant growth habit; and gross adaptation to prevailing temperatures, rainfall, day lengths, and levels of soil fertility. When generations have been advanced and seed increased in self-pollinated crops under the most convenient system for evaluating these qualitative traits, the advanced generations can be tested in appropriate cropping systems for quantitative traits such as yield potential, competitive ability with associated species, and stability of production.

Where heterosis is important as in maize and sorghum, some form of dynamic recombination and testing such as full-sib or half-sib family selection could be practiced. This allows testing of promising families for quantitative characters in each generation. This system is used extensively by CIMMYT in the international maize program. It is not known whether hybrids with the specificity of adaptation of traditional single crosses or double crosses could be applied to these complex and variable cropping systems which characterize multiple cropping. Prolificacy in maize and tillering in sorghum would appear to be traits which would greatly enhance their potential yields at low densities, while allowing light to penetrate to an understory crop. An adequate testing program over locations and systems would allow the identification of lines, as well as the evaluation of a number of promising hybrids, if this route appeared desirable. Distribution of existing maize hybrids in the tropics to large numbers of small farmers has been successful only in a few countries.

PRACTICAL SCREENING AND TESTING OF NEW CULTIVARS

The first critical step in the development of genotypes for multiple cropping systems is the decision of whether a separate breeding and testing program is necessary. If that decision is affirmative, the most efficient possible breeding scheme must be devised for rapid handling of large numbers. Promising new genotypes identified in the program must be tested both on the experiment station and on the farm; this is crucial before seed increase and wide-scale application of a new component of technology. Finally, the diversity of systems which characterize many small farmer zones presents some unique challenges in the transfer of technology. Each of these questions is explored.

Decision to Breed for Intercropping Systems

Results of trials conducted to date indicate no clear and generalized decision on whether or not a specific breeding program for intercropping systems is needed or justified. Factors which should be considered include the magnitude and nature of the correlations (significance of the G × S interactions), resources available for the total improvement program, similarity of traits and breeding objectives between the two (or more) breeding schemes under consideration, and relative importance of the two or more alternative cropping systems in the region into which improved genotypes are to be introduced.

The positive signs of almost all correlation coefficients in Tables 6.2 and 6.3 suggest that two separate breeding programs rarely would be justified. There generally is a positive association (though not always significant) between yields in the two contrasting systems. Prevalent insects and diseases likewise will affect each species in a region in almost any cropping system, although differences in severity between systems may dictate a change in relative priorities. Efficient response to applied fertility is vital in improved cultivars; but the modified natural fertility of an intercrop system which includes legumes, for example, may require less additional fertilizer for component cereal crops, and again may influence priorities. The relative importance of each crop in a system must be considered, as well as the contribution of each to total yield or income, and the interaction of yield of one crop with yield of the other. The most efficient combination of the two (or more) crops is the desired end product, with efficiency measured in yield, net income, nutrition, or other appropriate units.

Limited research personnel, facilities, and operating budget encourage the technician to make most efficient use of these scarce resources. With a fixed resource base in a breeding program, any dilution of funds to support two or more primary improvement activities would be expected to give less genetic progress in any specific direction than a single effort with one system and a relatively small number of breeding objectives. If a decision is made to focus entirely on a multiple cropping approach, due to the predominance of one or more related systems in a region, this would suggest a potential for rapid progress in yield potential and adaptation in that system. The division of germplasm and breeding activities into two separate and unrelated projects with no interchange between them rarely would be justified. It is possible to design an efficient combination for critical selection of parents, early generation screening for disease and insect resistance, and systematic testing in more than one cropping system. Examples used in several programs in the tropics are given in a later section.

Extremely important among these biological and other variables is the potential production to be achieved in multiple cropping systems in a region through introduction of improved genotypes, and whether over the short or medium term farmers are likely to preserve these systems or change to monoculture. A complex decision based on availability of relevant technology, information and capital, past experience, risk and other nutritional and economic factors, and the willingness and capacity of the farmer to modify his current system must be taken into consideration in the design of a crop improvement program for multiple cropping systems.

Phenotypic Traits Desirable for Intercropping

When the predominant cropping system or systems have been identified, and when the most critical limiting factors to production have been established and quantified, and a decision has been reached on which system or systems will be used in the improvement program, the next focus is on specific breeding objectives for each component crop species. Selection criteria vary with crop, cropping system, prevalent pathogens and insects, unique stress conditions in each region, and eventual use of the product, including relative prices and demand for component crops. An early decision within the cropping systems context is whether to breed improved genotypes that are specific to the farmer's current system, or whether some agronomic modifications—planting dates, densities, spatial arrangement, crop species, rotations—should be considered in the design of new components for the systems. Genetic improvement projects rarely are efficient in this context if not linked to a dynamic and imaginative activity in agronomy. Given the complex combination of factors summarized in Fig. 6.4, it is difficult to generalize about traits desirable for genotypes in a multiple crossing system. There are many reports in the literature about specific traits, but only a cursory treatment is available on this aspect in the previous review (Francis et al., 1976).

Photoperiod and Temperature Sensitivity

The genetic capacity to grow and mature in a given number of days, independent of day length, is a trait often associated with successful genotypes for intensive multiple cropping systems (Dalrymple, 1971; Jain, 1975; Moseman, 1966; Phrek et al., 1978;

Swaminathan, 1970). Photoperiod insensitivity has been among the important breeding criteria since the inception of rice improvement in IRRI (Coffman, 1977). This trait allows planting of a cultivar on any convenient date, with flowering and maturity controlled by genotype reaction to prevailing temperature patterns and to some degree to other cultural and natural fertility factors. Coupled with shorter duration cultivars, this capacity in rice and other crops encourages an intensification of the cropping system with additional crops during the same year. Photoperiod insensitivity is valuable in mungbeans (*Phaseolus aureus*) (Tiwari, 1978) and other legumes for intercropping, relay cropping, or double cropping with a principal cereal crop. In some specific situations, photoperiod sensitivity may be important in one component crop to assure that its major growth, flowering, and filling period do not coincide with another component of different seasonal duration. The suggestion that temperature insensitivity be incorporated (Tiwari, 1978) is useful in terms of broad adaptation and in tolerance to stress conditions (high and low extremes in temperature), but crop growth rates independent of temperature are biologically impossible.

Crop Maturity

Short crop maturity has been cited as a desirable trait in most reports (Moseman, 1966; Swaminathan, 1970) due to the potential this provides for intensification of the cropping system through addition of species or multiple plantings of the same species during the crop year. Short duration has been an important trait in most of the new rice cultivars released by IRRI, though genotypes currently being developed for low input agriculture include medium and long maturity alternatives (Coffman, 1977). Mungbeans (Catedral and Lantican, 1978; Gomez, 1976, 1977; IRRI, 1972; Phrek et al., 1978), soybeans, and cowpea (Gomez, 1977) are among the short cycle legume crops which fit well into these intensive cropping systems (Jain, 1975). Early and concentrated flowering to give uniform maturity is desirable in mungbean (Carangal et al., 1978). Sweet potato (IRRI, 1972) and maize (Gomez, 1977; Hart, 1977) cultivars with a short duration cycle likewise are desirable for intensive systems. Tall and long-cycle rice may fit better into some intercropping systems in Central America with maize of short duration, where the rice flowers and fills grain after the maize is doubled (Hart, 1977). In the international mungbean trials, Poehlman (1978) reported that vigorous late- and extended-flowering cultivars were most desirable for rainfed locations with low light intensity and higher temperatures, while short-cycle genotypes were best under high light conditions, irrigation, and cooler temperatures.

More important than the maturity characteristics of one component are the ways in which the two or more crops fit together in a system. Generally, the combination of an early and a late maturing crop is desirable, to better utilize available growth factors at different times (Andrews, 1972a, 1972b). Sorghum-millet intercrops at International Crop Research Institute for the Semi Arid Tropics (ICRISAT) (1977) were most successful when the earliest sorghum was combined with the latest millet, or when the earliest millet was combined with the latest sorghum, and these maturity differences were found to be more important than height differences of the two components. Selection of component crops with appropriate maturities is a critical part of the improvement program.

Plant Morphology

Short erect cereals have been developed for nitrogen responsiveness (Coffman, 1977), and this same trait is desirable for most multiple cropping applications. Medium to short cereal crop plants provide less competition to an understory legume or intercropped cereal of another species (Andrews, 1972a). Determinate growth habit and medium to short plant height are desirable in most legumes (Catedral and Lantican, 1978; Gomez, 1976, 1977; IRRI, 1972). In the maize-bean system, however, a climbing cultivar of beans appears to have greater yield potential than a bush type with simultaneous planting (Francis, 1978; Hart, 1977). Yield of a taller maize cultivar was less affected than yield of a dwarf hybrid by an associated climbing bean (CIAT, 1978), and which type is most desirable depends on total system yields and relative prices of the two crops (Francis and Sanders, 1978). Height differences between two components may be more important than the absolute height of each component (ICRISAT, 1977), and the interaction of component crop height with relative planting densities must be considered (Zandstra and Carangal, 1977). Leaf angle of crops affects the amount of light transmitted to lower components of a system, and influences distribution of light to different levels of leaf area within the canopy (Trenbath and Angus, 1975; Wien and Nangju, 1976).

Rooting Systems

Shallow rooted mungbean cultivars are most desirable for intercropping with a more permanent crop such as sugarcane to minimize competition for water and nutrients (Catedral and Lantican, 1978; Gomez, 1977). In general, the combination of two or more crops with different rooting patterns, such as a shallow-rooted species with a deep-rooted species, should give a better total water and nutrient extraction potential than either crop grown alone, or than the combination of two crops with similar rooting patterns (Krantz, 1974; Swaminathan, 1970).

Population Density Responsiveness

Component crops which respond to increased density give greater flexibility in the design of cropping systems with varied proportions of each crop in a mixture (Francis et al., 1978a; IRRI, 1973; Swaminathan, 1970). Optimum mixtures vary with the species, density response of each component, type of intercropping system, relative prices for the crops, and alternative schemes for the greatest total exploitation of the growth environment.

Early Seedling Growth

Particularly in low input cropping systems, early and competitive seedling growth is highly desirable to partially control weed growth (Catedral and Lantican, 1978; Gomez, 1977; IRRI, 1972). Growth of one species in a mixture also may be suppressed by allelopathy, an important interaction in weed/crop combinations or in multiple cropping systems (Trenbath, 1976). There is apparently genetic variation within some species tested for ability to alter weed growth (for cucumber [*Cucumis sativus*] example, see Putnam and Duke, 1974).

Insect and Disease Considerations

Pests that attack a given crop species in each region can be controlled or the attack modified to some degree by crop rotation and design of cropping systems (Altieri et al., 1978). Intercropping tall and short species may reduce attack on one or the other due to physical interference with insect movement (Chiang, 1978). Shorter duration crop cycles result in a shorter exposure time of each species to pests, and a longer total system cropping period may lead to higher population levels of naturally occurring biocontrol agents (Litzinger and Moody, 1976). Trenbath (1977) reviews the situation of reduced attack by insects on crops in mixed culture relative to monoculture, and, in a previous report, classified pest and disease interactions with crops according to several possible mechanisms: fly paper effect, compensation effect, and microenvironmental effects (Trenbath, 1975a). There is apparently less difference between intercropping and monoculture in the incidence of diseases, compared to insects, although the advantages of crop diversity for preventing widespread disease epidemics have been discussed (Day, 1973; Harlan, 1976). These insect and disease interactions with cropping systems are important because of the relative importance which must be placed on each resistance trait in a breeding program, and the stability which host plant resistance lends to these intensive cropping systems for the small farmer.

Screening Techniques for a Breeding Program

The first step in screening germplasm has involved large tests of available germplasm under alternative systems (see Tables 6.2 and 6.3). An example of the extension of this methodology to a double cropping system with rice was described by Carangal et al. (1978) for the evaluation of mungbean cultivars. Seven locations in Southeast Asia, as well as seven locations in the Philippines, were used to test a standard set of genotypes. 'Bhacti' was the best cultivar across countries, while 'CES-1D-21' was the best cultivar across locations in the Philippines. This is an international approach to cultivar choice; it is helpful in the identification of limiting factors and in the appropriate selection of parents for a breeding program.

Theoretical considerations for breeding of two or more species have been published by Hamblin and colleagues (Hamblin and Donald, 1974; Hamblin and Rowell, 1975; Hamblin, Rowell, and Redden, 1975). Comparisons of the use of pedigree breeding vs. bulk breeding are discussed, and a theoretical design is described for the simultaneous selection of two species for yield and ecological combining ability. Practical applications of the methods were not found in the literature.

The cowpea breeding program in IITA (IITA, 1976; Wien and Smithson, 1979) has focused on intercropping in the evaluation of some advanced lines. Tests in several locations in Nigeria during 1976 and 1977 revealed large differences among lines tested, and 'TVu1460' and 'TVu1593' were identified as cultivars with promise for this intercropped system.

Maize breeding in the ICTA program in Guatemala had concentrated on monoculture improvement in the lowlands until Poey and colleagues (1978) established a new and innovative selection and recombination program. They are farm testing 500 full-sib families in nonreplicated 5-m rows, with half of each row associated with beans. These results, analyzed using five locations (five different farms) as replications, lead to recombination of the best families on the experiment station and another cycle of farm testing. Two subregions—1500 to 1800 meters elevation and above 1800 meters elevation—are included, each with a separate set of families. Though no results are available yet for evaluation, this selection scheme appears likely to meet its objective of minimizing the genotype by environment interaction which has plagued highland maize improvement schemes in Central America and the Andean zone for years.

The climbing bean improvement program in CIAT can be used to illustrate a series of logical steps in the breeding process which leads from problem identification through agronomic testing, crossing, early generation, and advanced line evaluation. Recognition of the need for improved cultivars of climbing beans in association with maize (Mancini and Castillo, 1960; Francis et al., 1976) led to an early screening of available germplasm from the *Phaseolus* germplasm bank in Centro Internacional Agricultura Tropical (CIAT). This initial screening of almost 2000 accessions and seed increase on trellis supports in monoculture was followed by a screening of 500 promising lines in two cropping systems: intercrop with maize and monocrop on trellis (see line 8, Table 6.3). A subset of the best cultivars was tested in a replicated trial with the same two systems (see line 7, Table 6.3) in the next season. Concurrent agronomic trials explored the optimum planting dates for climbing beans with maize (Francis, 1978), densities of the two crops (Francis et al., 1978a), and the interaction of genotype by system with a small set of cultivars. Subsequent research on maize cultivars (CIAT, 1978) revealed that 'Suwan-1,' a cultivar with intermediate height, relatively narrow leaves, and lodging resistance, gave the greatest expression of differences in yield among bean cultivars.

Testing of 30 potential climbing bean parent materials in three highland locations (CIAT, 1978) showed highly specific temperature adaptation and a range of reaction in growth habit and flowering pattern to this range of environments. Preliminary evaluation of germplasm in association with maize is now conducted in hills which include three bean plants and three maize plants per bean progeny; this allows a cheap and effective evaluation of large numbers in a small area. Cultivars with good pod set, growth habit stability, apparent disease resistance, and a range of seed color were selected as parents and intercrossed. Because of the relatively high and positive correlations between intercrop and monoculture yields in climbers (Table 6.3), and because of the difficulty of testing for yield in early generations, these progeny were tested in early generations for resistance to rust, bean common mosaic virus and anthracnose, and given a preliminary yield evaluation in monoculture (CIAT, 1977). At least 1000 progeny per cross are evaluated (CIAT, 1978).

Advanced generation testing in four locations—monoculture in Popayan, intercrop with maize in Palmira and Obonuco, relay with maize in La Selva—recently was completed. Following seed increase in the first season of 1979, the best selections from this initial set of crosses will be entered by Davis and colleagues into the International Bean Yield and Adaptation Nursery for Climbers. This is the first time that an international trial has been developed by crossing, selecting, and testing specifically for use in a multiple cropping system. It supplements the 1978 international trials of climbing beans which were selected and increased from the germplasm bank.

The CIAT climbing bean improvement program concentrates on development of cultivars for simultaneous intercropping or relay systems, with sufficient testing at the different stages to identify superior types for monoculture. The bush bean improvement program, conversely, is directed toward efficient types for monoculture, or for double or relay cropping. The most promising bush selections likewise are tested in association with maize at regular intervals in the breeding process. Other bean plant ideotypes of a semi-determinate nature are being selected for relay and other intensive cropping systems (CIAT, 1977; Laing, 1978). Concentration of bush bean culture and a unique set of insect/disease problems in the lowlands, compared to the climbing beans and associated cropping systems in the highlands, simplifies this process somewhat in the Andean zone.

These breeding programs are all recent, and procedures have not been compared to alternative methods nor across several crops. They will give the first germplasm specifically developed for intensive systems. Over the next few years they should give relevant comparisons from which researchers in other regions, working on other crops, may be able to gain some perspective and save time and resources when designing new programs.

On-Farm Testing and Transfer of Technology

Critical to success in any crop improvement program is the testing of promising cultivars in the cropping systems and environment in which they will be grown by the farmer. Mechanisms for evaluation of new cultivars vary with the type of research organization and national policies on extension and agricultural development. Whatever the mechanism for validating new technology, several problems must be considered which are inherent in multiple cropping systems and the biological, economic, and cultural environment in which they are used.

As mentioned previously, it is difficult to generalize about multiple cropping systems and the farmers who use them. Many small farm regions are characterized by a multiplicity of systems, with much variation in planting systems, densities, species composition, and microclimatic influences. Planting dates often are dependent on rainfall patterns; thus they are somewhat uniform in a region. The number of species and major cropping systems also may be limited due to crop adaptations, markets, and preferred species in the diet and tradition. Thus it may be possible to identify a small and discrete number of systems in which to test cultivars in a zone.

If cultivars are to be introduced into existing systems, the procedure is less complicated than if cultivars are one component of a new or modified production package which includes other inputs and requires education of the farmer. Testing must be realistic, and any validation on the farm cannot depend on a transplanting of experiment station technology which is unrealistic or unavailable to the farmer. Enough replication throughout the region of application must be accomplished to assure an adequate evaluation over the range of possible soil and climatic conditions to be faced by a new cultivar. This replication of testing generally is limited by lack of resources, but many ingenious schemes have been devised which maximize farmer participation and input, and thus extend as much as possible the scarce funds available.

Although broad adaptation is desirable in improved cultivars from the breeder's or seed producer's point of view, the individual farmer's immediate concern extends only to his own range of cropping system variations and the soil and climatic conditions which he has experienced on his farm. If yield potential in specific systems and cultural conditions must be sacrificed for genetic adaptation to a wide range of conditions, some compromise must be attempted between these conflicting objectives.

Several examples of on-farm testing have been cited. The maize selection procedure in the Guatemalan highlands (Poey, 1978) involves farm tests during the initial stages of family evaluation, and presumably these same collaborators may be willing to test resulting populations or synthetics from the program. The Philippine program in collaboration with IRRI is sending new crop cultivars from the screening activity to additional sites in Southeast Asia. The CIAT bean program already has begun wide testing of bush cultivars in various systems, and has initiated through national research programs the first climbing bean trials in areas where these bean types are grown with maize and other support systems. There is no single scheme that is superior or to be recommended for this testing step; the most important issue is that adequate emphasis and resources should be put on this critical phase of research.

Economic and cultural factors may be more important than biological variables in the eventual adoption of new cultivars. Certainly the economic advantage of a new cultivar alone or as a part of a modified cultural system, as compared to the current cultivar, will be critical to success. Genetic characteristics such as seed color, size, taste, and cooking qualities may influence acceptability of a basic food crop cultivar. The nature and cost of the change in cropping system which may be needed for a new cultivar could negate any advantages of the technology, if these modifications are not understood and accepted by the farmer. If additional inputs are required, is the farmer capable of financing them, or is credit available to him at a realistic rate of interest? These questions plus others specific to each zone and crop must be considered in the design of new technology by the breeder who hopes to improve production through new cultivars and cropping systems.

POTENTIAL PRODUCTIVITY OF MULTIPLE CROPPING SYSTEMS

Traditional multiple cropping systems with unimproved cultivars and low levels of technology have been preserved by farmers to reduce risks, to provide a nutritious and varied diet, and to make better use of available land than would be possible with a comparable monoculture. With few outside inputs and traditional management, low crop densities, and limited moisture and/or plant nutrition, neither the combined crop components in a mixture nor a single species in monoculture is fully exploiting available resources, such as light, energy, and other nonlimiting growth factors. Thus it is not surprising that researchers have found in these low management systems that intercropping generally has an advantage over monoculture, sometimes up to 400 percent (Herrera and Harwood, 1973). When available components of technology—new cultivars, fertilizers, pest control, irrigation, density recommendations—which have been developed for monoculture are introduced into both systems, yields increase and the relative advantage of intercropping is reduced to levels of 10 to 40 percent in the better species combinations (Francis, 1978; Herrera and Harwood, 1973; Hiebsch, 1978; Lohani and Zandstra, 1977).

The potentials of a multiple cropping system depend on the competition of component crop species for available growth factors and some types of complementation between (among) species. Those cases in which overyielding (intercrop yield greater than yield of most productive component in monoculture) occurs are of greatest interest to the farmer, and this is the principal objective of crop improvement for multiple cropping systems. According to Andrews (1972b), overyielding of intercropping over monoculture occurs in those combinations in which: (1) intercrop competition is less than intracrop competition; (2) arrangement and relative numbers of the contributing crop plants affect the expression of the difference in competitive ability; (3) competition between crops is alleviated when their maximum demands on the environment occur at different times (either by choosing crops with different growth cycles or by planting at different times); (4) the seasonal period of growth is long enough to permit better total exploitation of this total season by two or more crops; and (5) legumes can be intercropped with nonlegumes under poorer fertility conditions.

The classic papers by de Wit (1960) and by Donald (1963) should be consulted for the basis of quantitative interactions between species. One of the most comprehensive recent reviews on crop interactions in multiple cropping is that of Trenbath (1976), to which the reader is referred for more complete treatment of the nature of competition in mixed culture. Only those aspects with direct relevance to crop improvement are discussed here.

Competitive Ability and Yield

Reports in the literature disagree on the association of competitive ability with yield in pure stand. Successful competition for scarce resources is critical to crop production by components of a mixture. An early report on barley and wheat cultivars (Suneson and Wiebe, 1942) found that survival in mixtures was unrelated to yield of component cultivars in pure stand. A good correspondence between high yields in monoculture and good competition in mixtures has been reported for barley (Allard and Adams, 1969; Harlan and Martini, 1938), for wheat (Allard and Adams, 1969; Jensen and Federer, 1965), and for maize (Kannenberg and Hunter, 1972). A negative relationship between yield in pure stand and competitive ability was reported in barley (Wiebe et al., 1963) and in rice (Jennings and de Jesus, 1968). Donald (1968) explored the theoretical basis for this negative association and suggested that a weak competitor in mixtures actually makes a minimum demand on resources per unit dry matter produced, and thus produces more in pure stand.

Competitive ability and the selective value of genotypes appear to be influenced strongly by environment, including the effects of neighboring plants (Allard and Adams, 1969; Hamblin, 1975). When genotype performance over a series of environments is plotted against the environmental mean yields (average of all genotypes in each environment, see Eberhart and Russell, 1966; Finlay and Wilkinson, 1963), the rate of response by individual genotypes to improving environments is variable (Hamblin, 1975). Likewise, it has been shown in the section on genotype by system interactions that in several species the relative performance of genotypes varies with the cropping system. Add to this the possible complication cited by Hamblin and Donald (1974) in barley, where yields of progeny in the F_3 were not correlated with yields in the F_5. There also was a negative correlation of F_5 grain yield with F_3 plant height and leaf lengths, factors favorable to competing ability.

If experience with one species suggests that the best yielding genotypes in a population will be eliminated by competition in early generations, then evaluation of spaced plants and a pedigree system probably is indicated (Hamblin and Rowell, 1975). If the individuals which compete well in a population are the best sources of germplasm for increased yields in later generations, then a bulk breeding scheme could be used in self-pollinated crops (Hamblin and Rowell, 1975). Further research on breeding methodology is badly needed to advance our understanding of which approaches are most appropriate and most efficient.

Species Interaction and Resource Utilization

Crop species present in the field at the same time—whether planted simultaneously or in relay pattern—interact by competing for available resources. There is rarely a competition for physical space, but rather for the light, water, nutrients, or CO_2 which that space receives or contains. Trenbath (1976) describes in detail the mechanisms by which genotypes compete for resources. Both field and greenhouse studies have attempted to quantify the nature of intra and interspecific competition and to determine how this division of resources is accomplished (for example, Donald, 1958;

Hall, 1974; Osiru and Willey, 1972; Trenbath, 1974b; Trenbath and Harper, 1973; Willey and Osiru, 1972).

The picture which emerges is of a dynamic and complex interaction in several dimensions among two or more crop species and their physical environment. Competition begins for an individual plant when growth and development is retarded or altered from what it would be if no other individuals were present. This interaction ends only when that plant is removed from the crop environment, or when it ceases to actively (in the case of root absorption of water and nutrients) or passively (in the case of light interception by a taller though mature plant) compete with neighboring plants of the same or a different species. The challenge to the plant breeder is to design each crop component to efficiently exploit resources in this environment with the maximum economic yield produced per unit of resources and per unit of time, and with a minimum effect on the same exploitation by the other species.

Total resource utilization is most complete when two or more species occupy different ecological niches within the cropping system (Loomis et al., 1971). Growth cycles of the intercropped species may be different (Lohani and Zandstra, 1977; Osiru and Willey, 1972), accomplished either through varied planting dates or choice of crops with different maturities. This complementary use of resources over time (Ludwig, 1950, as cited by Trenbath, 1976) holds potential in zones where two or more crops with shorter cycles and greater efficiency could occupy the field which now grows a single long cycle crop, or where a part of the potential cropping cycle is not utilized.

The complementarity of taller cereals and shorter legume species (Francis, 1978; Willey and Osiru, 1972) or of different component heights in related species (Khalifi and Qualset, 1974; Trenbath, 1975a) makes better use of light through the growing season. What has been called complementary competition for light describes the compensation for yield loss by one component by increased production in another. The nature of this compensation and complementary resource use will determine in part whether a mixture will overyield a monoculture. Spatial exploration of different layers by roots of two or more species is another type of complementation which may have advantages in deep soils (Trenbath, 1974a).

Differences in patterns of light interception or root exploration are useful not only in the choice of species and cultivars, but also in the conscious genetic selection of promising cultivars of each component species to best complement others in a multiple cropping system. The degree to which two or more species, and new cultivars of those species, can more efficiently and completely exploit total resources will determine the potential success of multiple cropping systems compared to high input monocultures.

CONCLUSIONS AND RECOMMENDATIONS

The potential for improving multiple cropping systems at any level of technology or management depends on the ability of the researcher to combine genetic advance with new agronomic methods. The realization of this potential to increase production on the farm requires viable tests of the best combinations and the eventual transfer of the new technology. Relevant research in these systems requires: (1) a thorough knowledge of existing cropping systems and the reasons for their popularity; (2) a comprehensive understanding of the nature and variation of prevailing climatic and soil conditions, and both the growth potentials and stress which they will impose on growing crops; (3) a practical experience with the limiting constraints to production in prevalent crop species; and (4) enough perspective on breeding, agronomy, plant protection, cropping systems, economics, and politics to make rational decisions on priorities in a breeding program. This is a colossal task!

The challenge will not be met successfully by an isolated individual working in a single academic discipline. Nor will the answers to these complex questions arise from any single basic research project or one applied development effort. A research focus that crosses departmental lines and the traditional disciplines, and that includes the spectrum from basic to applied activities, will have the best chance of success.

Who will train the geneticists and plant breeders to carry out these activities, and how should a program be organized? Is it possible to give scientists a relevant preparation for improving complex cropping systems in the tropics with an academic program and field apprenticeship in the temperate zone? And will the existing rules and organization of universities and research institutions in all parts of the world allow this to occur?

We as educators must accept this challenge. The responsibility for training young plant breeders from the developing world requires a better appreciation on the part of those doing the training in crops and cropping systems in these countries. A graduate study program should reflect the importance to future success of a broad technical preparation and capacity to communicate with specialists in other disciplines. The international centers have organized staff and research activities into interdisciplinary teams and can collaborate on thesis work for some students. Selection and development of thesis topics to give the best possible preparation for handling the complex types of biological problems that limit crop yields in the tropics are essential.

It is neither a small nor routine responsibility to design and direct a graduate program for a potential research leader in the developing world.

We as students and researchers have an even greater responsibility. The potential of crop species improvement for multiple cropping systems will not be realized without a concerted effort in the field, both on the experiment station and on the farm. We must explore the complexities of existing systems, sort out the components that are susceptible to research and improvement, and subject the most promising alternatives to rigorous evaluation under real world conditions. This is an application of the scientific method to solving practical problems of the farmer.

The world food crisis is upon us, and the geneticist and plant breeder have a significant role to play in its solution. Multiple cropping systems are complex, but they hold an exciting potential for increasing food production which has not yet been realized. Through the successful integration of our activities with those in other disciplines we can begin to focus on the complex problems which have limited the use of improved crop technology by most farmers in the tropics. The use of more intensive systems also opens a new and greater potential for increased production in temperate zones. We can each contribute in some way to a dynamic and imaginative research and training program which will make the greatest possible use of our talents in genetics and plant breeding to meet the challenge of increased world food production.

Discussion

R. K. CROOKSTON
R. M. LANTICAN

1. R. M. LANTICAN. Breeding work in a tropical setting is challenging because one has to deal with a myriad of situations associated with year-round seasonable variability; many options in the use of energy, including labor; and an array of established cropping systems. This challenge can tax the limited manpower and resources of national breeding programs, so a unified and systematic approach to problems in crop production becomes imperative to a plant breeding program. Certainly, breeding programs must be confined to major crop production systems prevailing in a geographic region that create the greatest impact on the food supply and socioeconomic situations.

In Asia, excluding China, Korea, and Japan, there are 39 million ha in the rain-fed wetlands that offer great potential for increased food output. This rain-fed area normally is used for growing a single crop of rice each year, and most of it remains idle for the rest of the year. With new early maturing rice cultivars, direct seeding on dry seedbeds replaces the traditional methods of transplanting seedlings, and as a result, cropping intensity can be doubled. The other major area available to crop production in the tropics is the space under plantation crops like coconuts, oil palm, and rubber, and between rows of sugarcane. In Asia, the coconut crop occupies 5 million ha. Thus, in tropical areas, the objectives when breeding plants for multiple or intercropping must be to adapt dryland crop strains for conditions of pre and postrice cultivation and to intercropping with plantation crops.

Postrice cultivation usually encounters low moisture supply, hot or cold temperatures, and poor soil granulation. Plantings under or between rows of plantation crops must contend with partial to full shading and competition for nutrients. So the breeding program must be done for a cropping and management system where the only available moisture is from rain; that is superior to traditional methods of second cropping.

In the Philippines, our first task was to establish the cropping system for which crop cultivars would be developed. So our objective became to develop cultivars of dryland crops for a postrice cropping system that is compatible with the soil puddling and seedling transplanting in rice cultivation. The second crop must rely on zero or minimum tillage. Seeds of the second crop are hand dibbled into the mud at the base of the rice stubble at a high density to make up for limited vegetative development of plants under stress. Sowing is done into the base of the rice stubble because that is where the residual moisture is located. Spaces between rice rows become cracked and dry quickly; thus they have no reserve moisture for germinating the second crop. Next, the soil is mulched to conserve residual moisture.

The second task is to screen crop species that will fit into the rice culture system. Dryland crop species researched to date are sorghum, mungbean, soybean, cowpeas, adzuki bean, rice bean, peanuts, and potatoes. Mungbeans produce especially good yields on residual moisture. Corn, watermelons, and tobacco are used by farmers traditionally, but they require supplemental irrigation.

For intercropping between rows of plantation crops, farmers generally use coffee, cacao, banana, papaya, pineapple, grain, and root crops. Once, I observed a farmer's intercropping which involved four species. The highest canopy was coconut trees, the second was papayas, the third was pineapples, and the fourth was sweet potatoes. The most successful species for intercropping with plantation crops are sorghum, peanuts, cowpeas, tomatoes, potatoes, ginger, and sweet potato. The most successful crops for growing between rows of a newly established ratoon crop of sugarcane are vegetables and grain legumes.

At issue is whether the establishment of a breeding program exclusively designed to select crops and genotypes for postrice cultivation is justified. My feeling is that it is not. Trials with elite lines of sorghum and dryland leguminous species that have been conducted under dryland cropping, cropping under shade, and paddy cultivation show that substantial degrees of cultivar by culture system interaction occur. However, high levels of yield are attainable already with elite cultivars of dryland crops grown under paddy conditions. Any additional yield increment that could be obtained by exploitation of cultivar specificity through an elaborate breeding program designed for a rice-based system would not be justified, considering the high cost of maintaining a program, limited availability of trained personnel, and low level of skill in and adaptation to using innovative technology by the average Asian farmer.

More appropriate is a unified two-stage breeding and evaluation program for all conditions. The first stage of selection would be done under optimum dryland environments to exploit known cultivar features that relate to general fitness and greater yield stability over a range of cropping systems and seasonal patterns. As an example, soybeans, which have erratic behavior in the Philippines, will produce quite stably if they: (1) mature in 80 to 95 days; (2) have a leaf area index of 3.0 to 5.0; (3) have a harvest index of 30 percent or greater; and (4) have a seed weight of 150 g or greater per 1000 seeds. These traits can be assessed readily in the first stage of selection, and genotypes that possess these levels of the traits are generally widely adapted. With the relatively small number of genotypes that survive first stage selection, the second stage evaluation should be initiated to exploit unique adaptiveness of tolerance that individual genotypes may have to stress situations such as: (1) shading and competition effects from intercropping; and (2) soil compaction and drought unique to paddy field cultivation. With this two-stage selection scheme, soybean breeding has been quite successful in developing cultivars for the tropics.

2. R. K. CROOKSTON. Reports from the tropics suggest that land usage has been improved from 10 to 40 percent as a result of intercropping. It is such information that has aroused interest in intercropping in the temperate zone. In western Minnesota, one intercropping pattern that was recently tested consisted of temporary wind breaks of corn and 12 rows of soybeans in a repeating pattern. Soybean yield per acre as a result of this arrangement was increased by 14 percent and the corn yields were a bonus. Apparently, the reason for the soybean yield increase was that the relative humidity over the soybean canopy was increased, and soybean plants were able to avoid water stress.

We have also researched two other intercropping patterns with corn and soybeans in Minnesota. For the first pattern we used a spacing of 75 cm between rows. We alternated single rows of corn and soybeans, and also sets of 3 rows of corn and 3 rows of soybeans, 6 rows and 6 rows, 12 and 12, and finally 24 consecutive rows of each crop served as a control. As we moved from the control to 12 and 12, 6 and 6, and the like, corn yields increased; but soybean yields decreased at the same time, so that our land efficiency ratio was held constant at a value of 1.0.

The second approach consisted for planting corn and soybeans with rows spaced 37.5 cm apart. The planting pattern was alternate single rows. Variables were maturity of the maize hybrids, maize planting date, and maize stand density. Planting date and stand density had dramatic effects on both maize and soybean yields, but all combinations of variables resulted in land usages less than the check.

My conclusion from these brief preliminary experiments is that more research is needed on management approaches before it is sensible to evaluate cultivars for their suitability in a multiple or intercropping system in a temperate region.

3. S. GALAL. When making a comparison of intercropping systems with monocultures for land equivalent ratios, it is important that optimum planting patterns and plant densities be used for both. In Minnesota, the pattern used for comparing solid planting and intercropping were the same, and under these circumstances it is unlikely that a land equivalent ratio greater than 1.0 could occur. In Egypt, when optimum but different planting patterns were used for monoculture and intercropping systems, a land equivalent ratio of 1.75 was obtained.

4. R. J. BAKER. Interactions among crop genotypes are of two types. One occurs when there are significant changes in ranks of the genotypes, and the second occurs without significant change in ranks. From a plant breeding point of view, only the first one is important. Unfortunately, the analysis of variance procedure fails to differentiate between these two types of interaction.

Dr. Francis computed correlation coefficients and argued that a high correlation was an indicator of no interaction. This argument is true, statistically, but the real question is, How high must that correlation be in order to conclude that indeed there is no interaction? It may be erroneous to conclude that interaction exists when a correlation is as low as 0.6 or 0.8. The low correlation may be due to random errors in one or both environments—not due to interaction.

The real caution, however, is involved when a researcher uses correlations to select one method over another. For example, one correlation is not significantly different from zero, whereas the second is; so the researcher concludes that the research methodology used to obtain the data for computing the second correlation is the best to use in future experimentation. But if the two correlations are not significantly different, there is no real basis for choosing among the methods of experimentation.

5. S. JANA. An aim of plant improvement for intercropping systems, of necessity, is to select genotypes that are good interspecies competitors. Several studies with barley and wheat have shown that genotypes that produce well in monoculture are not always the best synergists when grown in genotype mixtures. Is this also true for tropical legumes such as cowpeas in Nigeria or beans in Colombia?

6. C. A. FRANCIS. The cereal data are confusing. Correlations between competitive ability and yield in pure stand are variable, that is, some are positive, some are negative, and some are zero. So the central question remains, Are the traits that give yield potential to a cultivar in monoculture the same traits that provide for good value in associated croppings?

7. R. SHABANA. In Egypt, a positive correlation of 0.9 was found between intercropping tolerance of maize inbreds and their corresponding hybrids.

8. S. N. NIGUM. With intercropping, there are different planting patterns for different intercropping species and also for different cultivars. So should planting patterns be superimposed upon the experiments designed to select complementary genotypes of two species that will be components in an intercropping system?

9. C. A. FRANCIS. Probably the most efficient way to investigate this whole area is to work on one species at a time while holding the rest of the system, such as density, planting dates, and the like, constant. This means selecting among variable genotypes of one component crop in one field, selecting among variable genotypes of a second species in another field, and eventually bringing the selected cultivars of the component species together. Concurrent agronomic research can fine tune the intercropping system. Selecting within two species at the same time, plus agronomic variables in a factorial system, can result in so many treatments and interactions that the experiments become unmanageable and the data uninterpretable.

10. R. L. VILLAREAL. Panel members feel that there is no need to have a specific program to develop cultivars of crop species for multiple cropping, relay cropping, and intercropping systems. Actually, there is good reason *for* having such a specific program. Crop cultivars should be custom-tailored for a specific environmental condition or planting pattern, and in research at the AVRDC in Taiwan, this has actually been done. Tomato and sweet potato cultivars have been developed for use on fields that remain idle after a rice crop has been harvested. Land after rice has only residual moisture which is enough to germinate a seed crop but not to see the crop through to maturity. One sweet potato genotype that followed newly harvested rice under minimum input conditions, that is, no supplemental irrigation, low level fertility, and no pesticide, could survive exceptionally well and produce many roots and support these roots to maturity. The main difference between a poor and a good performing sweet potato was in the ability of the good performer to fill the already initiated roots to maturity. In the poor performer this trait was absent or not well developed.

Also, several breeding lines of tomatoes have been found in the Philippines that will germinate, survive, and give economic yield following a rice crop.

11. F. MARQUIS-SANCHEZ. Intercropping systems that now exist in tropical countries have evolved throughout the evolution of agriculture as heterogeneous systems, so doesn't it make sense that breeding programs carried out to produce genotypes for this system of farming should be initiated from the very beginning under an intercropping system instead of having one phase of selection in monoculture?

Such a breeding program would capitalize on the positive interactions among genotypes under the intercropping system.

12. C. A. FRANCIS. That is an excellent point. Doing the selection work under real farm conditions might give very good progress, at least for that specific intercropping system. If one particular cropping system is prevalent in a zone, perhaps it would be well to work with that system only.

This approach will narrow rapidly the potential applications of new cultivars, however. A compromise may be to conduct preliminary trials under the appropriate system in the experiment station, and carry out some advanced cycles under a range of on-farm conditions.

13. O. LELEJI. The concept of intercropping is much more complex than most researchers realize. In indigenous agriculture in Nigeria, a farmer is apt to use from five to ten crops in his intercropping system. However, as researchers, we assume that the interactions in intercropping are simple. For example, in no study are more than three or four crops grown together at the same time in a mixed planting. Further, indigenous intercropping varies from farm to farm, and from year to year on the same farm, depending on what is more profitable for the farmer. Therefore, plant breeders cannot develop genotypes specifically for an intercropping system because there is no typical intercropping system.

14. C. A. FRANCIS. As stated earlier, the location where research on multiple cropping should be conducted is open to question. Perhaps research results from experiment stations cannot be extended to all real farm situations, to all other cropping systems, to all soil types, and the like. But this criticism or skepticism can be made for any agricultural research, whether directed to agriculture in developing or developed countries. It must be accepted at the outset that the extension of results from research to an individual farm requires a bit of experimentation by the farmer himself.

15. F. AGBO. There is an intrinsic high correlation between economic and biological yields. Can harvest index be a reliable measure of economic productivity?

16. C. A. FRANCIS. The correlation between harvest index and economic yield is well established for a number of crops, but generally harvest index has not been explored as a means of increasing the efficiency of tropical food crops. Many tropical crops have extremely low harvest indexes, so if more emphasis were placed on the carbon partitioning process in plants, very rapid progress might be made in improving economic yields in some tropical crops.

17. E. A. CLARK. It is said that one reason why small farmers in the tropics prefer mixed cropping is that this practice confers yield stability over years and environmental fluctuations. Conversely, some reports indicate that large year to year and season to season variations occur in yields from mixed cropping, both in terms of absolute yield and relative to monoculture systems. If mixed cropping systems are inherently more stable than monoculture systems, what factors confer this stability?

18. C. A. FRANCIS. In a study of 80 comparisons that included monoculture maize, monoculture geans, and the bean-maize intercrop, the intercropping system was more stable both in production and in income. The main factor of yield stability is likely some kind of biologic buffering, that is, one condition is favorable for one crop and unfavorable for another, and vice versa. With this situation the intercrop somehow comes out better than the monoculture over years. The same holds on the economic side. As prices fluctuate, the more crops that are in the system, the better are the buffers from an economic standpoint. We need more information on nutrient cycling, root exploration, light interception, and other aspects of intimate crop associations.

19. K. DIESBURG. Perhaps insurance value or stability of intercropping would have its greatest advantage in marginal agricultural areas.

20. C. A. FRANCIS. Multiple cropping situations are just as susceptible to diversity of climate and soil as are monocultures, so we need to separate the advantages of diversity on a given farm from the advantages of multiple cropping. Diversity, even in monocrop systems, such as planting half a ha to each crop, gives the same buffering against economics as does multiple cropping. Unless there is a clear advantage in higher land efficiency ratio by putting crops together, as much diversity can be created within a given farm with monocultures of several crops as with multiple cropping.

21. J. GASKILL. Plant pathologists and entomologists know that growing the same crop year after year in the same field tends to cause a buildup in certain plant pathogens, insects, and nematodes with a consequent reduction in yield of that crop. Is the same tendency to be expected where a fixed mixture of crops is grown on the same land area year after year?

22. C. A. FRANCIS. Probably, over a long period, a simple intercropping system of two crops will succumb to disease problems, but much less rapidly than a single monoculture will. Apparently, the insect situation varies much more among cropping systems than the disease situation.

23. K. RAWAL. Across the continent of Africa, multiple cropping is indigenous to agriculture. But whenever introductions were made

of cash or commercial crops, such as cotton, peanuts, and cocoa, monocultures of these crops have developed. And such monocultures have been successful.

24. A. M. THRO. There has been some criticism of multiple cropping systems developed by experiment stations as being too labor intensive. Yet labor availability is purported to be no problem in developing countries. What is the labor situation for agriculture in developing countries?

25. R. LANTICAN. At the moment, developing countries tend to have an adequate supply of labor for the farming industry. In some countries such as Taiwan, where much manufacturing industry has developed, labor for farming is now in short supply and expensive, and cropping systems have had to adapt to this change. However, in most developing countries, there likely will continue to be some excess labor for some time into the future. This labor needs to be utilized, so why not utilize it in the farming industry?

26. O. LELEJI. That is a good point. Intercropping as practiced now is not amenable to mechanization. But how can we know that ten years from now, when our research is to be applied, that the developing countries will have an adequate labor supply to support multiple cropping as a way of agriculture? As changes occur in the society and economics of tropical countries, genotypes bred for today may no longer be useful.

27. R. K. CROOKSTON. Even in the midwestern USA, as land values continue to increase, it is becoming profitable for American farmers to use methods for intensifying the usage of their land. Intercropping or multiple cropping may be one way to accomplish this. In fact, this is the justification for inter and multiple cropping research in American experiment stations.

28. E. A. CLARK. It is commonly said that one of the primary limits to adoption of an intercropping system in the USA is that it cannot be mechanized. Is there any modification of existing machinery or new types of machinery that might permit mechanization of intercropping systems?

29. R. K. CROOKSTON. With conventional planters, there is no problem in designing the planter box arrangements for sowing crops

in an intercropping system. Herbicides are available that control weeds in a combination of corn and soybeans.

Harvesting is a real problem, however. In Georgia, two farmers grow corn and soybeans in combination, and the corn matures slightly ahead of the soybeans. It is a tall corn with the ears placed above the soybean canopy, so they drive through the field and harvest the ears of corn above the soybean plants. After the corn stalks have dried, they harvest the soybeans. This may not sound like a satisfactory approach. However, if agronomists can devise an intercropping system that will give a land efficiency ratio well above 1.0, engineers will invent a machine to take care of harvesting the component crops.

REFERENCES

Adams, M. W., A. H. Ellingboe, and E. C. Rossman. 1971. Biological uniformity and disease epidemics. Bioscience 21:1067-70.

Agboola, A. A., and A. A. Fayemi. 1971. Preliminary trials on the intercropping of maize with different tropical legumes in Western Nigeria. J. Agr. Sci., Cambridge 11:219-25.

Aiyer, A. K. Y. N. 1949. Mixed cropping in India. Indian J. Agric. Sci. 19 (pt. 4):439-543.

Allard, R. W., and J. Adams. 1969. Population studies in predominantly self-pollinated species. XIII. Intergenotypic competition and population structure in barley and wheat. Am. Natural. 103:621-45.

Allard, R. W., and P. E. Hansche. 1964. Some parameters of population variability and their implications in plant breeding. Adv. Agron. 16: 281-325.

Altieri, M. A., C. A. Francis, A. van Schoonhoven, and J. D. Doll. 1978. A review of insect prevalence in maize (*Zea mays* L.) and bean (*Phaseolus vulgaris* L.) polycultural systems. Field Crops Res. 1:33-49.

Andrews, D. J. 1972a. Intercropping with sorghum, pp. 545-56. *In* Roa, N. G. P., and L. House (eds.), Sorghum in seventies. Oxford and IBH Publ. Co., New Delhi.

Andrews, D. J. 1972b. Intercropping with sorghum in Nigeria. Exp. Agric. 8:139-50.

Andrews, D. J., and A. H. Kassam. 1976. Importance of multiple cropping in increasing world food supplies, pp. 1-10. *In* Multiple cropping. Am. Soc. Agron. Spec. Publ. 27.

Baker, E. F. I. 1975. Research on mixed cropping with cereals in Nigerian farming systems: A system for improvement, pp. 287-309. *In* Proc. Int. Workshop on Farming Syst., Int. Crops Res. Inst. Semiarid Trop., Hyderabad, India.

Blijenburg, J. G., and J. S. Sneep. 1975. Natural selection in a mixture of eight barley varieties grown in six successive years. 1. Competition between the varieties. Euphytica 24:305-15.

Borlaug, N. E. 1959. The use of multilineal or composite varieties to control airborne epidemic diseases of self-pollinated crop plants, pp. 12-27. *In* Proc. First Int. Wheat Genet. Symp., Univ. Manitoba, Winnipeg, Can.

Bradfield, R. 1970. Increasing food production in the tropics by multiple cropping, pp. 229-42. *In* Aldrich, D. G. Jr. (ed.), Research for the world food crisis. Am. Assoc. Adv. Sci., Washington, D.C.

Browning, J. A., and K. J. Frey. 1969. Multiline cultivars as a means of disease control. Annu. Rev. Phytopath. 7:355-82.

Buestan, H. 1973. Programa de leguminosas de grano. Inf. Anu. 1973. Estac. Exp. Boliche, Inst. Nac. de Invest. Agropecu., Guayaquil, Ecuador.

Carangal, V. R., A. M. Nadal, and E. C. Godilano. 1978. Performance of promising mungbean varieties planted after rice under different environments, pp. 120-24. *In* First Int. Mungbean Symp., Asian Veg. Res. Dev. Cent.

Catredal, I. G., and R. M. Lantican. 1977. Evaluation of legumes for adaptation to intensive cropping systems. II. Soybeans, *Glycine max*. Philipp. J. Crop Sci. 2:67-71.

Catedral, I. G., and R. M. Lantican. 1978. Mungbean breeding program of Univ. Philipp., Los Baños, Philipp., pp. 225-27. *In* First Int. Mungbean Symp., Asian Veg. Res. Dev. Cent.

Cent. Int. Agric. Trop. 1977. Annu. Rep. Cali, Colombia.

Cent. Int. Agric. Trop. 1978. Annu. Rep. Cali, Colombia.

Chiang, H. C. 1978. Pest management in corn. Annu. Rev. Entomol. 23: 101-23.

Chiappe, L., and J. Huamani. 1977. Oportunidad de siembra continuada de maiz sobre tres variedades de frijol. Univ. Agraria La Molina, Lima, Peru. Mimeogr.

Clark, A., R. Shibles, and D. R. Laing. 1978. Corn and bean interactions in mixed culture. Cent. Int. Agric. Trop. Mimeogr. (unpublished).

Coffman, W. R. 1977. Rice varietal development for cropping systems at IRRI, pp. 359-71. *In* Proc. Symp. on Cropping Syst. Res. and Dev. for the Asian Rice Farmer. Int. Rice Res. Inst., Los Baños, Philipp.

Crookston, R. K., C. A. Fox, D. S. Hill, and D. N. Moss. 1978. Agronomic cropping for maximum biomass production. Agron. J. 70:899-902.

Dalrymple, D. G. 1971. Survey of multiple cropping in less developed nations. Foreign Econ. Dev. Serv., USDA, Foreign Econ. Dev., Res. Publ. 12. 108 pp.

Day, P. R. 1973. Genetic variability of crops. Annu. Rev. Phytopath. 11: 293-312.

de Carvalho Prado, E., and C. Vieira. 1976. Yields of climbing bean varieties grown on trellises at three spacings. *In* Bean Imp. Conf. Newsl. 19:18-19.

de Wit, C. T. 1960. On competition. Versl. Landbouwk. Onderzoek. No. 66.8, Wageningen. 82 pp.

Dickinson, J. C. 1972. Alternatives to monoculture in the humid tropics of Latin America. Prof. Georgr. 24:217-32.

Dijkstra, J., and A. L. F. De Vos. 1972. The evaluation of selections of white clover (*Trifolium repens* L.) in monoculture and in mixture with grass. Euphytica 21:432-49.

Donald, C. M. 1958. The interaction of competition for light and for nutrients. Aust. J. Agric. Res. 9:421-35.

Donald, C. M. 1963. Competition among crop and pasture plants. Adv. Agron. 15:1-114.

Donald, C. M. 1968. The breeding of crop ideotypes. Euphytica 17:385-403.

Eberhart, S. A., and W. A. Russell. 1966. Stability parameters for comparing varieties. Crop Sci. 6:36-40.

Enyi, B. A. C. 1973. Effects of intercropping maize or sorghum with cowpeas, pigeon peas or beans. Exp. Agric. 9:83-90.

Finlay, K. W., and G. M. Wilkinson. 1963. The analysis of adaptation in a plant breeding program. Aust. J. Agric. Res. 14:724-54.

Finlay, R. C. 1974. Intercropping soybeans with cereals. Reg. Soybean Conf., Addis-Ababa. Mimeogr. 20 pp.

Francis, C. A. 1978. Multiple cropping potentials of beans and maize. Hortscience 13:12-17.

Francis, C. A. 1979. Small farm cropping systems in the tropics, Ch. 19. *In* Thorne, D. W., and M. D. Thorne (eds.), Soil, Water and Crop Production. AVI Press, Westport, Conn.

Francis, C. A., and J. H. Sanders. 1978. Economic analysis of bean and maize systems: Monoculture versus associated cropping. Field Crops Res. 1: 319-35.

Francis, C. A., C. A. Flor, M. Prager, and J. H. Sanders. 1978a. Density response of climbing beans in two cropping systems. Field Crops Res. 1:255-67.

Francis, C. A., C. A. Flor, and S. R. Temple. 1976. Adapting varieties for intercropping systems in the tropics, pp. 235-53. *In* Multiple Cropping, Am. Soc. Agron. Spec. Publ. 27.

Francis, C. A., M. Prager, and D. R. Laing. 1978b. Genotype × environment interactions in climbing bean cultivars in monoculture and associated with maize. Crop Sci. 18:242-47.

Francis, C. A., M. Prager, D. R. Laing, and C. A. Flor. 1978c. Genotype × environment interactions in bush bean cultivars in monoculture and associated with maize. Crop Sci. 18:237-42.

Francis, C. A., M. Prager, and G. Tejada. 1980. Genotype × environment interactions in maize cultivars in monoculture and associated with two types of beans. Crop Sci. (In press).

Fyfe, J. L., and H. H. Rogers. 1965. Effects of varying variety and spacing on yields and composition of mixtures of lucerne and tall fescue. J. Agric. Sci. 64:351-59.

Galal, S., Jr., L. H. Hindi, A. F. Ibrahim, and H. H. El-Hinnawy. 1974. Intercropping corn with soybean as a bio-assaying method for screening shade-tolerant corn stocks (*Zea Mays* L.). Z. Pflanzensucht. 71: 185-86.

Gomez, A. A. 1976. Varietal screening for intensive cropping. Progress rep. 3, Univ. Philipp., Los Baños-Int. Rice Res. Inst.-Int. Dev. Res. Cent. Prog., Univ. Philipp., Los Baños, Laguna, Philipp.

Gomez, A. A., 1977. Varietal screening for intensive cropping. Progress rep. 2, Univ. Philipp., Los Baños-Int. Rice Res. Inst.-Int. Dev. Res. Cent. Proj., Univ. Philipp., Los Baños, Laguna, Philipp.

Gomez, A. A., and H. G. Zandstra. 1977. An analysis of the role of legumes in multiple cropping systems. Symp. on Exploiting the Legume-Rhizobium Symbiosis in Trop. Agric., Hawaii. Coll. Trop. Agric. Misc. Publ. 145. Dep. Agron. and Soil Sci., Univ. Hawaii.

Guilarte, T. C., R. E. Perez-Levy, and G. M. Prine. 1974. Some double cropping possibilities under irrigation during the warm season in North and West Florida. Proc. Soil Crop Sci. Soc. Fla.

Hadfield, W. 1974. Shade in the north-east Indian tea plantations. I. The shade pattern. J. Appl. Ecol. 11:151-78.

Hall, R. L. 1974. Analysis of the nature of interference between plants of different species. I. Concepts and extension of the de Wit analysis to examine effects. Aust. J. Agric. Res. 25:739-47.

Hamblin, J. 1975. Effect of environment, seed size and competitive ability on yield and survival of *Phaseolus vulgaris* (L.) genotypes in mixtures. Euphytica 24:435-45.

Hamblin, J., and C. M. Donald. 1974. The relationships between plant form, competitive ability and grain yield in a barley cross. Euphytica 23:535-42.

Hamblin, J., 1979. (Pers. commun.).

Hamblin, J., and J. G. Rowell. 1975. Breeding implications of the relationship between competitive ability and pure culture yield in self-pollinated grain crops. Euphytica 24:221-28.

Hamblin, J., J. G. Rowell, and R. Redden. 1976. Selection for mixed cropping. Euphytica 25:97-105.

Harlan, H. V., and M. L. Martini. 1938. The effect of natural selection in a mixture of barley varieties. J. Agric. Res. 57:189-99.

Harlan, J. R. 1976. Diseases as a factor in plant evolution. Annu. Rev. Phytopath. 14:31-51.

Harper, J. L. 1967. A Darwinian approach to plant ecology. J. Ecol. 55: 247-70.

Hart, R. D. 1975a. A bean, corn and manioc polyculture cropping system. I. The effect of interspecific competition on crop yield. Turrialba 25: 294-301.

Hart, R. D. 1975b. A bean, corn and manioc polyculture cropping system. II. A comparison between the yield and economic return from monoculture and polycultural cropping systems. Turrialba 25:377-84.

Hart, R. D. 1977. Characteristicas de variedades que pueden tener potencial como componentes de los sistemas de cultivos en Yojoa, Honduras. Reunion Int. Colab. Tec. Cen. Agropicu. Trop. Invest. y Ensenyañca-Cen. Int. Agric. Trop.-Cent. Int. Mejoramiento de Maiz y Trigo-Inst. Int. Am. Cincias Agric., Turrialba, Costa Rica. Mimeogr. 3 pp.

Herrera, W. T., and R. R. Harwood. 1973. Crop interrelationships in intensive cropping systems. IRRI Semin. Unpublished Mimeogr. 24 pp.

Hiebsch, C. K. 1978. Interpretation of yields obtained in crop mixtures. ASA, Agron. Abstr., p. 41.

Inst. Cienc. y Tecnol. Agric. 1976. Programa de Prod. de Frijol, Inf. Anu. Guatemala.

Int. Crops Res. Inst. Semiarid Trop. 1977. Report of the cropping systems research carried out during the Kharif (moonsoon) and Rabi (post-monsoon) season of 1976. Int. Crops Res. Inst. Semiarid Trop. Farming Syst. Res. Program. Mimeogr.

Int. Inst. Trop. Agric. 1976. Annu. Rep. Ibadan, Nigeria.

Int. Inst. Trop. Agric. 1977. Annu. Rep. Ibadan, Nigeria.

Int. Rice Res. Inst. 1972. Annu. Rep. Los Baños, Philipp.

Int. Rice Res. Inst. 1973. Annu. Rep. Los Baños, Philipp.

Int. Rice Res. Inst. 1974. Annu. Rep. Los Baños, Philipp.

Jain, H. K. 1975. Breeding for yield and other attributes in grain legumes. Indian J. Genet. Plant Breed. 35:169-87.

Jain, H. K., and P. N. Bahl. 1975. Symposium recommendations. Indian J. Genet. Plant Breed. 35:304-5.

Jennings, P. R., and J. H. Cock. 1977. Centres of origin of crops and their productivity. Econ. Bot. 31:51-54.

Jennings, P. R., and J. de Jesus. 1968. Studies on competition in rice. I. Competition in mixtures of varieties. Evolution 22:119-24.

Jensen, N. F. 1952. Intra-varietal diversification in oat breeding. Agron. J. 44:30-34.

Jensen, N. F., and W. T. Federer. 1965. Competing ability in wheat. Crop Sci. 5:449-52.

Kannenberg, L. W., and R. B. Hunter. 1972. Yielding ability and competitive influence in hybrid mixtures of maize. Crop Sci. 12:274-77.

Kass, D. C. 1976. Simultaneous polyculture of tropical food crops with special reference to the management of sandy soils of the Brazilian Amazon. Ph.D. thesis, Cornell Univ. 265 pp.

Kass, D. C. L. 1978. Polyculture cropping systems: Review and analysis. Cornell Int. Agric. Bull. 32, Cornell Univ., Ithaca, N.Y.

Khalifa, M. A., and C. O. Qualset. 1974. Intergenotypic competition between tall and dwarf wheats. I. In mechanical mixtures. Crop Sci. 14:795-99.

Krantz, B. A. 1974. Cropping patterns for increasing and stabilizing agricultural production in the semi-arid tropics. ICRISAT Farming Syst. Workshop, Hyderabad, India. 43 pp.

Laing, D. R. 1978. Adaptability and stability of performance in common beans (*Phaseolus vulgaris* L.). Cent. Int. Agric. Trop., Cali, Colombia. Mimeogr.

Lantican, R. M. 1977. Field crops breeding for multiple cropping patterns. Proc. Symp. Cropping Syst. Res. and Dev. for the Asian Rice Farmer, Int. Rice Res. Inst., Los Baños, Philipp.

Lantican, R. M., and I. G. Catedral. 1977. Evaluation of legumes for adaptation to intensive cropping systems. I. Mungbean, *Vigna radiata* (L.) Wilczek. Philipp. J. Crop Sci. 2:62-66.

Litzinger, J. A., and K. Moody. 1976. Integrated pest management in multiple cropping, pp. 293-316. *In* Multiple Cropping, ASA Spec. Publ. 27.

Lohani, S. N., and H. G. Zandstra. 1977. Matching rice and corn varieties for intercropping. Int. Rice Res. Inst. Mimeogr. 14 pp.

Loomis, R. S., W. A. Williams, and A. E. Hall. 1971. Agricultural productivity. Adv. Agron. 22:431-68.

Ludwig, W. 1950. Zur theorie der konkurrenz: Die annidation (Einnischung) als funfter evolutionsfaktor. Zool. Anz. Ergangzungsband zu Band 145: 516-37.

Mancini, M. S., and D. M. A. Castillo. 1960. Observaciones sobre ensayos preliminares en el cultivo asociado de frijol de enredadera y maiz. Agric. Trop., Colombia 16:161-66.

Moseman, A. H. 1966. International needs in plant breeding research, pp. 409-20. *In* Frey, K. J. (ed.), Plant breeding. Iowa State Univ. Press, Ames, Ia.

Osiru, D. S. O., and R. W. Willey. 1972. Studies on mixtures of dwarf sorghum and beans (*Phaseolus vulgaris*) with particular reference to plant population. J. Agric. Sci., Cambridge 79:531-40.

Phrek, G., E. Methi, and J. Suthat. 1978. Multiple cropping with mungbean in Chiang Mai, Thailand, pp. 125-28. *In* First Int. Mungbean Symp., Asian Veg. Res. Dev. Cent.

Poehlman, J. M. 1978. What we have learned from the international mungbean nurseries, pp. 97-100. *In* First Int. Mungbean Symp., Asian Veg. Res. Dev. Cent.

Poey, F. 1978. (Pers. commun.). Guatemala.

Putnam, A. R., and W. B. Duke. 1974. Biological suppression of weeds: evidence for allelopathy in accessions of cucumber. Science 185: 370-72.

Rao, M. R., P. N. Rao, and S. M. Ali. 1960. Investigation on the type of cotton suitable for mixed cropping in the nothern tract. Indian Cotton Genet. Rev. 14(5): 384-88. (Field Crops Abstr. 1962. 15:377).

Sakai, K. 1955. Competition in plants and its relation to selection. Cold Spring Harbor Symp. Quant. Biol. 20:137-57.

Santa-Cecilia, F. C., and C. Vieira. 1978. Associated cropping of beans and maize. I. Effects of bean cultivars with different growth habits. Turrialba 28(1):19-23.

Saxena, M. C., and D. S. Yadav. 1975. Multiple cropping with short duration pulses. Indian J. Genet. Plant Breed. 35:194-208.

Schutz, W. M., and C. A. Brim. 1967. Inter-genotypic competition in soybeans. I. Evaluation of effects and proposed field plot design. Crop Sci. 7: 371-76.

Shia, F. Y., and T. P. Pao. 1964. On the yields of sugarcane interplanted with different varieties of sweet potato. Rep. Taiwan Sugarcane Exp. Stn. 35:55-63. (Field Crops Abstr. 1965, 18:306).

Singh, L. 1975. Breeding pulse crops varieties for inter and multiple cropping. Indian J. Genet. Plant Breed. 35:221-28.

Suneson, C. A. 1969. Survival of four barley varieties in a mixture. Agron. J. 41:459-61.
Suneson, C. A., and G. A. Wiebe. 1942. Survival of barley and wheat varieties in mixtures. J. Am. Soc. Agron. 34:1052-56.
Swaminathan, M. S. 1970. New varieties for multiple cropping. Indian Farming 20(7):9-13.
Tang, C. K. 1968. A study on interplanting sweet potato with sugarcane. I. Date of interplanting, variety of sweet potato and row width of autumn plant cane. Rep. Taiwan Sugarcane Exp. Stn. 31:27-55.
Tarhalkar, P. P., and M. G. P. Rao. 1975. Changing concepts and practices of cropping systems. Indian Farming 25(3):3-7, 15.
Tiwari, A. S. 1978. Mungbean varietal requirements in relation to cropping seasons in India. pp. 129-31. *In* First Int. Mungbean Symp., Asian Veg. Res. Dev. Cent.
Tiwari, A. S., L. N. Yadav, L. Singh, and C. N. Mahadik. 1977. Spreading plant type does better in pigeon pea. Trop. Grain Legume Bull., Int. Inst. Trop. Agric. No. 7:7-10.
Torregroza, M. 1978. (Pers. commun., Nat. Maize and Sorghum Program, Inst. Colombiano Agric., Cent. Natl. Inst. Agric.) Tibuitata, Bigoff, Colombia.
Trenbath, B. R. 1974a. Biomass productivity of mixtures. Adv. Agron. 26: 177-210.
Trenbath, B. R. 1974b. Neighbor effects in the genus *Avena*. II. Comparison of weed species. J. Appl. Ecol. 11:111-25.
Trenbath, B. R. 1975a. Diversify or be damned? Ecologist 5:76-83.
Trenbath, B. R. 1975b. Neighbor effects in the genus *Avena*. III. A diallel approach. J. Appl. Ecol. 12:189-200.
Trenbath, B. R. 1976. Plant interactions in mixed crop communities, pp. 129-69. *In* Multiple Cropping, ASA Spec. Publ. 27.
Trenbath, B. R. 1977. Interactions among diverse hosts and diverse parasites. Ann. N. Y. Acad. Sci. 287:124-50.
Trenbath, B. R., and J. F. Angus. 1975. Leaf inclination and crop production. Field Crop Abstr. 28:231-44.
Trenbath, B. R., and J. L. Harper. 1973. Neighbor effects in the genus *Avena*. I. Comparison of crop species. J. Appl. Ecol. 10:379-400.
Tuzet, R., V. L. Chiappe, and R. Sevilla. 1975. Comparacion de diferentes modalidades de siembra en el cultivo asociado maiz-frijol. Inf. del Maiz, Univ. Nac. La Molina, Lima, Peru. 8:10-11.
Vignarajah, N. 1977. Component technology: varietal requirements, pp. 347-48. *In* Cropping Syst. Res. and Dev. for the Asian Rice Farmer, Int. Rice Res. Inst. Symp. Proc.
Villareal, R. L. 1978. (Pers. commun., AVRDC).
Villareal, R. L., and S. H. Lai. 1976. Developing vegetable crop varieties for intensive cropping systems, pp. 373-90. *In* Cropping Syst. Res. and Dev. for the Asian Rice Farmer, Int. Rice Res. Inst. Symp. Proc.
Wiebe, C. A., F. G. Petr, and W. Stevens. 1963. Interplant competition between barley genotypes, pp. 546-57. *In* Stat. Genet. and Plant Breed. Natl. Acad. Sci. USA, Natl. Res. Coun. Publ. 982.
Wien, H. C., and D. Mangju. 1976. The cowpea as an intercrop under cereals. Symp. Intercropping Semi-Arid Areas, Morogoro, Tanzania. 17 pp.
Wien, H. C., and J. B. Smithson. 1979. The evaluation of genotypes for intercropping. Int. Intercropping Workshop, Jan. 10-13. Int. Crops Res. Inst. Semiarid Trop., Hyderabad, India.
Wiggans, R. G. 1935. Combinations of corn and soybeans for silage. Cornell Univ. Agric. Exp. Stn. Bull. 634. 34 pp.
Willey, R. W. 1979a. Intercropping: Its importance and research needs. Part I. Competition and yield advantages. Field Crop Abstr. 32:1-10.

Willey, R. W. 1979b. Intercropping: Its importance and research needs. Part II. Agronomy and research approaches. Field Crop Abstr. 32:73-85.

Willey, R. W., and D. S. O. Osiru. 1972. Studies on mixtures of maize and beans (*Phaseolus vulgaris*) with particular reference to plant population. J. Agric. Sci., Cambridge 79:517-29.

Wortman, S., and R. W. Cummings, Jr. 1978. To feed the world: The challenge and the strategy. Johns Hopkins Univ. Press, Baltimore. 440 pp.

Zandstra, H. G., and V. R. Carangal. 1977. Crop intensification for the Asian rice farmer. Agric. Mech. in Asia. Summer, 1977, pp. 21-30.

Zavitz, C. A. 1927. Forty years experiments with grain crops. Ontario Agric. Coll. Bull. 332.

CHAPTER 7

Breeding for Morphological and Physiological Traits

D. WILSON

SELECTION based on observable economic yield, health, and tolerance of climate of individual plants in breeding nurseries has led to most of our successful crop cultivars. However, there is no reason to suppose that this procedure detects all variation in all characteristics likely to contribute to yield and stability in a crop. Furthermore, yield assessment in one or more environments is time and space consuming and often restricts the range of genetic material that can be screened.

These objections might be avoided if we were able to identify suitable parents more accurately and at an early developmental stage by determining which plant characteristics contribute most to high economic yield under the range of climatic and management conditions likely to be encountered during the life of the crop. A group of characteristics could then be used to identify the crop ideotype (Jennings, 1964; Donald, 1968). Although physiological processes are common to all crops, the morphological and physiological diversity of crops is as great as the range in their economic end products and in the environments in which they grow. Therefore, there can be no universal ideotype in plant breeding, but rather many biological "models."

In some crops, we can now reasonably expect to improve certain aspects of yield by breeding for specific quantifiable traits which may often be measured in the young seedling (Wallace et al., 1972; Cooper, 1974; Wilson, 1976). Routine selection for such traits requires both significant additive genetic variation and simple rapid

Head, Department of Developmental Genetics, Welsh Plant Breeding Station, Aberystwyth, Wales, United Kingdom.

means of assessment. Fortunately, most morphological and anatomical characteristics of plants usually satisfy both of these requirements. The more common problem is determining the best combination among many traits, which may be correlated with one another, and the feasibility of combining them in one cultivar. The more direct measurement of dynamic physiological processes poses the problem that their accurate measurement is dependent on technological developments, and decisions on which processes to use depend on advances in knowledge of the physiological bases of crop production. However, genetic variation in such processes, and in plant morphology, is in itself an extremely powerful physiological tool which can lead to the development of hypotheses for initiating breeding programs.

Breeding for physiological and morphological traits can result in increase in crop or plant biomass (biological yield), in redistribution of assimilates to the economic product within the plant, in alleviating or avoiding the effects of some environmental constraint, or all three of these. Disease and pest tolerance may also be affected, but these are not considered here. Although increase in biomass can lead to increase in economic yield, in practice this is not always so, and increase in biomass can even occur with reduced economic yield (Goldsworthy, 1970). It is useful therefore to examine the main physiological and morphological traits which might be of use to breeders, in relation to crop processes and yield characteristics which they are most likely to affect.

BIOLOGICAL YIELD

All crop production ultimately depends on photosynthesis, although processes such as respiration, translocation, or the activity of metabolic "sinks" are all important (Evans, 1975). The capture and use of solar energy by crop communities takes place at several different organizational levels which may all be subject to genetic variation, and many of the photosynthetic units in a crop may suffer from limitations of various environmental factors whose effects can be ameliorated by genetically variable elements (Wilson, 1973).

Prolonged high levels of photosynthesis and efficient use of photosynthate depend on previous and current events in crop development; and, in general, growth is controlled at any point in time by the photosynthetic activity of individual leaves, the size of the photosynthetically active system, and the extent to which photosynthate is used for active growth (Cooper, 1976). The development and effectiveness of the assimilatory system and its effects on growth have been extensively studied (namely, Monsi and Saeki, 1953; Brown and Blaser, 1968; Monteith, 1969; Saeki, 1973). During the young seedling stage, photosynthesis and crop growth rate (CGR), are limited by the ability of the crop to intercept light. Also, the leaf area index (L, leaf area per unit area of ground) is very low, and the photosynthetic response curve to light is similar to that of an individual leaf (Brown et al., 1966). At this stage there is little shading; most or all leaves are photosynthesizing rapidly at or above light saturation level, and rate of photosynthesis per unit area of leaf (net assimilation rate, NAR), is high. As the crop develops and L increases, photosynthesis and hence CGR increase until a maximum L (L_{max}) is reached and the crop is intercepting all light falling upon it (de Wit, 1965). Although an increase beyond L_{max} in L may result in a decrease in crop photosynthesis (Black, 1964; Pearce et al., 1967), more usually, it effects little or no change (Ludwig et al., 1965; King and Evans, 1967; Alberda, 1971). Crop respiration follows a similar pattern; it is approximately proportional to gross photosynthesis and reaches a maximum when a ceiling yield of live plant tissue is reached (McCree, 1970; Robson, 1973). At complete light interception the mean NAR is low because illumination of individual leaves may range from complete light saturation at the top of the canopy to light compensation at the base.

It should therefore be possible to increase biological yield by (1) increasing photosynthetic capacity of the individual leaf, (2) improving light interception characteristics of the crop, or (3) reducing any wasteful respiration. Much breeding and research effort has been expended in examining these possibilities.

Photosynthesis

The Individual Leaf

The photosynthetic response of leaves to light shows an initial linear phase in which rate of photosynthesis is proportional to light intensity and photochemical processes are limiting. Eventually the rate of photosynthesis becomes strongly dependent on CO_2 diffusion or on biochemical processes, and the photosynthetic system becomes light saturated (P_{max}). Between low, completely limiting, and high, completely saturating, light intensities, a large transition range occurs where a number of environmental variables may limit photosynthesis simultaneously. Consequently, even within a crop canopy exposed to full sunlight, individual leaves may suffer from shortage of CO_2, light, or some combination of both; so that genetic variation in efficiency of utilization of both CO_2 and light may be important.

While differences in rates of photochemical processes or in photosynthesis in weak light are reported to exist both between and within species (Homann and Schmid, 1967; Brown, 1969; Wilson, 1973; May, 1975), the significance of such variation is not established. Narrow sense heritability of rate of light limited photosynthesis is reported to be low (17 percent) in ryegrass (*Lolium perenne*) (Wilson and Cooper, 1969a) and rice (*Oryza sativa*) (Wallace et al., 1972).

There are of course major differences in P_{max} and in responses to light and temperature between species depending on whether the first products of photosynthesis are C_3 or C_4 compounds (Wilson, 1973). In C_4 crops, photosynthesis may respond to light energies approaching full sunlight, although CO_2 can be an important limitation over much of the light intensity range (Wilson and Ludlow, 1970). Most C_3 crops appear to saturate at about one-quarter of full sunlight. This difference in photosynthetic pathway is further associated with differences in photorespiration; in leaf anatomy and chloroplast morphology; in rate of translocation; and in the efficiency of water use, which can have marked effects on yield (Hatch et al., 1971). The contrast between these groups of species is the basis for much interest in photorespiration (see below), the existence of which in C_3 species has been deprecated as a major reason for reduced P_{max} and biological yield compared with C_4 species (Zelitch, 1968).

Even so, large heritable differences in P_{max} per unit leaf area at normal CO_2 levels exist both between and within species and cultivars of these two major ecological groups (Hesketh and Moss, 1963; El-Sharkawy and Hesketh, 1965; Duncan and Hesketh, 1968; Wilson and Cooper, 1967; Dornhoff and Shibles, 1970; Asay et al., 1974). Furthermore, CO_2 enrichment during the storage phase has been shown to lead to increased yield in several crops (Evans, 1975); so it is at first sight surprising that there is usually either no relationship at all between genetic variation in P_{max} and yield, or it is negative (El-Sharkawy and Hesketh, 1965; Duncan and Hesketh, 1968; Delaney and Dobrenz, 1974; Kaplan and Koller, 1977). In general, it appears that it is only among certain seedling populations or where rate of leaf photosynthesis is an integrated expression (NAR) of all leaves on a plant that a good positive relationship is recorded (Morley, 1958; Wilson and Cooper, 1967; Treharne et al., 1971). Direct selection for P_{max} or related characters has not established the existence of any consistent effect on growth or dry-matter yield

(Wilson and Cooper, 1970; Moss and Musgrave, 1971; Hart et al., 1978). However, high grain-yielding semidwarf rice cultivars, have short, thick leaves, noted for their high P_{max} per unit leaf area (Wallace et al., 1972).

These results can be partly explained by the basis of expression of photosynthetic rate, its relationship with leaf anatomy and morphology, and the importance of adaptability for the individual leaf. Unit leaf area is of course a valid basis for assessing the effects of short-term fluctuations in environmental variables on photosynthesis. However, its use as a sole basis of comparing P_{max} of genotypes can be misleading (Charles-Edwards and Ludwig, 1975). A unit area of leaf can be thick or thin and comprise many or few mesophyll cells. Consequently, the interpretation of many comparative studies on photosynthesis is made difficult by large variations in the amount of photosynthetic tissue within unit leaf area. In many studies, rate of photosynthesis of unit leaf area has been related to leaf thickness and size and number of mesophyll cells (El-Sharkawy and Hesketh, 1965; Wilson and Cooper, 1967, 1969a; Charles-Edwards and Ludwig, 1975), and it has been suggested that correlations between specific leaf weight (SLW) and P_{max} (Pearce et al., 1969; Wilson and Cooper, 1969b; Dornhoff and Shibles, 1970; Delaney and Dobrenz, 1974) result from the effect of increasing cell numbers on SLW (Wilson and Cooper, 1969c). SLW per se need bear no relationship with P_{max} (Dunstone et al., 1973) since it is presumably also affected by supporting tissues. The rate of photosynthesis per mesophyll cell is much less variable (Eagles and Othman, 1974), and the cell could well be considered as the basis on which to express photosynthesis. However, this would be operationally difficult; so the most practical procedure might be to use several bases in addition to leaf area, such as fresh and dry weight.

A further difficulty is the often reported negative relationship between P_{max} and leaf size (Evans and Dunstone, 1970; Khan and Tsunada, 1970; Hanson, 1971) which can prevent the attainment of both high L_{max} and high P_{max} in one plant (see below) (Rhodes, 1971). However, large- and small-leaved barley (*Hordeum vulgare*) lines have been reported to exhibit the same rates of P_{max} (Berdahl et al., 1972), and in ryegrass selection for size of mesophyll cell need not alter leaf size (Wilson and Cooper, 1970). Even so, the most appropriate leaf design for high economic yields depends on the crop, and the one most suitable for high grain yields may not be the same as that leading to high forage yield. The short, thick, and erect leaf with high nitrogen content and high P_{max} proposed as most suitable for the rice plant (Ishizuka, 1971) and for sugarcane

(*Saccharum officinarum*) (Rosario and Musgrave, 1974) contrasts with the long-leaved type with relatively low P_{max} suggested for maximizing biological yield of temperate grasses (Rhodes, 1975) and which has resulted from selection for high grain yield in wheat (*Triticum aestivum*) (Evans, 1975).

There may be genetic variation in biochemical components of photosynthesis within both C_3 and C_4 species (Treharne et al., 1971; Nelson et al., 1975; Randall et al., 1977) but the significance of these differences is not clear. Even assuming it was possible to double P_{max} and retain the same canopy structure and light interception characteristics, Monteith (1977) has recently calculated that under North European conditions dry matter production would only increase by 30 percent during the summer (Fig. 7.1). He concludes that it might be difficult to increase crop yields much by breeding for small increases in P_{max}.

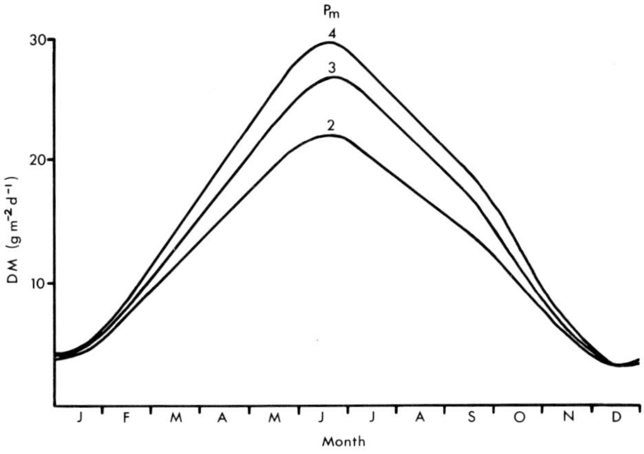

Fig. 7.1. Seasonal change of rate of dry matter production (DM) as function of P_{max}, maximum rate of leaf photosynthesis (g CH_2Om^{-2} day^{-1}) (K = 0.6). (Monteith, 1977.)

The significance of variations in stomatal traits for photosynthesis appears to differ between C_3 and C_4 species. Variation in stomatal resistance to diffusion (r_s) in unstressed plants is likely to affect CO_2 uptake of C_4 more than of C_3 species (Wilson, 1973). In the latter, genetic variation in r_s can affect transpiration more than photosynthesis, thus improving water use efficiency (WUE) (Miskin et al., 1972; Wilson, 1975a; Wilson et al., 1977). The use of stomatal traits in breeding for high r_s and improved water use in C_3 species is discussed later. However, in some circumstances a low resistance might be advantageous. These might prevail (1) where water is unlikely to be limiting, as in irrigated crops, so that any possible gain in CO_2 assimilation could be realized; and (2) in regions where supraoptimal temperatures may occur with some regularity, and the leaf cooling effect of high transpiration might be a useful attribute.

PHOTOSYNTHETIC ADAPTABILITY. A difficulty in interpreting comparative data on single leaf photosynthesis or of using it logically in breeding programs to increase yield is that the potential of a leaf for photosynthesis is dependent on both current environmental conditions and on those during its development, its position on the plant, and its age. Even within the range of climatic conditions encountered by crops growing in their normal environment the degree of physiological and morphological plasticity may be large, and the range of adjustment of photosynthesis to growth conditions may differ considerably among populations genetically adapted to different environments (Björkman, 1973). Increasing light intensity during leaf development, for example, often increases P_{max} (Björkman and Holmgren, 1963; Wilson and Ludlow, 1970) but there are strong species × light intensity (Dunstone et al., 1973) and genotype × light intensity (Björkman, 1968; Wilson and Cooper, 1969d) interactions. For example, diploid *Triticum* species respond to preceding light intensity whereas hexaploid species do not (Dunstone et al., 1973). Similarly, the temperature optimum for photosynthesis is greatly affected by temperature during leaf development (Björkman, 1973; Charles-Edwards et al., 1971; Treharne and Nelson, 1975), and genotype × temperature pretreatment interactions have been reported (Wilson and Cooper, 1969e). Many effects of environmental pretreatment do not correlate consistently with internal anatomical changes in leaves (Eagles and Treharne, 1969; Wilson and Cooper, 1969a, b; Treharne and Nelson, 1975; Peet et al., 1977) but may be associated with other, possibly biochemical, influences on mesophyll events and/or r_s (Holmgren, 1968; Hatch et al., 1969; Wilson and Ludlow, 1970). It may therefore be possible to breed for more adaptable photosynthetic apparatus, as suggested by Asay et al. (1974), without encountering problems of morphological association with photosynthetic activity.

LEAF AGE. A progressive decline in net photosynthesis with leaf age after maturity has been recorded in many species. The age at which this decline begins varies both between and within species

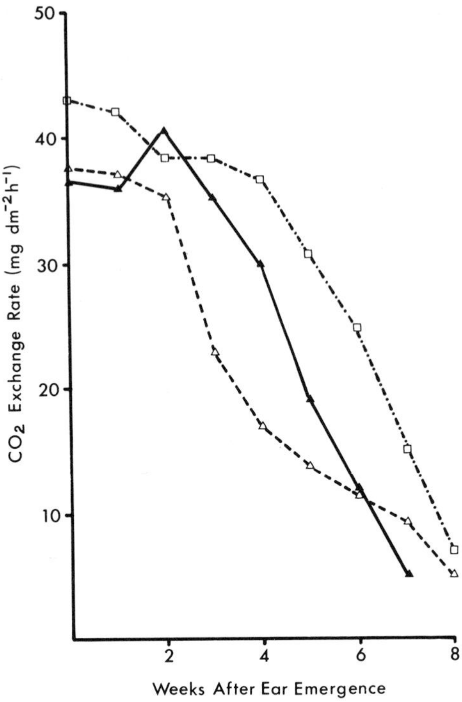

Fig. 7.2. CO_2-exchange rates of flag leaves of three spring-wheat cultivars after ear emergence (△ Glenleaf, ▲ Neepawa, □ Opal). 120 nEcm^{-2}s^{-1} photon flux density, 350 μ liters liter^{-1}CO_2. 5% LSD = 3.8 mg CO_2 dm^{-2}h^{-1}. (Adapted from Aslam and Hunt, 1978.)

(Hardwick et al., 1968; Wilson and Ludlow, 1970; Loach, 1970; Stoy, 1975; Aslam and Hunt, 1978). In temperate grasses the decline may not start until two or three weeks after full leaf expansion (Jewiss and Woledge, 1967); but in wheat the flag leaf often exhibits a maximum rate about one week after emergence, declines, then rises to a second maximum after the onset of rapid grain filling (Lupton, 1968; Evans and Rawson, 1970), a pattern which seems to reflect the demands of the grains for photosynthate (Rawson et al., 1976). Even so, genotypic differences in the aging pattern of flag leaf photosynthesis exist among wheat cultivars (Fig. 7.2). In sugar beet (*Beta vulgaris*) the highest yielding cultivars in terms of both total dry matter and sugar yield may be those which maintain largest NAR late in the season and exhibit least leaf senescence (Loach, 1970). Possible causes of differences in leaf senescence are a matter for conjecture, but a large part of the biochemical apparatus for senescence may preexist in a latent form (Thomas and Stoddart, 1977). If so, then selection criteria for reduced leaf senescence, measurable on young healthy leaves, could be developed.

The Crop Canopy

The main morphological traits which influence canopy structure are easily measurable and usually have usefully high narrow-sense heritabilities (Wallace et al., 1972; Rhodes, 1973). These include such leaf characteristics as length, width, rigidity, and angle, and in cereals, tiller angle. In soybeans (*Glycine max*), leaf directional response to light varies genetically (Wallace et al., 1972).

Variations in canopy structure affect the crop in two important ways: first, and most obviously, through light interception and distribution; and second, through correlated physiological, developmental, and morphological changes (Evans, 1975). Consequently, canopy design not only has to consider potential biological yield, but also crop management and economic end product.

The close relationship between light interception, crop photosynthesis, and dry matter production in the early stages of crop growth is well established (Biscoe and Gallaher, 1977). Both rate of attainment and level of L at which all light is intercepted are influenced by the geometry, presentation, and distribution of the component leaves and stems of the crop canopy. For example, in developing crops a prostrate leaf arrangement results in faster attainment of a closed canopy and hence more effective light interception and more rapid growth than an erect arrangement. Once a closed canopy has been achieved, however, a more erect leaf arrangement allows the incoming light to be distributed over a larger leaf area, leading to more efficient conversion and higher crop photosynthesis (Cooper et al., 1971). Theoretically, erect leaves are unlikely to be advantageous at L values less than 3 (Duncan, 1971).

These relationships can be expressed in terms of the light transmission coefficient K, which gives a quantifiable measure of the light trapping capabilities of a crop. In terms of leaf posture, a low value of K (0.3) indicates mainly erect leaves whereas with a high value (0.9) they are nearly horizontal (Monteith, 1965). K is also affected by leaf light transmission qualities and by their horizontal and vertical distribution within the canopy. Thus, lower values of K are achieved in communities with well separated and regularly distributed leaves than in those where the distribution is more random (Loomis and Williams, 1969).

The relationships between canopy architecture, light distribution, and crop production changes with time of day, weather, season, and stage of crop development. Productivity may also be affected by the influence of leaf arrangement on air circulation. The advantages of erectness, for example, are more distinct in high than in low light intensity environments so that breeders in continental climates are more likely to achieve a consistent effect from breeding for reduced K than are breeders in cloudier maritime climates (Monteith, 1977). Even so, a model has predicted that growth rate could be increased from about 17 to 29 g m^{-2} day^{-1} in mid-summer in the UK by changing from horizontal to erect leaves (Fig. 7.3). Such a difference has already been demonstrated between contrasting forage grasses (Sheehy and Cooper, 1973).

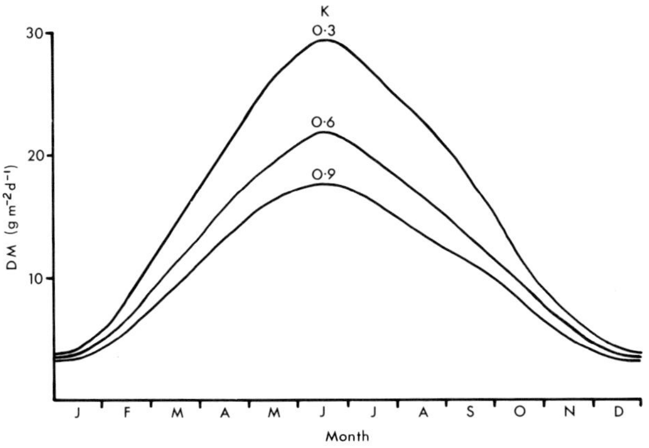

Fig. 7.3. Seasonal change of rate of dry matter production (DM) as function of K, light transmission coefficient of canopy assuming complete light interception (P_{max} = 3 g $m^{-2}h^{-1}$. (Monteith, 1977.)

In the outbreeding ryegrasses, Rhodes (1972, 1973, 1975) has shown that, (1) even within populations, considerable additive genetic variation exists for characteristics affecting K, and hence the value of L at which incoming light is completely intercepted (Fig. 7.4); (2) genotypes with contrasting expression of these characters can be detected at an early seedling stage; (3) selection at this stage can produce marked changes in subsequent crop dry matter production which are management dependent for their expression; (4) response to selection varies between populations

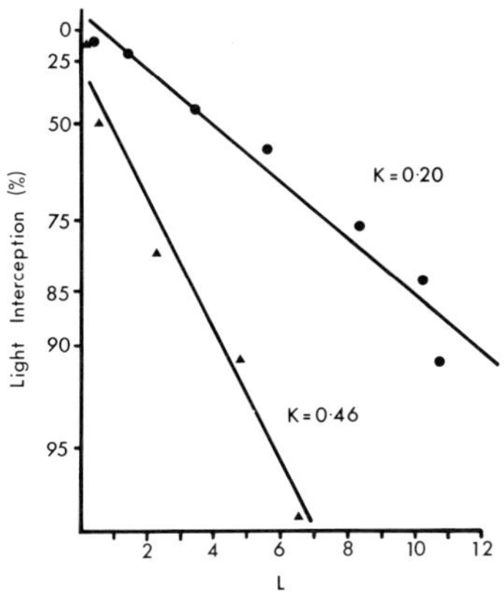

Fig. 7.4. Light interception, leaf area index (L) and $K_{(vis)}$ in two contrasting genotypes of *Lolium perenne* cv. S.321. (Rhodes, 1971.)

but may continue at least to the fourth generation. Since the harvested portion of forage grasses consists of virtually all the above ground biomass, theory would suggest that almost any strategy which heightens the canopy and increases L_{max} could improve light interception and yield. Consequently, Rhodes (1972, 1973, 1975) examined the effects of intracultivar selection for leaf length, leaf rigidity, leaf angle, and tiller angle, using the sixth seedling leaf as the basis of selection. Leaf length was found to be the most effective in influencing crop growth. Yield increases of up to 30 percent in dry-matter yield of infrequently cut plots, allowing complete development of L_{max}, have been recorded after four generations of selection for long leaves (Fig. 7.5). Although the potential biomass production of the short-leaf more prostrate type of plant is very much less, it is this type which is often most productive under frequent cutting management systems (Rhodes, 1973). Leaf size and rate of leaf appearance, which are negatively correlated with one another, also respond readily to selection in some ryegrass cultivars. Improved L_{max}, with corresponding increases in annual dry-matter yield under infrequent cutting systems, have resulted from selection either for large leaves or for slow rate of leaf appearance (Wonkyi-Appiah, 1970); so leaf size in itself often provides a useful indication of potential dry matter production in temperate forages.

In practice, the forage crop is not usually managed simply with

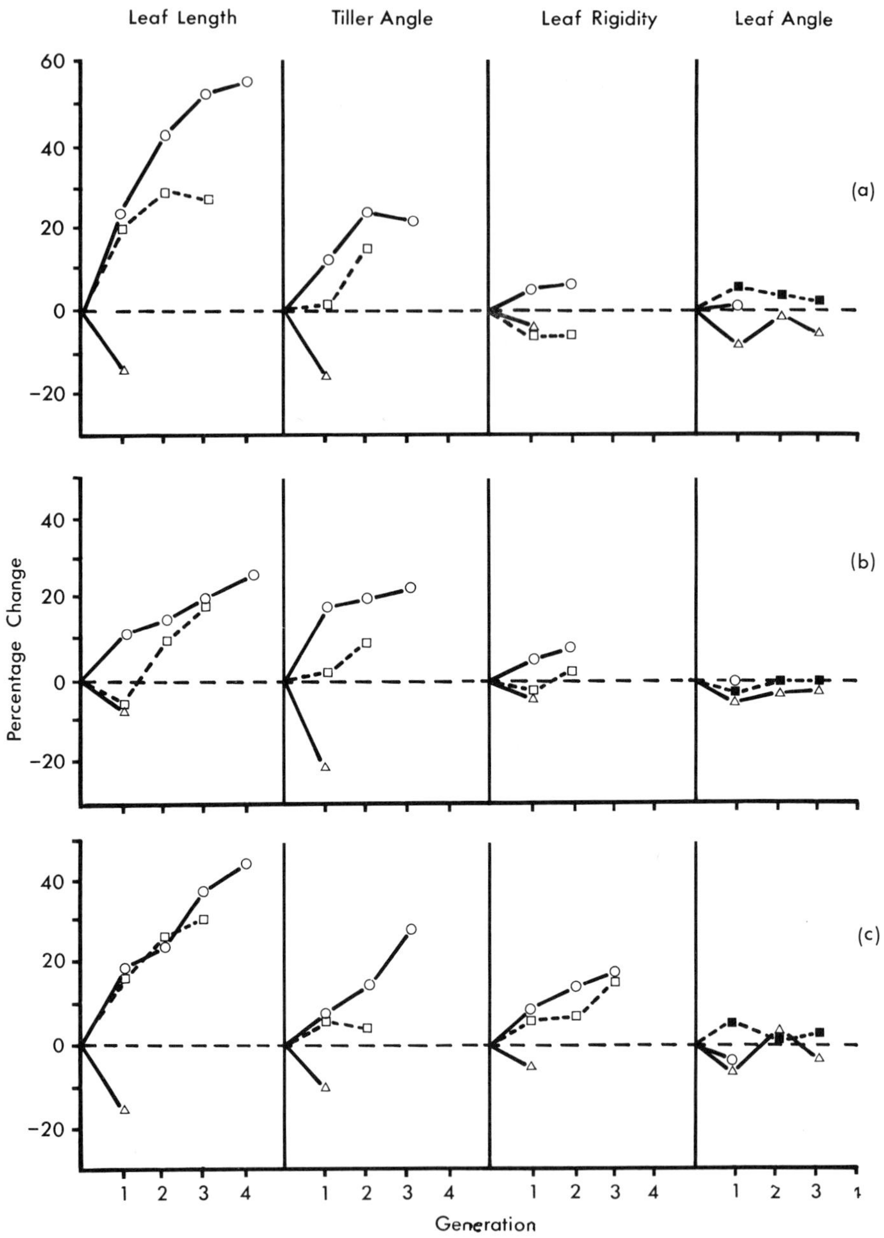

Fig. 7.5. Response to selection for canopy characters (○—○) and associated changes in dry matter yield (□---□) in three ryegrass cultivars (a, b, c). (Rhodes and Mee, 1979.)

a view to maximizing L. Efficient continuous production of forage must take into account short- and long-term requirements for crop maintenance. The appropriate cutting or grazing height for achieving highest annual yields and maintaining a vigorous stand or sward may differ at different seasons and may not even be closely related to concurrent growth rates (Brown and Blaser, 1968). Grazing or frequent cutting may be necessary at times to promote tiller development. In some species such as the stoloniferous *Cynodon dactylon*, the morphology may be such that defoliation intensity makes little difference in yield (Brown and Blaser, 1968).

Breeding to maximize the benefits (N-fixation, superior feed value) of clover in grass/clover pastures involves consideration of the canopy of the community as affected by its constituent parts. Clover stature is important in controlling ability to coexist with associate grasses, particularly at high L (Donald, 1963). In perennial pastures, long petioles and large leaves, which are associated with high intrinsic clover yield, lead to strong competitive ability in the establishment year. However, in subsequent years the decreased investment in stolon tissue often associated with this type of plant leads to reduced ability to compete with grass (Rhodes and Harris, 1979). Even so, it may well be possible to combine greater investment in stolon tissue with the long petiole trait (Rhodes, pers. commun.).

It has been suggested that an erect canopy is likely to be more beneficial for crops with axillary inflorescences such as soybeans, peas (*Pisum sativa*), and cotton (*Gossypium hirsutum*), which bear fruit at many nodes, than for grain crops, because of the prime importance of the upper leaves for grain filling (Evans, 1975). However, there is little evidence of variation in canopy structure affecting K, L_{max} and canopy photosynthesis of soybeans, although cultivar differences in canopy photosynthesis independent of any effects of L have been reported (Shibles et al., 1975). In the pea crop, factors other than efficiency of light interception have hitherto been considered more important by breeders (Pate, 1975).

Morphological designs for grain crop canopies (Pendleton et al., 1968; Ishizuka, 1971; Wallace et al., 1972; Yoshida, 1972; Mock and Pearce, 1975; Austin and Jones, 1976) emphasize structure at the time of grain filling and, particularly in wheat and barley (*Hordeum vulgare*), the predominant role of the flag leaf and ear in supplying assimilate for the developing grain. In these two crops the spike exists in a favorable light environment and may contribute a substantial proportion of the total carbohydrate in the grain (Evans,

1975). This contribution is greater in awned than awnless cultivars (Evans and Rawson, 1970), and it has been suggested that rapid progress in developing genotypes with enhanced spike photosynthesis is likely to come from selecting for more awn tissue (Johnson et al., 1975). However, awns may be disadvantageous at times because of shading effects on leaves or of their effect in increasing spike weight and hence susceptibility to wind. Awns may, however, be of some advantage in water stress conditions (Evans et al., 1975).

In contrast to wheat and barley, the rice panicle has low photosynthetic activity, and it is important that panicles are situated to avoid shading leaves. Most modern rice cultivars have their ears at the top of the canopy, and it has been suggested that cultivars with their ears at lower levels are preferable (Murata and Mitshusima, 1975). Leaves below the flag leaf are relatively more important to grain filling in this crop than in wheat and barley (Yoshida, 1972). Consequently, leaf posture is particularly important in rice breeding for improving the light environment of the lower leaves. Recently produced high-yielding cultivars have short, erect leaves as well as short culms and profuse and upright tillers. Maize (*Zea mays*) tassels also show low photosynthetic activity, and Mock and Pearce (1975) have suggested that the maize ideotype should have erect leaves above the ear, although there has been no significant effect on grain yield from selecting for four generations for erect leaves in maize (Ariyanayagam et al., 1974).

The potential benefits of the erect leaf habit have been quantified in models (Monteith, 1965; de Wit, 1965; Duncan et al., 1967) which predict that for L of 4-5, the maximum difference in photosynthesis to be expected between canopies with erect and those with horizontal leaves would be roughly 10-15 percent. In practice, at this L the erect leaf habit in barley has been shown to increase both dry matter production and grain yield only by permitting the crop to respond to increase in plant density (Angus et al., 1972). A similar density dependent relationship has been reported in sugarcane (Bull and Glasziou, 1975). Even so, the beneficial effects of erect leaves on wheat crop photosynthesis at high L (Fig. 7.6) may not only be direct through improved light distribution, but may also be indirect through altered senescence patterns (Austin et al., 1976). However, in that experiment the erect leaf type was only marginally more productive, since the lax cultivar produced more grains and appeared to fill them by drawing on a reserve of assimilate in the stem. Similar transfers of photosynthates to grain from stem have been reported by others (Stoy, 1965), and in extreme situations, up to

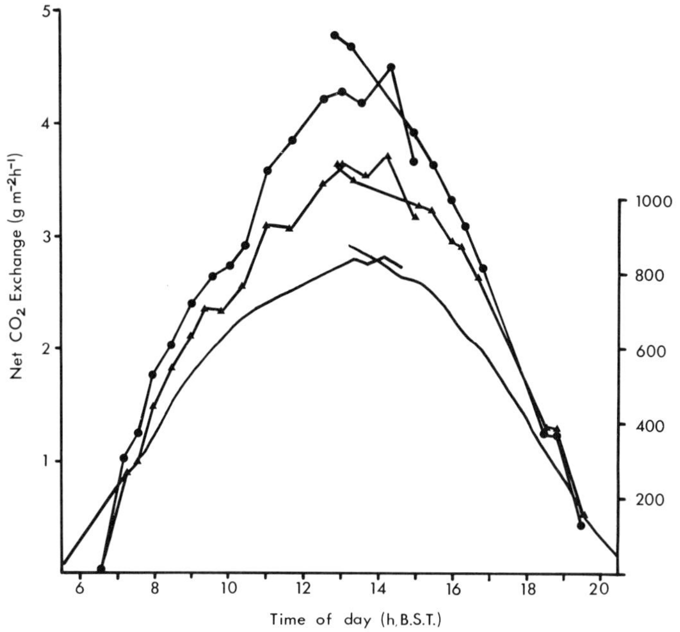

Fig. 7.6. Diurnal change in net CO_2 exchange of canopies of erect (•) and lax (▲) leaved wheat genotypes on 20-21 June 1974, Cambridge, England. Continuous line indicates irradiance. (Austin et al., 1976.)

70 percent of the contribution to grain filling in barley can be from assimilate stored prior to anthesis (Gallaher et al., 1975). Although lower leaves are usually of lesser importance at the time of grain filling, penetration of light down the canopy may be advantageous in promoting root growth and therefore water and nutrient uptake, and in influencing spikelet number and grain set (the size of the "sink") (Evans et al., 1975). In some circumstances increases in preanthesis photosynthetic activity can have greater benefit to final grain yield than any postanthesis changes. In one experiment using CO_2-enrichment to increase photosynthesis of a field crop of barley during the preanthesis period, grain yield was increased by 50 percent compared with only 25 percent when the same treatment was applied postanthesis (Gifford et al., 1973).

Short-strawed rice and wheat cultivars have met with success because of their ability to withstand lodging and to respond to increased fertilizer nitrogen without collapse of the canopy (Wallace et al., 1972; Yoshida, 1972). Nevertheless, theory would suggest that tall plants should reach greater L_{max} and be capable of producing greater biomass; and a positive genetic relationship has been shown between height and grain yield in barley (Riggs and Hayter, 1975), oats (*Avena sativa*) (Rosielle and Frey, 1975), sorghum (*Sorghum bicolor*) (Campbell et al., 1975), and wheat (Knott and Kumar, 1975; Law et al., 1978) (Fig. 7.7). To exploit this relationship in wheat it has been suggested that lodging resistance might be achieved in other ways than excessively reducing

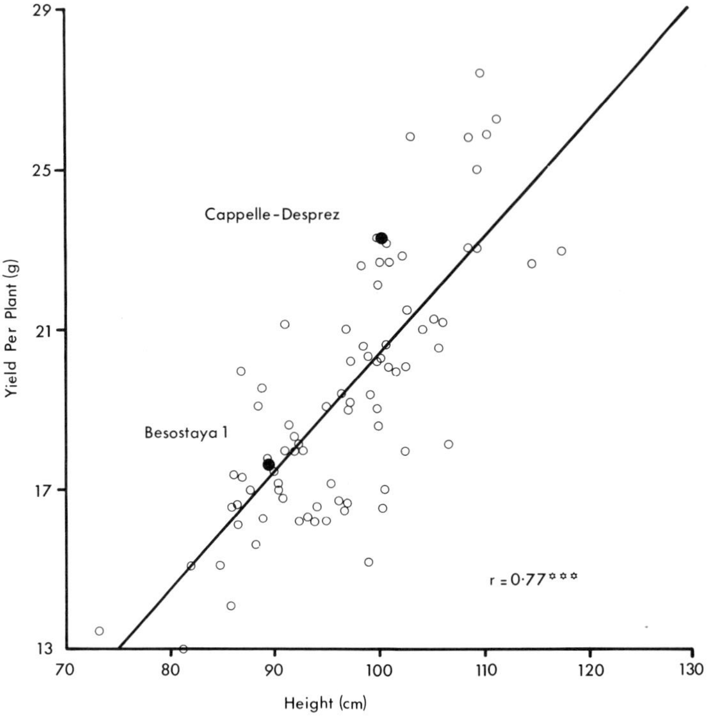

Fig. 7.7. The relationship between grain yield per plant and height of F_3 lines from the cross of Cappelle-Desprez with Besostayal wheat. (Law et al., 1978.)

height. Selecting positively for height together with yield components has been proposed to produce "tall-dwarfs," which already tend to be higher yielding than "short-dwarfs" (Law et al., 1978). Yoshida (1972) has suggested the existence of an optimum plant height for a given species and that the present short stature is not necessarily the optimum for rice and wheat. However, in crops such as cereals, overinvestment of assimilate in stem could occur and might cause reduction in harvest index (see below).

The characteristics which confer lodging resistance and which might be important in breeding "tall-dwarf" cereals include culm density, the dry weight per unit length of culm, and the number of vascular bundles per culm (Pinthus, 1973). In grasses, genetic variation in the amount or arrangement of structural components of the leaf that can affect rigidity and genetic differences in leaf structure tend to be paralleled by those in stem structure (Sant and Rhodes, 1970; Selim, 1976) so that leaf traits might be used to select for lodging resistance. In cereals, it would seem desirable to obtain flag leaf rigidity without severe reduction in length. Long rigid leaves are attainable in certain forage grasses (Selim, 1976), but whether this is a realistic objective in cereals is not known. In maize, selection for erect leaves has not influenced leaf length but has reduced plant height (Ariyanayagam et al., 1974).

The less obvious effects of selecting for canopy structure traits are often the most important. In forage grasses, animal products are the economic "yield" and it is the readily digestible shoot dry matter from the plant that is important. Fractions such as lignin which lend rigidity to leaves, so improving light interception, are waste products to the animal. Even cellulose is only slowly digested. Therefore, the greater amounts of dry matter which might be produced by improving light interception could be of less use to the animal. For example, in tall fescue (*Festuca elatior*), strong, erect-leaved families have appreciably lower digestibility, harsher leaves, and more cellulose in both leaves and stems, than the lax-leaved families selected from the same cultivar (Selim, 1976). However, in ryegrass, leaf rigidity can be achieved by having more structural units across the width of the leaf rather than changing the relative proportions of leaf tissues (Sant and Rhodes, 1970). In that species, erect-leaved populations have now been produced which have similar digestibilities to lax-leaved types (Rhodes, pers. commun.). A high quality erect-leaved grass may also be more efficiently harvested because it is more accessible to the animal (Jackson, 1974).

Strong, erect-leaved tall fescue selections are also sparse tillering and have reduced tolerance to subzero temperatures, while lax types are profuse tillering and have improved cold tolerance (Selim, 1976). Because of their few tillers, erect-leaved cultivars may only be suitable for conservation or for leniently grazed management systems. The association between erect habit and sparse tillering may well be general in temperate forage grasses (Rhodes, pers. commun.) but the relationship with cold tolerance does not seem to be related to erect habit per se, since strong- and weak-leaved ryegrass selections which have no clear habit differences are less and more cold tolerant, respectively (Table 7.1).

Another less obvious effect of canopy structure changes is that if these result from increasing the total size of the shoot system relative to the root system then this may adversely affect the nutrient or water economy of the plant under some circumstances (Throughton and Whittington, 1969).

Table 7.1. Effect of selecting for leaf tensile strength on cold tolerance of seedling populations of perennial ryegrass cv S. 23. (Wilson, 1978.)

Population	Mean Leaf Strength ($g\ mg^{-1}dw$)	Second Leaf Damage at −9° C	Plant Kill at −7° C
		%	%
Strong leaf	81	63	34
Weak leaf	50	28	13
5% LSD	15	8	5

Respiration

The main aspects of respiration that have attracted the attention of breeders are the possibilities that, (1) photorespiratory losses of carbon from C_3 crops can be eliminated or reduced by selective breeding; and (2) crop energetic efficiency can be improved by reducing the proportion of the energy budget, as measured by dark respiration, which is expended on processes not directly contributing to growth.

Photorespiration

Much of the greater efficiency of the C_4 pathway for carbon fixation over that of the C_3 at the single leaf level (Cooper, 1975) can be attributed to photorespiratory losses of CO_2 from C_3 species (Troughton, 1975). Although the advantage of the C_4 pathway is much reduced at the crop level (Gifford, 1974), even after allowing for the different environmental conditions in which the two groups of species normally grow, crops of C_4 species are about 40 percent more efficient than those of C_3 (Monteith, 1978). Therefore, although the main ecological niche of most C_4 plants is in hot arid environments where their more efficient use of water and high optimal temperatures are advantageous (Björkman, 1973; Laetsch, 1974; Ehleringer and Björkman, 1977), it seems reasonable to suggest that breeding for reduced photorespiratory losses from C_3 crops might increase potential crop photosynthesis. However, this has not been clearly demonstrated, and hybridization and selection experiments have emphasized the genetic and physiological complexities of incorporating complete C_4 systems into C_3 crops (Björkman, 1976) or of fixing reduced photorespiration into C_3 populations by selection (Zelitch, 1976). Even so, the value of low photorespiration in C_3 crops is not clear since recent biochemical experiments suggest that energy dissipation by photorespiration protects the photosynthetic apparatus from damage from photooxidative processes during periods of CO_2 shortage, such as might be caused by stomatal closure at high light intensities (Tolbert and Ryan, 1976; Krause et al., 1978; Lorimer et al., 1978). If so, then high photorespiration might even be a useful breeding objective in some conditions.

For technical reasons, variation is usually assessed by measuring photorespiration-related phenomena such as the CO_2 compensation concentration ($[CO_2]$), rate of CO_2 evolution into (or O_2 uptake from) CO_2-free air in the light, photosynthetic response to low oxygen, and presence or absence of the typical C_4 Kranz anatomy. C_4 plants characteristically have low or zero levels of these traits whereas C_3 plants generally have $[CO_2]_c$ greater than 40 ppm, exhibit relatively greater CO_2 evolution and O_2 uptake in the light and show enhanced photosynthesis at low oxygen (Zelitch, 1971).

No major C_3 crop has yet been shown to contain individuals with these C_4 traits (Cannell et al., 1969; Moss and Musgrave, 1971; Dvorak and Nátr, 1971; Hall, 1972; Wilson, 1972a). Furthermore, the *Atriplex* hybridization experiments at the Carnegie Institute indicate that it is unlikely that such an individual might arise by mutation because the complete integration of all biochemical and anatomical C_4 characteristics, each largely under separate genetic control, seems necessary for full expression of the C_4 advantage (Björkman, 1976). At least fifteen genera, none of which include important crops, contain both C_3 and C_4 species (Björkman, 1976); two species, *Panicum milioides* (Quebedeaux and Chollet, 1977) and *Mollugo verticillata* (Sayre and Kennedy, 1978) seem to be examples of intermediate behavior; and in one, *Spartina townsendii*, the relatively low temperature optima for photosynthesis of the C_3 pathway coexists with C_4 biochemistry and anatomy (Long and Woolhouse, 1978). Therefore, it seems possible that integration might be achieved by interspecific hybridization in genera which contain species with both major photosynthetic pathways. However, the prospects of introducing the C_4 pathway into agriculturally important C_3 species by genetic manipulation do not look promising.

Genetic variation in photorespiration-related phenomena do exist within C_3 species (Zelitch and Day, 1968, 1973; Carlson et al., 1971; Wilson, 1972a; Garrett, 1977) although the extent to which any single trait provides an accurate indication of photorespiration is questionable (Martin et al., 1972; Schrader, 1976). Nevertheless, these differences between genotypes are often highly repeatable, but no consistent effects on net photosynthesis have been observed. In some experiments, selected plants with reduced apparent photorespiration have greater net photosynthesis (Zelitch and Day, 1973). In others, there has been either no association (Carlson et al., 1971; Samish et al., 1972; Wilson, 1975b), a positive correlation (Wilhelm and Nelson, 1978), or all possible combinations depending on which photorespiration-related trait was measured (Martin et al., 1972). In view of the lack of a simple and wholly satisfactory method of directly measuring photorespiration, and of the present uncertainties surrounding it, there seems little point in breeders attempting to breed for low apparent photorespiration at present.

Dark Respiration

The development of the hypothesis that dark respiration can be compartmentalized into a "biosynthetic" and a "maintenance" component (McCree, 1970, 1974; Thornley, 1970, 1971; Penning de Vries, 1972, 1974; McCree and Van Bavel, 1977) has focussed attention on the relative importance of these activities in the developing crop. This in turn has raised the possibility that some plants may be more "efficient" than others and that productivity might be raised by reducing the energy requirement for maintaining the status quo of existing plant material (McCree, 1974; Penning de Vries, 1974). Although the concept may be an oversimplification (Thornley, 1977), it does prove a framework within which crop respiratory losses can be considered. In early stages of crop growth much of the energy expenditure is concerned with cell division and expansion. However, as the crop develops and live weight and L increase, so does the proportion of fully developed, degradable, plant structure requiring "maintenance" (Fig. 7.8), and the maintenance respiration becomes proportionately greater than the biosynthetic component. Loss of carbohydrate by respiration plays a major role in determining seasonal changes in the efficiency with which crops store solar energy. In cotton, for example, some 30-40 percent of

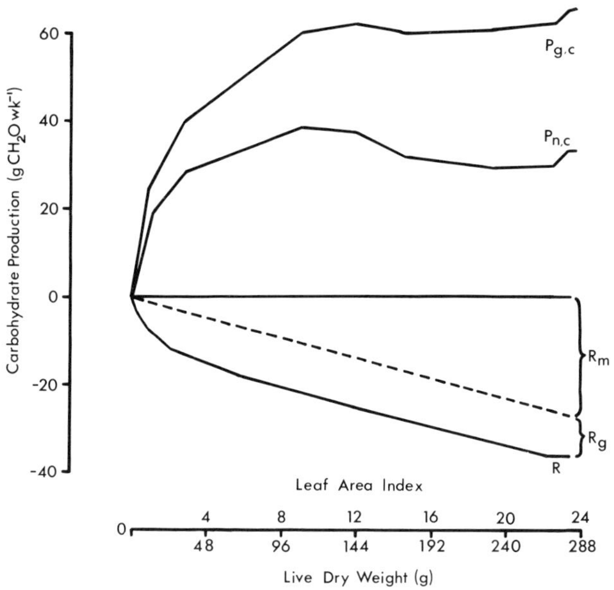

Fig. 7.8. The relationship between gross photosynthesis ($P_{g.c.}$), net photosynthesis ($P_{n.c.}$) and respiration (R) on the one hand, and leaf area index and dry weight on the other, determined experimentally in young ryegrass swards assuming that respiration consists of a maintenance component (R_m), and a growth component (R_g). (Robson, 1973.)

the photosynthate transported to the growing organ can be lost through respiration (Baker et al., 1972). Carbon budget studies of a developing barley crop in England have shown that the total amount of assimilate "lost" by maintenance respiration increased from about 7 percent in May to nearly 65 percent at the stage of grain filling in July (Biscoe et al., 1975). In the undefoliated grass sward at high L, maintenance respiration may reach up to two and a half times the biosynthetic component (Robson, 1973).

McCree (1974) has calculated maintenance coefficients (m) by taking the efflux of CO_2 after 48 h in the dark and plotting this against plant dry weight to obtain the proportionality constant. Much of the variation in grain yielding capacity among five grain sorghum cultivars growing at high L has been attributed to variation in m (Fernandez, 1977).

Some studies have demonstrated a negative relationship between dark respiration rate of fully grown tissues and growth of different cereal species and cultivars (Scheibe and Meyer zu Drewer, 1959), contrasting corn inbreds (Heichel, 1971a), ryegrass genotypes (Wilson, 1975b), and barley cultivars (Kolawole, 1978). Genetic variation in rate of dark respiration of fully developed leaves or roots might partly reflect variation in losses of CO_2 not associated with synthesis of new tissue. If so, then any inherent variability in this trait may have little effect in the early stages of crop development but should influence dry matter production at later stages when there is a large standing biomass. In ryegrass, selection for slow and for rapid dark respiration of mature leaves gave a significant response to selection in both directions (h^2 ca 0.45). Subsequent growth studies revealed that selection had not affected growth of young plants, but once L exceeded 5 the slow respiration families developed a greater L_{max} and plant dry weight than the fast respiration group (Wilson, 1975b; 1976). In subsequent cutting and regrowth of mature plots, there was a consistently greater crop growth rate in the slow than fast respiration families (Fig. 7.9). In the field in Wales, this procedure has increased annual dry matter production by about 7 percent (Wilson, 1977). In this species the rate of dark respiration

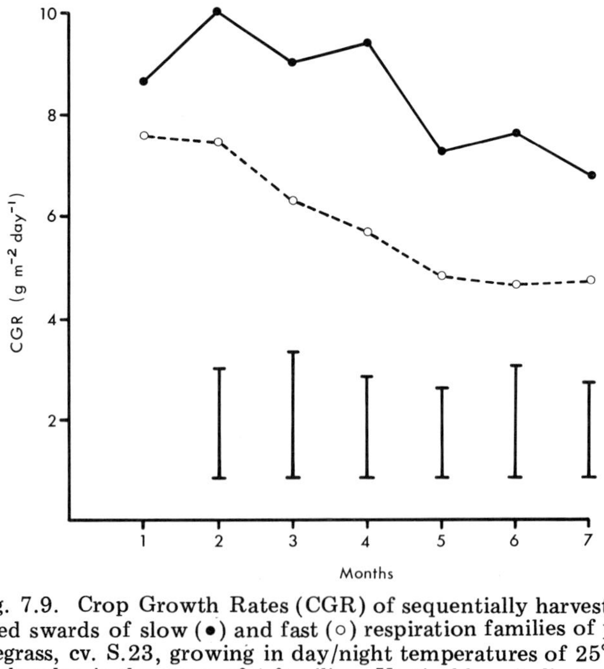

Fig. 7.9. Crop Growth Rates (CGR) of sequentially harvested simulated swards of slow (•) and fast (○) respiration families of perennial ryegrass, cv. S.23, growing in day/night temperatures of 25°C/20°C. Each value is the mean of 4 families. Vertical bars indicate 5% LSD. (Data from Wilson, 1976.)

does not appear to be correlated with leaf morphology or anatomy so that it may be possible to combine slow leaf respiration with any required canopy structure. The advantage of a reduced respiration is likely to be greater at high than low temperatures (Wilson, 1972a; Ryle et al., 1976).

Although the possibility of genetic variation in the biosynthetic component of respiration has been considered slight (Penning de Vries, 1974), genetic differences in root growth respiration have been reported (Lambers and Steingrover, 1978), and seedling respiration has been positively related to seedling growth (Whalley and McKell, 1967; Abernethy et al., 1977).

On a more biochemical level, the so-called mitochondrial efficiency, the ratio of adenosine diphosphate (ADP) to oxygen (ADP:O ratio) has been suggested as an indicator of dry-matter yield potential (McDaniel, 1969, 1972; Schneiter et al., 1974; Doney et al., 1975a). However, experimenters have found it difficult to predict sugar beet yield on the basis of ADP:O ratios (Doney et al., 1975b).

ECONOMIC YIELD

The capacity for high biological yield sets the framework for manipulating plant structure so that the economically useful part forms as large a proportion of the plant as is consistent with vigor and health. The importance of harvest index in cereals, its relationship with grain yield, its independence from biological yield, and its use as a crude selection criterion have recently been discussed by Rosielle and Frey (1975) and Donald and Hamblin (1976). In these crops, attempts to increase yield by altering the relative contribution of the different components of grain yield (ears per plant, grains per ear, grain size) have usually proved unsuccessful because increase in one component tends to be accompanied by reduction in another (Adams, 1967). In contrast, increases in the proportion of grain to vegetative parts which have usually accompanied yield increase are common to most cereals (Van Dobben, 1962; Syme, 1970; Donald and Hamblin, 1976; Lawes, 1977; Mohamed et al., 1978).

The particular plant form which gives a high harvest index in one competitive situation is likely to differ from that which gives high values in another, giving rise to familiar difficulties when extrapolating from the breeding nursery to the field (Donald and Hamblin, 1976). However, with adequate information about the relationships between the various morphological traits which are likely to influence the harvest index and economic yield in the field, detection of appropriate parents when growing as spaced individuals in the field should be possible. Variations in such traits seldom influence only the harvest index. Indeed the prime reason for their choice may be concerned with some other effect, and the associated improvement in the index may be considered an additional bonus. In temperate cereals, high harvest index is usually associated with early heading and shorter leaves and stems, although the extent of these relationships is affected by soil nitrogen status and species (Donald and Hamblin, 1976).

The potential for yield improvement by breeding for morphological traits to increase harvest index obviously depends on the crop and on the remaining scope for improvement. In crops where the end product is a reproductive organ almost any proportional reduction in leaf or stem might increase the index (Table 7.2), but efficient development and growth of the organ requires the preceding growth of enough foliage to form a complete canopy. Conversely, lack of sufficient storage capacity in the form of metabolically active "sinks" may prevent use of the full photosynthetic potential of a crop (Neales and Incoll, 1968; Evans, 1975). There is little point in breeding for high photosynthetic capacity unless the plant can handle the large quantities of assimilate which may be produced. Either, or both, storage and photosynthetic capacity may limit yield, emphasizing the necessity for parallel development of the two (Evans, 1975; Austin and Jones, 1976).

In tuberous crops such as potato (*Solanum tuberosum*), where harvest indexes already exceed 0.8 (Watson, 1971) it may not be possible to increase the index much more. In some short-strawed cereals with index values in the range 0.5-0.6 (Bingham, 1971; Lawes, 1977), further increases may concern the extent to which almost complete reliance for grain filling can be placed on current photosynthesis. Assimilate stored in the stem can be valuable in seasons of poor growth and during periods of stress. Although the contribution of reserves accumulated prior to anthesis may not be large in most crops in good growing conditions (5-10 percent in unstressed wheat [Evans et al., 1975], 20 percent in maize [Hume and Campbell, 1972]), it is highly variable with season. For example, in fields of 'Proctor' barley in England, the stem contribution to increase in grain weight between anthesis and harvest ranged from 2-74 percent in three contrasting years although yield of grain was consistent (Gallaher et al., 1975). Although the extent of genetic variation in this characteristic, or its relationship with any morphological

Table 7.2. The contribution to biological yield (tops only) of various plant parts and the correlation between harvest index and those plant parts in genotypes of three cereals[a]. (Donald and Hamblin, 1976.)

	Winter Wheat		Spring Barley		Oats	
	Biol. yield	Corr. of wt. of plant part (gm^{-2}) with H.I.	Biol. yield	Corr. of wt. of plant part (gm^{-2}) with H.I.	Biol. yield	Corr. of wt. of plant part (gm^{-2}) with H.I.
	%		%		%	
Leaves	9	-0.18	6	-0.72*	7	-0.47
Stems	33	-0.36*	28	-0.83*	34	-0.57
Veg. tillers	8	-0.29	5	-0.35	3	-0.27
Chaff	10	-0.18	10	-0.46	15	-0.58
Grain	40	0.62**	51	0.66*	41	0.50
Height (cm)		0.60**		-0.38		0.96**

[a]Fifteen centimeter row spacing.

traits, is not known, variation between wheat genotypes in ability to relocate assimilate to grains has been reported (Austin et al., 1977). In unstressed wheat cultivars, the lag period between cessation of height growth and onset of rapid grain growth appears to be important in building up reserves, and it has been shown that cultivars with a more pronounced lag set many grains and reach high grain yields (Rawson and Evans, 1971).

Tillering in cereals needs to be considered in relation to plant survival and compensation for effects of variations in numbers of plants per unit area (Bingham, 1971; Austin and Jones, 1976). Although highest harvest index may be achieved with fewest tillers (Donald, 1968), in practice the need for sufficient tillering to ensure plant survival dictates a compromise.

In crops such as sugarcane, sugar beet, and potatoes, the breeder is most concerned with maximizing storage and retention of the products of photosynthesis in vegetative organs. The main limitation to yield, particularly in high-yielding cultivars, may lie in the storage capacity of the plant. In sugar crops, where the storage organ is a container for sucrose, yield might be increased by either the size of the container, the concentration of sugar within it, or both (Milford, 1973; Oworu et al., 1977). The ideal beet might be one in which cell multiplication dominates over cell expansion for a longer developmental period to produce a large root of many small cells (Milford, 1973).

In sugarcane, selection indexes comprising measurements of stalk length, diameter, and number have been proposed as a simple convenient procedure which is 90 percent as successful as direct measurement of total cane (Miller et al., 1978). Another approach is to adjust the leaf blade:joint weight ratio. Clones with a high sugar content but low leaf blade:joint weight ratio would not only improve harvest index but would also allow more canes to survive per unit area of ground because of reduced interplant competition for light (MacColl, 1977).

Breeding for increase in the shoot:root ratio may also offer scope for improving the harvest index and for increasing total dry matter production by increasing the size of the assimilatory system. Under good growing conditions, high shoot:root ratio is often associated with rapid growth (Troughton, 1965; Taylor et al., 1971). Many root characteristics, including shoot:root ratio, show at least as high heritabilities as shoot morphological traits (Troughton and Whittington, 1969). Any advantage to shoot dry-matter yield which may accrue from an increase in the ratio will of course depend on the ability of the reduced root system to continue to supply adequate nutrients and water to the shoot, and in practice this may be the most important consideration for the breeder. However, this ability may depend as much, or more, on root morphology, anatomy, and distribution as on the weight of roots (Troughton and Whittington, 1969).

In forage crops, opportunities for increasing economic yield by structural reorganization of the shoot system, rather than increase in biomass, occur at two levels: (1) the relative proportions of the different plant organs (leaves, leaf sheaths, stems), and (2) the anatomy and physical structure of these organs. Changes in these can influence both the proportion of the plant which can be digested by the animal (digestibility) and the rate of ingestion (voluntary intake). In these crops, breeding for increased dry matter production may often be insufficiently sensitive, by itself, to increase animal production. For example, in recent trials at Aberystwyth in which growth of steers grazing on two closely related perennial ryegrass cultivars were compared, liveweight gains over a 5-month period were 14 percent greater in animals fed on the cultivar which produced 4 percent less dry matter (Evans et al., 1978). Measurement of digestibility *in vitro* has now become routine in many programs of forage crop breeding and evaluation (Jones, 1975). However, differences in whole plant digestibility between or within species can be a consequence of differences in the relative proportion of plant parts rather than any which are inherent to the anatomy or morphology of the plant. Selection for digestibility of whole plant samples may therefore result in change in agronomic type. In many species, such as ryegrass, leaves tend to be more digestible than tillers so that a leafier plant would result; but in others, such as orchardgrass (*Dactylis glomerata*), stems may be more digestible at certain times so that an increase in proportion of stem tissue would follow (Terry and Tilley, 1964; Mansat, 1972; Walters and Evans, 1974). The corollary is that the effect of selecting for leaf:stem ratio on digestibility depends on the crop. In alfalfa (*Medicago sativa*), leaves may be 50 percent more digestible than stems at maturity (Terry and Tilley, 1964). Furthermore, differences in digestibility dependent on differences in tillering of spaced plants may largely disappear in the sward environment (Walters and Evans, 1974). A more satisfactory procedure, which avoids these difficulties, is to select for digestibility or some major component of it, or for specific plant organs, leaves, or stems. In this way whole plant digestibility can be improved with minimum change in the proportions of plant parts (Selim, 1976).

Quite large differences in animal production can occur between forages of the same digestibility as a result of differences in the level of cell wall components, the physical arrangement of fibers, and external morphological features (Ulyatt, 1970; Walters, 1971; Laredo and Minson, 1973; Jones, 1975). Hanna and his coworkers (1974) have drawn attention to the more rapid digestion (and rate of water loss) of bloomless as against bloomed sorghum, and pubescent as against glaborous pearl millet (*Pennisetum glaucum*). Both traits are controlled by single genes. In temperate grasses, cellulose is particularly important (Jones, 1975), and intracultivar selection in high digestibility ryegrass has shown that small (7 percent) but consistent improvements in animal liveweight gain, attributable to intake differences, can be achieved after one generation of selection for low-leaf cellulose (Wilson, 1965; Lancashire and Ulyatt, 1975). In species with initially low digestibility, this may be improved by selection for low cellulose alone (Selim, 1976). Physical traits which influence rate of fiber disintegration play a major role in determining the rate of passage of feed through the rumen, and intake differences have been ascribed to energy consumption in milling (Chenost, 1966; Laredo and Minson, 1973; Jones et al., 1974), ease of maceration under laboratory conditions (Troelson and Bigsby, 1964), and leaf tensile strength (Evans, 1964), all of which can be used in breeding. Genotypic differences in rate of breakdown of fresh leaf tissues subjected to attack by ruminal microorganisms (Monson et al., 1972) and by cellulolytic enzymes (Selim et al., 1975) also indicate potential for improving voluntary intake and hence animal production and economic yield.

CLIMATIC CONSTRAINTS

Breeding for consistency and uniformity of yield is often more important than a high potential. Most breeding programs take into account various unidentified stresses occurring during the normal course of field assessment (Breese, 1969). However, if known climatic constraints are likely to occur during growth of a crop then faster progress in breeding may be made by making use of specific adaptive traits in selection. These constraints range from those which might normally limit the seasonal or geographical range over which a crop or cultivar can be grown to the occasional occurrence of climatic extremes causing severe stresses, and the common transient difficulties such as reduced water supply which most crops encounter at some time. Whether it is worthwhile breeding specifically for ability to withstand or avoid these problems will of course depend on their frequency and severity, on the existence of some economic means of agronomic amelioration, and the possibilities of growing a satisfactory alternative crop.

Day Length

Response to day length is one of the most important controlling mechanisms of reproductive behavior, and therefore yield, in many crops. In those which are day-length sensitive, it determines times of flowering and seed production. The existence of genetic variation in response within species allows breeders to use it to ensure that the reproductive or vegetative phase of growth coincides with the most favorable environmental conditions. These responses, which are controlled by many genes, are often modified by temperature and many long-day perennial plants from high latitude may have a vernalization requirement for cold and/or short days before they will respond to long days (Cooper, 1961, 1963). Indeed, breeding for degree of this requirement can also be used to manipulate flowering behavior.

Delayed flowering can be of value in crops such as sugar beet, tobacco (*Nicotiana tabacum*), or grazed pasture, where maximum vegetative development is required. Breeding for particular day-length responses is a useful means of stress avoidance (below), and in temperate-zone cereals a requirement for long days can be used to delay floral initiation until risk of frost injury is past (Kirby, 1969). Delayed initiation may also improve grain yield since rapid initiation tends to promote the terminal spikelet at the expense of lateral ones (Rawson, 1970).

In some crops, indifference to day length is a major objective, allowing normally short-day species such as maize, sorghum, and soybean to be cultivated at higher latitudes (Evans, 1973). Although maize, for example, is widely adapted, photoperiodic sensitivity is a critical factor limiting the range of adaptation of many individual genotypes (Francis, 1970). Genotypes most suited to short growing seasons must be insensitive to photoperiod for places in the temperate zone and must have a very strong tendency to flower (Aitken, 1977). However, there is variation in maize for both these traits (Francis et al., 1970).

Water Shortage

Tolerance and/or avoidance of drought, and the range and relative importance of traits which are likely to be involved in the exploration for and utilization of water have been much discussed (Levitt, 1972; Moss et al., 1974; Boyer and McPherson, 1975; Jones,

1976, 1977; Passioura, 1976). However, definitive evidence of the direct effects of breeding or selecting for such traits is sparse, and understanding of the mechanisms of stress tolerance is far from complete. Even so, it is possible to propose traits that are likely to be worthwhile breeding for in particular environments. In some, it may be more appropriate to aim primarily for drought avoidance, in others drought tolerance, or even some combination of the two. The requirements for regions with a definite dry season differ from those with short and irregular periods of deficit where drought avoidance mechanisms that maintain high leaf-water potentials and photosynthesis may be all that is required. In most crops, productivity is usually the first consideration so that water-use efficiency may be more important than simple survival. In general the traits which influence these aspects of the water relations of crops do so through their effects on (1) the timing of crop development; (2) the efficiency of the root system in extracting and transporting water; (3) the effectiveness of the shoot system in controlling transpiration, and the relationship between transpiration and photosynthesis; and (4) the ability of plants to endure stress.

Timing of Crop Development

In determinate crops growing in climates with a highly predictable dry season the most useful strategy may well be rapid development and early maturation. For example, the use of early maturing cotton cultivars in Texas has led to 20 percent less use of water but greater yield than conventional cultivars growing in the same region (Namken et al., 1974). In a study of factors affecting drought tolerance among spring wheats in Australia, 40-90 percent of the observed variation in tolerance could be accounted for by earliness of maturation date (Derera et al., 1969).

Roots

The characteristics of the root system that might be of use in breeding for ability to avoid physiological drought depend on whether supplies of moisture are likely to be available at greater depths. Where reserves exist at depth, ability to produce a deep extensive rooting system which responds positively to a declining water table is advantageous (Troughton and Whittington, 1969; Hurd, 1974). Although efficiency of extraction of water seems to be less at depth, the rate of uptake of deep roots is reportedly greater than shallow ones (Willat and Taylor, 1978). Where moisture reserves are confined to the upper layer of soil it has been shown in practice that rooting depth is not important; but the number of nodal roots may be associated with yield, lodging, and tiller survival in wheat (Derera et al., 1969). Where genetically uniform grain crops are growing mainly on stored water it may be an advantage to have small root systems with high hydraulic resistance, so that water can be conserved during early growth to leave a sufficiency for later grain filling. Breeding for small xylem vessels in the seminal roots has been suggested as a means of increasing the resistance (Passioura, 1972). Similarly, a uniform forage crop with small root relative to shoot growth, which is likely to produce plants with a high yield potential, might be able to withstand short periods of water deficit better than one with more root, although more roots would be suitable if any root competition were likely (Troughton, 1974).

In practice the most useful root system over a range of environments seems likely to be one in which the ability to produce deep roots is combined with an efficient extraction system in the surface layers, where most nutrients are concentrated, but which does not invest a high proportion of the total weight of the plant in roots. In addition, the ability to readily produce new root axes, which has shown genetic variation in ryegrass (Troughton, 1978), could be important in recovery from drought.

Control of Transpiration

The rate of soil water depletion, and therefore the time of onset and rate of intensification of drought, is determined by the rate of evaporation from the crop. Consequently, traits which affect evapotranspiration are likely to be important in breeding programs. The appropriate overall strategy of control of water use will be affected by the genetic uniformity of the crop. In pure stands likely to encounter periods of water shortage, a conservation strategy would be suitable provided photosynthesis was not impaired in normal conditions. In mixed stands, rapid water use to grow as fast as possible while water is available, followed by measures to severely restrict further loss (stomatal closure, leaf rolling) may be advantageous for individual plants (Passioura, 1976). Any trait which reduces crop transpiration or increases photosynthesis will increase water-use efficiency (WUE). Over a more extended period, reduced night respiration might also be expected to increase WUE.

In grain crops, increasing harvest index, for example, by breeding for restricted tillering could reduce wasteful use of water in dry conditions (Hurd, 1968; Blum, 1973; Jones and Kirby, 1977). In crops where the leaves are the harvested portion, control of transpiration at the leaf surface may be the only feasible approach to drought control by the shoot system. Such control may be stomatal or cuticular, and species and cultivars show much variation in stomatal frequency, length, and behavior (Dobrenz et al., 1969; Heichel, 1971b; Miskin et al., 1972; Wilson, 1972b, 1975a; Ciha and Brun, 1975; Sheriff and Kaye, 1977) and in morphological and anatomical traits affecting stomatal and cuticular transpiration (Hall et al., 1965; Wilson, 1975c; Ebercorn et al., 1977; Wilson et al., 1977). However, the value of these in field environments is not always clear.

If cuticular resistance can be increased then this would reduce transpiration without interfering with photosynthesis (Moreshet, 1970; Hanna et al., 1974). In this respect, the recent development of a rapid colorimetric method for measuring epicuticular wax could prove a useful technique for breeders (Ebercorn et al., 1977). Specific adaptations such as narrow, erect, shiny leaves, sunken stomata, or even awns in barley (Johnson et al., 1975) may well be worthwhile traits where water is certain to be limited but, for many crops, stomatal control is likely to be the most efficient and flexible means of influencing the water economy (Jones, 1976). This control can be exercised by the size, distribution and siting of the stomata, the morphology and physiology of the leaf, and the adaptive responses of all these to water stress.

The effect of stomatal frequency on transpiration and photosynthesis depends on stomatal dimensions. Frequency is often negatively correlated with length, but genetic variation in frequency may (Miskin et al., 1972; Wilson, 1972b; Yoshida, 1978; Kolawole, 1978) or may not (Jones, 1977; Kolawole, 1978) be associated with variation in transpiration depending on correlations with other dimensions. In ryegrass, selection for relatively high stomatal resistance to water loss, based mainly on low stomatal frequency, can reduce potential transpiration and improve WUE of plants in pots (Wilson, 1975a) and in field plots during the establishment phase (Fig. 7.10). In this species, as in barley (Miskin et al., 1972) and *Dactylis* (Wilson et al., 1977), quite large variations can occur at the same rate of photosynthesis. In the ryegrass experiments, water loss from stressed plants was most affected by stomatal aperture which was influenced by leaf surface morphology (Wilson, 1975c). Selecting for a flatter leaf surface led to more responsive stomata and a "water-saver" type of plant which controlled soil moisture better and gave greater yields in pure stands in the field than the original cultivar (Fig. 7.10). Heritable leaf surface variations can also account for differences in water-use efficiency in pot-grown *Dactylis* selections (Wilson et al., 1977).

Selection for these anatomical and morphological traits can often be conducted on seedling populations (Wilson, 1973;

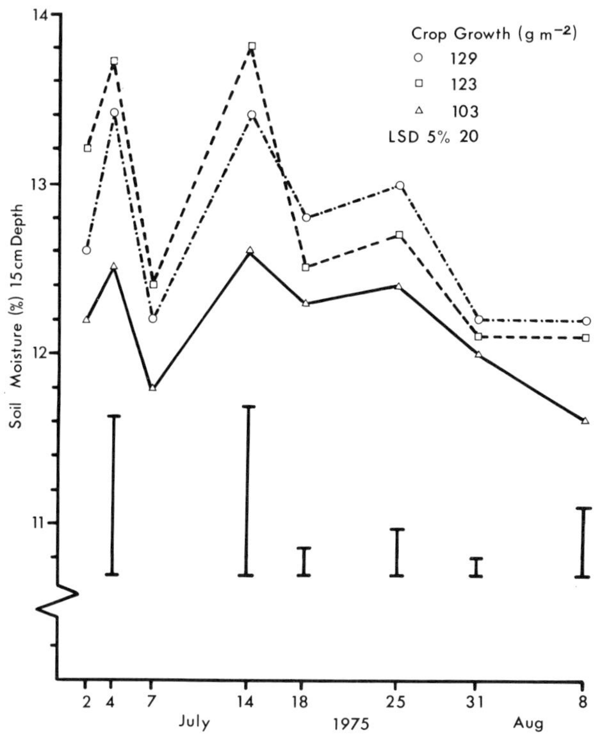

Fig. 7.10. Growth and soil moisture use of plots of perennial ryegrass experimental populations selected for either low theoretical stomatal conductance (○–·–○) or flatter leaf surface (□---□), compared with the original cultivar (△——△). (Data from Wilson, 1976.)

Kolawole, 1978). With advances in instrumentation it should also now be possible to base selection on direct measurement of transpiration under controlled environmental conditions, so that differences in physiologically based stomatal response and perhaps more importantly, stomatal adaptability, might be detected. Genetic reduction in potential transpiration may not always be beneficial because of the high temperatures commonly associated with drought and the possibility of overheating as a result of reduced evaporative cooling.

Stress Tolerance

Mechanisms affecting tolerance, as against avoidance, of water stress are even less clear (Hsiao and Acevedo, 1974) so that it is difficult to propose specific criteria, although techniques using osmotic solutions can be used to measure seedling response to stress directly (Johnson and Asay, 1978). The ability to accumulate free proline, a characteristic of stressed plants, has been suggested as an indication of tolerance, but its usefulness in breeding is uncertain (Hanson et al., 1977). More elastic cell walls may aid in the maintenance of leaf turgidity when leaf water potential is declining (Weatherly, 1965), and differences in drought tolerance between sorghum genotypes have been explained on this basis (Blum, 1974), although those results have been questioned (Jones and Turner, 1978). In ryegrass, selecting for reduced leaf tensile strength greatly improved (> 100 percent) yields in dry field conditions compared with the parent cultivar (Lancashire et al., 1977). Similar tall fescue selections have shown the ability to continue producing leaves of high specific leaf area under stress conditions in which those of the parent cultivar had successively lower values (Silcock, 1978). However, the extent to which these phenomena are related to drought tolerance or tolerance of some other stress such as high temperature is not known.

Temperature

For most C_3 species, the temperature optimum for growth is around 20-25°C, whereas that for most C_4 species lies between 30-35°C (Cooper and Tainton, 1968), although the exact temperature is subject to phenotypic adaptation (Charles-Edwards et al., 1971). However, there is considerable variation within each of these groups and species which occur over different climatic regions usually comprise a number of ecological races genetically adapted to the local environment (Turesson, 1922; Mooney and Billings, 1961). Temperature adaptation provides the means of extending the latitudinal or altitudinal limits or the seasonal growth patterns of important crop species. In the temperate maritime climates of much of New Zealand and Britain, for example, hybridization of summer active North European populations of forage legumes and grasses with winter active Spanish and Portuguese types has produced cultivars with broader temperature optima than either parent and which combine both good early spring growth and good main season activity (Barclay, 1970; Borrill et al., 1973). In these species, the rate of seedling leaf area expansion can be used to measure temperature adaptation, and improvement in spring growth can even be achieved by selection for rapid expression from within outbreeding grasses growing in low temperatures and short days in controlled environments. However, such material often has reduced cold hardiness so that the potential advantage of early spring growth may only be realized in regions not subject to severe cold (Cooper, 1974). This negative relationship between growth at low temperature and cold hardiness represents a major unresolved challenge to plant scientists.

For breeding purposes, identification of plants tolerant to temperature extremes generally has to be made on the basis of their growth reactions, either in the field or in a controlled environment (Breese and Foster, 1971). No physiological or morphological traits have been shown to exhibit consistent correlated responses with heat or cold tolerance. However, it is possible that selection for reduced dark respiration might improve heat tolerance (Biscoe and Gallaher, 1977), and provided soil moisture supplies are adequate, rapid transpiration might be useful in hot climates (above). Although the relationship between reduced leaf strength and cold tolerance has been found in two forage grass species only (above), it may well be more widespread. In grain crops the effect of high temperature on floret formation and grain set is particularly important, and morphologically similar wheat cultivars have been shown to differ in the way that different characters respond to heat stress so that it has been suggested that these differences might be used as a basis for fitting cultivars to different temperature zones (Bagga and Rawson, 1977).

CONCLUSIONS

Knowledge of the direct effects of breeding for specific physiological or morphological traits is still fragmentary, and much of the available information has been obtained by inference from experiments or breeding programs designed to answer more general questions. It is often difficult to make valid conclusions about the effect of differences in a single trait on yield of cultivars which may also differ in many others. Any difference in duration of growth, for example, may totally obscure the effects of other traits. The best evidence comes from those experiments in which the populations compared have a common genetic background and only exhibit contrasting expression of the trait in question, apart from close linkages (Jennings and Herrera, 1968; Wilson and Cooper, 1970; Rhodes, 1973).

The stage at which a trait can most suitably be introduced into a breeding program will partly be determined by its ease of measurement and heritability. In outbreeding crops, selection from within existing high yielding cultivars has already led to significant improvements in biological yield, but it may well be that earlier introduction into the program would be beneficial, particularly where a radically new model is envisaged. Wallace and his colleagues (1972) have discussed operational considerations in breeding for physiological components of yield and suggest that for complex characters of low h^2 selection in self-fertile crops should not begin until the F_3 so that maximum response can be achieved.

In the long-term it should be possible to provide highly complex blueprints for crops, but the use of these in practice will likely be to supply a number of key traits for application in standard breeding programs, possibly in the seedling stage. In the foreseeable future it seems particularly important to (1) define the optimum canopy structure for each crop and determine the most appropriate management to maximize its effect on yield, (2) define the strategies and forms of genetic adaptation to environmental variables and devise methods of easily assessing genetic variation in phenotypic adaptability, (3) determine the optimum harvest index of major crops, (4) resolve the puzzle of photorespiration, (5) examine the wider application of breeding for reduced unnecessary dark respiratory losses, and the effects on temperature and drought tolerance. In certain conditions the most serious limitation to crop production may be some highly specific stress, for which usable genetic variation in tolerance exists. These include water logging (Austin and Jones, 1976), herbicides (Faulkner, 1974), and soil mineral stress (Wright, 1977).

It will become increasingly important to aim for efficiency of use of scarce resources, and it may well be that in many regions we should now be concerned with productivity per unit of water or phosphorus, rather than unit of land. It has been estimated, for example, that at the present rate of increase in use, workable deposits of phosphorus, a nonrenewable resource, will run out in 100 years (Evans, 1975). Efficiency of recovery and use of nitrogen, which has not been specifically considered here, becomes a more important consideration in nonlegumes as that element becomes increasingly expensive. Genetic variation in use of nitrogen has been known for some time (Vose, 1963; Vose and Breese, 1964). However, because high yields are possible only under high levels of nitrogen supply, characteristics that confer high-yielding ability are often concerned with response to nitrogen. In legumes there are encouraging signs from current research that nitrogen-fixation might in the future be improved by breeding of both host plant and the symbiont partner (Hollaender, 1977; Mytton et al., 1977).

Whatever the emphasis in crops or crop production, increasing knowledge of the physiological basis of yield and adaptation will continue to improve our ability to more accurately identify the most suitable genetic material to be parents of the cultivars of the future.

Discussion

J. W. HANOVER
R. W. F. HARDY
D. C. RASMUSSON

1. R. W. F. HARDY. The objectives of plant breeding and agrichemical research are common—to increase crop productivity, improve quality, and facilitate production processes. In the agrichemical area, we believe that improved plant protectants which now represent an $8 billion world market will continue to increase in importance. Some of us visualize that plant growth regulants, which are currently a trivial $0.1 billion world market, will become a most important generation of agrichemicals possibly equivalent to plant protectant chemicals. Certainly, a fundamentally based approach will be critical for the successful development of major plant growth regulators, and several steps are common for plant growth regulator and plant breeding research—identification of the limitations or what's wrong with the natural system; assessment of the quantitative significance of what's wrong; and development of simple, specific, high capacity screening systems to select solutions. I will give several examples of the identification of what's wrong and assessments of their significance for carbon and nitrogen input.

Carbon Input: The CO_2-Fixing Enzyme and Photorespiration

The rapid expansion of our fundamental understanding of ribulose 1, 5-bisphosphate carboxylase has enabled the tabulation of what's wrong and the opportunities for improvement of the CO_2-fixing enzyme. The "what's wrong" list includes 65 limitations at the biochemical, genetic, physiological, and agronomic levels. The most significant limitation so far recognized is the promiscuity of carboxylase for O_2 as well as CO_2 coupled with the low and high affinities of the enzyme for CO_2 and O_2 relative to their respective ambient concentrations. And the relative affinities get worse with increasing temperature. These characteristics are the primary cause of photorespiration. Alteration of the CO_2:O_2 ratio around the crop canopy has been used to assess the quantitative significance of decreasing photorespiration in field-grown legumes. Yield increases are 50-100 percent for peanuts, peas, beans, and soybeans under conditions where photorespiration is substantially decreased. There is no known essential function for photorespiration, and there is substantial information that argues against any essential function for it. I am optimistic that within a reasonable time chemical effectors and/or genetic modifications will correct the photorespiration error.

Carbon Input: Photosynthesis Regulation and Sugar Translocation

What regulates the activity of the CO_2-fixing enzyme and related processes of starch synthesis and translocation of sugar from sources to sinks? Molecular understanding is developing. For example, vectorial ATPase producing a proton gradient provides the driving force for export of sucrose. Such information may provide the insight to deregulate the photosynthetic system and optimize export.

Carbon Input: Assimilate Partitioning

The surprising observation that the partitioning of assimilates between vegetative and reproductive tissue is regulated by an O_2-process provides the first fundamental probe of the regulation of reproductive sink activity. Subambient O_2 decreases reproductive growth and supraambient O_2 increases the translocation of sugar to reproductive sinks. Will chemists or geneticists be able to exploit this system to maximize harvest index of forage and grain crops?

Nitrogen Input

The nitrogen input for cereal grains and legumes over the next 20 years may be provided by a variety of systems, but only systems that are compatible with increasing yields should be considered. The primary driving force in crop production is increasing yield, and the secondary one is the understanding of biological N_2 fixation. This phenomenon has 86 "what's wrongs" recognized so far at the mathematical, chemical, biochemical, genetic, physiological, and agronomic levels. The major limitation is the excessive requirement of energy by nitrogenase—25 ATPs/N_2 molecule fixed—or a minimum of 10 gm of carbohydrate/gm N_2 fixed in recent measurements comparing nodulated and nonnodulated isolines of soybeans. Increased provision of photosynthesis to the N_2-fixing nodule by CO_2 enrichment of the aerial environment of field-grown legumes markedly increases N_2 fixation. There are several approaches that may enhance biological N_2 fixation, especially in legumes, but the energy problem must be solved before extension of biological N_2 fixation to cereals should be considered.

In my judgement, our rapidly expanding fundamental understanding of key processes such as carbon input and nitrogen input is providing the basis for a more efficacious fundamental approach to develop agrichemical and genetic solutions for increased crop productivity.

2. J. W. HANOVER. The principal considerations in analyzing physiological and morphological traits for yield improvement in agronomic crops also apply to tree crops. In fact, the application of physiological methods to breeding trees is one means for increasing the efficiency of tree breeding. Early detection of superior parents to use in a tree breeding program is vital, but elite tree selection is usually ineffective except for highly heritable traits. Progeny testing is necessary, and it can be accelerated only by using techniques that induce early flowering to shorten generation intervals. For some traits, indirect selection can be applied by utilizing the genetic correlation between some physiological process and the trait under selection. As an example, blue foliage color is a trait for which selection is practiced in the blue spruce trees. The blue spruce tree may have either blue or green leaves, and this color trait has a chemical basis and a physiological function. With blue foliage, epidermal waxes cover the leaf surface and occlude the stomata, whereas with green leaves, the waxes are in the stomata and not on the leaf surface. The function of leaf waxes is to prevent solarization and heat buildup in the leaves. There is no differential effect on photosynthesis or transpiration between the wax situations of blue and green leaves, but selection for foliage color gives simultaneous improvement in growth traits. As a second example, growth analyses which tell the rates of leaf area development are good indicators of growth potential of deciduous trees. Leaf photosynthetic rate is not a good indicator of growth potential in screening genotypes, but high leaf growth rate and high photosynthetic rate are both used as selection criteria.

The greatest use that tree breeders make of physiologic processes in the tree breeding programs is in the manipulation of environmental factors that influence the plants' physiology in such a way that the growth cycle is expedited. As an example, when breeding blue spruce and aspen, systematic seed selections are made from natural populations for use in a progeny testing program. Progeny screening is done first in greenhouses, where the use of long photoperiods converts the phytochrome system to its far red form. This and growth control procedures, such as carbon dioxide enrichment, permit early evaluation of some progeny traits such as foliage color, and they stimulate early flowering in many species. Tree seedlings are then transplanted to gardens for further evaluation and use in hybridization. Early flower induction allows breeding to be done rapidly. Using these procedures, it takes only five years to obtain interspecies hybrids, and to produce F_2 and backcross progenies for testing and advanced generation breeding. Even more rapid progress is possible with some deciduous trees such as aspen.

Some species, such as poplars can be induced to grow to 8 m in height in three years. Others, such as white birch and certain conifers, can be induced to flower at one year of age by applying the techniques of physiological genetics and growth analyses intensively.

3. D. C. RASMUSSON. There are four items that I would like to emphasize: (1) placing greater emphasis on breeding for yield potential, (2) the use of morphological traits to improve yield, (3) the question of who should be involved in breeding for greater yield potential, and (4) a research strategy that might be used.

In most plant breeding programs, the selection pressure that is placed on yield potential is modest. Two factors contribute to this—parental lines are selected because they possess disease resistance, desirable seed characteristics, lodging resistance, or shattering resistance; and a genotype selected to be a new cultivar must possess numerous desirable traits. In this type of combination breeding, the likelihood of sizable gain in yield potential is low. We ought to find ways to put greater selection pressure on yield potential. One way is breeding for the optimum expression of morphological plant characteristics. There is direct and indirect evidence that morphological traits can influence yield and many of them can be selected visually without recourse to a replicated field test.

Traits that should be related to yield are plant size, height, and size of stem and leaf. These have to do with capture, transport, and partitioning of the canopy energy. Duration of grain filling and the morphological components of yield, inflorescence number, seed number, and seed weight determine sink size and strength. Harvest index, biological yield, and tillering are other morphological traits that must be associated with yield.

Who should be involved in attempting to improve yield via use of morphological characters? I suggest that all plant breeders who aspire to enhance yield should research yield related characters. Applied breeders should try to achieve genetic diversity for yield related characters, make crosses, and do selection in imaginative ways. This seems to be reasonable, based on what we know about how yield is achieved. Consider what has happened in wheat, rice, and sorghum.

But the options are oftentimes limited in individual crops because of a lack of knowledge about the role of various traits in determining yield and because genetic variation for traits is not available in suitable parental stocks. Therefore, I suggest that a few breeders for each crop team up with plant physiologists to obtain information and to develop the prerequisite germplasm. I recommend a four-step strategy or procedure. The first step is selection of characters for study. This should be a team judgement, based upon the anticipated impact the trait will have on yield and on the ease of handling the trait in the breeding program. The second step is a search for genetic diversity for the trait. The third step involves incorporation of the genetic diversity for the trait into the adapted genetic background. Often the necessary diversity is found in unimproved stocks; hence there is need for a breeding program which may require more than one cycle of crossing and selection. The fourth step involves evaluation of the worth of the contrasting states of a trait relative to its impact on yield. If it has a positive effect, further experimentation is needed to determine whether the trait interacts with genetic backgrounds and environments. This last phase is demanding and expensive and optimally involves the cooperative effort of a plant physiologist and a breeder.

Some may question doing any breeding work before learning the value of a trait. Yet the real value of a trait in determining yield must always be determined in field trials or in a farmer's field. And gains in yield potential can be evaluated only with good quality germplasm.

4. A. FASSOULAS. If one can effectively select for high yield on a single plant basis, how can morphological or physiological traits be used as selection criteria to further increase yield?

5. D. WILSON. If yielding ability can be measured accurately on spaced plants in the nursery, then the benefits of using specific

productivity related traits may be confined to convenience and to speeding up the selection process in those cases where the trait can be measured rapidly on young plants. However, for many crops, conditions in the breeding nursery, where selection takes place, are different from farm conditions in climate, soil, and plant competetion and this leads to a point where yields in the two situations are not correlated. Consequently, selection of specific traits which affect productivity or adaptability on the farm, but which are measurable in the nursery, makes more accurate identification of superior genotypes possible.

6. I. EDWARDS. To many breeders of cereal crops such as wheat in continental climates with high radiation, a shortage of plant photosynthate per se is of lesser importance than the limitations of sink size, or specifically spikelet number. Is there a possibility to increase the period of spikelet formation in the plant, and thereby, increase spikelet number?

7. D. WILSON. Although the available evidence supports the fact that photosynthesis itself is often not the limiting factor in yield of grain crops, both photosynthetic capacity and sink size need to be taken into account in long-term advances. Certainly a major adjustment between these two factors has taken place with the selection for high harvest index which has led to major increases in grain yield. I know of no intrinsic reason why the period of spikelet formation might not be increased. Whether such a change in a specific component of the harvest index would give further yield advances would depend on its relationships with other components and on whether the optimum index was thereby exceeded.

8. Z. WICKS. Is it reasonable to expect that a mutant may be found in a crop species that would totally eliminate photorespiration? And if so, what methodology would be used to find it?

9. D. WILSON. Data from the Carnegie Institute indicate that single mutations that would eliminate photorespiration are highly unlikely to occur because several processes are involved in photorespiration and each of these is under separate genetic control. Thus, several favorable mutations would need to occur simultaneously in one plant. Furthermore, even when most of the observable traits of the C_4 photosynthetic pathway are combined in hybrids between C_3 and C_4 species, the resultant genotypes are apt to have

photorespiration and not to exhibit the agronomically useful C_4 traits.

10. J. W. HANOVER. To have any hope of success at incorporating nonphotorespiration into a C_3 plant would require the use of a tissue culture system that would permit screening of very large numbers of cells or tissues for this trait.

11. R. W. F. HARDY. Photorespiration arises from errors from the viewpoint of crop productivity in the CO_2-fixing enzyme—ribulose 1, 5-bisphosphate carboxylase. These errors are its substrate promiscuity where it reacts with O_2 as well as CO_2 coupled with its low affinity for CO_2 relative to atmospheric CO_2 concentration. It is a reasonable speculation to hope to produce a mutant in which the above errors have been significantly reduced. In fact, it is reported that a polyploid cultivar of ryegrass may be such an example. Many tests exist—possibly the compensation point would be an appropriate initial test. Attempts to incorporate the CO_2-pumping capability of C_4 plants into C_3 plants is another approach but may be less attractive than direct mutation of the CO_2-fixing enzyme.

12. R. NELSON. Dr. Wilson showed that selection for decrease in respiration caused increased yield in forages, and Dr. Hardy showed that superambient O_2 increased seed yield in soybeans. Does this mean that the desired direction to manipulate respiration rate would depend on whether the economic yield was vegetative or seed? Also, what are the techniques used in screening for respiration rate in breeding populations?

13. R. W. F. HARDY. The oxygen process that has to do with reproductive growth is totally unrelated to mitochondrial based dark respiration. Mitochondrial based dark respiration saturates at 1 percent oxyen, and the effects in our studies occurred between 5 and 21 percent oxygen. This is a unique oxygen process that occurs within reproductive structures of all crops, and it has nothing to do with photorespiration. It is an oxygen process that is unique to assimilate unloading in reproductive structures. Whether it involves a physical or chemical process is unknown.

14. D. WILSON. Selecting for respiration is not an easy technique to use. High errors are associated with the technique, so extreme care and many measurements must be made on a plant. Respiration rates of fully developed tissues were measured, and a standard

respirometer was used to measure oxygen uptake.

15. J. MAC KEY. For its growth, the plants need solar energy, carbon dioxide, water, and nutrients. In attempting to guarantee the right balance between these basic elements, the plant tends to develop some kind of mirror image between its shoot and root. For example, in wheat there exists a positive correlation between plant height and root depth, between number of tillers and of crown roots, as well as between dry weight of shoot and of root prior to grain filling. Morphological changes moderated for photosynthetic consideration such as development of dwarf wheats will thus pleiotropically influence the development of the roots as well. Dwarf wheats have shallow root systems and need irrigation. To what extent do we really understand the correlative growth pattern for different plant parts in our attempts to build proper ideotype concepts? At least in connection with cereal breeding for high grain yield, there has been a tendency to model types from photosynthetic considerations without fully understanding the risks of such a one-sided approach.

16. D. WILSON. Your point is a good one. No ideotype is set up in our program. When a physiological trait is chosen for study, the question is asked, What effect is it going to have on the plant and the crop at different levels of expression, namely, how does it effect shoot:root ratio, water economy, and the like? These questions are answered by breeding experimental populations and looking at the physiology and agronomy of them. It is not until these questions are answered that ideotypes are constructed. With this type of program, high yielding cultivars are used as a starting point; so sometimes even a higher yielding cultivar results if the trait studied happens to be associated with yield.

17. R. L. COOPER. It has been said that to evaluate the worth of a trait, it needs to be put into an isogenic background. There is a limitation to that approach. That is, the genetic background which complements the normal trait may not complement the new trait being researched. For example, a 2x photosynthetic rate incorporated into a genetic background that complements an x rate may not be beneficial. The way to achieve the proper genetic background to complement the new trait requires simultaneous selection for the trait and yield. Then, yield genes will be accumulated to complement the new trait.

18. S. S. CHASE. The greatest single problem facing the tree breeder is generation time. What techniques are available for manipulating flowering induction time in trees? Controlled induction of first season flowering would be a tremendous accomplishment.

19. J. W. HANOVER. Early flower induction in trees involves the use of continuous light and carbon dioxide to accelerate phase change and then a cold treatment which stimulates flowering response in birches and spruces. And there are many hormonal treatments that can be applied in certain tree species to cause early flowering.

20. J. J. MOCK. This morphological-physiological breeding aspect needs to be put into a more practical point of view. Much of improved grain yield in cereals accomplished via breeding and selection has resulted from changing the distribution of photosynthate among plant parts. A plant has only a certain amount of photosynthetic output and it must be distributed to roots, stem, leaves, and grain. Selection to distribute more photosynthate to one organ causes other organs to suffer. For example, with maize, selection for increased grain yield oftentimes causes stalk quality to suffer. So it seems evident that the morphological or physiological characteristics that can increase yield without causing deterioration of other plant parts must be those that increase photosynthate production, such as increased photosynthetic rate per unit area, increased leaf area per plant, or increased duration of active leaf area during critical stages of development. Selecting for one or more of these traits and yield simultaneously probably is the best way for obtaining high yielding agronomically acceptable genotypes.

REFERENCES

Abernethy, R. H., L. Neal Wright, and K. Matsuda. 1977. Association of seedling respiratory metabolism and adenylate energy charge with seed weight of *Panicum antidotale* Retz. Crop Sci. 17: 563-66.

Adams, M. W. 1967. Basis of yield component compensation in crop plants with special reference to the field bean, *Phaseolus vulgaris*. Crop Sci. 7:505-10.

Aitken, Y. 1977. Evaluation of maturity genotype-climate interactions in maize (*Zea mays* L.). Z. Pflanzenzuchtg. 78:216-37.

Alberda, Th. 1971. Potential production of grassland, pp. 151-71. *In* Wareing, P. F., J. P. Cooper (eds.), Potential Crop production. Heinemann,London.

Angus, J. F., R. Jones, and J. H. Wilson. 1972. A comparison of barley cultivars with different leaf inclinations. Aust. J. Agric. Res. 23:945-57.

Ariyanayagam, R. P., C. L. Moore, V. R. Carangal. 1974. Selection for leaf angle in maize and its effect on grain yield and other characters. Crop Sci. 14:551-56.

Asay, K. H., C. J. Nelson, and G. L. Horst. 1974. Genetic variability for net photosynthesis in tall fescue. Crop Sci. 14:571-74.

Aslam, M., and L. A. Hunt. 1978. Photosynthesis and transpiration of the flag leaf in four spring wheat cultivars. Planta 141:23-28.

Austin, R. B., and R. B. Jones. 1976. The physiology of wheat. Camb. Plant Breed. Inst. Rep. for 1975, pp. 20-73.

Austin, R. B., J. A. Edrich, M. A. Ford, and R. D. Blackwell. 1977. The fate of the dry matter, carbohydrates and ^{14}C lost from the leaves and stems of wheat during grain filling. Ann. Bot. 41:1309-21.

Austin, R. B., M. A. Ford, J. A. Edrich, and B. E. Hooper. 1976. Some effects of leaf posture on photosynthesis and yield in wheat. Ann. Appl. Biol. 83:425-46.

Bagga, A. K., and H. M. Rawson. 1977. Contrasting responses of morphologically similar wheat cultivars to temperatures appropriate to warm temperate climates with hot summers: A study in controlled environment. Aust. J. Plant Physiol. 4:877-87.

Baker, D. N., J. D. Hesketh, and W. G. Duncan. 1972. Simulation of growth and yield in cotton: 1. Gross photosynthesis, respiration, and growth. Crop Sci. 12:431-35.

Barclay, P. C. 1970. Some aspects of the development and performance of grasslands 4700 white clover. Proc. N. Z. Grassl. Assoc. Meet. for 1969.

Berdahl, J. D., D. C. Rasmusson, and D. N. Moss. 1972. Effect of leaf area on photosynthetic rate, light penetration, and grain yield in barley. Crop Sci. 12:177-80.

Bingham, J. 1971. Plant breeding: Arable crops, pp. 273-94. *In* Wareing, P. F., and J. P. Cooper (eds.), Potential crop production. Heinemann, London.

Biscoe, P. V., and J. N. Gallaher. 1977. Weather, dry matter production, and yield, pp. 75-100. *In* Landsberg, J. J., and C. V. Cutting (eds.), Environmental effects of crop physiology. Academic Press, London.

Biscoe, P. V., R. K. Scott, and J. L. Monteith. 1975. Barley and its environment. III. Carbon budget of the stand. J. Appl. Ecol. 12:269-93.

Björkman, O. 1968. Further studies on differentiation of photosynthetic properties in sun and shade ecotypes of *Solidago virgaurea*. Physiol. Plant. 21:84-99.

Björkman, O. 1973. Comparative studies on photosynthesis in higher plants. Photophysiology 8:1-63.

Björkman, O. 1976. Adaptive and genetic aspects of C_4 photosynthesis, pp. 287-310. *In* Burris, R. H., and C. C. Black (eds.), CO_2 metabolism and plant productivity. University Park Press, Baltimore.

Björkman, O., and P. Holmgren. 1963. Adaptability of the photosynthetic apparatus to light intensity in ecotypes from exposed and shaded habitats. Physiol. Plant. 16:889-914.

Black, J. N. 1964. An analysis of the potential production of swards of subterranean clover (*Trifolium subterraneum* L.) at Adelaide, South Australia. J. Appl. Ecol. 1:3-18.

Blum, A. 1973. Components analysis of yield responses to drought of sorghum hybrids. Exp. Agric. 9:159-67.

Blum, A. 1974. Genotypic responses in sorghum to drought stress. 1. Response to soil moisture stress. Crop Sci. 14:361-64.

Borrill, M., B. F. Tyler, and M. Kirby. 1973. The evaluation and development of cocksfoot introductions. Rep. Welsh Plant Breed. Stn. for 1972, pp. 37-42.

Breese, E. L. 1969. The measurement and significance of genotype-environment interactions in grasses. Heredity 24(1):27-44.

Breese, E. L., and C. A. Foster. 1971. Breeding for increased winter hardiness in perennial ryegrass. Rep. Welsh Plant Breed. Stn. for 1979, pp. 77-86.

Boyer, J. S., and H. G. McPherson. 1975. Physiology of water deficits in cereal crops. Adv. Agron. 27:1-24.

Brown, K. W. 1969. A model of the photosynthesizing leaf. Physiol. Plant. 22:620-37.

Brown, R. H., and R. E. Blaser. 1968. Leaf area index in pasture growth. Herb. Abstr. 38(1):1-9.

Brown, R. H., R. E. Blaser, and H. L. Dunton. 1966. Leaf area index and apparent photosynthesis under various microclimates for different pasture species, pp. 108-113. Proc. 10th Int. Grassl. Congr., July 7-16, Helsinki, Finl.

Bull, T. A., and K. T. Glasziou. 1975. Sugar cane, pp. 51-72. *In* Evans, L. T. (ed.), Crop physiology: Some case histories. Cambridge Univ. Press.

Campbell, L. G., A. J. Casady, and W. J. Cook. 1975. Effects of a single height gene (dw 3) of sorghum on certain agronomic characters. Crop Sci. 15: 595-97.

Cannell, R. Q., W. A. Brun, and D. N. Moss. 1969. A search for high net photosynthetic rate among soybean genotypes. Crop Sci. 9:840-41.

Carlson, G. E., R. B. Pearce, D. R. Lee, and R. H. Hart. 1971. Photosynthesis and photorespiration in two clones of orchardgrass. Crop Sci. 11:35-37.

Charles-Edwards, D., and L. J. Ludwig. 1975. The basis of expression of leaf photosynthesis activities, pp. 37-43. *In* Marcelle, R. (ed.), Environmental and biological control of photosynthesis. Dr. W. Junk, The Hague.

Charles-Edwards, D. A., J. Charles-Edwards, and J. P. Cooper. 1971. The influence of temperature on photosynthesis and transpiration in ten temperate grass varieties grown in four different environments. J. Exp. Bot. 22:650-62.

Chenost, M. 1966. Fibrousness of forages: Its determination and its relation to feeding value, pp. 406-11. Proc. 10th Int. Grassl. Congr., July 7-16, Helsinki, Finl.

Ciha, A. J., and W. A. Brun. 1975. Stomatal size and frequency in soybeans. Crop Sci. 15:309-13.

Cooper, J. P. 1961. Selection and population structure in *Lolium*. V.Continued response and associated changes in fertility and vigour. Heredity 16: 435-53.

Cooper, J. P. 1963. Species and population differences in climatic response, pp. 381-404. *In* Evans, L. T. (ed.), Environmental control of plant growth. Academic Press, New York.

Cooper, J. P. 1974. The use of physiological criteria in grass breeding. Rep. Welsh Plant Breed. Stn. for 1973:95-102.

Cooper, J. P. 1975. Control of photosynthetic production in terrestial systems, pp. 593-621. *In* Cooper, J. P. (ed.), Photosynthesis and productivity in different environments. Cambridge Univ. Press.

Cooper, J. P. 1976. Photosynthetic efficiency of the whole plant, pp. 107-26. *In* Duckham, A. N., J. G. W. Jones, and E. H. Roberts (eds.), Food and production consumption: The efficiency of human food chains and nutrient cycles. North-Holland Publ. Co., Amsterdam.

Cooper, J. P., and N. M. Tainton. 1968. Light and temperature requirements for the growth of tropical and temperate grasses. Herb. Abstr. 38:167-76.

Cooper, J. P., I. Rhodes, and J. E. Sheehy. 1971. Canopy structure, light interception, and potential production in forage grasses. Rep. Welsh Plant Breed. Stn. for 1970:57-69.

Delaney, R. H., and A. K. Dobrenz. 1974. Morphological and anatomical features of alfalfa leaves as related to CO_2 exchange. Crop Sci. 14:444-47.

Derera, N. F., D. R. Marshall, and L. N. Balaam. 1969. Genetic variability in root development in relation to drought tolerance in spring wheats. Exp. Agric. 5:327-37.

Dobrenz, A. K., L. Neal Wright, A. B. Humphrey, M. A. Massengale, and W. R. Kneebone. 1969. Stomate density and its relationship to water-use efficiency of blue panicgrass (*Panicum antidotale* Retz.). Crop Sci. 9: 354-57.

Donald, C. M. 1963. Competition among crop and pasture plants. Adv. Agron. 15:1-118.

Donald, C. M. 1968. The breeding of crop ideotypes. Euphytica 17:385-403.

Donald, C. M., and J. Hamblin. 1976. The biological yield and harvest index of cereals as agronomic and plant breeding criteria. Adv. Agron. 28:361-405.

Doney, D. L., R. E. Wyse, and J. C. Theurer. 1975a. Mitochondrial efficiency and growth rate in sugar beet. Crop Sci. 15:5-7.

Doney, D. L., J. C. Theurer, and R. E. Wyse. 1975b. Absence of a correlation between mitochondrial complementation and root weight heterosis in sugar beets. Euphytica 24:387-92.

Dornhoff, G. M., and R. M. Shibles. 1970. Varietal differences in net photosynthesis of soybean leaves. Crop Sci. 10:42-45.

Duncan, W. G. 1971. Leaf angles, leaf area and canopy photosynthesis. Crop Sci. 11:482-85.

Duncan, W. G., and J. D. Hesketh. 1968. Net photosynthetic rates, relative leaf growth rates, and leaf numbers of 22 races of maize grown at eight temperatures. Crop Sci. 8:670-74.

Duncan, W. G., R. S. Loomis, W. A. Williams, and R. Hanau. 1967. A model for simulating photosynthesis in plant communities. Hilgardia 38:181-205.

Dunstone, R. L., R. M. Gifford, and L. T. Evans. 1973. Photosynthetic characteristics of modern and primitive wheat species in relation to ontogeny and adaptation to light. Aust. J. Biol. Sci. 26:295-307.

Dvorak, J., and L. Na'tr. 1971. Carbon dioxide compensation points of *Triticum* and *Aegilops* species. Photosynthetica 5:1-5.

Eagles, C. F., and B. O. Othman. 1974. Regulation of leaf expansion in *Dactylis*. Rep. Welsh Plant Breed. Stn. for 1973:12-13.

Eagles, C. F., and K. J. Treharne. 1969. Photosynthetic activity of *Dactylis glomerata* L. in different light regimes. Photosynthetica 3:29-38.

Ebercorn, A., A. Blum, and W. R. Jordan. 1977. A rapid colorimetric method for epicuticular wax content of sorghum leaves. Crop Sci. 17:179-80.

Ehleringer, J., and O. Björkman. 1977. Quantum yields for CO_2 uptake in C_3 and C_4 plants. Plant Physiol. 59:86-90.

El-Sharkawy, M., and J. Hesketh. 1965. Photosynthesis among species in relation to characteristics of leaf anatomy and CO_2 diffusion resistances. Crop Sci. 5:517-21.

Evans, L. T. 1973. The effect of light on plant growth, development and yield, pp. 21-36. *In* Slatyer, R. O. (ed.), Plant response to climatic factors. UNESCO, Paris.

Evans, L. T. 1975. The physiological basis of crop yield, pp. 327-55. *In* Evans, L. T. (ed.), Crop physiology: Some case histories. Cambridge Univ. Press.

Evans, L. T., and R. L. Dunstone. 1970. Some physiological aspects of evolution in wheat. Aust. J. Biol. Sci. 23:725-41.

Evans, L. T., and H. M. Rawson. 1970. Photosynthesis and respiration by the flag leaf and components of the ear during grain development in wheat. Aust. J. Biol. Sci. 23:245-54.

Evans, L. T., I. F. Wardlaw, and R. A. Fischer. 1975. Wheat, pp. 101-49. *In* Evans, L. T. (ed.), Crop physiology: Some case histories. Cambridge Univ. Press.

Evans, P. S. 1964. A study of leaf strength in four ryegrass varieties. N.Z. J. Agric. Res. 7:508-13.
Evans, W. B., J. M. M. Munro, and R. V. Scurlock. 1978. Comparative pasture and animal production from cocksfoot and perennial ryegrass varieties under grazing, pp. 5.7-.8. *In* Grazing: Sward production and livestock output. Brit. Grassl. Soc. Hurley, Berkshire, Engl.
Faulkner, J. S. 1974. Heritability of paraquat tolerance in *Lolium perenne* L. Euphytica 23:281-88.
Fernandez, C. J. 1977. Differences in carbon economy among five grain sorghum cultivars. Texas A & M Univ. Thesis abstr.
Francis, C. A. 1970. effective day lengths for photoperiod sensitive reactions in plants. Agron. J. 62:790-92.
Francis, C. A., D. Sarria, D. D. Harpslead, and C. Cassalett. 1970. Identification of photoperiod insensitive strains of maize (*Zea mays* L.). II. Field tests in the tropics with artificial lights. Crop Sci. 10:465-68.
Gallaher, J. N., P. V. Biscoe, and R. K. Scott. 1975. Barley and its environment. V. Stability of grain weight. J. Appl. Ecol. 12:319-36.
Garrett, M. K. 1977. Control of photorespiration at the level of RuDP carboxylase in *Lolium*, p. 123. *In* Coombs, J., compiler. Fourth Int. Congr. Photosyn. (abstr.). UKISES, London.
Gifford, R. M. 1974. A comparison of potential photosynthesis, productivity and yield of plant species with differing photosynthetic metabolism. Aust. J. Plant Physiol. 1:107-17.
Gifford, R. M., P. M. Bremner, and D. B. Jones. 1973. Assessing photosynthetic limitation to grain yield in a field crop. Aust. J. Agric. Res. 24:297-307.
Goldsworthy, P. R. 1970. The growth and yield of tall and short sorghums in Nigeria. J. Agric. Sci. 75:109-22.
Hall, A. E. 1972. Photosynthesis in the genus *Beta*. Crop Sci. 12:701-2.
Hall, D. M., A. I. Matus, J. A. Lamberton, and H. N. Barber. 1965. Infraspecific variation in wax on leaf surfaces. Aust. J. Biol. Sci. 18:323-33.
Hanna, W. W., W. G. Monson, and G. W. Burton. 1974. Leaf surface effects on *in vitro* digestion and transpiration in isogenic lines of sorghum and pearl millet. Crop Sci. 14:837-38.
Hanson, A. D., C. E. Nelsen, and E. H. Everson. 1977. Evaluation of free proline accumulation as an index of drought resistance using two contrasting barley cultivars. Crop Sci. 17:720-26.
Hanson, W. D. 1971. Selection for differential productivity among juvenile maize plants: Associated net photosynthetic rate and leaf area changes. Crop Sci. 11:334-39.
Hardwick, K., M. Wood, and H. W. Woolhouse. 1968. Photosynthesis and respiration in relation to leaf age in *Perilla frutescens* (L.) Britt. New Phytol. 67:79-86.
Hart, R. H., R. B. Pearce, N. J. Chatterton, G. E. Carlson, D. K. Barnes, and C. H. Hanson. 1978. Alfalfa yield, specific leaf weight, CO_2 exchange rate and morphology. Crop Sci. 18:649-53.
Hatch, M. D., C. R. Slack, and T. A. Bull. 1969. Light-induced changes in the extent of some enzymes of the C_4-dicarboxylic acid pathway of photosynthesis and its effect on other characteristics of photosynthesis. Phytochemistry 8:697-708.
Hatch, M. D., C. B. Osmond, and R. O. Slatyer (eds.). 1971. Photosynthesis and photorespiration. Wiley-Interscience, New York and London.
Heichel, G. H. 1971a. Confirming measurements of respiration and photosynthesis with dry matter accumulation. Photosynthetica 5(2):93-98.
Heichel, G. H. 1971b. Genetic control of epidermal cell and stomatal frequency in maize. Crop Sci. 11:830-32.

Hesketh, J. D., and D. N. Moss. 1963. Variation in the response of photosynthesis to light. Crop Sci. 3:107-11.

Hollaender, A. (ed.). 1977. Genetic engineering for nitrogen fixation. Plenum Press, New York.

Holmgren, P. 1968. Leaf factors affecting light-saturated photosynthesis in ecotypes of *Solidago virgaurea* from exposed and shaded habitats. Physiol. Plantarum 21:676-98.

Homann, P. H., and G. H. Schmid. 1967. Photosynthetic reactions of chloroplasts with unusual structures. Plant Physiol. 42:1619-32.

Hsiao, T. C., and E. Acevedo. 1974. Plant responses to water deficits, water-use efficiency, and drought resistance. Agric. Meteorol. 14:59-84.

Hume, D. J., and D. K. Campbell. 1972. Accumulation and translocation of soluble solids in corn stalks. Can. J. Plant Sci. 52:363-68.

Hurd, E. A. 1968. Growth of roots of seven varieties of spring wheat at high and low moisture levels. Agron. J. 60:201-5.

Hurd, E. A. 1974. Phenotype and drought tolerance in wheat. Agric. Meteorol. 14:39-55.

Ishizuka, Y. 1971. Physiology of the rice plant. Adv. Agron. 23:241-315.

Jackson, D. K. 1974. Efficiency in grass production for grazing. Rep. Welsh Plant Breed. Stn. for 1973:111-16.

Jennings, P. R. 1964. Plant type as a rice breeding objective. Crop Sci. 4:13-15.

Jennings, P. R., and R. M. Herrera. 1968. Studies on competition in rice. II. Competition in segregating populations. Evolution 22:332-36.

Jewiss, O. R., and J. Woledge. 1967. The effect of age on rate of apparent photosynthesis in leaves of tall fescue (*Festuca arundinacea* Schreb.). Ann. Bot. N.S. 31:661-67.

Johnson, D. A., and K. H. Asay. 1978. A technique for assessing seedling emergence under drought stress. Crop Sci. 18:520-22.

Johnson, R. R., C. M. Willmer, and D. N. Moss. 1975. Role of awns in photosynthesis, respiration and transpiration of barley spikes. Crop Sci. 15: 217-21.

Jones, D. I. H. 1975. Some recent developments in techniques for assessing the digestibility and intake characteristics of grasses. Rep. Welsh Plant Breed. Stn. for 1974:128-33.

Jones, D. I. H., R. J. K. Walters, and E. L. Breese. 1974. The evolution of herbage breeding programmes for improved voluntary intake and other nutritive characteristics, pp. 111-20. *In* Aberg, E. (ed.), Quality of herbage. Proc. 5th Gen. Meet. Eur. Grassl. Fed., 1973, June 12-15, Uppsala.

Jones, H. G. 1976. Crop characteristics and the ratio between assimilation and transpiration. J. Appl. Ecol. 13:605-22.

Jones, H. G. 1977. Transpiration in barley lines with differing stomatal frequencies. J. Exp. Bot. 28:162-68.

Jones, H. G., and E. J. M. Kirby. 1977. Effects of manipulation of number of tillers and water supply on grain yield in barley. J. Agric. Sci. 88:391-97.

Jones, M. M., and N. C. Turner. 1978. Osmotic adjustment in leaves of sorghum in response to water deficits. Plant Physiol. 61:122-26.

Kaplan, S. L., and H. R. Koller. 1977. Leaf area and CO_2-exchange rate as determinants of the rate of vegetative growth in soybean plants. Crop Sci. 17:35-38.

Khan, M. A., and S. Tsunada. 1970. Growth analyses of cultivated wheat species and their wild relatives with special reference to dry matter distribution among different plant organs and to leaf area. Tohoku J. Agric. Res. 21:47-59.

King, R. W., and L. T. Evans. 1967. Photosynthesis in artificial communities of wheat, lucerne and subterranean clover plants. Aust. J. Biol. Sci. 20: 623-36.

Kirby, E. J. M. 1969. The effects of daylength upon the development and growth of wheat, barley and oats. Field Crop. Abstr. 22:1-7.

Knott, D. R., and J. Kumar. 1975. Comparison of early generation yield testing and a single seed descent procedure in wheat breeding. Crop Sci. 15: 295-99.

Kolawole, K. B. 1978. Genetic variation in some physiological and agronomic characters in barley (*Hordeum vulgare* L.). Univ. Coll. of Wales, Aberystwyth. Ph.D. Thesis. 317 pp.

Krause, G. H., G. H. Lorimor, U. Heber, and M. R. Kirk. 1978. Photorespiratory energy dissipation in leaves and chloroplasts, pp. 299-310. *In* Hall, D. O., J. Coombs, and T. W. Goodwin (eds.), Photosynthesis '77: Proceedings of the fourth international congress on photosynthesis, 1977, Sept. 4-9. The Biochemical Soc., Reading U.K. and London.

Laetsch, W. M. 1974. The C_4 syndrome: A structural analysis. Annu. Rev. Plant Physiol. 25:27-52.

Lambers, H., and E. Steingrover. 1978. Growth respiration of a flood-tolerant and a flood-intolerant *Senecio* species. Physiol. Plant. 43:219-24.

Lancashire, J. A., and H. M. J. Ulyatt. 1975. Live-weight gains of sheep grazing ryegrass pastures with different cellulose contents. N.Z. J. Agric. Res. 18: 97-100.

Lancashire, J. A., D. Wilson, R. W. Bailey, M. J. Ulyatt, and P. Singh. 1977. Improved summer performance of a "low cellulose" selection from "Grasslands Ariki." N.Z. J. Agric. Res. 20:63-67.

Laredo, M. A., and D. J. Minson. 1973. The voluntary intake, digestibility, and retention time by sheep of leaf and stem fractions of five grasses. Aust. J. Agric. Res. 24:875-88.

Law, C. N., J. W. Snape, and A. J. Worland. 1978. The genetical relationship between height and yield in wheat. Heredity 40:133-51.

Lawes, D. A. 1977. Yield improvement in spring oats. J. Agric. Sci. 89:751-57.

Levitt, J. 1972. Plant Responses to Environmental Stresses. Academic Press, New York. 697 pp.

Loach, K. 1970. Analysis of differences in yield between six sugarbeet varieties. Ann. Appl. Biol. 66:217-23.

Loomis, R. S., and W. A. Williams. 1969. Productivity and the morphology of crop stands: Patterns and leaves, pp. 27-47. *In* Eastin, J. D., F. A. Haskins, C. Y. Sullivan, and C. H. M. van Bavel (eds.), Physiological aspects of crop yield. ASA and CSSA, Madison, Wis.

Long, S. P. and H. W. Woolhouse. 1978. The responses of net photosynthesis to light and temperature in *Spartina townsendii* (sensulato), a C_4 species from a cool temperate climate. J. Exp. Bot. 29:803-14.

Lorimer, G. H., K. C. Woo, J. A. Berry, and C. B. Osmond. 1978. The C_2 photorespiratory carbon oxidation cycle in leaves of higher plants: Pathway and consequences, pp. 311-22. *In* Hall, D. O., J. Coombs, and T. W. Goodwin (eds.), Photosynthesis '77: Proceedings of the fourth international congress on photosynthesis, 1977, Sept. 4-9. The Biochemical Soc., Reading, U.K. and London.

Low, S. B., and L. A. Wilson. 1974. Comparative analysis of tuber development in six sweet potato (*Ipomoea batatas* (L.) Lam) cultivars. 1. Tuber initiation, tuber growth and partition of assimilate. Ann. Bot. 38:311-18.

Ludwig, L. J., T. Saeki, and L. T. Evans. 1965. Photosynthesis in artificial communities of cotton plants in relation to leaf area. 1. Experiments with progressive defoliation of mature plants. Aust. J. Biol. Sci. 18: 1103-18.

Lupton, F. G. H. 1968. The analysis of grain yield of wheat in terms of photosynthetic ability and efficiency of translocation. Ann. Appl. Biol. 61: 109-19.

MacColl, D. 1977. Growth and sugar accumulation of sugarcane. III. Development of commercial clones and their progenies in single row plots. Exp. Agric. 13:161-67.

Mansat, P. 1972. Improving the quality of forage crops: Modification of plant type and selection for a biochemical character, pp. 197-206. *In* Lupton, F. G. H., G. Jenkins, and R. Johnson (eds.), The way ahead in plant breeding. Proc. Sixth Congr. Eucarpia. Cambridge Univ. Press.

Martin, F. A., J. L. Ozbun, and D. H. Wallace. 1972. Intraspecific measurements of photorespiration. Plant Physiol. 49:764-68.

May, D. S. 1975. Genetic and physiological adaptation of the Hill reaction in altitudinally-diverse populations of *Taraxacum*. Photosynthetica 9(3): 293-98.

McCree, K. J. 1970. An equation for the rate of respiration of white clover plants grown under controlled conditions, pp. 221-30. *In* Setlik, I. (ed.), Prediction and measurement of photosynthetic productivity. Cent. Agric. Publ. and Doc., Wageningen.

McCree, K. J. 1974. Equations for the rate of dark respiration of white clover and grain sorghum, as functions of dry weight, photosynthetic rate, and temperature. Crop Sci. 14:509-14.

McCree, K. J., and C. H. M. van Bavel. 1977. Respiration and crop production: A case study with two crops under water stress, pp. 199-216. *In* Landsberg, J. J., and C. V. Cutting (eds.), Environmental effects on crop physiology. Academic Press, London.

McDaniel, R. G. 1969. Relationships of seed weight, seedling vigour and mitochondrial metabolism in barley. Crop Sci. 9:823-27.

McDaniel, R. G. 1972. Mitochondrial heterosis and complementation as biochemical measures of yield. Nature New Biol. 236:190-91.

Milford, G. F. J. 1973. The growth and development of the storage root of sugarbeet. Ann. Appl. Biol. 73:427-38.

Miller, J. D., N. I. James, and P. M. Lyrene. 1978. Selection indices in sugarcane. Crop Sci. 18:369-72.

Miskin, K. E., D. C. Rasmusson, and D. N. Moss. 1972. Inheritance and physiological effects of stomatal frequency in barley. Crop Sci. 12:780-83.

Mock, J. J., and R. B. Pearce. 1975. An ideotype of maize. Euphytica 24: 613-23.

Mohamed, A. M. A., S. O. Okiror, and D. C. Rasmusson. 1978. Performance of semidwarf barley. Crop Sci. 18:418-22.

Monsi, M., and T. Saeki. 1953. Uber den lichtfaktor in den pflanzengesellschaften und seine bedeutung fur die stoffproduktion. Jap. J. Bot. 14: 22-52.

Monson, W. G., J. B. Powell, and G. W. Burton. 1972. Digestion of fresh forage in rumen fluid. Agron. J. 64:231-33.

Monteith, J. L. 1965. Light and crop production. Field Crop Abstr.18:213-19.

Monteith, J. L. 1969. Light interception and radiative exchange in crop stands, pp. 89-111. *In* Eastin, J. D., F. A. Haskins, C. Y. Sullivan, and C. H. M. van Bavel (eds.), Physiological aspects of crop yield. ASA and CSSA, Madison, Wis.

Monteith, J. L. 1977. Climate and the efficiency of crop production in Britain. Philos. Trans. R. Soc. London, Sect. B. Vol. 281, pp. 277-94.

Monteith, J. L. 1978. Reassessment of maximum growth rates for C_3 and C_4 crops. Exp. Agr. 14:1-5.

Mooney, H. A., and W. D. Billings. 1961. Comparative physiological ecology of arctic and alpine populations of *Oxyria digyna*. Ecol. Monogr. 31:1-28.

Moreshet, S. 1970. Effect of environmental factors on cuticular transpiration resistance. Plant Physiol. 46:815-18.

Morley, F. H. W. 1958. Effects of strain and temperature on the growth of

subterranean Clover (*T. subterraneum*). Aust. J. Agric. Res. 9:745-53.
Moss, D. N., and R. B. Musgrave. 1971. Photosynthesis and crop production. Adv. Agron. 23:317-36.
Moss, D. N., J. T. Woolley, and J. F. Stone. 1974. Plant modification for more efficient water use: The challenge. *In* Stone, J. F. (ed.), Plant modification for more efficient water use. Agric. Meteorol. 14:311-20.
Murata, Y., and S. Matshushina. 1975. Rice, pp. 73-100. *In* Evans, L. T. (ed.), Crop physiology: Some case histories. Cambridge Univ. Press.
Mytton, L. R., M. H. El-Sherbeeny, and D. A. Laws. 1977. Symbiotic variability in *Vicia faba*. 3. Genetic effects of host plant, Rhizobium strain and of host x strain interaction. Euphytica 26:785-91.
Namken, L. N., C. L. Wiegand, and W. O. Willis. 1974. Soil- and air-temperatures as limitations to more efficient water use. Agric. Meteorol. 14:169-81.
Neales, T. F., and L. D. Incoll. 1968. The control of leaf photosynthesis rate by the level of assimilate concentration in the leaf: A review of the hypothesis. Bot. Rev. 34:107-25.
Nelson, C. J., K. J. Treharne, and E. J. Lloyd. 1975. Genetic variation in enzyme activity of tall fescue leaf blades. Crop Sci. 15:771-74.
Oworu, O. O., C. R. McDavid, and D. MacColl. 1977. The anatomy of the storage tissue of sugar-cane in relation to sugar uptake. Ann. Bot. 41: 401-4.
Papendick, R. I., P. A. Sanchez, and G. B. Triplett (eds.). 1976. Multiple cropping. Am. Soc. Agron. Spec. Publ. 27. 378 pp.
Passioura, J. B. 1972. The effect of root geometry on the yield of wheat growing on stored water. Aust. J. Agric. Res. 23:745-52.
Passioura, J. B. 1976. The control of water movement through plants, pp. 373-80. *In* Wardlaw, I. F., and J. B. Passioura (eds.), Transport and transfer processes in plants. Academic Press, New York.
Pate, J. S. 1975. Pea, pp. 191-224. *In* Evans, L. T. (ed.), Crop physiology: Some case histories. Cambridge Univ. Press.
Pearce, R. B., R, H. Brown, and R. E. Blaser. 1967. Photosynthesis in plant communities as influenced by leaf angle. Crop Sci. 7:321-24.
Pearce, R. B., G. E. Carlson, D. K. Barnes, R. H. Hart, and C. H. Hanson. 1969. Specific leaf weight and photosynthesis in alfalfa. Crop Sci. 9:423-26.
Peet, M. M., J. L. Ozbun, and D. H. Wallace. 1977. Physiological and anatomical effects of growth temperature on *Phaseolus vulgaris* L. cultivars. J. Exp. Bot. 28:57-69.
Pendleton, J. W., G. E. Smith, S. R. Winter, and T. J. Johnston. 1968. Field investigations of the relationships of leaf angle in corn (*Zea mays* L.) to grain yield and apparent photosynthesis. Agron. J. 60:422-24.
Penning de Vries, F. W. T. 1972. Respiration and growth, pp. 327-48. *In* Rees, A. R., K. E. Cockshull, D. W. Hand, and R. G. Hurd (eds.), Crop processes in controlled environments. Academic Press, London.
Penning de Vries, F. W. T. 1974. Substrate utilization and respiration in relation to growth and maintenance in higher plants. Neth. J. Agric. Sci. 22: 40-44.
Pinthus, M. J. 1973. Lodging in wheat, barley, and oats: The phenomenon, its causes and preventive measures. Adv. Agron. 25:209-63.
Quebedeaux, B., and R. Chollet. 1977. Comparative growth analyses of *Panicum* species with differing rates of photorespiration. Plant Physiol. 59:42-44.
Randall, D. D., C. J. Nelson, and K. H. Asay. 1977. Ribulose bisphosphate carboxylase: Altered genetic expression in tall fescue. Plant Physiol. 59: 38-41.
Rawson, H. M. 1970. Spikelet number, its control and relation to yield per ear in wheat. Aust. J. Biol. Sci. 23:1-15.

Rawson, H. M., and L. T. Evans. 1971. The contribution of stem reserves to grain development in a range of wheat cultivars of different height. Aust. J. Agric. Res. 22:851-63.
Rawson, H. M., R. M. Gifford, and P. M. Bremner. 1976. Carbon dioxide exchange in relation to sink demand in wheat. Planta 132:19-23.
Rhodes, I. 1971. The relationship between productivity and some components of canopy structure in ryegrass (*Lolium* spp.). II. Yield, canopy structure and light interception. J. Agric. Sci. 77:282-92.
Rhodes, I. 1972. Yield, leaf-area index and photosynthetic rate in some perennial ryegrass (*Lolium perenne* L.) selections. J. Agric. Sci. 78:509-11.
Rhodes, I. 1973. Relationship between canopy structure and productivity in herbage grasses and its implications for plant breeding. Herb. Abstr. 43: 129-33.
Rhodes, I. 1975. The relationship between productivity and some components of canopy structure in ryegrass (*Lollium* spp.). IV. Canopy characters and their relationship with sward yields in some intra-population selections. J. Agric. Sci. 84:345-51.
Rhodes, I., and W. Harris. 1979. The nature and basis of changes in the composition and productivity of ryegrass-white clover mixtures, pp. 55-60. *In* Charles, A. H., and R. J. Haggar (eds.), Changes in sward composition and productivity. Occas. Symp. 10 Br. Grassl. Soc.
Rhodes, I., and S. S. Mee. 1980. Changes in dry matter yield associated with selection for canopy characters in ryegrass. Grass and Forage Sci. (in press).
Riggs, T. J., and A. M. Hayter. 1975. A study of the inheritance and interrelationships of some agronomically important characters in spring barley. Theor. Appl. Genet. 46:257-64.
Robson, M. 1973. The growth and development of simulated swards of perennial ryegrass. II. Carbon assimilation and respiration in a seedling sward. Ann. Bot. 37:501-18.
Rosario, E. L., and R. B. Musgrave. 1974. The relationship of sugar yield and its components to some physiological and morphological characters. Proc. Int. Soc. Sugar Cane Technol. 15:1011-20.
Rosielle, A. A., and K. J. Frey. 1975. Estimates of selection parameters associated with harvest index in oat lines derived from a bulk population. Euphytica 24:121-31.
Ryle, G. J. A., J. M. Cobby, and C. E. Powell. 1976. Synthetic and maintenance respiratory losses of $^{14}CO_2$ in uniculum barley and maize. Ann. Bot. 40:571-86.
Saeki, T. 1973. Distribution of radiant energy and CO_2 in terrestrial communities, pp. 297-322. *In* Cooper, J. P. (ed.), Photosynthesis and productivity in different environments. Cambridge Univ. Press.
Samish, Y. B., J. E. Pallas, Jr., G. M. Dornhoff, and R. M. Shibles. 1972. A re-evaluation of soybean leaf photorespiration. Plant Physiol. 50:28-30.
Sant, F. I., and I. Rhodes. 1970. A note on the relationship between leaf rigidity and leaf anatomy of *Lolium perenne* L. J. Br. Grassl. Soc. 25: 233-35.
Sayre, R. T., and R. A. Kennedy. 1977. Ecotypic differences in the C_3 and C_4 photosynthetic activity in *Mollugo verticillata*, a C_3-C_4 intermediate. Planta 134:257-62.
Scheibe, A., and H. Meyer zu Drewer. 1959. Vergleichende untersuchungen zur atmungsintensitat der worzeln unterschiedlicher genotypen bei getreidearten. Z. Acker-u. Pflanzenbau 108:223-52.
Schneiter, A. A., R. G. McDaniel, A. K. Dobrenz, and M. H. Schonhorst. 1974. Relationship of mitochondrial efficiency to forage yield in alfalfa. Crop Sci. 14:821-24.

Schrader, C. E. 1976. CO_2 metabolism and productivity in C_3 plants: An assessment, pp. 385-96. *In* Burris, R. H., and C. C. Black (eds.), CO_2 matabolism and plant productivity. University Park Press, Baltimore.

Selim, O. I. 1976. Variation in leaf anatomy in relation to nutritive value of *Festuca arundinacea* Schreb. and other temperate grasses, Univ. Coll. of Wales, Aberystwyth. Ph.D. Thesis. 135 pp.

Selim, O. I., D. Wilson, and D. I. H. Jones. 1975. A histological technique, using cellulolytic enzyme digestion, for assessing nutritive quality differences in grasses. J. Agric. Sci. 85:297-99.

Sheehy, J. E., and J. P. Cooper. 1973. Light interception, photosynthetic activity, and crop growth rate in canopies of six temperate forage grasses. J. Appl. Ecol. 10:239-50.

Sheriff, D. W., and P. E. Kaye. 1977. Responses of diffusive conductance to humidity in a drought avoiding and a drought resistant (in terms of stomatal response) legume. Ann. Bot. 41:653-55.

Shibles, R. M., I. C. Anderson, and A. H. Gibson. 1975. Soybean, pp. 151-89. *In* Evans, L. T. (ed.), Crop physiology: Some case histories. Cambridge Univ. Press.

Silcock, R. G. 1978. Transpiration and water-use efficiency as affected by leaf characteristics of *Festuca* species. Welsh Plant Breed. Stn., Aberystwyth. Ph.D. Thesis. 269 pp.

Stoy, V. 1965. Photosynthesis, respiration and carbohydrate accumulation in spring wheat in relation to yield. Physiol. Plant Suppl. 4:1-125.

Stoy, V. 1975. Use of tracer techniques to study yield components in seed crops, pp. 43-45. *In* Proceedings of a symposium: Tracer techniques for plant breeding. Vienna Int. Atomic Energy Agency.

Syme, J. R. 1970. A high-yielding Mexican semi-dwarf wheat and the relationship of yield to harvest index and other varietal characteristics. Aust. J. Exp. Agric. Anim. Husb. 10: 350-53.

Taylor, A. O., J. A. Rowley, and N. M. Jepsen. 1971. Factors regulating the growth rate of *Lolium perenne* L. cv. 'Grasslands Ruanui' and *L. multiflorum* Lam. cv. 'Grasslands Tama' a tetraploid. 1. Seeds, photosynthetic rates, photosynthetic products, translocation, and proportion of plant parts. N.Z. J. Bot. 9:504-18.

Terry, R. A., and J. M. A. Tilley. 1964. The digestibility of the leaves and stems of perennial ryegrass, cocksfoot, timothy, tall fescue, lucerne and sainfoin, as measured by an *in vitro* procedure. J. Br. Grassl. Soc. 19: 390-92.

Thomas, H., and J. L. Stoddart. 1977. Biochemistry of leaf senescence in grasses. Ann. Appl. Biol. 85:461-62.

Thornley, J. H. M. 1970. Respiration, growth and maintenance in plants. Nature 227:304-5.

Thornley, J. H. M. 1971. Energy, respiration and growth in plants. Ann. Bot. 35:721-28.

Thornley, J. H. M. 1977. Growth, maintenance and respiration: A reinterpretation. Ann. Bot. 41:1191-1203.

Tolbert, N. E., and F. J. Ryan. 1976. Glycolate biosynthesis and metabolism during photorespiration, pp. 141-59. *In* Burris, R. H., and C. C. Black (eds.), CO_2 metabolism and plant productivity. University Park Press, Baltimore, Md.

Treharne, K. J., and C. J. Nelson. 1975. Effect of growth temperature on photosynthetic and photo-respiratory activity in tall fescue, pp. 61-69. *In* Marcelle, R. (ed.), Environmental and biological control of photosynthesis. Dr. W. Junk, The Hague.

Treharne, K. J., A. J. Pritchard, and J. P. Cooper. 1971. Variation in photosynthesis and enzyme activity in *Cenchrus ciliaris* L. J. Exp. Bot. 22: 227-38.

Troelson, J. E., and F. W. Bigsby. 1964. Artificial mastication—a new approach for predicting voluntary forage consumption by ruminants. J. Anim. Sci. 23:1139-42.

Troughton, A. 1965. Intra-varietal variation in *Lolium perenne.* Euphytica 14: 59-66.

Troughton, A. 1974. The development of leaf water deficits in plants of *Lolium perenne* in relation to the sizes of the root and shoot systems. Plant and Soil 40:153-60.

Troughton, A. 1978. Shoot:root relationships. Rep. Welsh Plant Breed. Stn. for 1977:147-49.

Troughton, A., and W. J. Whittington. 1969. The significance of genetic variation in root systems, pp. 296-313. *In* Whittington, W. J. (ed.), Root growth. Fifteenth Proc. Nottingham Easter School, Butterworth, London.

Troughton, J. H. 1975. Photosynthetic mechanisms in higher plants, pp. 357-91. *In* Cooper, J. P. (ed.), Photosynthesis and productivity in different environments. Cambridge Univ. Press.

Turesson, G. 1922. The genotypical response of the plant species to the habit. Hereditas 3:211-350.

Ulyatt, M. J. 1970. Factors contributing to differences in the quality of short-rotation ryegrass, perennial ryegrass, and white clover, pp. 709-13. *In* Proc. 11th Int. Grassl. Congr. Apr. 13-23, Queensland, Aust.

Van Dobben, W. H. 1962. Influence of temperature and light conditions on dry-matter distribution, development rate in arable crops. Neth. J. Agric. Sci. 10:377-89.

Vose, P. B. 1963. Varietal differences in plant nutrition. Herb. Abstr. 33:1-13.

Vose, P. B., and E. L. Breese. 1964. Genetic variation in the utilization of nitrogen by ryegrass species *Lolium perenne* and *L. multiflorum.* Ann. Bot. 28:251-70.

Wallace, D. H., J. L. Ozbun, and H. M. Munger. 1972. Physiological genetics of crop yield. Adv. Agron. 24:97-146.

Walters, R. J. K. 1971. Variation in the relationship between *in vitro* digestibility and voluntary dry matter intake of different grass varieties. J. Agric. Sci. 76:243-52.

Walters, R. J. K., and E. M. Evans. 1974. Herbage quality. Rep. Welsh Plant Breed. Stn. for 1973:42-44.

Watson, D. J. 1971. Size, structure and activity of the productive systems of crops, pp. 76-88. *In* Wareing, P. F., and J. P. Cooper (eds.), Potential crop production. Heinemann, London.

Weatherly, P. E. 1965. The state and movement of water in the leaf. The state and movement of water in living organisms. Symp. 19 Soc. Exp. Bio., pp. 157-184.

Whalley, R. D. B., and C. M. McKell. 1967. Interrelations of carbohydrate metabolism, seedling development, and seedling growth rate of several species of *Phalaris.* Agron. J. 59:223-26.

Wilhelm, W. W., and C. J. Nelson. 1978. Irradiance response of tall fescue genotypes with contrasting levels of photosynthesis and yield. Crop Sci. 18:405-8.

Willat, S. T., and H. M. Taylor. 1978. Water uptake by soya-bean roots as affected by their depth and by soil water content. J. Agric. Sci. 90: 205-13.

Wilson, D. 1965. Nutritive value and the genetic relationships of cellulose content and leaf tensile strength in *Lolium.* J. Agric. Sci. 65:285-92.

Wilson, D. 1972a. Variation in photorespiration in *Lolium.* J. Exp. Bot. 23: 517-24.

Wilson, D. 1972b. Effect of selection for stomatal length and frequency on the theoretical stomatal resistance to diffusion in *Lolium perenne* L. New Phytol. 71:811-17.

Wilson, D. 1973. Physiology of light utilization by swards, 2:57-101. *In* Butler, G. W., and R. W. Bailey (eds.), Chemistry and biochemistry of herbage. Academic Press, London.
Wilson D. 1975a. Leaf growth, stomatal diffusion resistances and photosynthesis during droughting of *Lolium perenne* populations selected for contrasting stomatal length and frequency. Ann. Appl. Biol. 79:67-82.
Wilson, D. 1975b. Variation in leaf respiration in relation to growth and photosynthesis of *Lolium*. Ann. Appl. Biol. 80:323-28.
Wilson D. 1975c. Stomatal diffusion resistances and leaf growth during droughting of *Lolium perenne* plants selected for contrasting epidermal ridging. Ann. Appl. Biol. 79:83-94.
Wilson, D. 1976. Physiological and morphological selection criteria in grasses, pp. 9-18. *In* Dennis, B. (ed.), Breeding methods and variety testing in forage plants. Rep. Meet. Fodder Crops Sect., Eucarpia, Sept. 7-9, Roskilde, Den.
Wilson, D. 1977. Dark respiration. Rep. Welsh Plant Breed. Stn. for 1976: 126 (Table 4).
Wilson, D. 1978. Cold-hardiness in relation to leaf quality in *Festuca arundinacea*. Rep. Welsh Plant Breed. Stn. for 1977: 160.
Wilson, D., and J. P. Cooper. 1967. Assimilation of *Lolium* in relation to leaf mesophyll. Nature 214:989-92.
Wilson, D., and J. P. Cooper. 1969a. Diallel analysis of photosynthetic rate and related leaf characters among contrasting genotypes of *Lolium perenne*. Heredity 24(4):633-49.
Wilson, D., and J. P. Cooper. 1969b. Apparent photosynthesis and leaf characters in relation to leaf position and age, among contrasting *Lolium* genotypes. New Phytol. 68:645-55.
Wilson, D., and J. P. Cooper. 1969c. Effect of light intensity and CO_2 on apparent photosynthesis and its relationship with leaf anatomy in genotypes of *Lolium perenne* L. New Phytol. 68:627-44.
Wilson, D., and J. P. Cooper. 1969d. Effect of light intensity during growth on leaf anatomy and subsequent light-saturated photosynthesis among contrasting *Lolium* genotypes. New Phytol. 68:1125-35.
Wilson, D., and J. P. Cooper. 1969e. Effect of temperature during growth on leaf anatomy and subsequent light-saturated photosynthesis among contrasting *Lolium* genotypes. New Phytol. 68:1115-23.
Wilson, D., and J. P. Cooper. 1970. Effect of selection for mesophyll cell size on growth and assimilation in *Lolium perenne* L. New Phytol. 69:233-45.
Wilson, D., I. B. Abdullah, and S. A. Trickey. 1977. Variation in transpiration rate in *Dactylis*, pp. 71-79. *In* Proc. 13th Int. Grassl. Congr., May 18-27, Leipzig.
Wilson, G. L., and M. M. Ludlow. 1970. Net photosynthetic rate of tropical grass and legume leaves, pp. 534-38. *In* Proc. 11th Int. Grassl. Congr., Apr. 13-23, Queensland, Aust.
de Wit, C. T. 1965. Photosynthesis of leaf canopies. Versl. Landbouwkd. Onderz 663:1-57.
Wonkyi-Appiah, J. B. 1970. Selection for leaf growth in ryegrass and its influence on potential production. Univ. Coll. of Wales, Aberystwyth. Ph.D. Thesis. 133 pp.
Wright, M. J. (ed.). 1977. Plant adaptation to mineral stress in problem soils. Proc. of workshop, Nov. 22-23. Cornell Univ. Agric. Exp. Stn. 420 pp.
Yoshida, S. 1972. Physiological aspects of grain yield. Annu. Rev. Plant Physiol. 23:437-64.
Yoshida, T. 1978. On the stomatal frequency in barley. V. The effect of stomatal size on transpiration and photosynthetic rate. Jap. J. Breed. 28(2):87-96.

Zelitch, I. 1968. Investigations on photorespiration with a sensitive ^{14}C-assay. Plant Physiol. 43:1829-37.

Zelitch, I. 1971. Photosynthesis, photorespiration and plant productivity. Academic Press, New York.

Zelitch, I. 1976. Biochemical and genetic control of photorespiration, pp.343-58. *In* Burris, R. H., and C. C. Black (eds.), CO_2 metabolism and plant productivity. University Park Press, Baltimore.

Zelitch, I., and P. R. Day. 1968. Variation in photorespiration. The effect of genetic differences in photorespiration on net photosynthesis in tobacco. Plant Physiol. 43(11):1838-44.

Zelitch, I., and P. R. Day. 1973. The effect on net photosynthesis of pedigree selection for low and high rates of photorespiration in tobacco. Plant Physiol. 52:33-37.

CHAPTER 8

Breeding for Insect Resistance

J. N. JENKINS

IN A RECENT REVIEW, Kennedy (1978) stated that between 1966 and 1977 over 200 papers were published that dealt with insect resistance in fruit and vegetable crops in North America. He indicated that insect resistance was investigated in 9 fruit crops with 19 different crop-insect interactions and 21 vegetable crops with 70 crop-insect associations. Maxwell et al. (1972), in reviewing the literature on insect resistance from 1958 to 1971, found 1,400 research papers, of which they cited 555 in discussing 14 species in the three field-crop families—Leguminosae, Gramineae, and Malvaceae. They also cited eight horticultural and two miscellaneous families.

Breeding programs for host plant resistance exist for alfalfa (*Medicago sativa*), clover (*Trifolium* spp.), peanuts (*Arichis hypogaea*), soybeans (*Glycine max*), cotton (*Gossypium hirsutum*), rice (*Oryza sativa*), wheat (*Triticum* spp.), oats (*Avena sativa*), barley (*Hordeum* spp.), rye (*Secale cereale*), maize (*Zea mays*), sorghum (*Sorghum bicolor*), and sugarcane (*Saccharum officinarum*), and there may be others. It is difficult to catalogue all of the different host-plant resistance programs in the USA, and it is even more difficult to catalogue those in other countries. There is an international group of researchers on host-plant resistance, which on an informal basis, meets biannually for a two-day session and which publishes an annual newsletter. In the 1978 "Host-Plant Resistance Newsletter," 307 USA and 169 foreign researchers indicated that they were working in host-plant resistance. Contributions from 64 USA

J. N. Jenkins, Research Geneticist, Agricultural Research, Science and Education Administration, USDA and Adjunct Professor of Agronomy, Mississippi State University, Mississippi State.

researchers in 20 states and 34 scientists from nine other countries were included in the 1978 newsletter. Obviously, many researchers other than plant breeders are involved in insect resistance work, with nearly every breeding program having an entomologist and a plant breeder working as a team. Several programs also have biochemists studying the nature of resistance. Hedin et al. (1977) cited 258 papers in a limited review of research on the chemical basis for insect resistance. Compounds that have been isolated from plants exhibit biological activity with more than 100 species of insects. The book entitled *Host-Plant Resistance to Pests* (Hedin, 1977) has 16 chapters and 34 authors.

The number of research programs mentioned in this brief introduction indicates an intense scientific endeavor in the field of host-plant resistance to insects. Thus, it is appropriate and timely to discuss this subject at a symposium on plant breeding. In this chapter selected specific examples have been reviewed to illustrate principles involved in breeding programs for insect resistance. The following topics are covered: (1) significance of host-plant resistance to insects in plant breeding programs, (2) breeding methodology in host-plant resistance to insects, (3) how resistance type affects the breeding program, and (4) special problems related to permanence of resistant cultivars.

SIGNIFICANCE OF HOST-PLANT RESISTANCE TO INSECTS

Painter (1951) in his classic textbook cites several examples of early work in host-plant resistance to insects. The apple (*Pyrus malus*) cultivar 'Winter Majetin,' was first reported to be resistant to the woolly aphid (*Eriosoma lanigerum* (Hausm.)) in 1831. The first extensive observations on resistance of wheat to the Hessian fly (*Mayetiola destructor* (Say)) were between 1886 and 1892 in California. Another well known example of host-plant resistance is to grape phylloxera (*Phylloxera vitifoliae* (Fitch)). Phylloxera resistant stocks were exported from the USA to France about 100 years ago to combat that insect, and those vines still form an important means of control for the insect in France.

To show the significance of breeding for insect resistance, examples will be discussed in four crops—alfalfa, wheat, maize, and cotton.

Alfalfa

Alfalfa is an important forage crop, and the spotted alfalfa aphid (*Therioaphis maculata* (Buckton)) which attacks this crop seriously was accidentally introduced into the USA in 1954. Following major crop losses from this insect, much research effort was devoted to breeding alfalfa cultivars with resistance to it. This effort has led to the release of more than 30 alfalfa cultivars with resistance to spotted alfalfa aphid. Many of them have multiple pest resistance combined with high forage yields (Nielson and Lehman, 1980). The success attained in breeding for spotted alfalfa aphid was so spectacular that nearly all current alfalfa breeding programs include a primary objective of breeding for insect resistance. In addition to the spotted alfalfa aphid, resistance to the pea aphid (*Acyrthosiphon pisum* (Harris)), the blue alfalfa aphid (*A. kondoi* (Shinji)), the alfalfa weevil (*Hypera postica* (Gyllenhal)), the Egyptian alfalfa weevil (*H. brunneipennis* (Boheman)), the potato leafhopper (*Empoasca fabae* (Harris)), the meadow spittlebug (*Philaenus spumarius* (Linnaeus)), the alfalfa seed chalcid (*Bruchophagus roddi* (Gussakovsky)), and lygus bugs (*Lygus hesperus* (Knight) and *L. lineolaris* (Palisot de Beauvois)) are being sought in various alfalfa breeding programs in the USA.

Much resistance to the aphid species in alfalfa cultivars is due to high degrees of antibiosis. This places an intense selection pressure upon the insect population, so many aphid biotypes have developed. As a consequence, a succession of cultivars has been required to maintain sufficient resistance to grow alfalfa economically in the areas where aphids are serious pests. Government imposed restrictions upon the use of insecticides on alfalfa fed to cattle have intensified the need for insect resistance as a cultivar characteristic.

Wheat

In breeding of small grains, and particularly wheat, insect resistance has played a major role. In the early twentieth century differences were noted in wheat cultivars for degree of susceptibility to the Hessian fly. Painter (1951) summarized the results of a classification of about 400 cultivars, selections, and hybrids of wheat for resistance to the Hessian fly. One of the first papers to characterize insect resistance on the basis of genetic factors in the host was that by Cartwright and Wiebe (1936) for the Hessian fly in two wheat crosses.

Maxwell et al. (1972) reported that over four million ha in 34 states were planted with 23 Hessian fly resistant cultivars of wheat, and the annual value of increased yield resulting from those resistant cultivars was estimated at $238 million. The primary types of resistance to Hessian fly are antibiosis and tolerance, and the insect has developed several biotypes. The Great Plains race is most prevalent west of central Kansas, whereas races A, B, C, D, and E are found in the eastern soft wheat region.

The wheat stem sawfly (*Cephus cinctus* (Norton)) has long been known as a pest of wheat. Much of the resistance to this insect is due to a solid stem character in wheat; however, solid stem often is associated with low yielding capacity, and for this reason, resistant cultivars are not grown universally in areas where the insect is a pest.

A long standing breeding effort has been concerned with resistance of wheat to the greenbug (*Toxoptera graminum* (Rond.)), and recently, some wheat breeding programs have included work on resistance to the cereal leaf beetle (*Oulema melanopa* (L.)).

Maize

Painter (1951) gave an extensive survey of the early studies for resistance of maize to the European corn borer (*Ostrinia nubilalis* (Hubner)), and recently, Brindley and Dicke (1963), and Brindley et al. (1975) reviewed comprehensively all research on this insect since its introduction in the USA. Early breeding work for resistance to European corn borer in maize used natural infestations of the insect to screen lines for resistance, but this was a very unsatisfactory technique. Now, however, researchers are able to rear the European corn borer on artificial media in sufficient quantities to obtain millions of egg masses for artificial infestation of maize plants. This breakthrough has greatly increased the sureness and precision of the research, and the number of maize lines that can be evaluated is almost unlimited. Corn Belt hybrids carry resistance to the first brood European corn borer, and in the late 1960s, efforts were begun to develop resistance to second brood borers. Most major maize seed companies have ongoing programs in breeding for resistance to this pest.

The southwestern corn borer (*Diatraea grandiosella* (Dyar)) is an important pest of maize in many areas of the USA. Research has been done on plant resistance to this insect since the mid-1960s, and a few inbred lines that exhibit resistance have been released by experiment stations. The research effort of the USDA, Science and Education Administration, Agricultural Research, and the Mississippi Agricultural and Forestry Experiment Station have led to the development of genotypes resistant to southwestern corn borer, as well as in promoting interest in breeding for resistance (Davis, 1976).

In most of the southern USA, the fall armyworm (*Spodoptera frugiperda* (J. E. Smith)) is a limiting factor in the production of late planted maize. Recently, Williams et al. (1978) reported resistance to this insect in breeding lines developed in Mississippi.

Cotton

Cotton is grown under the protection of large amounts of insecticides in much of the U.S. Cotton Belt, but also, there has been a substantial research effort in breeding for insect resistance in this crop. The earliest use of escape as a mechanism for reducing damage from an insect was applied in the breeding for earliness of maturity in cotton during the initial years of the boll weevil (*Anthonomus grandis* (Boheman)) attack. The boll weevil spread rapidly across the Cotton Belt between 1892 and 1920, and even though cotton breeding was just beginning during that period it was recognized that earliness was an important factor in escaping boll weevil damage. Several resistant cultivars were developed rapidly. This use of increased earliness represented an indirect approach to insect resistance in cotton.

An excellent discussion of the early development of jassid (*Empoasca* spp.) resistance was provided by Painter (1951). Genetic resistance to the jassid, initially developed over 50 years ago in South Africa (Parnell, 1925) and India (Hutchinson, 1962), was the first case of success in using resistant cultivars to control a pest of cotton. Growing pilose cultivars has essentially eliminated jassid as a major cotton pest in tropical Africa. The general consensus among research workers is that plant hairiness and resistance to jassids are associated, but opinions vary about the degree to which other plant traits influence jassid resistance.

Beginning around 1960, increased emphasis was placed upon research on host-plant resistance to insects in cotton because of the experience with the boll weevil developing resistance to chlorinated hydrocarbon insecticides and the increased cost of controlling cotton insects with insecticides. Since that time, numerous researchers have investigated host-plant resistance to bollworm (*Heliothis* spp.), pink bollworm (*Pectinophera gossypiella* (Saunders)), *Lygus* spp., and spider mites (*Tetrancychus urticae* (Koch)). Sources of resistance to each of these arthropods are available presently in most public and private cotton breeding programs in the USA. No proprietary commercial cultivars of cotton are advertised as being resistant to insects, but one commercial cultivar of cotton is nectariless, and this trait reduces populations of lygus, *Heliothis*, and pink bollworm. Two cultivars, 'Gumbo' and 'Pronto,' have modified leaf shape which increases resistance to banded wing whitefly (*Trialeurodes abutilonea* (Haldeman)). With the high cost of insect control in cotton and the many breeding programs emphasizing host-plant resistance, cultivars with good levels of resistance to the major insects will be available soon.

Recent Efforts with New Insect-Crop Situations

Several recently initiated research programs involve breeding for insect resistance in relatively new insect-crop situations. In sorghum, the recently developed resistance to the greenbug (*Schizaphis graminum* (Rondani)), biotype C, and the release of commercial hybrids with this resistance marks an important step forward for insect control in this crop. Several stations are developing resistance in soybeans to leaf-feeding beetles and Lepidopterous insects. The advents of alfalfa weevil and Egyptian alfalfa weevil have caused an increased effort to breed alfalfa cultivars with resistance to these pests, but so far these programs have not met with the same degree of success as those breeding for aphid resistance. These weevils are of importance in some areas; thus, if breeding for resistance is not successful, alfalfa production may be impaired in these areas. Commercial companies are now rearing southwestern corn borers so they can have eggs to use in screening maize genotypes for resistance. Cotton breeders at universities and with the USDA now have a good base of experience and cotton germplasm to build on for resistance to *Heliothis* spp., boll weevil, *Lygus* spp., and bollworm, and commercial breeders also are becoming more interested in using these basic sources of resistance. Recently, earliness of a different level of magnitude from that used as an escape mechanism in the boll weevil era has been used by producers and seed companies in the mid-South and Texas. Several early-maturing determinant and indeterminant cultivars are being grown in the Cotton Belt, and new crop production systems built around these early cotton strains are being developed in the more arid areas of the Cotton Belt.

Chemistry of Insect Resistance

Considerable research has been done on the chemistry of insect resistance in cotton, and this work has been reviewed by Waiss et al. (1977), Stipanovic et al. (1977), and Hedin et al. (1977).

Resistant Cultivars as Major Means of Control

In several crops, host-plant resistance is the major means of insect control. As examples, for alfalfa, host-plant resistance is the major means of control for the spotted alfalfa aphid and the pea aphid; for wheat, the Hessian fly is controlled primarily by resistant cultivars. Resistance to the European corn borer has come to play an important role in maize production in the USA also.

Resistant Cultivars Used in Integrated Pest Management

Host-plant resistance is utilized as a component in an integrated program of insect control in several crops. For example, in cotton, the traits of earliness and nectariless are valuable components in the management of cotton insects. For alfalfa, moderate levels of resistance to the alfalfa weevil are available, but these levels of resistance must be supplemented with insecticide application to give good control of this insect. A good level of resistance to the greenbug is available in sorghum, but insect control is obtained only when resistant hybrids are grown where predators can have a significant impact upon the greenbug population. In each above mentioned crop, an increase of host-plant resistance will occur as new germplasm sources become available.

Potential Programs in Host-Plant Resistance to Insects

Several crops or crop-insect situations have potential for a greatly expanded use of host-plant resistance, but for these, essentially no resistant cultivars are being used. Examples are: the rice water weevil (*Lissorhoptrus oryzophilus* (Kuschel)) and the stinkbug (*Oebalus pugnax* (Fab.)) in rice; *Heliothis* spp., *Lygus* spp., and pink bollworm in cotton; and southwestern corn borer and fall armyworm in maize. For each of these crops, the presently available germplasm shows a great potential for developing cultivars with resistance.

Regulatory Measures Affecting Use of Resistant Cultivars

When considering the value of resistant cultivars in future breeding programs, we should be aware that the Environmental Protection Agency (EPA) is placing more and more restrictions on the use of pesticides. With food crops, the Food and Drug Administration (FDA) regulations also will have an effect upon what pesticides can be used, as well as the type of resistant cultivars developed. For example, for crops that come under FDA regulations, toxins in the plant genotype to bring about insect resistance must be below certain levels before that crop can be sold for food. This further complicates breeding programs. Because of this complication, there is a need to begin cooperation with chemists to determine the chemical bases of resistance.

Integrated Pest Management and Resistant Cultivars

Integrated pest management (IPM) programs can reduce pesticide use on crops in many situations, and it may allow some crops to be produced in areas where pesticides alone have failed or become excessively expensive. IPM should permit the use of lower levels of resistance in crop cultivars than are necessary when plant resistance is used alone. This then becomes a very important aspect of breeding for insect resistance. Because lower levels of resistance may be usable in IPM programs, the number of cultivars available for commercial use may increase. Also, the use of resistant cultivars in an IPM program should reduce the selection pressure for the development of pesticide-resistant biotypes of insects.

BREEDING METHODOLOGY FOR HOST-PLANT RESISTANCE TO INSECTS

Gene Control of Host-Plant Resistance to Insects

The literature on the inheritance of resistance to insects in plants shows that some resistance traits are controlled qualitatively by dominant genes, others by recessive genes, and yet others show quantitative inheritance. There are examples of resistance due to morphological plant traits and examples of resistance due to physiological characteristics. All in all, the breeding strategy to be used when developing cultivars with insect resistance depends upon the crop species and other breeding objectives in the program, with due emphasis given to selection for resistance.

Special Considerations in Breeding for Insect Resistance

Several special aspects must be considered in a breeding program for insect resistance, particularly when breeding for insect resistance is integrated into traditional breeding programs. The first set of considerations deals with insect-plant interactions. There are some similarities, yet some marked differences, between insect-plant interactions and pathogen-plant interactions. For example, insects can and do exercise choice. Their choices vary with the situation under which the insect is placed, such as monoculture vs. a varied crop culture. Another aspect to be considered is the life cycle of the insect. Some insects feed on both crop and weed species. The generation time and the population dynamics of the insect species must be considered also. Some insects have several generations per year, whereas others have only one. Obviously, this has a profound effect on how a breeding program is structured. Some insects have a very low potential for population size increase, whereas others have "explosive" potential; for these two situations, the levels and types of resistance needed may be very different.

Finally, with some crops, several insect species are usually involved simultaneously, and resistance to one particular insect may affect other pest species of that crop.

A second consideration in breeding plants for insect resistance involves people-to-people interactions; that is, by necessity, an entomologist must become a member of the research team. This scientist has a vital role to play in the breeding program. It is mandatory for the entomologist and plant breeder to work as a team and to share equally in credit that results from their cooperative work. Plant breeders have a long history of cooperation with plant pathologists, and in a few breeding programs, there is a history of cooperation with entomologists, but generally, this type of team cooperation represents a new experience for entomologists. The entomologist, because it represents a new endeavor for him, must develop knowledge and appreciation for plant breeding, and likewise, the plant breeder must develop an appreciation for and understanding of the general field of entomology. Areas that are frequently new to plant breeders are the population dynamics of the insect and the ability of many insects to attack more than one host crop.

A third special aspect of breeding for insect resistance in host plants deals with having an insect supply for infesting plant breeding nurseries. Either natural or artificial infestations can be used, but often natural infestations fail to give good differentiation among plant genotypes because the insects are insufficient in number or they do not infest a whole nursery uniformly. In this situation, it may be possible to enhance natural infestations. An example in cotton is the "nurse-crop principle" with the tarnished plant bug (Laster and Meredith, 1974), whereby garden mustard (*Brassica juncea*) is interplanted with cotton to serve as a nurse crop for the natural rearing of populations of the tarnished plant bug early in the year. With this technique, large populations of this insect can be obtained in the field immediately surrounding the cotton plants, and the critical factors of numbers, uniformity, and consistency can be met. If enhancement of natural infestations is not possible, the entomologist must innovate a laboratory rearing program to provide insects for artificial infestations. For example, European (Guthrie et al., 1965) and southwestern corn borers (Davis, 1976) can be reared and induced to lay eggs in the laboratory. These masses are placed on maize and sorghum plants to give uniform, high-level infestations. This technique also permits the timing of infestation so that evaluation of plants can be done at the convenience of the breeder and entomologist; also, it permits the planting and evaluation of plant material at times in the season that would not be possible under natural infestation.

The procedure for evaluating insect resistance in plant genotypes must meet certain criteria:

1. A good supply of eggs, larvae, or adult insects must be available for infesting plants in a breeding nursery. If the insects are reared in the laboratory, they must represent the wild population in vitality, biotype composition, and genetic structure, and they must be nourished so that their behavior and population capabilities are similar to those in the wild.

2. A rapid, repeatable rating scale must be developed that is related to the development of the insect, or to the economic damage done by the insect. Sometimes both items are used in scoring plant genotypes. The effects of the plant upon the insect, namely, insect weight, fecundity, growth interval of instars, can be measured, or the effects of the insect upon the plant, such as a damage rating scale or a recovery rating, can be assayed. Under any circumstances, the evaluation ratings must be rapid and repeatable.

The speed and cost of a screening procedure can materially affect the breeding methodology used in an insect resistance program.

1. In evaluating resistance, the degree of damage that the insect does to the plant, the uniformity of the infestation, and the inheritance pattern for resistance will determine whether single plants or progeny-row evaluations need to be utilized.

2. Certain stages of plant growth are more suited to evaluation for insect resistance than are others. For some insects, seedlings can be evaluated, whereas with others, ratings must be made on more mature plants. One type of resistance to insects involves feeding site specificity. For example, first and second broods of the southwestern (Davis et al., 1972) and European corn borers (Russell et al., 1974) feed on different plant tissues, and good resistance to the first brood does not necessarily confer resistance to the second one. Also, the first generation of southwestern corn borer does not diapause, whereas the second generation does; thus, the different physiology of the two generations may affect insect behavior and feeding.

3. Insects differ in the types of damage they to to the plant

and this fact must be considered in the evaluation scale. If the insect kills the plant or chews off parts of the leaves, a damage rating scale is easy to develop. However, where insects damage fruit, reduce yields, or delay maturity of the crop, a rating scale that measures resistance is difficult to construct. For example, if the insect damages a fruit or vegetable, it is necessary to test the insect on the fruit or vegetable itself, and this requires special evaluation techniques. It suffices to say that a primary responsibility of the entomologist is to develop rapid, economic, consistent techniques for evaluating plant genotypes for reaction to an insect pest and this requires the availability of a dependable and adequate supply of healthy insects. It also requires the continuous interaction of the plant breeder and entomologist to assure that the techniques developed can be used efficiently and that they can be effectively integrated into a breeding program.

When host-plant resistance to an insect is discovered, provision must be made to assure that the cultivar, inbred line, or germplasm source is in adequate supply and is made available to all breeders, public and private alike, of this crop. Usually, the breeder on a breeder-entomologist team takes this responsibility. By all means, the breeding team that works on host-plant resistance in a public institution must foster close cooperation with commercial companies that have breeding programs on the crop to catalyze the use of the host-plant resistance in their breeding programs. Sometimes the breeder-entomologist team must develop rearing, infestation, and selection techniques adaptable to the needs of commercial companies as well as public research institutions.

In summary, inheritance patterns for insect resistance in plants are no different from those for other plant traits, but special aspects must be considered. These deal with insect-plant interactions; the need for entomologists, plant breeders, and chemists to work as a team; appropriate supplies of insects and evaluation techniques for screening plant progenies; and close cooperation with plant breeding companies, so that the technology and germplasm needed for breeding for insect resistance are used.

HOW RESISTANCE TYPE AFFECTS THE BREEDING PROGRAM

Resistant Cultivars as the Only Method of Control

Resistant cultivars can be used as the only method for controlling insects or they may be used as one component in an integrated pest management program (IPM). Where host-plant resistance is the only method used for control, the breeding team must determine what degree of resistance gives economically acceptable insect control. For example, vegetables are quite different from alfalfa in the degree of insect damage that can be tolerated in the salable product from the crop. Also, the breeding team must determine the relation of the resistant cultivar to insect biotype selection. When the insect is an obligate feeder on the crop in question, or the crop is grown on large hectarages, or when the type of resistance utilized places a high selection pressure upon the insect population, biotypes of the insect are likely to be differentiated. And of course, if biotype selection does occur, the agricultural production life of any one cultivar is apt to be quite short, that is, two to five years. Therefore, the team must consider whether the resources available to it will permit the development and release of a new cultivar that often. If not, a type of resistance should be used that will minimize biotype selection. Tolerance or multiline cultivars may be satisfactory choices in such situations.

Resistant Cultivars as a Component in IPM

Two factors that affect the use of host-plant resistance as one of the components in IPM are the degree of resistance required and whether the insect has single or multiple generations per year on the crop.

For insect populations that build up from small parental stocks over several generations, a low level of resistance may compound over generations to lengthen the time required to reach an economically damaging level. A resistant cultivar may enhance chemical and natural-enemy control of insects in IPM, that is, a resistant cultivar may serve as a base component in IPM. In fact, a cultivar with low-level resistance to an insect is quite useful as a base component in an IPM program, whereas it has limited value when used as the only method of control. For example, earliness and determinancy in cotton (Walker et al., 1978) are quite successful in an IPM program involving all aspects of cotton production. And, in this setting, the other components of the IPM program may enhance the resistant cultivar's capacity to cause biotype selection.

Mechanisms of Resistance

Three mechanisms of host-plant resistance to insects have been defined. The term "antibiosis" was proposed by Painter (1951) and defined as those factors of a resistant plant that cause adverse effects on the insect life cycle when the insect uses that plant for food. He defined "tolerance" as resistance in which the plant shows an ability to grow and reproduce or to repair insect injury to a marked degree in spite of supporting an insect population nearly equal to that damaging a susceptible host. Kogan and Ortman (1978) developed the term "antixenosis" for plant characters that cause differential oviposition and habitation by the insect on the host plant. Antixenosis means that the plant is avoided because it is an undesirable host. Very good examples of each of these three mechanisms of resistance can be given; although, in most cases more than one mechanism may be operating.

If the team has the opportunity to choose the type of resistance that will be utilized, it should select the one that places the least selection pressure on the insect population for new biotypes while providing useful protection to insects in the crop production system. Where possible, two or more types of resistance should be combined into the same cultivar; however, for some insects that feed on multiple hosts, a single mechanism of resistance will prove sufficient.

The production system in which the resistant cultivar is grown may affect the value of its resistance type. Antixenosis may give resistance where the insect has several different plants to choose for food or oviposition, whereas it would not give resistance when used in monoculture. This type of resistance is useful in strip cropping for controlling insect damage.

Antibiosis is the most striking mechanism of insect resistance. High levels of antibiosis usually place great selection pressure on the insect for new biotypes, especially if the insect is a primary or obligate feeder on one crop. Again, the effect of the plant on the insect's life history and its population dynamics is much more important than the art of classifying the mechanism of resistance.

PERMANENCE OF RESISTANT CULTIVARS: SPECIAL PROBLEMS

Number of Biotypes in the Population

Several factors affect the length of time a resistant cultivar remains resistant to a particular insect species. Of primary importance is the number of biotypes that comprise the original field population of insects. Usually, the number is not known initially, and it only becomes obvious after resistant cultivars have been widely grown. In some instances, a resistance source has been used for a number of years without the selection of biotypes that can damage the resistant plants. Examples are the woolly aphid on the resistant apple, 'Winter Majetin,' and the use of resistance to first brood European corn borer in Corn Belt hybrids. On the other hand, the breeding history of aphid resistant alfalfa cultivars has caused a succession of biotype selections that can survive on formerly resistant cultivars, and Hessian fly biotypes are strongly selected by resistant wheat cultivars. The probability of biotype selection by the resistant cultivars should be considered carefully when a research team is choosing the intricacies for a breeding program. Failure to do so can lead to very short periods of commercial usefulness for cultivars developed by the team.

Selection Pressure for Biotype Development

The degree of permanence of a resistance source, which is really mediated via selection pressure among insect biotypes, can be influenced by those factors in the environment of the insect that favor the survival of insects that can reproduce on the resistant cultivar. Biotype selection is more probable in some insects than others. For example, insects that reproduce parthenogenetically, are obligate feeders on a single host, or feed on plant species that have high levels of antibiosis, are more subject to severe selection for biotypes than are those that do not meet one of these criteria. Generally, when antixenosis or tolerance is used as an insect resistance mechanism there is some selection pressure on the insect, but the pressure may not be as great as with antibiosis. For example, a resistant cotton cultivar with the trait, frego bract, causes the boll weevil female to oviposite fewer eggs on a cotton flower. The females spend more time moving about on the plants in a state of frustration with the result that each produces fewer progeny (Mitchell et al., 1973). This type of resistance is very usable, but it places very little selection pressure on the insect population. A recent example where a very low selection pressure is applied on the insect population is provided with resistance to the greenbug in sorghum (Teetes et al., 1974), which is of the tolerance type; namely, the greenbug populations are not reduced drastically on the resistant hybrids, but the plants are not damaged to the same degree as in susceptible hybrids.

Additional factors in biotype selection are the number of insect generations per year, which can affect the speed of selection and host range of the insect. Some insects feed primarily on one host-plant species, whereas others feed on a wide range of cultivated and wild species of plants. It is quite clear that an insect species that feeds on a number of plant species has less pressure for biotype selection placed upon its population by a resistant cultivar than does a species that feeds on only one plant species. And of course, the proportions of land area grown to a single crop and the relative proportions of this hectarage sown to resistant and susceptible cultivars can affect the degree of biotype selection pressure placed on an insect population. Even with an insect that is specific to one plant species, selection pressure is reduced if the entire area is not planted with the resistant cultivar. Annual geographic migration of insects also is an important factor in biotype selection. Usually, there is a general mixing of populations of insects due to migration from one area to another. Some insects move to a new location after each generation, whereas others tend to stay near the location where they emerged. Sedentary insects would be more subject to biotype selection than would migratory ones.

Management of Resistant Cultivars

How insect resistant cultivars are managed will have a pronounced effect on the permanence of the resistance. In a cropping sequence where resistant cultivars are only one component of an IPM program, all factors of the program place different selection pressures on the insect; so there is less likelihood for rapid selection of a new biotype that can attack the resistant cultivar than where the major or only means of control is the resistant host cultivar. For example, strip-cropping sorghum in cotton significantly reduces *Heliothis* spp. damage because predators of the *Heliothis* spp. build up in the sorghum (Massey and Young, 1975). This cultural practice greatly reduces selection pressure of the cotton cultivar used in this production scheme upon the *Heliothis* population.

Each host-plant resistance research team must choose production strategies to increase the permanence of their resistant cultivars. Nearly all strategies revolve around reducing biotype selection pressure to delay the emergence of new dominant biotypes that can infest a resistant cultivar. All factors that have been discussed should be considered in developing this strategy. Two strategies that have been used in managing resistance genes to control plant diseases should be considered also for insect resistance, that is, multiline cultivars and gene deployment. Both have possibilities for use with insect resistant cultivars, but multiline cultivars probably have greater potential than gene deployment for increasing the permanence of insect resistance. Gene deployment appears less useful for insect resistance than for disease resistance because the geographic migration of most insects is usually less than that of airborne disease organisms. In general, however, the population parameters that are considered when choosing between multiline cultivars, pure-line cultivars, and gene deployment are the same for disease resistance and insect resistance genes. Thus, in some situations multiline cultivars could work well, but in others, not so well. There may even be situations where geographic deployment of resistance genes would fit.

SUMMARY

In summary, host-plant resistance to insects is a significant element in crop breeding, as shown by the examples of alfalfa, wheat, maize, and cotton. Recently expanded programs of breeding for insect resistance in sorghum, soybeans, alfalfa, maize, and cotton have occurred. In some cases, host-plant resistance is the major means for controlling insects, whereas in others resistant cultivars were used as a part of an IPM. Resistant cultivars as a base component in IPM programs is a new and important concept. And there is potential for expanded use of resistant cultivars in several crops. Government regulation of levels of toxins in edible crops may dampen the breeding of resistant cultivars, so it is important to learn more about the chemistry of insect resistance. Thus, the chemical basis of resistance is an area receiving increased emphasis.

The inheritance of insect resistance is similar to that of other plant traits, but there are several special considerations that affect plant breeding programs where insect resistance is a goal. These are insect-plant interactions, the team approach to breeding cultivars, and developing adequate supplies of insects for infestation, and good evaluation criteria. A good insect supply and a rapid repeatable rating scale are necessities, and resistance sometimes involves feeding site specificity.

How insect resistant cultivars are used in agricultural production can have significant consequences on their value. Where resistant cultivars are used as the only means of insect control, the effects of the resistant cultivar on insect biotype selection are considerable. When resistant cultivars are used as a basic component in an IPM program, the cultivar may enhance the total insect control and the impact that the IPM program has on insect biotype selection.

The permanence of resistant cultivars is influenced by the

number of biotypes originally in the field populations, the type of insect reproduction, the effect that the plant exerts on the insect, the number of insect generations per year, the host range of the insect, the proportions of resistant and susceptible cultivars grown, and insect migration between generations and between years.

The management of resistant cultivars within an IPM program can enhance the longevity of resistant cultivars. Multiline cultivars and regional deployment of genes for insect resistance may be valuable strategies for making insect resistance more permanent. Certainly, the parameters that are used to determine the value of these strategies for controlling diseases are the same ones to use when deciding their value for controlling insects.

REFERENCES

Brindley, T. A., and F. F. Dicke. 1963. Significant developments in European corn borer research. Annu. Rev. Entomol. 8: 155-76.

Brindley, T. A., A. N. Sparks, W. R. Showers, and W. D. Guthrie. 1975. Recent research advances in the European corn borer in North America. Annu. Rev. Entomol. 20: 221-39.

Cartwright, W. B., and G. A. Wiebe. 1936. Inheritance of resistance to the hessian fly in the wheat cross Dawson x Poso and Dawson x Big Club. J. Agric. Res. 52: 691-95.

Davis, F. M. 1976. Production and handling of eggs of southwestern corn borer for host plant resistance studies. Miss. Agric. and Forestry Exp. Stn. Tech. Bull. 74. 11 pp.

Davis, F. M., C. A. Henderson, and G. E. Scott. 1972. Movements and feeding of larvae of the southwestern corn borer on two stages of corn growth. J. Econ. Entomol. 65: 519-21.

Guthrie, W. D., E. S. Raun, F. F. Dicke, G. R. Pesho, and S. W. Carter. 1965. Laboratory production of European corn borer egg masses. Iowa State J. of Sci. 40(1): 65-83.

Hedin, P. A., (ed.). 1977. Host-Plant Resistance to Pests. Am. Chem. Soc. Symp. Ser. 62.

Hedin, P. A., J. N. Jenkins, and F. G. Maxwell. 1977. Behavioral and developmental factors affecting host plant resistance to insects. Am. Chem. Soc. Symp. Ser. 62: 231-75.

Hutchinson, Joseph. 1962. The history and relationships of the world's cottons. Endeavour 11(81): 5-15.

Kennedy, G. G. 1978. Recent advances in insect resistance of vegetable and fruit crops in North America: 1966-1977. Entomol. Soc. Am. Bull. 24(3): 375-84.

Kogan, M., and Ortman, E. F. 1978. Antixenosis: A new term proposed to define Painter's "Nonpreference" modality of resistance. Entomol. Soc. Am. Bull. 24: 175-76.

Laster, M. L., and W. R. Meredith, Jr. 1974. Evaluating the response of cotton cultivars to tarnished plant bug injury. J. Econ. Entomol. 67: 686-88.

Massey, W. B., Jr., and J. H. Young. 1975. Linear and directional effects in predator populations, insect damage, and yield associated with corn and sorghum. Environ. Entomol. 4: 637-41.

Maxwell, F. G., J. N. Jenkins, and W. L. Parrott. 1972. Resistance of plants to insects. Adv. in Agron. 24: 187-265.

Mitchell, H. C., W. H. Cross, W. L. McGovern, and E. M. Dawson. 1973. Behavior of boll weevil (*Coleoptera: Curculionidae*) on frego bract cotton. J. Econ. Entomol. 66: 677-80.

Nielson, N. W., and W. F. Lehman. 1980. Breeding approaches in alfalfa, pp. 277-312. *In* Maxwell, F. G., and P. R. Jennings (eds.). Breeding plants for resistance to insects. Wiley-Interscience, New York.

Painter, R. H. 1951. Insect resistance in crop plants. McMillan Co., New York.

Parnell, F. R. 1925. The breeding of jassid-resistant cottons: Report for the season 1924-25. Empire Cott. Grow. Rev. 2: 330-36.

Russell, W. A., W. D. Guthrie, and R. L. Grindeland. 1974. Breeding for resistance in maize to first and second broods of the European corn borer. Crop Sci. 14: 725-27.

Stipanovic, R. D., A. A. Bell, and M. J. Lukefahr. 1977. Natural insecticides from cotton (*Gossypium*). *In* Host plant resistance to pests. Am. Chem. Soc. Symp. Ser. 62: 197-214.

Teetes, G. L., C. A. Schaefer, J. W. Johnson, and D. T. Rosenow. 1974. Resistance in sorghums to the greenbug: Field evaluation. Crop Sci. 14: 706-8.

Waiss, A. C., Jr., G. G. Chan, and C. A. Elliger. 1977. Host plant resistance to insects. *In* Host plant resistance to pests. Am. Chem. Soc. Symp. Ser. 62: 115-28.

Walker, J. K., R. E. Frisbie, and G. A. Niles. 1978. A changing perspective: *Heliothis* in short-season cottons. Entomol. Soc. Am. Bull. 24: 385-91.

Williams, P. W., F. M. Davis, and G. E. Scott. 1978. Resistance of corn to leaf-feeding damage by fall armyworm. Crop Sci. 18: 861-63.

CHAPTER 9

Disease Resistance in Plants and Its Consequences for Plant Breeding

J. E. PARLEVLIET

IN MODERN AGRICULTURE, the dynamic nature of the host-pathogen relationship is evident through the frequency by which the pathogen overcomes the sources of resistances introduced into agricultural use. Loss of resistance was observed as early as 1916 (Kommedahl et al., 1970), but it took some 40 years before the seriousness of this phenomenon was fully realized. Van der Plank (1963, 1968, 1975) must be credited for developing a general hypothesis to explain the dynamics of the host-pathogen relationship. His ideas have generated a considerable discussion (Nelson, 1975; Robinson, 1976; Parlevliet and Zadoks, 1977; Parlevliet, 1977) including controversies. Basically, Van der Plank classified host-plant resistances as horizontal and vertical resistance, but this approach is too simple. Clifford (1975) stated this reservation: "In common with other workers, the author accepts the convenience of cataloguing resistance into two types. Nature, I am sure, never intended this division." There is a vast number of host-pathogen interaction systems in various stages of coevolution, so any classification of resistance must be a rough one. Nevertheless, to gain insight into the way host-pathogen systems evolve, which is a prerequisite for developing an effective host breeding program, the coevolution of the two organisms must be analyzed and described in a generalized way.

Defense of host plants against parasites may be due to either avoidance or resistance mechanisms (Parlevliet, 1977). Avoidance reduces the chance of contact between the prospective host tissue and the parasite, whereas resistance operates when host tissue and

Institute of Plant Breeding, Agricultural University, Wageningen, The Netherlands.

the parasite come into contact. Mechanisms of avoidance may be (1) the production of volatile repellent compounds (such as pyrethrins in *Chrysanthemum cinerariaefolium*); (2) mimicry (such as the resemblance in leaf shape of *Passiflora* species, host of *Heliconius* larvae, to sympatric species inedible to *Heliconius* [Harper, 1977]); (3) features of the leaf surface like the presence of hairs, thorns, or resin ducts, and composition and texture of the leaf surface; (4) sparse and scattered occurrence of the host; and (5) the susceptible tissues being present for only a short time during the host's life cycle (Feeny, 1976). Most avoidance mechanisms are morphological and are designed against animal parasites, but some do operate against viral, bacterial, and fungal pathogens.

Resistance mechanisms often are chemical in nature, and they may be naturally occurring or induced. Naturally occurring resistance compounds are present in host tissues prior to its contact with the pathogen, whereas induced compounds occur only after such contact. Chemical defense to feeding by animals is mostly naturally occurring, whereas chemical defense to pathogens often is induced (Levin, 1976). Therefore, induced chemical resistances as defense mechanisms to pathogens will be emphasized in this chapter.

ASSESSMENT OF RESISTANCE

Resistance refers to the ability of the host to interfere with the normal growth and/or development of the pathogen; so to assess resistance, the growth and development of the parasite in or on the host must be measured. This is only possible for ectopathogens like powdery mildew and some other biotrophics like rusts. In most cases, disease symptoms are assessed as if they reflect the quantitative growth of the pathogen in the host.

The assessment of disease resistance is done in various ways:

1. Disease incidence is defined as the proportion of plant units infected, that is, percentage of diseased plants or ears.

2. Disease severity is defined as the proportion of the total area of plant tissue affected by disease (James, 1974).

3. More detailed evaluations measure the number and size of successful infections. For pathogens that spread according to the compound interest law (Van der Plank, 1963), disease severity is the cumulative result of infection frequency (proportion of spores that result in sporulating lesions), latent period (time from infection to spore production), spore production (spores produced per lesion

or per unit area of tissue per unit of time), and infectious period (period of sporulation).

A different way of assessment is used with many biotrophic leaf pathogens, such as rusts and powdery mildews, to which the host may react to give high resistance. With this reaction type, a halo of necrotic or chlorotic tissue develops around the infection point, and the pathogen growth is hindered qualitatively (no sporulation) or quantitatively (restricted sporulation). The disease development is rated by infection types (IT), where IT 0 equals a necrotic or chlorotic fleck without sporulation, and IT 4 equals a sporulating pustule without chlorosis or necrosis. IT 1, 2, and 3 describe pustules surrounded by necrotic or chlorotic tissue with increasing intensities of sporulation. Generally IT 3 and 4 are classed as susceptible or high IT reactions, and IT 0, 1, and 2 are classed as resistant or low IT reactions. Resistances of the low IT type may be expressed throughout the plant's life cycle (seedling resistance) or in the adult-plant stage only (adult plant resistance). Low IT type resistance is often race-specific and simply inherited (Parlevliet, 1978d).

THE HOST-PATHOGEN SYSTEM

The natural ecosystems from which our crops and their pathogens originated were extremely complex and both evolved with dependence upon the physical and biological environments. One part of this ecosystem was a coevolution involving the mutual interactions between hosts and pathogens, resistance and pathogenicity being the principal reflections of it, respectively.

Evolution and Coevolution

In such an ecosystem, the pathogen was confronted by a great diversity in host species and a diversity of vegetative tissues in each plant (such as roots, tubers, stems, leaves, fruits, and the like). Also the pathogen had to adapt to the growth cycles of its host, which means that the pathogen's resting stages had to coincide with the host's period of dormancy.

Adaptation to such factors requires a certain specialization on the part of the pathogen over and above adaptation to the defense mechanisms of the host.

In this adaptation process, the pathogen may evolve to either a lower or a higher level of specialization. With lower specialization, the pathogen has many hosts, each of which can be exploited less efficiently. With higher specialization, the pathogen becomes more an obligate parasite on a few closely related hosts that are parasitized very efficiently. All of this leads to a coevolution where host and pathogen are interdependent on each other.

As an initial situation from which coevolution began, we might consider a number of host species all equally parasitized by several unspecialized leaf pathogens each from a different species. It is a realistic assumption that all host and pathogen species were genetically heterogeneous. The genotypic variation within each host species means that some plants were more and some were less easily attacked by the pathogens than the average host plant. Those less easily attacked would have a reproductive advantage and the trait responsible for their reaction to the pathogens would become more abundant in the plant population, that is, the defense mechanisms that would tend to be generally effective to all parasitic species would have the greatest survival value. The parasites were gradually faced with less susceptible host species, but the defense mechanisms probably would differ from host species to host species because the defense mechanism of each species developed independently. Of course, concomitantly, the pathogens also had genetic variations, and genotypes which could attack host plants with a certain defense mechanism had a reproductive advantage on the host employing that defense system. This adaptation would tend to be exploited by this particular pathogen, and the pathogen would become adapted to the host species using this defense mechanism. Other pathogens would develop other adaptations to the same or other defense mechanisms. This process automatically led to increased specialization of the pathogens to the various hosts, or rather to their defense mechanisms. Each pathogen became more and more adapted to some of the many defense mechanisms in use, and, therefore became more and more tightly linked to a small range of host species employing these defense mechanisms.

In coevolution, the first defenses developed by the host were general in nature, but when pathogens became increasingly specialized, new defense mechanisms were directed toward more specialized pathogens. Ultimately, a situation arose where all host species employed various forms of general defense mechanisms and diverse types of specialized defense mechanisms against pathogens. And all pathogens developed adaptations to some defenses, but the level of adaptations and the defense mechanisms to which they adapted varied from pathogen to pathogen. The result was a complex system of host-pathogen interactions that we know today.

Rates of coevolution vary. Some host defense mechanisms are more likely to be eroded by adaptation processes of the pathogen than are others. Chemical defenses are more readily neutralized than are morphological ones, but no defense mechanism is sacrosanct (Harper, 1977; Parlevliet, 1977).

With increasing specialization, the pathogen tends to become obligate on its host as exemplified by mildews, rusts, and smuts which cannot exist without their hosts. In turn, the requirement to conserve their hosts keeps the level of pathogenicity of the pathogens in balance, and hosts and pathogens coexist.

Specificity of Defense Mechanisms

Specificity with respect to the pathogen can be considered on various levels. The closed-flowering habit of some barley (*Hordeum vulgare*) cultivars renders them inaccessible to pathogens that enter through the open flowers, such as loose smut (caused by *Ustilago nuda*) (Pedersen, 1960). This avoidance mechanism is nonspecific because it excludes other ecologically similar pathogens as well, such as ergot (caused by *Claviceps purpurea*) which can parasitize barley (Darlington, et al., 1977). Most avoidance mechanisms are nonspecific by avoiding groups of parasites with similar ecological requirements. Specific avoidance mechanisms may exist, such as parasite-specific repellents.

Resistance can be nonspecific also. The host-specific phytoalexins produced by many plants are highly nonspecific. Many other resistances, however, are highly pathogen specific or even race specific. The partial resistances of maize (*Zea mays*) to the rusts (caused by *Puccinia sorghi, P. polysora*) are independent (Robinson, 1976). And, with barley, race-nonspecific resistance to leaf rust (caused by *Puccinia hordei*) occurs in the 'Vada' and 'Berac' cultivars, both of which are highly susceptible to yellow rust (caused by *P. striiformis*) (Parlevliet, 1978d). So race-nonspecific resistances can be highly pathogen specific. Resistance breeding deals predominantly with resistances that occur within a host-pathogen system that has both race-specific and race-nonspecific manifestations. The variation in degree and frequency of race specificity has dominated the discussions on resistance breeding for years.

Specificity within Host-Pathogen Systems

Resistances or pathogenicities can be classified as to their level of specificity only when both host and pathogen populations vary for resistance and pathogenicity, respectively. The degree of resistance or pathogenicity is measured by the disease incidence, disease severity, or another method of disease assessment. When a number of host genotypes (cultivars) are tested against a number of pathogen genotypes (isolates), the genetic variation in the resulting disease assessment scores is dependent or independent. Dependent variation in resistance results from variation in pathogenicity, that is, the ranking of cultivars for resistance depends on the pathogen isolate used, and a significant interaction occurs between cultivars and isolates. With the independent genetic variation, the rankings of cultivars is the same for all isolates, so the host resistance varies independently of the pathogenicity in the pathogen. Dependent genetic variation is specific variation, and independent genetic variation is nonspecific. This can be referred to as race-specific and race-nonspecific resistance and cultivar-specific and cultivar-nonspecific pathogenicity. The terms vertical and horizontal resistance and virulence and aggressiveness, as used by Van der Plank (1963), have the same meanings.

An example that illustrates specificity and nonspecificity is

Table 9.1. $\sqrt{\text{Lesion area in mm}^2}$ of eight rice cultivars inoculated with four isolates of *Xanthomonas oryzae* by the needle-pricking method. (After Yamamoto et al. 1977; Ezuka et al., 1975.)

Cultivar	*Xa*-resistance[a] gene	Isolate[b] T-7174	T-7147	T-7133	Xo-7323
Nikisakae	. . .	10	13	13	12
Asahi 1	. . .	23	21	24	21
Pelita I/1	*Xa-1*	1	8	7	7
Norin 27	*Xa-1*	1	26	29	27
TKM 6	*Xa-1, Xa-2*	1	1	16	9
Tadukan	*Xa-1, Xa-2*	1	3	32	39
Nagomasari	*Xa-3*[c]	2	2	1	13
Chugoka 45	*Xa-3*[c]	2	1	2	24

[a]A number of *Xa*-resistance genes conferring race-specific resistance have been recognized.

[b]On the left side of the line, resistant host-pathogen combinations; on the right, susceptible host-pathogen combinations.

[c] Formerly named *Xa-w*.

that of bacterial leaf blight (caused by *Xanthomonas oryzae*), a serious disease on rice (*Oryza sativa*). The host has a number of resistance genes (*Xa*-genes). Some give almost complete protection to some but not all isolates of the pathogen (Table 9.1), whereas others give a race-nonspecific pattern. 'Nikisakae' cultivar is more resistant than 'Asahi 1' for all four bacterial isolates. Similar differences in resistance occur among cultivars carrying *Xa*-genes. When the *Xa*-genes exert their effect, the race-nonspecific resistance is masked (left of the stepwise line in Table 9.1). When no effects of the *Xa*-genes occur, race-nonspecific resistance expresses itself (right of the stepwise line in Table 9.1). Likewise, isolates vary in pathogenicity. The four isolates in Table 9.1 vary for virulence genes that correspond to the *Xa*-genes (assuming a gene-for-gene system). Xo-7323 carries virulence to *Xa-1*, *Xa-2*, and *Xa-3*, whereas T-7174 carries no virulence genes. This illustrates cultivar-specific pathogenicity. For nonspecific pathogenicity, the four isolates do not differ in their reactions on the cultivars. But, other isolates of the pathogen do differ in a cultivar-nonspecific way for levels of disease severity that they cause (Table 9.2). Isolate Xo-7306 is more pathogenic on all cultivars tested than Xo-7323, irrespective of the presence or absence of *Xa*-genes (Table 9.2).

In such a host-pathogen system where major, race-specific resistance genes operate, race-specific resistance can be separated from other resistance easily. This other resistance generally shows a more continuous, quantitative type of variation among host cultivars and it is assumed to be race nonspecific. This, however, may

Table 9.2. $\sqrt{\text{Lesion area in mm}^2}$ of six rice cultivars inoculated with two isolates of *Xanthomonas oryzae* by the needle-pricking method. The two isolates are virulent on cultivars carrying *Xa-1*, *Xa-2*, or *Xa-3*. (After Yamamoto et al., 1977.)

Cultivar	*Xa*-resistance gene	Isolate Xo-7323	Isolate Xo-7306
Ketan Bengawan	*Xa-3*	10	19
Ketan Jahe	. . .	12	28
Hijau Gadring	*Xa-3*	14	30
Cicih Jambu	. . .	18	35
Cicih Baat	*Xa-3*	25	44
Soba Enim	. . .	26	45

not be quite true. Small, but significant deviations from race-nonspecific patterns in quantitative resistance have been reported for several host-pathogen interactions, such as potato (*Solanum tuberosum*) and late blight (caused by *Phytophthora infestans*) (Caten, 1974; Latin et al., 1978), barley and leaf rust (Clifford and Clothier, 1974; Parlevliet, 1978b), wheat (*Triticum aestivum*) and leaf rust (caused by *P. recondita*) (Kuhn et al., 1978), wheat and Septoria blight (caused by *Septoria tritici*) (Ziv and Eyal, 1978), and barley and scald (caused by *Rhynchosporium secalis*) (Habgood, 1976). The rice-*Xanthomonas oryzae* system, discussed above, too seems to show such small race-specific effects as shown in Table 9.3.

Table 9.3. $\sqrt{\text{Lesion area in mm}^2}$ of six rice cultivars inoculated with two isolates of *Xanthomonas oryzae* by the needle-pricking method. The two isolates are virulent on cultivars carrying *Xa-3*. (After Yamamoto et al., 1977.)

Cultivar	*Xa*-resistance[a] gene	Isolate Xo-7323	Isolate Xo-7306
Ketan Jahe	. . .	12	28
Ketan Menuh	. . .	14	21
Cicih Selem	. . .	14	32
Selem Gempel	. . .	18	29
Cicih Godangan	Xa-3	18	43
Ketan Bulu	Xa-3	30	45

The ranking order of the cultivars varies somewhat with the two isolates, and pairs of cultivars show small interactions. 'Selem Gempel' and 'Cicih Godangan' show the same rating to isolate Xo-7323 but quite different ratings to Xo-7306. However, isolate Xo-7306 attacks Cicih Godangan and 'Ketan Bulu' equally, but Xo-7323 does not.

The Genetics of Specificity

Specific or vertical and nonspecific or horizontal effects are expressed in terms of population dynamics, not in terms of genes. Resistances scored qualitatively, such as race-specific effects, can be interpreted via simple genetic ratios, but resistances that are assessed quantitatively are more difficult to interpret in terms of genetic patterns. These two cases are contrasted in Table 9.4. Assume that a host population and a pathogen population possess two loci

Table 9.4. Disease assessments of 16 combinations between two alleles at each of two host loci and two alleles at each of two pathogen loci. In Model A the assessment is of a qualitative nature, in Model B of a quantitative nature. The host is assumed to be homozygous diploid with resistance being dominant, the pathogen being haploid with p representing virulence and P avirulence. The host and pathogen genes operate on a gene-for-gene basis.

Model A[a]

Host	Pathogen P_1P_2	p_1P_2	P_1p_2	p_1p_2
$r_1r_1r_2r_2$	S	S	S[b]	S
$R_1R_1r_2r_2$	MR	S	MR	S
$r_1r_1R_2R_2$	MR	MR	S	S
$R_1R_1R_2R_2$	R	MR	MR	S

Model B[c]

	P_1P_2	p_1P_2	P_1p_2	p_1p_2	$\bar{x}$
$r_1r_1r_2r_2$	80	80	80	80	80
$R_1R_1r_2r_2$	40	80	40	80	60
$r_1r_1R_2R_2$	40	40	80	80	60
$R_1R_1R_2R_2$	0	40	40	80	40
$\bar{x}$	40	60	60	80	60

[a] Assessment as infection types: S = susceptible, MR = moderately resistant, R = resistant.

[b] Differential interaction.

[c] Assessment as a disease incidence or severity (in percentages).

each for resistance and pathogenicity, respectively, and that the host and pathogen operate on a gene-for-gene basis. Assume that each resistance gene has an incomplete effect, but together the resistance is complete. The effects of the host genes are additive. With the qualitative assessment (Model A in Table 9.4), vertical resistance would be assumed, whereas with the quantitative assessment, the conclusion would be quite different. A statistical analysis shows

67 percent of all variance in Model B to be caused by differences between cultivars and between isolates (main effects), whereas 33 percent is due to cultivar × isolate interactions. The variance for main effects would be assigned to horizontal effects and the interaction variance to vertical effects. So when two genes that give incomplete resistance interact additively, and operate on a gene-for-gene basis with two pathogenicity genes, the larger part of the genetic variation is horizontal or nonspecific in nature. With five minor genes for resistance and pathogenicity, each variance becomes so small that it cannot be distinguished from the interaction error variances (Parlevliet and Zadoks, 1977); so the researcher concludes that the differences in resistance and pathogenicity are truly horizontal in nature.

When resistance genes are nonspecific, that is, they do not operate on a gene-for-gene basis with the pathogenicity genes, the situation is not very different in reality from that described above. This situation is illustrated in Table 9.5. Each pathogenicity allele neutralizes the effect of one resistance allele, either R_1 or R_2. With both assessments, there are three categories of reaction (namely, S, M, and R or 80, 40, and 0). In Model B, 78 percent of the total variance is due to main effects (horizontal effects) and 22 percent is due to interaction (vertical effects). So whether the gene-for-gene situation operates or not has little effect on the occurrence of horizontal and vertical effects. Thus this method of analysis is not a very good way for separating the two types of gene actions. The only qualitative difference between the two systems is the ranking order. In Table 9.4, where the resistance of the host cultivars varies with the isolate used, there is a differential interaction that is equal to the effect of one gene (40 percent; Table 9.4). Such a differential reaction does not occur in Table 9.5. Van der Plank (1968, 1975) recognized this difficulty, and he emphasized that differential interactions are needed as evidence of vertical effects.

With polygenic systems the effects of individual resistance genes are small; thus the possible differential interactions are small also. If the interaction effects are of the same order of magnitude as the experimental error effects, it will be difficult, if not impossible, to discover the presence of such interactions. Some small interactions mentioned previously were differential interactions which suggests that at least in some polygenic systems, the polygenes in the host operate on a gene-for-gene basis with polygenes in the pathogen.

Discerning unambiguously between vertical and horizontal effects is easy only when two or more host and two or more

Table 9.5. Disease assessments of 16 combinations between two alleles at each of two host loci and two alleles at each of two pathogen loci. In Model A the assessment is of a qualitative nature, in Model B of a quantitative nature. The host is assumed to be homozygous diploid with resistance being dominant, the pathogen being haploid with p representing virulence and P avirulence. The host and pathogenicity genes do not operate on a gene-for-gene basis.

Model A

Host	Pathogen			
	P_1P_2	p_1P_2	P_1p_2	p_1p_2
$r_1r_1r_2r_2$	S	S	S	S
$R_1R_1r_2r_2$	MR	S	S	S
$r_1r_1R_2R_2$	MR	S	S	S
$R_1R_1R_2R_2$	R	MR	MR	S

Model B

	P_1P_2	p_1P_2	P_1p_2	p_1p_2	$\bar{x}$
$r_1r_1r_2r_2$	80	80	80	80	80
$R_1R_1r_2r_2$	40	80	80	80	70
$r_1r_1R_2R_2$	40	80	80	80	70
$R_1R_1R_2R_2$	0	40	40	80	40
$\bar{x}$	40	70	70	80	65

pathogen genes with large effects are involved in the interaction. With only one gene in the host and one in the pathogen, no differential effects are possible; and with minor genes, differential effects usually cannot be discerned from error effects. One may even go further and wonder whether dividing resistance into vertical (VR) and horizontal (HR) types has any use. Only clear cut VR can be recognized as such, with all other cases being difficult to classify unambiguously. Van der Plank connected race nonspecificity with stability, a viewpoint that is not necessarily true (Parlevliet and Zadoks,

1977). If host genes operate nonspecifically against the pathogen population, then there should be nonspecific pathogen genes operating against the host population. Thus, if so-called HR is difficult to separate from VR, and if it is not more stable, perhaps the effort to distinguish between these two resistance types is meaningless.

Hypotheses about Horizontal Resistance

Van der Plank (1963) introduced the concepts of VR and HR, and Van der Plank (1975, 1978) and Robinson (1976) assumed that the resistance genes involved in HR do not operate on a gene-for-gene basis with genes in the pathogen. Supposedly, HR gives stable resistance and usually is controlled by several additive genes; thus, HR genes act in a way completely different from VR genes. Robinson (1976) assumed that the pathogen, to adapt to HR, must change its ecological requirements drastically. But, without a serious loss of fitness, the pathogen cannot make this drastic change in a short time span. This may be valid for many mechanisms for avoidance of disease. For example, to adapt to closed-flowering barley cultivars, loose smut would need to change its infection pattern profoundly. And certainly, the assumption of Robinson (1976) is not a good explanation for the HR or race-nonspecific resistance of maize to *P. sorghi* and *P. polysora.* These two rusts have rather similar ecological requirements, but the HR for one does not operate against the other (Robinson, 1976). So here, the race-nonspecific resistance to one pathogen is highly pathogen specific when different, but related and ecologically similar, pathogen species are compared. HR to one pathogen has been overcome by the other without profound differentiation of their ecological requirements. Other authors have suggested that HR represents the accumulated "ghost" or "residual" effects of resistance genes that have succumbed to corresponding virulence genes (Riley, 1973; Clifford, 1975; Boukema and Garretsen, 1975) with these residual effects being race nonspecific. Indeed, Martin and Ellingboe (1976) observed that a wheat line carrying the resistance gene *Pm4*, which gave a susceptible IT 4 with mildew races carrying the corresponding virulence allele, permitted less abundant pathogen growth than a near-isogenic line that possessed the allele for susceptibility, *pm4*. They, however, concluded that the ghost effect was race specific, but that it did contribute to the phenomenon known as HR.

The viewpoint of Nelson (1975), where he assumed that VR and HR genes are one and the same, is similar to the ghost gene hypothesis. He maintained that VR and HR are not the actions of different genes, but rather they are expressions of the same genes in different genetic backgrounds. There are no major or minor genes. He concluded that resistance genes function vertically when separate and horizontally when together.

Abdalla (1970) proposed that specialized polygenes confer HR, and major switching genes, if present, determine the specificity of the resistance by directing the polygenic actions against specific races.

Eenink (1976) equated HR with stable resistance and assumed that the stability of a resistance was determined by the genetics of pathogenicity. If pathogenicity was inherited polygenically, the resistance was stable, whether inherited via oligo or polygenes.

Parlevliet and Zadoks (1977) concluded that all resistance genes in the host population and virulence genes in the pathogen population form on integrated system. Major and minor resistance genes interact on a gene-for-gene basis with major and minor virulence genes. When only a few genes with major effects operate, differential interactions can be detected easily, and this is called VR. When several to many genes with small effects determine the resistance/pathogenicity pattern collectively, it is difficult to separate the small differential interactions from experimental errors. This resistance behaves as HR and appears to be race nonspecific.

All of these hypotheses present the same difficulty. From knowledge gained with only a restricted group of host-pathogen systems, a hypothesis is formed to cover all host-pathogen systems. And most hypotheses have come from data collected from highly specialized biotrophic pathogens such as the rusts. These generalizations, however, probably are not valid. As an example, animal parasites and pathogens differ greatly in that parasites have sensorial capacities and are mobile, whereas pathogens lack such faculties. As a result, avoidance mechanisms are more effective against animal parasites than against pathogens. Also, the level of specialization of the parasite or pathogen may be of great importance. The highly specialized biotrophic pathogens, such as the mildews, rusts, and smuts, are obligate organisms, whereas less specialized parasites, such as many insects, feed on several to many host species. Resistance mechanisms employed by the host against these two classes may depend upon their differing degrees of specialization.

Defense mechanisms, to protect against less specialized parasites because of the parasite's more general nature, may be race nonspecific or horizontal. Specialized parasites are likely to have overcome such barriers; thus resistances to these parasites must be more specific. It is within such specialized parasite-host relationships that the integrated system of Parlevliet and Zadoks (1977) is thought to operate.

The Genetics of Resistance

Several excellent reviews have been produced on the inheritance of host-plant resistance to pathogens (Hooker and Saxena, 1971; Person and Sidhu, 1971; Simons, 1972; Day, 1974; Eenink, 1976; Nelson, 1978), so only a short summary on this subject will suffice herein.

Resistance may be controlled by any number of genes, and the effects of resistance genes vary from large to minute. Resistance genes may interact epistatically or additively. For relationships between biotrophic parasites and host plants, resistance and virulence genes often operate on a gene-for-gene basis (Flor, 1971; Day, 1974).

Vertical or race-specific resistance is inherited through oligogenes with relatively large effects. Resistance genes such as *Sr*-genes, *Lr*-genes, and *Pm*-genes in wheat that give a hypersensitive or low infection type of reaction for resistance to stem rust (caused by *Puccinia graminis tritici*), leaf rust, and powdery mildew (*Erysiphe graminis tritici*), respectively; the *Pa*-genes and *Ml*-genes in barley for resistance to leaf rust (caused by *Puccinia hordei*) and powdery mildew (caused by *Erysiphe hordei*), respectively; the *Pi*-genes and *Xa*-genes in rice for resistance to rice blast (caused by *Pyricularia oryzae*) and bacterial leaf blight, respectively; the *Dm*-genes in lettuce (*Lactuca sativa*) for resistance to downy mildew (caused by *Bremia lactucae*); and resistance genes in many other biotrophic pathogen-host plant combinations, are typically of a vertical nature. For each case, many (8-30) resistance genes have been identified with each giving a medium to high level of resistance to some but not all pathogen races. All of these host-pathogen systems are proven or assumed to operate on a gene-for-gene basis. Such resistance genes, by virtue of their large effects, differentiate races carrying different virulence genes. Zadoks (1966), therefore, defined a race as a "taxon" within the pathogen species characterized by a specific combination of virulence genes. The differential series of cultivars used to identify races, in fact, identify combinations of virulence genes.

The inheritance of HR is more complicated. Defense mechanisms that are pathogen nonspecific, such as some avoidance mechanisms, can be expected to be race nonspecific. The resistance of closed-flowering barley cultivars to loose smut (Pedersen, 1960), and the higher resistance that tall wheat cultivars versus short ones have to glume blotch (caused by *Septoria nodorum*) (Bronnimann et al., 1973), are examples of such nonspecific avoidance mechanisms. The inheritance of such general defenses may be governed by any number of genes (Parlevliet, 1977). The lack of adaptation responses from the pathogen to such defense mechanisms is due to the fact that it has to change for a number of genes to overcome such barriers.

The resistance mechanisms operating against specialized pathogens are, in general, highly pathogen-specific even when they are race nonspecific. Avoidance mechanisms or general resistance (resistance operating to whole groups of pathogens) do not play an important role in this group of host-pathogen combinations. The highly specialized pathogens are of prime importance for resistance breeding because the greater part of crop disease losses is caused by them. Therefore, an understanding of what HR means within a host highly specialized pathogen system is essential.

HR can arise in two ways: (1) when the host genes do not operate in a gene-for-gene way with the pathogen genes, no differential interactions are possible (Van der Plank, 1975, 1978; Parlevliet and Zadoks, 1977); (2) also, when several to many host genes with small effects operate on a gene-for-gene basis with an equivalent number of genes in the pathogen population, differential effects are so small as to be undetectable, and the result appears to be HR (Parlevliet and Zadoks, 1977). Polygenic resistance always appears to give HR whether the genes involved operate on a gene-for-gene basis or not. The experimental evidence based on small differential interactions suggests that the gene-for-gene system operates at least in some cases of polygenic HR (Parlevliet and Zadoks, 1977; Parlevliet, 1978b).

Assuming that gene-for-gene relationships are common in host-specialized pathogen systems, VR and HR are extremes of a continuum. With few genes operating, each with large effects, differential interactions are easily discernable and the result is VR. With more genes operating, each with smaller effects, differential interactions are less easy to discern, and the result is a mixture of VR and HR. With many genes of small effects, differential interactions cannot be recognized, and the resistance is predominantly HR.

This concept of VR and HR explains well a number of host-pathogen relationships that do not fit Van der Plank's concept. The wheat-yellow rust (caused by *Puccinia striiformis*) combination serves as an example. The effects of the many resistance genes vary from small to large (Robbelen and Sharp, 1978). Zadoks (1972), in trying to fit this host-pathogen system in Van der Plank's concept of VR and HR, observed a continuum between instances of near-HR and instances of clear VR. In a second case, Darlington et al. (1977) infected 12 male-sterile barley and wheat cultivars with seven isolates of ergot collected from different grass hosts. There was a great variation in pathogenicity among the isolates and in resistance among the cultivars. They concluded that there was no clear-cut evidence for the existence of specific races and that the resistance-pathogenicity pattern was a quantitative one. Their data do not differentiate whether the resistance is of a vertical or a horizontal nature (Table 9.6). Cultivar 'Betzes msg 8' is more resistant than 'Club Mariot' to all isolates, and isolate 26 is less pathogenic than isolate 82 on all barley cultivars. For these two cultivars and two isolates, there is a constant ranking of the counterparts, so this would be considered HR. Differential interactions, however, do occur as well (see within broken lines sections of Table 9.6). These data could result from the action of a few resistance genes, each with medium effects that are additive, operating on a gene-for-gene basis with corresponding virulence genes in the pathogen.

Table 9.6. Percentage florets of four male sterile barley cultivars infected by four isolates of ergot (*Clavicips purpurea*). (After Darlington et al., 1977.) Differential interactions are given within broken lines.

Cultivar	Isolate				
	36	26	43	82	$\bar{x}$[a]
Betzges msg 8	16	24	9	39	19
Hoodless Beardless	3	11	28	47	19
Compana msg, c	51	36	52	44	33
Club Mariot	34	37	57	75	39
$\bar{x}$[b]	27	33	38	56	

[a]Mean over seven isolates.

[b]Mean over eight cultivars.

Are There Two Gene-for-Gene Systems?

The flax (*Linium usitatissimum*)-flax rust (caused by *Melampsora lini*) relationship and the maize-northern corn blight (caused by *Helminthosporium turcicum*) relationship are clearly different, with

Table 9.7. In the quadratic check four combinations occur between two alleles at a host locus and two alleles at a pathogen locus. Only one combination leads to a specific interaction which gives resistance (– in I) or susceptibility (+ in II).

Host	I, Pathogen		II, Pathogen	
	P_1	p_1	P_1	p_1
R_1	–	+	–	–
r_1	+	+	–	+

the first operating on a gene-for-gene system (Flor, 1956), and the second perhaps operating on such a system (Lim et al., 1974). With one locus in each of the host and pathogen, resistance occurs only in one of four possible interaction combinations for the flax-flax rust relationship (Table 9.7 I) and in three of four (Table 9.7 II) in the maize-northern leaf blight relationship. The specific interaction is for incompatibility (resistance) in the first illustration and for compatibility (susceptibility) in the second. This suggest two gene-for-gene systems, one as described by Flor (1956) and a second which is its reverse (Wheeler, 1975; Ellingboe and Gabriel, 1977); Loegering, 1978). Wheeler (1975) and Loegering (1978) concluded that these systems are not basically different. Loegering (1978) said that there can only be one gene-for-gene system because the interaction results from genes in both organisms; and Wheeler (1975) reasoned that, if two loci are operating in both host and pathogen, the two systems tend to become similar. This is true as shown in Table 9.8. In Model II, the reversed gene-for-gene system is assumed to operate. If one deals with the host genotypes $r_1r_1r_2r_2$ and $r_1r_1R_2R_2$ and the pathogen genotypes P_1p_2 and p_1p_2 one obtains the same quadratic check as in Table 9.7 I, while the two host and the two pathogen genotypes segregate for one gene only. Ellingboe and Gabriel (1977) assumed that the reversed gene-for-gene mechanism might be involved with the variation for basic compatibility.

Considering the known facts, one must conclude the existence of two systems that differ in the result of the specific interactions. In Model I (the gene-for-gene relationship as observed with many biotrophic pathogens), the specific interaction between the avirulence allele of the pathogen and the resistance allele of the host prevents the establishment of a compatible relationship between host

Table 9.8. Resistance (–) or susceptibility (+) of 16 combinations between two alleles at each of two host loci and two alleles at each of two pathogen loci when the specific interaction is for resistance (Model I) or for susceptibility (Model II).

Model I

Host	Pathogen			
	P_1P_2	p_1P_2	P_1p_2	p_1p_2
$r_1r_1r_2r_2$	+	+	+	+
$R_1R_1r_2r_2$	–	+	–	+
$r_1r_1R_2R_2$	–	–	+	+
$R_1R_1R_2R_2$	–	–	–	+

Model II

	P_1P_2	p_1P_2	P_1p_2	p_1p_2
$r_1r_1r_2r_2$	–	+	+	+
$R_1R_1r_2r_2$	–	–	+	+
$r_1r_1R_2R_2$	–	+	–	+
$R_1R_1R_2R_2$	–	–	–	–

and pathogen—the phenomenon variously called resistance, incompatibility, hypersensitivity, or low infection type reaction. In Model II (the gene-for-gene relationship assumed to occur with heterotrophic pathogens) the specific interaction between the virulence allele of the pathogen and the susceptibility allele of the host gives susceptibility or compatibility.

The situation in Model II is easy to understand in terms of host-pathogen evolution. Pathogen genotype P_1P_2 is nonpathogenic and could only exist as a saprophyte. Allele p_1 would produce a toxin for pathogenicity, the effect of which can be neutralized by the product of R_1. To restore pathogenicity, another toxin produced by p_2 is needed, and this, in turn, can be neutralized by a second product from gene R_2. This system seems to operate for various host-*Helminthosporium* relationships. Strobel (1975) observed that susceptibility in the sugarcane-*H. sacchari* combination arises when the toxin produced by the pathogen is bound to a specific protein located in the plasma membrane of the host (r). If the protein is changed somewhat (R), the recognition and binding capacities are lost, and the host reacts resistant. Resistance to Victoria blight (caused by *H. victoriae*) in oats (*Avena sativa*) and to Helminthosporium leaf blight (caused by *H. carbonum*) in maize is inherited monogenically. Pathogenicity is determined monogenically also. Each pathogen produces a host-specific toxin, and isolates that do not produce the host-specific toxin are nonpathogenic (Scheffer et al., 1967). In maize, the resistance gene *Ht1* governs resistance to race 1, gene *Ht2* to races 1 and 2 of the northern leaf blight, and virulence to *Ht1* is inherited monogenically (Lim et al., 1974). Lim et al. (1974) concluded that the maize-northern corn leaf blight system fits the gene-for-gene model, but Van der Plank (1978) countered that race 2 produces so much of the same toxin that resistance allele *Ht1* cannot neutralize it all with the result that this gene is no longer effective. Since Strobel (1975) found that it is the protein from the susceptibility allele that recognizes and binds the toxin and not that from the resistance allele, the explanation of Van der Plank (1978) cannot be true.

Model I is less easy to fit into a host-pathogen coevolution system. The interaction, too, is between two alleles with a positive function, and the interaction leads to resistance. Ellingboe (quoted from Sequeira, 1978) concluded that, in most systems where the genetics of the interaction between host and biotrophic parasite have been examined in detail, resistance appears to be a positive function, that is, the product of a functional gene (R_1 or R_2 in Table 9.8 I). Likewise, avirulence is a positive function, the product of a functional gene (P_1 or P_2 in Table 9.8 I). Compatibility arises when either the host or the pathogen or both produce the wrong product or nothing at all (Samborski, 1978). Absence of the R, r locus altogether results in susceptibility (Loegering, 1978). The use of wheat monosomics to locate rust and mildew resistance genes on certain chromosomes makes use of this principle because the absence of the chromosome arm carrying the resistance gene gives susceptibility. Similarly, absence (deletion) of the avirulence locus gives virulence in flax-flax rust systems (Flor, 1960). The fact that resistance genes often occur as multiple allelic series and virulence genes do not (Flor, 1971; Robbelen and Sharp, 1978) indicates that resistance results from a functional and virulence from a nonfunctional or a malfunctional allele. Allelic series can be expected only among slightly different functional alleles.

In summary, with both systems of host-pathogen relationships, the product of the host allele and the product of the pathogen allele, when in contact, recognize and interact with each other. In Model I the reaction leads to incompatibility, whereas in Model II the reaction results in compatibility.

The fact that these two systems coexist is in line with the discussion of Esser and Blaich (1973) who distinguished two kinds of incompatibility in plants and animals. Within biological species, incompatibility regulates the coexistence of different populations, different individuals, and different tissues that are in close proximity. When incompatibility between two genotypes results from the meeting of identical alleles or their products, the reaction is called homogenic; when it results from nonidentical alleles, heterogenic incompatibility occurs. In the latter case, compatibility results when identical alleles or their identical products meet. Within fungal species both systems may occur, homogenic incompatibility to regulate sexual mating and heterogenic incompatibility to control vegetative hybridization (Esser and Blaich, 1973).

The two host-pathogen systems that lead to incompatibility (Model I) or to compatibility (Model II) when the products of host and pathogen alleles recognize each other resemble the homogenic and heterogenic incompatibility systems, respectively, remarkably well. This resemblance is not accidental because the coexistence of the pathogen with its host requires a precise adjustment, especially for biotrophic pathogens. Similarly, a precise adjustment is needed within the mating system of the pathogen population so that adapted genotypes can coexist without losing evolutionary flexibility. Such precise adjustments result from using a combination of two systems, one determining incompatibility, the other compatibility.

Evolutionary Meaning of the Two Gene-for-Gene Systems

Intuitively, the two gene-for-gene resistance systems (discussed in the last section) could have different evolutionary meanings. System I, where incompatibility arises from allelic products recognizing each other, could be a phenomenon superimposed upon system II. System II could be required for basic pathogenicity. The pathogen has genes for pathogenicity, and the host has genes to neutralize the effects of such pathogenicity genes, namely, they confer resistance. An example is take-all (caused by *Gaeumannomyces graminis*), a soilborne disease of wheat and barley (Table 9.9). Oats is resistant because it produces avenacin which prevents the growth of the take-all pathogen (Turner, 1953). In Wales, a take-all form was found that overcomes the resistance of oats because it produces avenacinase, an enzyme that hydrolyzes avenacin into products that are less toxic to the pathogen (Turner, 1961). Now, the resistance gene or genes code for a product, and the virulence gene or genes code for a product that neutralizes the resistance. In this way the host and the pathogen coevolved, each trying to be a step ahead of the other. Coevolution of this system is slow when measured in terms of our agricultural evolution. The take-all form virulent on oats was found only once, so apparently it does not arise frequently. Similarly, resistance genes in maize to the various Helminthosporium pathogens are not easily overcome.

Table 9.9. Resistance (–) or susceptibility (+) of wheat and oats for *Gaeumannomyces graminis*, var. *graminis* and var. *avenae*.

Host	*G. graminis*	
	var. *graminis* (toxin)	var. *avena* (toxin and avenacinase)
Wheat	+	+
Oats (avenacin)	–	+

The resistance resulting from system I is more difficult to rationalize. This gene-for-gene system is restricted predominantly to biotrophic parasites. It differs from system II in that the products from the resistance allele in the host and the avirulence allele in the pathogen react to give resistance. There are two reasons why the pathogen would give a product that can be used by the host to give resistance: (1) the product of the avirulence allele has other essential functions, or (2) the product of the avirulence allele provokes an incompatibility reaction to regulate the coexistence of the pathogen with its host.

With reason (1) the virulence allele should produce a similar product which would maintain its essential function but not be recognized by the product of the host resistance allele. This is not compatible with the observations of Flor (1956) that deletion of the avirulent locus gives virulence, and it does not explain the ease with which avirulence mutates to virulence.

Reason (2) fits the research data better. When the avirulence allele has the sole function to provoke incompatibility, a change to virulence is quite easy. Any nonsense or loss mutation prevents the incompatibility reaction between host and pathogen, and compatibility is restored. This system can operate only if it is superimposed upon basic pathogenicity. When virulence is nothing more than restored compatibility, the ease by which many monogenic resistances against biotrophic pathogens are overcome is explained. There is a definite advantage for the pathogen population to carry avirulence alleles, such as alleles that provoke incompatibility. The biotrophic pathogen is highly specialized and depends totally upon its host for existence; so endangering the host endangers the pathogen. Because the host plant competes with other plants, a slight reduction in its fitness, and thus in its competitive ability, could result in a serious decline of the host population. A biotrophic pathogen does influence the fitness of its host populations considerably, and maximizing its pathogenicity could endanger its host and thus itself.

The balance between the Myxoma virus with its rabbit host is a good example (Fenner, 1965). The Brazilian Myxoma virus which is highly pathogenic, was introduced into the Australian rabbit population in 1950, and it killed 99.5 percent of the rabbits. By 1958, this strain was completely replaced by less pathogenic ones, and the rabbits had become moderately resistant. This coevolution for a decreased pathogenicity and an increased resistance which resulted in a survaval rate of over 75 percent in the rabbit population occurred in only eight years. The Phleum mottle virus in Wales (Catherall and Chamberlain, 1977) provides another example. Many grasses were symptomless carriers for the mild common strain of the virus. The strain that caused severe symptoms was quite rare.

Large fluctuations in pathogenicity can equally endanger the host population. When strong pathogenicity and favorable conditions for the pathogen coincide, the host population could be endangered. A system that regulates and stabilizes the level of pathogenicity at an intermediate level would be selected, because genotypes carrying such a system would have a better chance to survive over the long run. The gene-for-gene relationship for incompatibility provides such a system. With a substantial number of different avirulence alleles at varying frequencies, the pathogen population has a feedback system that keeps the "mean population pathogenicity" stabilized around an intermediate level. Within the host population the frequencies of R-genes tend to increase because they confer an advantage to the plants carrying them, but when the frequencies of R-genes become too high, that is, the pathogen population too small, the frequencies of the corresponding virulence alleles increase to restore the balance (Mode, 1958). Because a change from avirulence to virulence occurs with ease, there is no danger to the pathogen that the host population will become too resistant.

Consequences for Modern Agriculture

The pathogen-host equilibrium, maintained by the gene-for-gene system for incompatibility, operates satisfactorily in nature, where the genetically heterogeneous host and pathogen populations are exposed to a heterogeneous environment. In a natural environment (Parlevliet and Zadoks, 1977) it is of little use to differentiate between HR and VR; the system and population, not the individual genotype, are important to survival.

Modern agriculture, however, represents a totally different environmental situation. Genotypic variation of the host population has been reduced enormously. Plant breeders, when breeding for resistance to biotrophic pathogens, have selected and used the incompatibility genes rather than the genes that confer resistance to basic pathogenicity. These incompatibility genes have been exposed to the pathogen one by one. Because the function of these genes was not to protect the host from the pathogen, but the pathogen from becoming too aggressive, it is not surprising that these genes when not embedded in the natural system, capitulated to the pathogen population quickly. The manifestation has been the numerous cases of resistance genes losing their effectiveness and the rapid evolution of pathogen races. The finely tuned balance that existed in natural host-pathogen populations has been lost; the pathogen population has been allowed to specialize to a high degree on a host which was made uniform and abundant by man. The pathogen has become more virulent because this no longer endangers the existence of the host since man multiplies it.

It is within modern agriculture that the need for stable resistance arose and that concepts like HR and VR can have a useful meaning.

The hypotheses developed in the previous two sections suggest that two types of resistance genes exist, one being easy, and the other being difficult to overcome. These two types of resistance do not necessarily coincide with VR and HR, respectively, because resistances belonging to the second type (difficult to overcome) can be vertical (Table 9.9) as well.

The unstable incompatibility genes do have a use in plant breeding. As soon as such genes are embedded in a kind of integrated system, albeit simple compared with the natural one, their durability increases. The use of resistance genes in an integrated system will be discussed in the next section.

RESISTANCE BREEDING

When developing plant cultivars with improved disease resistance and other traits, the breeding methodologies and selection methods used depend upon the mating system of the plant species and heritabilities of selected traits. Disease resistance is treated just as any other trait in the breeding process, but breeding for it takes a special position for two reasons: (1) resistance can be assayed only by diseasing the plants, that is, employing another living and variable organism, and (2) resistance may appear elusive, that is, the resistance may break down.

The following subsections, therefore, concentrate on screening for resistance and how to use resistance genes more effectively.

Selection for Disease Resistance

Sources of Resistance

Resistance may be found both within and outside of the primary gene pool of the crop being bred. At first, resistance is sought in closely related material, such as local and foreign commercial cultivars and local land cultivars, because the less related the resistant source is to the material being improved, the more difficult it is to transfer the resistance without transferring simultaneously undesirable genes or gene complexes. If resistance cannot be found in closely related genotypes, it is necessary to search for it in primitive cultivars and weedy relatives, and as a last resort, related species and genera.

Resistances from distantly related host genotypes are no more durable than resistances from closely related ones. For example, the yellow rust resistance in the Dutch winter wheat cultivar 'Clement,' derived from rye, and the stem rust resistance in the Australian wheat cultivar 'Mengavi,' introduced from *Triticum timophevi*, both broke down within two years. The R-gene resistances of potato to late blight derived from *S. demissum*, and the *Cf*-gene resistances of tomato to *Cladosporium fulvum*, transferred from the wild species *Lycopersicon pimpinellifolium*, *L. hirsutum*, and *L. peruvianum* lost their effectiveness quickly.

Another source of resistance is via mutation induction. In mint (*Mentha piperita*), the clone 'Mitcham,' an allohexaploid hybrid of the species *M. aquatica* and *M. spicata*, is highly susceptible to Verticillium wilt (caused by *Verticillium albo atrum*). This precise hybrid, but enriched with resistance to Verticillium wilt, cannot be reconstituted from the original parental species, but through mutation induction and selection on a large scale, several Mitcham-like clones with resistance to *V. albo atrum* were produced (Murray, 1971). Producing a crop cultivar with improved resistance via mutation breeding is laborious and time-consuming because the rare mutants for resistance seldom occur without negative changes in other traits.

Screening for Resistance

Escape, dependency upon the natural races, interaction of various diseases, and even the absence of the pathogen, make screening for resistant genotypes under natural field conditions unreliable. So breeders and pathologists have developed many artificial screening methods. A good screening technique must discriminate between resistant and susceptible genotypes (high heritability), be relatively easy and cheap to apply, and select resistance that is effective in agricultural production. Screening methods are developed to fit the specificity of a given host-pathogen relationship. The various general steps and conditions needed for a successful screening procedure are shown in Fig. 9.1. In a few instances, screening for resistance can be done in the absence of the pathogen by exposing the host plants to the toxin produced by the pathogen. Sorghum genotypes can be screened for resistance to *Periconia circinata* by placing the seedling roots in a toxin-containing filtrate (Schertz and Tai, 1969).

Over the past 60 to 70 years, procedures for recognizing resistant plants unambiguously, easily, and cheaply have been improved enormously. Generally, these procedures are most effective in selecting resistances that are monogenically inherited and cause near immunity. And, with biotrophic pathogens, such genes generally condition incompatibility and are easily neutralized by the pathogen. When the breeder screens for other types of resistances, it must be done under conditions most favorable for the selected resistances and in the absence of resistances that break down easily. In the biotrophic pathogens, such other types of resistances often are polygenically inherited. To discern resistance levels quantitatively, the levels of disease infection must be moderate as would occur under natural conditions. Otherwise, differences in resistance will disappear. Because polygenic resistances are expressed best in adult

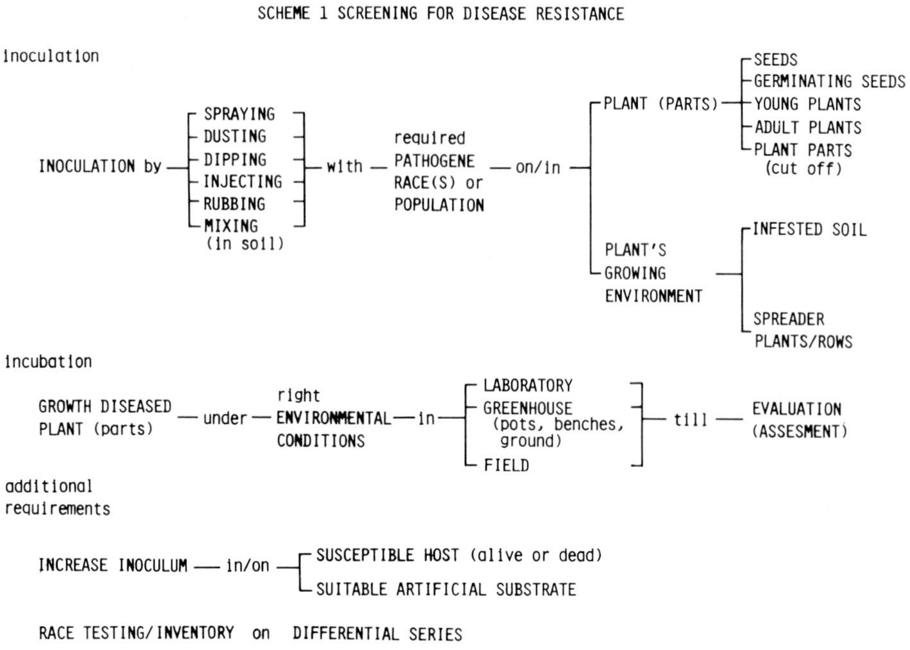

Fig. 9.1. Screening for disease resistance.

plant stages, testing for them should not be done on seedlings. If major gene resistances, which express a hypersensitive or low infection type reaction, are present in the host population, the genotypes that show these infection types or no disease at all should be discarded because they either carry the major genes or they are escapes. Genotypes to be saved are those with moderate amounts of disease.

Wind-borne pathogens such as rusts and powdery and downy mildews can spread radially from plot to plot (Van der Plank, 1968). This can be a serious problem when partial or medium resistances are being evaluated. Parlevliet and van Ommeren (1975) studied this problem with the barley-leaf rust relationship. Cultivars with varying levels of partial resistance to leaf rust were compared in plots of varying sizes and isolated by winter wheat (no interplot interference) or adjacent to each other (strong to very strong interplot interference). The ranking order of the cultivars for resistance was not affected much by the interplot interference, but the differences between cultivars in percentage leaf area affected (measuring the resistance) were greatly influenced. The difference in leaf areas affected between the most susceptible and the most resistant cultivar amounted to a factor of 2000 in 1973 and a factor of 1000 in 1974 in the absence of plot interference. With the cultivars adjacent to each other, with plot widths ranging from 0.25 m (1 row) to 2.0 m (8 rows), this difference was reduced to a factor varying between 15 and 50. In adjacent plots the differences in resistance were strongly underestimated.

With pathogens that spread horizontally with difficulty (*Septoria nodorum* in wheat) or build up slowly (many smuts, bunts, and soil pathogens), there is little or no problem of plot interference, but generally, such a pathogen spreads over the trial fields heterogeneously. Of course, this interferes with a good differentiation among quantitative differences in resistance. Frequent replication of a cultivar with known resistance can aid in the assessment of genotypes for resistance somewhat, but uniform inoculation is required for a good assessment. Also, assessment must be done at the right moment, when differences in resistance are maximal. This optimal stage usually is found empirically because it varies with host-pathogen systems.

Selection under Natural Conditions

Efficient screening methods are available for only a restricted number of host-pathogen systems. Each agriculturally important crop is attacked by many pathogens that do not cause significant economic losses, and therefore, do not warrant special sophisticated screening techniques. The breeder, however, can improve the crop's resistance to such minor diseases by removing susceptible plants or lines in the field when infections occur. This approach is quite effective. An example of such a minor disease is leaf rust of barley. Breeders in Western Europe generally remove lines that are severely affected, but they make no attempts to inoculate barley nurseries artificially. Parlevliet (unpublished) studied 40 barley cultivars of recent Western European origin in 1978, a year when leaf rust developed moderately. Leaf rust readings were made four weeks after heading. 'Akka,' an extremely susceptible cultivar, had 25 percent of its leaf area affected. Five cultivars had more than 5 percent leaf area affected, 22 had less than 1 percent, and 3 had less than 0.1 percent. This shows that considerable levels of partial resistance to leaf rust were present in recently released cultivars. The resistance probably was race nonspecific. These results emphasize that mild selection under natural field conditions is effective in obtaining considerable levels of resistance of a race-nonspecific type.

Results of Selection for Disease Resistance

Van der Plank (1963, 1968) reasoned that selection for race-specific resistance would be accompanied by a simultaneous loss of polygenic, partial resistance. He demonstrated this "Vertifolia effect" by comparing potato cultivars with and without race-specific (R) genes for resistance to *Phytophthora infestans* with races virulent on the R genes. The epidemic built up faster on cultivars with race-specific than on those with race-nonspecific resistance, that is, cultivars with R genes carried fewer race-nonspecific genes. His reasoning seems logical, but the Vertifolia effect, although occurring in some cases, is not a universal phenomenon. For example, Parlevliet and Kuiper (1977) reported that 'Cebada Capa' barley carried the race-specific resistance gene *Pa7* and a high level of polygenic, partial resistance to leaf rust. A second example involves powdery mildew disease of barley for which European breeders have utilized one race-specific resistance gene after another. Parlevliet (unpublished data) compared seven cultivars carrying different and no longer effective race-specific resistance genes with lines from the barley composite cross XXI that had not been selected for mildew resistance. The percentage leaf area affected (three weeks after heading) by mildew races virulent on the R genes varied from 8 to 23 percent with a mean of 15 percent for the seven cultivars, whereas of the unselected composite cross many lines had 80 percent leaf area affected with some approaching 100 percent. Clearly, the European barley cultivars carry a considerable amount of partial resistance, so there was no evidence of a Vertifolia effect.

A third example where the Vertifolia effect does not operate is provided by the Dutch lists of crop cultivars. Many had complete or nearly complete resistances (scores of 10, 9, or 8) which broke down after one or more years. The level of resistance remaining after the breakdowns of these R genes varied greatly among cultivars, being generally in the range from 6 to 3. (A score of 3 denotes extreme susceptibility.) The fact that most cultivars showed from some to a considerable rust resistance does not support the Vertifolia effect.

Probably the Vertifolia effect does not occur universally for two reasons: (1) very susceptible genotypes are constantly removed from breeding nurseries, and in fact, parents used in crosses likely have some genes for partial resistance; and (2) major race-specific genes often have an incomplete expression in the field with their expression becoming more complete when backstopped by genes for partial resistance (Parlevliet and Kuiper, 1977). Thus, complete resistance would occur when R and partial resistance genes are selected together. In other words, intense selection for disease resistance would tend to produce lines with major genes in backgrounds with a varying level of partial resistance. The major genes tend to be of a race-specific and the partial resistance of a race-nonspecific type.

For diseases of less importance, selection consists of removing highly susceptible types, rather than selecting the resistant ones, and this type of selection tends to accumulate genes for race-nonspecific resistance. This agrees with the observations of Fishbeck (1973) that cultivars developed in a given area are more resistant to minor diseases in that area than cultivars developed elsewhere.

Strategies in the Use of Resistance Genes

Stability of Resistance

Two types of resistance genes are assumed to occur: (1) the highly unstable incompatibility genes often expressed as a hypersensitive or low infection type reaction to biotrophic pathogens, and (2) the more stable resistance genes that affect the basic compatibility processes.

Examples of the unstable incompatibility genes occur in many host-pathogen relationships, such as oats—*Puccinia coronata,* flax—*Melampsora lini*, wheat—*Ustilago nuda* f. sp. *tritici aestivi*, tomato—*Cladosporium fulvum*, potato—*Phytophthora infestans*, maize—*Puccinia sorghi*, apple—*Venturia inaequalis*, and others (Flor, 1971; Day, 1974). All of the pathogens involved in these interactions are biotrophic.

It is difficult to classify resistances that affect basic pathogenicity processes because so little is known about the mechanisms by which they operate. To use stability of a resistance gene as the basis for classification is chancy because incompatibility genes differ in stability (Eenink, 1976). Major resistance genes operating in the oats—*Helminthosporium victoriae*, maize—*Helminthosporium maydis, H. turcicum*, and *H. carbonum*, cabbage—*Fusarium oxysporum*, and the sorghum—*Periconia circinata* systems may affect the basic pathogenicity processes. The apparent stability of the partial resistances of the potato—*Phytophthora infestans* (Black, 1970), maize—*Puccinia sorghi* (Hooker, 1969; Kim and Brewbaker, 1977), and barley—*Puccinia hordei* (Parlevliet, 1978a) systems may be due to their polygenic inheritance. Possibly, these genes interfere with basic pathogenicity. Not all polygenes for resistance, though, affect the basic pathogenicity processes. Some incompatibility genes that have small effects behave as minor genes. Thus, polygenic resistance may result from either incompatibility genes or genes that affect basic pathogenicity; in either case, each gene has a small effect.

There is no easy method for distinguishing the two types of resistance.

For breeding strategies, it is of course important to know whether the breeder is dealing with unstable or durable resistance. The best guidance in this situation is past experience; and resistance, once shown to be durable, can be introduced and selected in the same way as other traits are incorporated into new cultivars. To protect crops by means of incompatibility genes requires a different strategy. These genes need "protection," so they must be combined with other genes so that the combination interferes with the development of new, adapted races of the pathogen.

Stabilizing Selection

Van der Plank (1963, 1968) introduced the concept of stabilizing selection which says that unnecessary virulence genes per se reduce the fitness of the pathogen race carrying them. Stabilizing selection favors simple races, namely, races with the least number of unnecessary virulence genes. In nature there are many cases that contradict the concept of stabilizing selection. In several crop-rust interactions, pathogen races were found to carry virulence genes that would not be needed to permit these races to parasitize the host crop (Flor, 1971; Clifford, 1975; MacKey, 1976). Most *Puccinia recondita* races in the old world carry from several to ten virulence genes (Boskovic, 1976) even though the wheat cultivars grown do not carry that many corresponding resistance genes. Dixon and Wright (1978) observed a similar situation in the lettuce—*Bremia lactucae* relationship in England. Lettuce cultivars carry from none to four *Dm* resistance genes, whereas pathogen races with up to 10 virulence factors are quite common. Wolfe et al. (1976), analyzing the barley—*Erysiphe graminis* system in Western Europe, concluded that many virulence genes showed no stabilizing effects, but some did. Perhaps it is the general situation that stabilizing selection operates weakly or not at all with a majority of virulence genes and to some extent with a minority. Overall, stabilizing selection probably is not an important factor in preventing the development of complex races.

Absence of a Strategy

The most common strategy used in breeding a crop for disease resistance actually appears to be no strategy at all. The primary objective is to develop good resistant cultivars, and often this results in most cultivars developed for a given growing area carrying the same resistance gene because it represents the best source of resistance. The breeding for barley resistant to powdery mildew, wheat to yellow rust, tomato to *Cladosporium fulvum*, and lettuce to *Bremia lactucae* in Western Europe are typical examples of this strategy. The situation is characterized by a fairly rapid turnover in cultivars used and where only a few race-specific resistances are exposed to the pathogen population at any point in time. The most common reason for a cultivar to be eliminated from production is because its resistance has succumbed to a new pathogen race. (The resistance is said to have broken down.) Among the resistance breeding strategies, this one of releasing one, two, or only a few resistance genes at a time is the easiest to apply, and simultaneously, the easiest for the pathogen to adapt to. The Western European breeders, at present, wonder if and how they can continue with this approach in the host-pathogen systems mentioned above.

Resistance Genes Used One at a Time

Another strategy in resistance breeding is to have a controlled plan for the release and use of resistance genes. One or a few resistance genes are used in agricultural production of the crop at any point in time, and the virulence-gene composition of the pathogen population, especially relative to the resistance gene or genes being used, is monitored annually on a differential series of host genotypes that carry different resistance genes singly or in various combinations. Such information permits the breeder to detect trends of change in the frequencies with which virulence genes occur, and especially increases in the frequencies of those that can parasitize the cultivar carrying the resistance genes currently in use.

As soon as a new race that is virulent on the currently used resistance gene appears, new cultivars that carry another effective gene are released. That is, breeding for resistance tries to keep a step ahead of the pathogen. The breeding of flax in North America represents a good example of this strategy. From the 1930s to date, flax has been protected from flax rust by a succession of resistant cultivars. Consecutively, the resistance genes *L9*, *P*, *M*, *L*, and *N1* were introduced and each broke down to a new race; their periods of effectiveness ranged from 5 to 13 years. Presently, the resistance genes *L6* and *L11* are in use (Flor and Comstock, 1971; Zimmer and Hoes, 1974). This strategy worked for flax production in North America because the crop was grown predominantly in only one continuous area and the breeding of resistant cultivars was coordinated centrally. In agricultural production areas where the farming is spotty or discontinuous because of topographic or political reasons (such as many small countries in Europe), and/or the breeders operate independently, this strategy cannot be applied. Actually, it is slightly better than "no strategy at all."

Multiple Resistance Genes

A third strategy in resistance breeding involves the placement of two, three, or more new and still effective resistance genes into a new cultivar so that the pathogen population has a barrier of several resistances presented to it simultaneously. This should be an effective strategy because a new race, to overcome the multiple resistance genes, must have two or three simultaneous changes toward virulence, whereas a new race needs only one virulence gene change to overcome a single gene for resistance (Flor and Comstock, 1971). This strategy can operate satisfactorily only when the breeding is coordinated centrally and when the production area is isolated from other areas where the system is not applied. If some resistance genes are simultaneously released singly, in separate cultivars, the effectiveness of this approach will be reduced materially.

In North America, this strategy has been tried with flax (Flor and Comstock, 1971). Flax cultivars carrying the rust resistance genes *L6* and *M3* and others with *L6*, *M3*, and *N1* were developed and released. *N1* became ineffective in 1973 (Zimmer and Hoes, 1974), and as *L6* occurs also singly in some cultivars, the multiple-gene barrier has been weakened considerably.

Breeding wheat for stem rust resistance in Australia has followed this strategy also. During the 1930s, 1940s, and early 1950s, the genes *Sr6*, *Sr11*, *Sr9b*, and *SrTt* were used singly in succession to protect wheat from the rust pathogen. But since the late 1950s, two or more *Sr*-genes have been "pyramided" into the same cultivar, and at present, the breeding and selection for multiple-gene resistance is well coordinated. The parents for crosses are chosen on the basis of their reactions to rust infections in international nurseries, and each breeder selects within and among progenies in his nursery for rust resistance against naturally occurring races of the pathogen. Promising wheat lines are tested centrally in a greenhouse against anticipated rust races from the mutagenesis program. These races have considerably wider virulence spectra than races found in the field, so it is possible to select wheat genotypes with a multiple resistance barrier (Watson, 1977).

Polygenic Resistance

A fourth strategy in resistance breeding is to accumulate polygenes in a single genotype to provide a multiple-resistance barrier against the pathogen. This strategy assumes that the virulence system of the pathogen is inherited in a polygenic way also. Because such resistance genes cannot be recognized individually, the breeder likely will not know whether oligo or polygenically controlled resistance is being selected. If, as seems true, partial resistances tend to be inherited polygenically, there is a question about whether it is possible to select this type of resistance; and if it is possible, how should the selection proceed? Breeding for partial resistance should be approached as with any other polygenically inherited trait. Usually, it is not possible to concentrate all of the required genes into one genotype in a single breeding cycle; so a breeding method somewhat akin to recurrent selection should be used so that the improvement of resistance is a continuous process. Selection for partial resistance in potato to *Phytophthora infestans* (Black, 1970. Umaerus, 1970), in barley to leaf rust (Parlevliet and van Ommeren, 1975), and in maize to *Puccinia sorghi* (Hooker, 1969) has been shown to be effective. To show that the removal of the more susceptible genotypes can be effective for increasing polygenic resistance, the following example will serve. In 1976, about 5000 plants of each of two heterogeneous barley populations were screened under natural field conditions for partial resistance to powdery mildew. Only the 30 percent most susceptible plants were removed, and in 1977, the results from this selection were assessed. In the first population (created by several cycles of intercrossing among progenies of seven barley cultivars that did not carry effective race-specific resistance genes), the mean percentage of diseased leaf area decreased from 15 percent to 12 percent from selection, and in the second (barley composite cross XXI), the mean was reduced from 44 to 37 percent.

Accumulating polygenes for resistance seems quite effective in many host-pathogen systems. When partial resistance has a very low heritability (that is, for many soil pathogens), it is not yet clear how to select for resistance of this type.

Mosaic Patterns of Cultivars in Time and Space

Use of a number of cultivars, each carrying a different resistance gene, simultaneously or in succession in a production area, is another strategy proposed to provide a diverse pathogen environment which would decrease the erosion of the host resistance. This strategy may be successful if the resistance genes are known, the racial composition of the pathogen population is monitored closely, and the mosaic pattern of cultivar use is controlled. In principle, it is similar to regional deployment of resistance genes. However, without a fairly organized pattern of cultivar use, this strategy is almost certain to fail. In fact, the four host-pathogen systems, barley—powdery mildew, wheat—yellow rust, tomato—*Cladosporium fulvum*, lettuce—*Bremia lactucae*, as they operate in Western Europe, resemble a mosaic pattern because at any moment, a few to several resistance genes were used in different cultivars, but there is no indication that the pathogen population has been hindered in its evolution of new virulent races.

Multilines and Cultivar Mixtures

The use of host populations that are heterogeneous for resistance genes is a long recognized strategy as a buffering system against disease loss. However, differences of opinion about its usefulness and practical problems have limited its use in modern agricultural production (Browning and Frey, 1969). The durability of the resistance of multiline cultivars depends on the rate at which complex pathogen races, such as races that carry virulence to most or all resistance genes in the multiline cultivar, evolve. Various factors influence this rate of evolution (Parlevliet, 1979).

1. Stabilizing selection. Van der Plank (1968) assumed that stabilizing selection would prevent the development of complex pathogen races in a multiline. Marshall and Pryor (1978) concluded that the strength of this phenomenon must be considerable against all virulence genes to keep the pathogen races simple. And since stabilizing selection seems to operate weakly or not at all with most virulence genes, the development of races with many virulence genes and the erosion of the multiline resistance are probably not retarded greatly by this phenomenon.

2. Number of component lines or cultivars in a mixture. Several authors have reasoned that the greater the heterogeneity of a crop cultivar or blend, the better its buffering effect is against erosion of resistance genes (Browning, 1974; Parlevliet and Zadoks, 1977). Marshall and Pryor (1978) showed that the strength of stabilizing selection needed per virulence gene to prevent the development of complex races was reduced as the number of component genotypes in the cultivar or blend increased. Of course, practical considerations in handling the multiline and the few resistance genes available restrict the number of components in a multiline cultivar or blend. Certainly, the limited number of components used in modern multiline cultivars, together with a low degree of stabilizing selection, are insufficient to keep the pathogen population composed of simple races.

3. Epidemiological situation. The epidemiological situation may affect the durability of a multiline cultivar. Factors that are important in an epidemiological sense are (1) whether the pathogen overwinters or oversummers on the multiline, (2) the proportion of the total acreage of the epidemiological unit occupied by the multiline, and (3) what resistance genes are used in the cultivars grown on the remaining acreage. When the multiline is grown extensively in both the overwintering area or season and in the area or season where the epidemic develops, and when the simultaneously grown pure-line cultivars carry resistance genes also present in the multiline, the situation is set for a rapid evolution of complex races. Also, the greater the difference in disease reaction level between fully virulent and fully avirulent races, the greater the chance that a new race with increased virulence will become established in the pathogen population. Thus, a high disease pressure will tend to increase the rate of erosion of the multiline resistance.

4. Replacement of components. The advantage that a new more complex race may have in the pathogen population can be annulled somewhat by replacing the multiline component(s) that has become susceptible with one carrying a resistance gene still effective against the new race. This procedure is likely to increase the durability, not only of the multiline as a whole, but also of the individual resistance genes. The dynamic pathogen population should be confronted with a dynamic host population.

Undoubtedly, the multiline strategy is an improved method for managing resistance genes. However, it entails benefits and costs. The primary benefits are the longer useful life for individual resistance genes and the delay in disease buildup that occurs in fields planted to multiline cultivars. These benefits lead to greater crop yields in the long and short run, respectively. The costs are mainly that multiline breeding is a conservative strategy relative to improvement of agronomic traits and the development of isolines for a multiline cultivar involves considerable extra work of a technical nature. An important complication derives from the many race-specific resistance genes needed for present-day and future multiline cultivars. These genes must be effective against most virulence genes in the pathogen population. However, because some virulence on each gene is expected, resistance genes of no use for the conventional breeding program because they have broken down can be used in the multiline program. Needless to say, a well organized system is essential to assess the virulence and resistance genes in the pathogen and host populations, respectively. The maximum benefit comes from the multiline strategy by using such cultivars as dynamic entities in time and space; in time, by replacing components when appropriate, and in space, by using multilines with different sets of resistance genes in regions where the pathogen overwinters and where it builds up to epidemics. This, of course, amounts to gene deployment and the multiline strategy really entails gene deployment on a microscale.

Development of Resistant Genes

An additional strategy for using host resistance to control diseases is called "gene deployment." Gene deployment refers to using different sets of resistance genes in the various subareas of an epidemiological unit. This technique can interfere with the annual development of epidemics and with the formation of complex races. As an example, Frey et al. (1973) suggested subdivision of the Puccinia Path in central North America into three areas to aid in controlling crown rust disease of oats: (1) the most southern where overwintering of the *Puccinia coronata avenae* occurs, (2) the far northern area where oats is a full season crop and epidemics can occur every year, and (3) the central region where many oats are grown and which serves as a transmittal zone for the pathogen from south to north and north to south in spring and fall seasons, respectively. Different sets of resistance genes are allocated to each of these areas. Similarly, different sets of powdery mildew resistance genes could be introduced into winter and spring barley in Western Europe, because the pathogen overwinters on the winter crop and develops epidemics on spring barley in summer.

All of these strategies, which are aimed at prolonging the period of effectiveness of the race-specific resistance genes, require cooperation among breeders, institutions, and countries, and a more thorough knowledge of the host-pathogen systems involved. The cooperation at various levels is an especially difficult problem to realize.

Resistance Genes in Integrated Disease Control

Browning (1974) concluded that agro-ecosystems should be diversified as much as possible without causing undue sacrifices in yield and product quality. This could be realized best by using resistance as a component of a completely integrated control program for the most serious pathogens. This diversity along with the other components, proper agronomic and phytosanitary measures, a variety of fungicides, and different types of resistances would reduce the pathogen evolution toward more virulence considerably. Actually, the use of single race-specific resistance (incompatibility) genes as a sole method for protecting crops from their pathogens is an unrealistic approach.

Tolerance

Tolerance, like avoidance and resistance, helps the host to cope with the parasite. If the host cannot avoid or restrict the parasite, it can tolerate (endure) its presence, that is, it suffers relatively little biological damage from the parasite. To evaluate tolerance in a series of host genotypes it is necessary to compare the damage inflicted by the parasite at equal amounts of parasite present at the same developmental stage of each host genotype (Schafer, 1971). Tolerance is difficult to measure because it is confounded with partial resistance and disease escape. For example, assume two barley cultivars are equally affected by powdery mildew at various assessment dates, which on the surface would indicate similar amounts of pathogen on both. However, if cultivar A is one week later than cultivar B, it likely would suffer more damage because a given level of the disease occurred at an earlier stage in the cultivar's ontogeny, and as a result, the disease would have a longer time in the plant's life cycle to become severe and cause damage. So cultivar B, which might be considered tolerant relative to cultivar A, really is not tolerant; it simply escapes damage due to its earliness.

An example of true tolerance probably occurs in the 'Proctor' barley cultivar which suffers less damage from powdery mildew than other cultivars (Little and Doodson, 1972). Nine cultivars were compared at many sites over a period of three years. In eight trials where no other diseases occurred, mildew caused a 13 percent yield loss with Proctor, whereas 'Zephyr' and 'Sultan' had a 20 percent loss even though all three had nearly the same amount of the disease. 'Midas' and 'Julia' had significantly less disease than Proctor, but they registered a similar 12 percent yield loss. The reduced damage of Proctor apparently was not due to partial resistance or escape, because it is a very late cultivar.

Another factor that may be mistaken for tolerance involves the energy invested by the host to bring about the resistance reaction. 'Akka' barley cultivar is extremely susceptible to leaf rust, whereas 'Vada' has a high level of partial resistance due to a combination of a reduced number of pustules, a longer latent period, a reduced spore production per pustule, and a reduced infectious period (Neervoort and Parlevliet, 1978). To compare the tolerance of these two cultivars, equal levels of pathogen, that is, equal leaf areas invaded, must be present during the epidemiological development. This requires that Vada must receive many times more inoculum than Akka during the epidemic period. With equal levels of leaf rust, Vada might show a considerably greater yield reduction than Akka. The energy needed to resist the pathogen, and not a lack of tolerance, might be the cause of the larger yield reduction by Vada.

These problems make it difficult to measure and select for tolerance. Of course, breeders likely have selected for tolerance unconsciously when yield measured in the presence of disease was a major selection criterion. One of the elements contributing to higher yields would be tolerance, and genes for it would be accumulated gradually.

The value of tolerance in limiting the damage caused by pathogens is undoubtedly so important that much effort should be put into understanding and recognizing it. Tolerance should be a stable, permanent type of protection because, theoretically, it does not exert a selection pressure on the pathogen. However, it probably does not operate for all host-pathogen interactions. When the quality of product depends on the absence of disease symptoms, tolerance would be of little value as a mechanism of protection from disease.

Virologists use the concept of tolerance to mean lack of symptom expression. Tolerant cultivars show few or no disease symptoms even though the virus is present. The reduced level of disease symptoms is caused by either a lower concentration of virus (resistance to virus multiplication), or to real tolerance to symptom expression, or to a combination of both. Tolerance to symptom expression does not necessarily mean tolerance to damage also (Kooistra, 1968). Tolerance to symptom expression can be strain-specific (Wiersema, 1972).

CONCLUSIONS

The plant host uses several mechanisms to limit biological damage from parasites. They are avoidance, resistance, and tolerance. Avoidance operates before and resistance and tolerance operate after contact between host tissue and the parasite. Avoidance and tolerance are assumed to operate generally in a race-nonspecific way; resistance may vary from race specific (monogenically inherited) to race nonspecific (quantitatively inherited). Race-specific resistances may be relatively stable resulting from genes that operate within the basic resistance-pathogenicity system or highly unstable which is caused by genes that act within an incompatibility system and is characteristic of highly specialized biotrophic pathogens. This incompatibility system superimposed upon basic pathogenicity serves as a genetic feedback mechanism to regulate the coexistence of pathogen and host at intermediate levels of pathogenicity.

Screening techniques to permit selection of resistant host genotypes have been developed for many host-pathogen systems. These screening techniques permit easy recognition of major gene resistances, but for polygenic resistances (partial resistance), field screening with selection against susceptibility is recommended. If major incompatibility genes and polygenes for partial resistance are present together in the host genotypes, those carrying the major genes can often be recognized by their low infection types and be removed. Thus, the major gene resistance will not inhibit the selection for accumulation of polygenes. Actually, such mild selection against susceptibility can cause a fairly rapid buildup of resistance.

To increase the durability of resistance, various strategies can be used such as multiple resistance genes, polygenic resistance, mosaic planting of cultivars in time and space, multilines, gene deployment, integrated disease control, and tolerance.

In nature and in agriculture, host-pathogen systems are dynamic. The rate at which the pathogen adapts to the resistant host, however, can be reduced considerably by using diversified resistances. This diversification should be controlled relative to the virulence composition of the pathogen which requires a regular and thorough monitoring of the pathogen population.

Discussion

J. A. BROWNING
W. D. GUTHRIE
F. G. MAXWELL
R. R. NELSON

1. F. G. MAXWELL. Certain experiences with pest management and national issues have made it necessary to put greater emphasis on breeding plants resistant to insects and diseases. The history of breeding for insect resistance is short and meager when compared to that of breeding for resistance to pathogens.

Three decades ago, it appeared that insect control via insecticides was simple and adequate. But three factors have kept chemical insecticides from being a panacea for controlling insect pests adequately. First, over time, with the extensive and intensive use of insecticides, high levels of resistance have been selected in populations of arthropods. Second, serious regulatory restrictions have been placed on insecticide use by government agencies because these chemicals have brought about environmental or health hazards, and the trend for limiting or ending the use of insecticides will continue. Third, for economic reasons and the difficulty of getting a new insecticide approved for use, industry is reluctant to undertake the research on and production of new effective insecticides. Whereby it cost two or three million dollars to produce and market an insecticide two decades ago, now it costs from 10 to 15 million. The high costs of developing an insecticide for marketing and the greatly increased costs of fossil energy, which is used so extensively in insecticide production, are going to very materially increase the on-farm cost of insecticides. This cost impact will be especially hard on developing countries.

Considering the uncertainties of insecticides for the future, it is logical to turn to host-plant resistance to insects as an effective alternative for controlling insect damage to our commercial crops. And insect resistance in crop plants is a substantial part of Integrated

Pest Management (IPM), a program that promotes the utilization of several pest suppression tactics to keep populations below economic thresholds. In fact, the initial factor in IPM is pest resistance in the plant cultivars.

Breeders can select resistant cultivars without knowing the chemical basis for resistance in the plant, but in an effort to give breeders a better screening tool for selecting resistant plant genotypes, chemists have tried to learn the chemical basis for this resistance. Further, if the toxins or chemical factors responsible for resistance could be identified, maybe these chemicals, if not hazardous to humans, could be synthesized and used as insecticides. However, there are too few chemists working on research in this field.

2. J. A. BROWNING. A goal of Plant Breeding Symposium II (PBS II) was "to assess progress made from plant breeding research, especially over the past 15 years," that is, since PBS I. Much significant research has occurred on disease resistance and epidemiology in the last 15 years, and major credit for stimulating this research must go to J. E. Van der Plank. It was his thought-provoking concepts of vertical and horizontal resistance, stabilizing selection, and strong and weak genes, that stimulated scientists to test his theories. The accumulated thinking and research enabled Parlevliet and Zadoks to develop a unifying theory of resistance.

Van der Plank defined vertical and horizontal resistances genetically but projected their effects epidemiologically, and this has led to confusion. Thus, a basic problem today in resistance breeding is one of semantics. A simple solution is to separate terms for genetic and epidemiologic concepts of resistance. Browning et al. (1977) proposed retaining "specific" and "general" resistance to express genetic concepts of resistance. They proposed the term "dilatory resistance" for the epidemiologic concept of the resistance that delays pathogen development regardless of the means of genetic control.

Seldom has a pathogen been excluded successfully from a host population, so the important question becomes, How much *genetic* resistance must a population of plants possess to achieve an adequate level of *epidemiologic* dilatory resistance?

Research since PBS I has shown that it does not take as much resistance to protect a population as previously thought, and that this resistance can be specific or general (or preferably both) genetically. As an example, multiline cultivars of oats have been used in Iowa and Texas to study their effects on the epidemiology of crown rust in large field plots. They have done a very effective job of controlling crown rust in Iowa and in experimental plots in Texas. We found that if ca. 1/3 of the plants in a field individually had specific resistance, the population of plants was protected from crown rust by an adequate level of epidemiologic dilatory resistance. General resistance in the background of the specific resistance gave a higher level of dilatory resistance. The degree of diversity needed in the population to hold a given level of dilatory resistance depends on the favorableness of the environment, however.

The adequateness of this 1/3 resistance as shown by multiline oat cultivars has been corroborated by data from a wild ecosystem in Israel. There, ca. 1/3 specific resistance in wild oat (*Avena sterilis*) stands has been adequate to protect the wild populations from the ravages of crown rust.

Another recent example of where a diverse population of plants controls a disease has been researched at the Plant Breeding Institute, Cambridge, England. There, mixtures of three barley cultivars, each cultivar carrying a different gene for specific resistance, give a level of dilatory resistance that is as effective in controlling powdery mildew as four applications of a fungicidal spray. This barley mixture was grown commercially on ca. 2000 ha in England last year. Each cultivar in the mixture is agronomically sound and possesses a unique gene for incompatibility to powdery mildew. The cultivars are mixed in equal proportions so that, in this case, 2/3 of the plants have specific resistance and the population has excellent resistance of a dilatory type.

A fourth example is provided by potatoes in New York (Fry, 1978). These studies showed that clone L-521-7, with dilatory resistance, required only 1/3 the fungicidal protection as did the susceptible check to completely control late blight.

Thus, depending on the host and the disease, incomplete genetic resistance of either a specific or general nature may give an adequate level of dilatory protection to the host population. Heterogeneity finally is being recognized as an effective mechanism for adequately and effectively reducing the rate of disease development in host populations. 'Miramar 63' and 'Miramar 65,' multiline cultivars of wheat, were released from the Rockefeller Foundation Agricultural Program in Colombia in the sixties. 'Multiline M' and 'Multiline E' cultivars of oats have been produced in the midwestern USA for a dozen years. Three multiline cultivars of wheat are being released in India this year. Cultivar mixtures of barley are showing promise in extensive tests in England. And most important, the logic of this approach for disease management is corroborated by data from the indigenous ecosystem in Israel.

3. W. D. GUTHRIE. It is essential to have good screening techniques when breeding plants for resistance to insects. The technique must provide for an adequate number of insects and a uniform level of infestation, and usually this requires artificial rearing of insects. Many species of insects, such as aphids, leaf hoppers, plant hoppers, and Hessian flies can be reared in large numbers in a greenhouse or insectary on seedling plants. Also, insects of the order Lepidoptera often can be reared on artificial media. In 1978, 10.5 million egg masses of the European corn borer, 23 million larvae of the southwestern corn borer, 5.5 million eggs of the fall army worm, 4.7 million masses of the corn ear worm, and 2.4 million larvae of the sugarcane borer were produced by insects reared on artificial media. These egg masses, eggs, and larvae can be distributed at any density and uniformly on plants in a nursery dedicated to breeding for insect resistance, and subsequent ratings on the basis of feeding on the plants is a guide to what genotypes to save as resistant. Certainly, much of the very significant progress that has been made in breeding crop plants for resistance to Lepidopterous insects can be traced to entomologists' capability to rear insects of this order on artificial media. This provides an assured source of eggs and larvae with which to infest populations of plants.

4. R. R. NELSON. Disease resistance can be classified into two types. One type is based on a resistance to the infection process and usually results in a hypersensitive reaction. It is known as vertical or race-specific resistance. The other resistance is designed to resist the growth and reproduction of the pathogen after infection has taken place. This resistance is aimed at disease management to some acceptable level. It is known as horizontal or nonspecific resistance.

Researchers, early on, concluded that the two types of resistance were conditioned by different kinds of genes; race-specific and nonspecific resistances are controlled by major genes and minor genes, respectively. However, some years ago, I proposed the concept that the genes controlling vertical and horizontal resistance are the same genes. There are no major or minor genes for disease resistance; there are only genes. Genes will act as vertical resistance genes if they are present alone, and the same genes will act as horizontal resistance genes if they are present together in the same genotype. My concept further suggests that the so-called minor genes controlling horizontal resistance were at one time so-called major genes, which after being rendered susceptible by a new race of a pathogen, continued to exert a ghost effect. Supposedly they were retained in the native populations of plants because they were of some value to the genotype in retarding disease increase. In contrast, when a resistance gene in a modern cultivar is overcome, it usually is removed and replaced by another gene, thus preventing any opportunity for resistance genes to function in an additive manner.

There are numerous examples which suggest that wild hosts and their parasites have evolved to genetic equilibrium. They have evolved into equilibrium due to the fact that each had accumulated a large number of genes, perhaps one by one, over time. All wild hosts in their natural ecosystems sustain some levels of disease, and usually they do not react in a hypersensitive manner. They sustain a few small lesions, sporulation is retarded, and disease increase is nominal. They appear exactly like modern day cultivars that have horizontal resistance. In evolution, the last gene added by either the host or the parasite did not result in a massive disruption of the equilibrium because the new gene was in the presence of many other genes that had coevolved.

Some published observations lend credence to my concept. They can be summarized in one way as follows: major genes are modified by minor genes, and minor genes are modified by major genes. Major genes together exhibit additive effects, and minor genes together exhibit additive effects. Major genes and minor genes together express modifying and additive effects. Major genes mask minor genes, and minor genes mask major genes. Major gene resistance is enhanced by minor genes, and minor genes resistance is enhanced by minor genes. One dose of a major gene may confer susceptibility, whereas two or more doses result in increasing levels of resistance. Finally, genes may be major in one background and minor in another.

The practical implication of this is that for developing a type of resistance designed to manage disease at some acceptable level, one should gather all genes used and discarded in the past, all currently used ones, and some new and unused ones, and pyramid all of these into a single host genotype.

5. E. AYEH. Nelson says major and minor genes for disease resistance are the same, but they have differential interactions with the environment. Also, in view of the high specificity of gene-to-gene reaction that confers either a compatible or incompatible reaction, how would alleles of the minor genes interact with alleles of a major gene?

6. R. R. NELSON. This is a question of the action of a gene and how it is influenced by its own genetic background. A single resistance gene in a background with no other resistance genes will act one way to the parasite and the environment. The same gene in a background with other genes will react in a different way. Its role in a multigene background is additive whether one is talking about a reaction to the environment or to the parasite.

7. J. G. HAWKES. The experience with potatoes in Mexico would substantiate what Nelson has said about the evolution of disease resistance in indigenous plant populations. One can find single genes that give a hypersensitivity reaction to the late blight organism (vertical resistance) and genes that simply delay the growth of the mycelia (horizontal resistance) both in the same species, *Solanum demissum.* They have evolved these two types of resistance in response to *Phytophthora* that was also evolving in the same area. The dual occurrence of vertical and horizontal resistance can also be found in *S. verrucosum*, *S. stoloniferum*, and several other species. Initially, plant breeders used one of these vertical resistance genes at a time, but the value of such a resistance gene was soon lost. After many experiences with this type of failure, they turned to the use of genes that give horizontal resistance, such as genes that delayed incubation and other aspects of pathogen development. The latter is a different type of effect altogether. Its effect is to retard growth of the fungus. Could the panel give its views on the interrelationships between these two types of resistance?

8. J. E. PARLEVLIET. Likely, genes that give a hypersensitive reaction in *Solanum* and those that give partial resistance have quite different modes of affecting the pathogen. It is not likely that the partial-resistance genes represent shadow effects of useless hypersensitivity genes, because if this were the case the plant would not have evolved the Vertifolia effect that occurs in *Solanum.* My assumption is supported by research done on potatoes in Germany. Six cultivars, each carrying a gene for the hypersensitive type of resistance to late blight that had been overcome by a virulence gene in the pathogen, were compared with six cultivars that had no hypersensitivity genes. The average susceptibility of the cultivars with overcome-resistance genes was considerably higher than the average for cultivars carrying no R genes. If the R genes had a shadow effect, at least there should have been no difference. This suggests that there is no shadow effect, which leads me to think the

hypersensitivity genes have a function that is different from that of the partial-resistance genes. Further, there are no signs that this partial resistance erodes or breaks down, and in species where both hypersensitivity and partial-resistance genes to a disease coexist, they act independently of one another.

9. R. R. NELSON. Minor genes are considered to be minor genes because there are no genotypes of the parasite to differentiate them as major. An example is provided by the wheat cultivar 'Redcoat.' It has some horizontal resistance to mildew. Recently, an isolate of the mildew organism was obtained that caused a fleck reaction on Redcoat and a highly susceptible reaction on a cultivar having no known genes for resistance. This would suggest that Redcoat carries a major gene for resistance which remained undetected as a result of the apparent masking effect of horizontal resistance.

10. R. MCBROOM. Dr. Nelson, did I understand correctly that you advocate pyramiding resistance genes for a crop species into a single cultivar?

11. R. R. NELSON. Yes, I advocate putting all resistance genes into one genetic background of the host. This is based, in part, on a hedge that the superbiotype will not occur in nature. The superbiotype, of course, is one that can parasitize all the resistance genes that have been pyramided into one genotype. The real question is concerned with genetic probabilities. That is, what is the liklihood that a genotype of the parasite will acquire and retain the first virulence gene in its population until it acquires a second virulence gene and holds those two genes in the same genotype until it adds a third gene, and so on. It is not a matter of whether all virulence genes are present in nature. What matters is whether they are all present in one genetic background of the pathogen. Now, if by some rare genetic probability, one genotype of the parasite does acquire all the virulence genes to match all the resistance genes, my concept suggests that the ensuing confrontation would not result in a total overcome of the host. Likely, it would result in slow rusting, slow blighting, and so forth; and a new host and parasite equilibrium would result, with the host appearing to possess horizontal resistance. This concept is being tested by pyramiding four mildew resistance genes into a single winter wheat genotype. An isolate of the mildew fungus that has all four virulence genes to overcome these four resistance genes is available. By this time next year, we will

know whether the pyramid will be successful. No one knows how many genes must be pyramided to create equilibrium.

12. J. J. MACKEY. The success of pyramiding genes for resistance in the host plant greatly depends on the ability of the parasite to accumulate the matching genes for virulence into one genotype. Irrespective of whether this happens by recombination or mutation, a genetic background adaptation must occur to restore general fitness to the pathogen. The possibility for recombination within the parasite population thus is highly decisive, and this ability is different under different ecological conditions. Nelson's way of explaining how pyramiding resistance works will fit particularly under conditions of asexual reproduction of the parasite combined with high assortive selection pressure causing overall genetic erosion. True haploidy is another important prerequisite.

13. D. HILLERISLAMBERS. Multilines may provide a good population system of resistance for diseases, but infection spores of a pathogen are sedentary once they have been deposited on a plant tissue. However, insects are actively mobile through a crop, and if an insect is incompatible with one plant of a multiline, it could move to the next plant, and so on. In view of these facts, would one expect multilines to be less successful for insect resistance than for resistance to pathogens?

14. J. A. BROWNING. Suneson, in California, got good control of Hessian fly by using diverse populations of wheat. Probably, if an insect or nematode dissipates much of its energy in searching for a suitable host, this will be expressed in some level of protection for the host population.

15. F. G. MAXWELL. Perhaps the mobility of insects would have no effect on the value of multiline cultivars for controlling them. For example, an insect is drawn to a plant because the plant contains a chemical that serves as an attractant. Once there, the insect may lay eggs, and the progeny would have no choice about leaving that host plant. The plant may have an antibiotic factor that would be detrimental to the young insect, so it would die. There are many mechanisms operating to give plants resistance to insects, and the number and type operating in a multiline may or may not make such a population system of resistance effective.

16. J. N. JENKINS. The gene-to-gene relationship of host-insect interaction probably operates for host-specific insects but not for insects that have many hosts. For example, grasshoppers feed on a wide range of plants, whereas Hessian flies infect only wheat or barley. The two types of insects will react differently to multiline or even multicrop plant populations. So when deciding whether to use a multiline cultivar or gene deployment to protect against insects, the same questions must be asked as would be asked for a disease. When the parameters are put together, one finds that in some situations multilines would be successful, and in others, not.

17. O. MYERS. Blends or multiline cultivars of crops give yield stability against yield losses from excessive pathogen buildup, but do multilines have sufficient reproductive or genetic stability that they can be used without reconstructing them each generation?

18. J. A. BROWNING. Because epidemics of the disease a multiline cultivar is meant to control cannot develop in a multiline, it is hard to see how the composition of the cultivar could be shifted by the disease. And, if the composition of the multiline cultivar is not changed by the disease, there should be no need to reconstitute it each year.

At Iowa State, an experiment was conducted with a five-isoline multiline cultivar that was propagated for five consecutive generations under rust and nonrust conditions. Some minor changes did occur in the isoline composition of this multiline, but they were multidirectional and not at all related to the crown rust races used in the experiment.

19. I. KIBIRIGE-SEBUNYA. It is argued that horizontal resistance delays the epidemic by reducing the rate at which the epidemic proceeds. Others have suggested that this should be called rate-reducing resistance.

20. J. A. BROWNING. In multiline cultivars, genes for specific resistance are used in isolines to form a diverse population, and an epidemiological result is obtained that Van der Plank would attribute to horizontal resistance. Yet no component of the multiline may carry horizontal resistance. It is much simpler to just separate the genetic and epidemiological concepts. In each isoline is a gene for specific resistance or incompatibility, but the population of plants has an epidemiological response that retards the rate of development

of the epidemic, and this we attribute to dilatory resistance. In this example it resulted from the use of specific resistance in mixtures. It also is the epidemiologic expression of pure-line general resistance that reduces the rate of pathogen development.

21. W. DEWEY. The successful examples of multiline use all deal with the foliar diseases, such as rust, mildew, and late blight, that recur in cycles within the same season. Would multilines be as successful against smut disease of cereals that has only one cycle per season?

22. J. A. BROWNING. True, work has been done using diversity against foliar pathogens. However, the diversity principle must hold for other diseases as well. When two targets, in this case susceptible plants,are farther apart, the dispersion effect will reduce the amount of inoculum that reaches the target, in this case the inolulation site. Therefore, a reduced amount of disease will probably result.

In wild populations of small grains in Israel, one sees smut occurrence in every wild host species, but there never is a very high incidence of the disease. In populations of wild barley, loose smut occurs in every floret of an infected plant, yet the population seems to be well protected against loose smut.

At Iowa State, we investigated the value of multiline cultivars for reducing disease damage from *Helminthosporium victoreae*, a soilborne pathogen. The multilines gave significant protection from it. This is exactly what one would expect, because the susceptible roots were spaced farther apart, and it took the pathogen longer to get from one susceptible root to another.

23. G. BOWERS. Are multilines useful in breeding for multiple disease resistance?

24. J. A. BROWNING. Certainly, heterogeneous wild populations must have a considerable level of multiple-disease resistance. Also, the first two multiline wheat cultivars released by the Rockefeller Foundation, 'Miramar 63' and 'Miramar 65,' had protection against both stripe rust and stem rust. And they were both quite successful.

25. K. S. GILL. Multiline wheat cultivars released in India are resistant to both stripe rust and leaf rust.

26. N. BORLAUG. The multiline wheat cultivar developed in India

on the background of 'No. 8165' must carry many sources of resistance to both loose smut and bunt. Also, it carries some genes for resistance to mildew.

27. T. S. COX. In the literature on disease tolerance, true tolerance and the effects from partial resistance have not been effectively separated. If tolerance is so difficult to detect, how can a breeder select for it in a practical breeding program?

28. M. D. SIMONS. One method is simply to select for host genotypes that yield best in the presence of disease. This, of course, says what difference does it make whether it is tolerance or some obscure form of resistance.

29. J. A. BROWNING. Probably if these tolerant populations were put into an epidemiological test, each would have a form of dilatory resistance.

30. E. LAMB. Cultivars with low levels of insect resistance are useful within a system of integrated insect control. So would it be useful to include other components of the integrated insect control plan, such as biological, chemical, or cultural factors, in screening programs for resistant genotypes in order to produce cultivars better adapted to an integrated pest management system?

31. J. N. JENKINS. That is a very good suggestion, and such a plan was tried in Mississippi in 1979. A group of plant genotypes is being assayed to see how they fit into established systems of biological control, cultural control, and the like.

32. P. ROWE. In most cases, the mechanisms of resistance to insects or disease are not known, but are there any documented cases where resistance to a pest has been associated with a characteristic that made the resistant plants undesirable for human or animal consumption?

33. F. G. MAXWELL. Yes. One is gossypol in cotton. It is toxic to higher animals, and it is associated with resistance to *Heliothis* bollworm. The boll weevil, through the evolutionary process, has developed a large tolerance to gossypol.

34. J. J. MACKEY. A clear interconnection was recently demonstrated between human throat cancer and the habit of eating

sorghum with a high content of catechin tannin bred for bird repellence. Since tannins are protective substances not only against birds but also insects, viruses, and the like, there exists an obvious conflict of interest.

When considering the breeding strategy to use for developing cultivars with disease resistance, one must first understand the reproductive strategy of the pathogen. For example, here in North America along the *Puccinia* Path, stem rust and particularly the virulent races are almost entirely reproduced via the uredinial stage. This implies little recombination, that is, little chance occurs for reshuffling virulence genes and/or background genes. Under such circumstances, race-specific resistance is subjected to dramatic gene erosion, and the surviving population will be exhausted in its ability to adapt. This is probably why the multiple-resistance strategy in Australia and the multiline and gene deployment concept in North America are successful. In parts of Europe where survival of the rusts depends on host alternation interconnected with a sexual phase, the recombination of different categories of genesis allows the pathogens to store what Van der Plank calls unnecessary genes for virulence, and thus, to develop complex races.

35. I. KIBIRIGE-SEBUNYA. Some very virulent races of pathogens have been created in the laboratory in Australia, and it has been recommended that breeders should send breeding lines there to be tested for reaction to these races. But how can a breeder be sure that the mutant that may arise in his region in the future will resemble the Australian laboratory mutant of the pathogen?

36. J. E. PARLEVLIET. What the Australians are trying to do is create races that will attack a resistance gene that is effective to all known races. They treat the most virulent race with a mutagen and test mutant pathogen clones on the omnipotent resistance gene. A mutant race that will attack this resistance gene represents a new race. Now any plant that is resistant to this new race must carry a hitherto unknown gene for resistance in addition to the original resistance gene. So this resistant plant carries two resistance genes effective against all races in the field. The mutant race is produced on the basis of known resistance genes. And, in that sense, the mutant is a future race, because the resistance genes used in the field will select for races with virulence genes that can overcome these resistance genes.

37. R. L. GALLUN. This research is somewhat analogous to what

we have done with Hessian flies. Hessian fly biotypes have been crossed and entirely new biotypes have segregated from these crosses. These new biotypes are used to screen for new sources of resistance which can be bred into new wheat cultivars.

38. R. R. NELSON. The method has been used in my research programs with northern maize leaf blight and powdery mildew of winter wheat.

REFERENCES

Abdalla, M. M. F. 1970. Inbreeding, heterosis, plasmon differentiation and phytophthora resistance in *Solanum verrucosum* Schlechtd., and some interspecific crosses in *Solanum*. Agric. Res. Rep. 748: 1-213.

Black, W. 1970. The nature and inheritance of field resistance to late blight (*Phytophthora infestans*) in potatoes. Am. Potato J. 47: 279-88.

Boskovic, M. M. 1976. International pathogenicity survey of *Puccinia recondita* f. sp. *tritici*. Proc. Fourth Eur. and Mediterr. Cereal Rusts Conf., Interlaken, Switz., pp. 75-78.

Boukema, I. W., and F. Garretsen. 1975. Uniform resistance to *Cladosporium fulvum* Cooke in tomato (*Lycopersicon esculentum* Mill.). 2. Investigations on F_2's and F_3's from diallel crosses. Euphytica 24: 105-16.

Bronnimann, A., A. Fossati, and F. Hani. 1973. Ausbreitung von *Septoria nodorum* Berk. und Schadigung bei kunstlich induzierten Halmlangemutanten der Winterweizensorte 'Zenith' (*Triticum aestivum*). Z. Pflanzensuecht. 70: 230-45.

Browning, J. A. 1974. Relevance of knowledge about natural ecosystems to development of pest management programs for agro-ecosystems. Proc. Am. Phytopathol. Soc. 1: 190-99.

Browning, J. A., and K. J. Frey. 1969. Multiline cultivars as a means of disease control. Annu. Rev. Phytopathol. 7: 355-82.

Browning, J. A., M. D. Simons, and E. Torres. 1977. Managing host genes: Epidemiologic and genetic concepts. *In* Horsfall, J. G., and E. B. Cowling (eds.), Plant disease. Vol. I.

Caten, E. E. 1974. Inter-racial variation in *Phytophthora infestans* and adaptation to field resistance for potato blight. Ann. Appl. Biol. 77: 259-70.

Catherall, P. L., and J. A. Chamberlain. 1977. Relationships, host ranges and symptoms of some isolates of Phleum mottle virus. Ann. Appl. Biol. 87: 145-57.

Clifford, B. C. 1975. Stable resistance to cereal diseases: Problems and progress. Rep. Welsh Plant Breed. Stn. 1974: 107-13.

Clifford, B. C., and R. B. Clothier. 1974. Physiologic specialization of *Puccinia hordei* on barley. Trans. Br. Mycol. Soc. 63: 421-30.

Darlington, L. D., D. E. Mathre, and R. H. Johnston. 1977. Variation in pathogenicity between isolates of *Claviceps purpurea*. Can. J. Plant Sci. 57: 729-33.

Day, P. R. 1974. Genetics of host-parasite interaction. W. H. Freeman, San Francisco, 238 pp.

Dixon, G. R., and I. R. Wright. 1978. Frequency and geographical distribution of specific virulence factors in *Bremia lactucae* populations in England from 1973 to 1975. Ann. Appl. Biol. 88: 287-94.

Eenink, A. H. 1976. Genetics of host-parasite relationships and uniform and differential resistance. Neth. J. Plant Pathol. 82: 133-45.

Ellingboe, A. H., and D. W. Gabriel. 1977. Induced conditional mutants for studying host/pathogen interactions, pp. 36-46. *In* Induced mutations against plant disease. IAEA, Vienna.

Esser, K., and R. Blaich. 1973. Heterogenic incompatibility in plants and animals. Adv. Genet. 17: 107-52.

Ezuka, A., O. Horino, K. Toriyama, H. Shinoda, and T. Morinaka. 1975. Inheritance of resistance of rice variety Aikoku 3 to *Xanthomonas oryzae.* Bull. Tokai-Kuki Natl. Agric. Exp. Stn. 28: 124-30.

Feeney, P. 1976. Plant apparancy and chemical defense. *In* Wallace, J. W., and R. L. Mansell (eds), Biochemical interaction between plants and insects. Recent Adv. Phytochem. 10: 1-40.

Fenner, F. 1965. Myxoma virus and Oryctolagus cuniculus: Two colonizing species, pp. 485-99. *In* Baker, H. G., and G. L. Stebbins (eds.), Genetics of colonizing species. Academic Press, New York and London.

Fishbeck, G. 1973. Methodische grundlagen der Resistenzzuchtung. Vortrage fur Pflanzenzuecht. 13: 4-33.

Flor, H. H. 1956. The complementary genic systems in flax and flax rust. Adv. Genet. 8: 29-54.

Flor, H. H. 1960. The inheritance of X-ray induced mutations to virulence in a urediospore culture of race 1 of *Melampsora lini.* Phytopath. 50: 603-5.

Flor, H. H. 1971. Current status of the gene-for-gene concept. Annu. Rev. Phytopathol. 9: 275-96.

Flor, H. H., and V. E. Comstock. 1971. Flax cultivars with multiple rust-conditioning genes. Crop Sci. 11: 64-66.

Frey, K. J., J. A. Browning, and M. D. Simons. 1973. Management of host resistance genes to control diseases. Z. Pflanzenzuecht. 80: 160-80.

Fry, W. E. 1978. Quantification of general resistance of potato cultivars and fungicide effects for integrated control of potato late blight. Phytopathology 68: 1650-55.

Habgood, R. M. 1976. Differential aggressiveness of *Rhynchosporium secalis* isolates towards specified barley genotypes. Trans. Br. Mycol. Soc. 66: 201-4.

Harper, J. L. 1977. Population biology of plants. Academic Press, London, New York, San Francisco. 892 pp.

Hooker, A. L. 1969. Widely based resistance to rust in corn. Field Crops. Spec. Rep. Ia. Agric. Home Econ. Exp. Stn. 64: 28-34.

Hooker, A. L., and K. M. S. Saxena. 1971. Genetics of disease resistance in plants. Annu. Rev. Genet. 5: 407-24.

James, W. C. 1974. Assessment of plant disease and losses. Annu. Rev. Phytopathol. 12: 27-48.

Kim, S. K., and J. L. Brewbaker. 1977. Inheritance of general resistance in maize to *Puccinia sorghi* Schw. Crop Sci. 17: 456-61.

Kommedahl, T., J. J. Christensen, and R. A. Frederiksen. 1970. A half century of research in Minnesota on flax wilt caused by *Fusarium oxysporium.* Tech. Bull. Minn. Agric. Exp. Stn. 272: 35.

Kooistra, E. 1968. Significance of the non-appearance of visible disease symptoms in cucumber (*Cucumis sativus* L.) after infection with *Cucumis* virus 2. Euphytica 17: 136-40.

Kuhn, R. C., H. W. Ohm, and G. E. Shaner. 1978. Slow leaf-rusting resistance in wheat against twenty-two isolates of *Puccinia recondita.* Phytopathology. 68: 651-56.

Latin, R. X., D. R. MacKenzie, and H. Cole. 1978. A significant host/pathogen interaction determined among apparent infection rates. Proc. 35th Annu. Meet. Potomac Div. Am. Phytopathology Soc., 1978. Phytopathology News 12: 70-71 (abstr.).

Levin, D. A. 1976. The chemical defenses of plants to pathogens and herbivores. Annu. Rev. Ecol. Syst. 7: 121-59.

Lim, S. M., J. G. Kinsey, and A. L. Hooker. 1974. Inheritance of virulence of *Helminthosporium turcicum* to monogenic resistant corn. Phytopathology 64: 1150-51.

Little, R., and J. K. Doodson. 1972. The reaction of spring barley cultivars to mildew, their disease resistance rating and an interim report on their yield response to mildew control. J. Natl. Inst. Agric. Bot. 12: 447-55.

Loegering, W. Q. 1978. Current concepts in interorganismal genetics. Annu. Rev. Phytopathol. 16: 309-20.

MacKey, J. 1976. Genes of virulence and their general adaptability in oat stem rust, p . 35. *In* Proc. Fourth Eur. and Mediterr. Cereal Rusts Conf., Interlaken, Switz.

Marshall, D. R., and A. J. Pryor. 1978. Multiline varieties and disease control. I. The dirty crop approach with each component carrying a unique single resistance gene. Theor. Appl. Genet. 51: 177-84.

Martin, T. J., and A. H. Ellingboe. 1976. Differences between compatible parasite/host genotypes involving the Pm4 locus of wheat and the corresponding genes in *Erysiphe graminis* f. sp. *tritici*. Phytopathology 66: 1435-38.

Mode, C. J. 1958. A mathematical model for the co-evolution of obligate parasites and their hosts. Evolution 12: 158-65.

Murray, M. J. 1971. Additional observations on mutation breeding to obtain verticillium resistant strains of peppermint, pp. 171-95. *In* Mutation breeding for disease resistance. IAEA, Vienna.

Neervoort, W. J., and J. E. Parlevliet. 1978. Partial resistance of barley to leaf rust, *Puccinia hordei*. V. Analysis of the components of partial resistance in eight barley cultivars. Euphytica 27: 33-39.

Nelson, R. R. 1975. Horizontal resistance in plants: Concepts, controversies and applications, pp. 1-20. *In* Galvez, I. E. (ed.), Proceedings of the seminar on horizontal resistance to the blast disease of rice. CIAT Publ. Ser. C. E.-9, Cali, Colombia.

Nelson, R. R. 1978. Genetics of horizontal resistance to plant disease. Annu. Rev. Phytopathol. 16: 359-78.

Parlevliet, J. E. 1977. Plant pathosystems: An attempt to elucidate horizontal resistance. Euphytica 26: 553-56.

Parlevliet, J. E. 1978a. Screening for partial resistance in barley to *Puccinia hordei* Otth., pp. 153-56. *In* Proc. Fourth Eur. and Mediterr. Cereal Rusts Conf.., Interlaken, Switz.

Parlevliet, J. E. 1978b. Race-specific aspects of polygenic resistance of barley to leaf rust, *Puccinia hordei*. Neth. J. Plant Pathol. 84: 121-26.

Parlevliet, J. E. 1978c. Further evidence of polygenic inheritance of partial resistance in barley to leaf rust, *Puccinia hordei*. Euphytica 27: 369-79.

Parlevliet, J. E. 1978d. Aspects of and problems with horizontal resistance. Crop Improv. 5(1): 1-10.

Parlevliet, J. E. 1979. The multiline approach in cereals to rusts; aspects, problems, and possibilities. Proc. satellite symposium on use of multilines for reducing rust epidemics. Indian J. Genet. Plant Breed. 39 (1): 21-28. New Delhi.

Parlevliet, J. E., and A. van Ommeren. 1975. Partial resistance of barley to leaf rust, *Puccinia hordei*. II. Relationship between field trials, micro-plot tests and latent period. Euphytica 24: 293-303.

Parlevliet, J. E., and H. J. Kuiper. 1977. Partial resistance of barley to leaf rust, *Puccinia hordei*. IV. Effect of cultivar and development stage on infection frequency. Euphytica 26: 249-55.

Parlevliet, J. E., and J. C. Zadoks. 1977. The integrated concept of disease resistance; a new view including horizontal and vertical resistance in plants. Euphytica 26: 5-21.

Pedersen, P. N. 1960. Methods of testing pseudo-resistance of barley to in-

fection by loose smut, *Ustilago nuda* (Jens). Acta Agric. Scand. 10: 312-32.
Person, C., and G. Sidhu. 1971. Genetics of host-parasite interrelationships, pp. 31-38. *In* Mutation breeding for disease resistance. IAEA, Vienna.
Riley, R. 1973. Genetic changes in hosts and the significance of disease. Ann. Appl. Biol. 75: 128-32.
Robbelen, G., and E. L. Sharp. 1978. Mode of inheritance, interaction and application of genes conditioning resistance to yellow rust. Adv. Plant Breed., Suppl. 9. Verlag Paul Parey, Berlin and Hamburg. 88 pp.
Robinson, R. A. 1976. Plant pathosystems. Springer Verlag, Berlin, Heidelberg, New York. 184 pp.
Samborski, D. J. 1978. Concepts dealing with specificity in host-parasite systems. Proc. Third Int. Congr. Plant Pathol., Munich. Abstr. 220.
Schafer, J. W. 1971. Tolerance to plant disease. Annu. Rev. Phytopathol. 9: 235-52.
Scheffer, R. P., R. R. Nelson, and A. J. Ullstrup. 1967. Inheritance of toxin production and pathogenicity in *Cochliobolus carbonum* and *Cochliobolus victoriae.* Phytopathology 57: 1288-91.
Schertz, K. F., and Y. P. Tai. 1969. Inheritance of reaction of *Sorghum bicolor* (L.) Moench to toxin produced by *Periconia circinata* (Mang.) Sacc. Crop Sci. 9: 621-24.
Sequeira, L. 1978. Lectins and their role in host-pathogen specificity. Annu. Rev. Phytopathol. 16: 453-81.
Simons, M. D. 1972. Polygenic resistance to plant disease and its use in breeding resistant cultivars. J. Environ. Qual. 1: 232-40.
Strobel, G. A. 1975. A mechanism of disease resistance in plants. Sci. Am. 232: 81-88.
Turner, E. M. C. 1953. The nature of resistance of oats to the take-all fungus. J. Exp. Bot. 4: 264-71.
Turner, E. M. C. 1961. An enzymic basis for pathogenic specificity in *Ophiobolus graminis.* J. Exp. Bot. 12: 169-75.
Umaerus, V. 1970. Studies on field resistance to *Phytophthora infestans.* 5. Mechanisms of resistance and application to potato breeding. Z. Pflanzenzuecht. 63: 1-23.
Van der Plank, J. E. 1963. Plant diseases: Epidemics and control. Academic Press, New York and London. 349 pp.
Van der Plank, J. E. 1968. Disease resistance in plants. Academic Press, New York and London. 206 pp.
Van der Plank, J. E. 1975. Principles of plant infection. Academic Press, New York, San Francisco, and London. 216 pp.
Van der Plank, J. E. 1978. Genetic and molecular basis of plant pathogenesis. Springer Verlag, Berlin, Heidelberg, New York. 167 pp.
Watson, I. A. 1977. The national wheat rust control programme in Australia. Fac. of Agric. Univ. of Sydney, Sydney, 24 pp.
Wheeler, H. 1975. Plant pathogenesis. Springer Verlag. Berlin, Heidelberg, New York. 106 pp.
Wiersema, H. T. 1972. Breeding for resistance, pp. 174-87. *In* de Bokx, J. A. (ed.), Viruses of potatoes and seed-potato production. Pudoc, Wageningen.
Wolfe, M. S., J. A. Barrett, R. C. Shattock, D. S. Shaw, and R. Whitbread. 1976. Phenotype-phenotype analysis: Field application of the gene-for-gene hypothesis in host-pathogen relations. Ann. Appl. Biol. 82: 369-74.
Yamamoto, T., H. R. Hifni, M. Machmud, T. Nishizawa, and D. M. Tantera. 1977. Variation in pathogenicity of *Xanthomonas oryzae* (Uyeda et Ishiyama) Dowson, and resistance of rice varieties to the pathogen. Contr. Centr. Res. Inst. Agric. Bogor, No. 28. 22 pp.

Zadoks, J. C. 1966. Problems in race identification of wheat rusts. *In* Fifth Yugoslav Symp. on Res. in Wheat (Novi Sad, 1966). Contemp. Agric. 11/12: 299-305.
Zadoks, J. C. 1972. Modern concepts of disease resistance in cereals, pp. 89-98. *In* Lupton, F. G. H., G. Jenkins, and R. Johnson (eds.), The way ahead in plant breeding. Proc. Sixth Congr. Eucarpia. Cambridge Univ. Press.
Zimmer, D. E., and J. A. Hoes. 1974. Race 370, a new and dangerous North American race of flax rust. Plant Dis. Rep. 58: 311-13.
Ziv, O., and Z. Eyal. 1978. Assessment of yield component losses caused in plants of spring wheat cultivars by selected isolates of *Septoria tritici.* Phytopathology 68: 791-96.

CHAPTER 10

Breeding for Improved Nutritional Quality

J. D. AXTELL

HISTORICALLY, plant breeding has contributed enormously to increased production of total nutrients as well as to the increased efficiency of nutrient production per unit area. So, in this sense, all plant breeders address the problem of how this profession can continue to elevate the status of human nutrition through the application of the science of plant breeding to crop improvement. We should not lose sight of this overriding reality as we consider the problems and opportunities of breeding for improved nutritional quality of specific crop species.

To make most progress, however, in improving the nutritional quality of our food crops, plant breeders must become more thoroughly acquainted with nutritional issues. In a somewhat similar setting, H. J. Muller asked, in the early 1950s at the dawn of an era when genetics was entering the molecular age, "Must we become biochemists, physiologists and bacteriologists in addition to the already numerous disciplines required of a geneticist?" His answer was a resounding "Yes," because the science of genetics in his view would benefit enormously from the interdisciplinary cross-fertilization which would result. In some sense, plant breeders also must become knowledgeable in the problems of human nutrition if they are to fully exploit the opportunities that our science of plant breeding offers to humankind.

The recent World Food and Nutrition Study by the National Academy of Science is replete with recommendations for more

Professor of Genetics, Department of Agronomy, Purdue University, Lafayette, Indiana.

Journal Paper No. 7834, Purdue University Agricultural Experiment Station. Research supported in part by U.S. Contract AID/ta-c-1212.

research in human nutrition to fill important knowledge gaps and to provide information on the long-term effects of diet on human health. Field studies on nutritional problems in developing countries are badly needed. It is imperative that plant breeding become an important tool in the arsenal of the human nutritionist when attacking the problems of undernutrition and malnutrition in the world. Human nutritionists must provide clear-cut nutritional objectives to the plant breeder. Plant breeders, in turn, must provide information on what changes in nutrient composition are possible within a given crop species, and also what, if any, trade-offs are involved. Then, a strategy for nutritional improvement can be developed; namely, Can the plant breeder provide a cheaper solution to a nutritional problem than the nutritionist can provide by nutrient fortification or alteration in dietary habits? Of course, answers to these questions will vary between countries and between socio-economic groups within a country. We have only begun to explore this new dimension in plant breeding.

Herein, this subject will be approached from two dimensions—first, from the point of view of the plant breeder, and second, from the perspective of the human nutritionist.

THE BREEDER'S VIEWPOINT

For the human population to be adequately nourished it is necessary to assure nutritional quality in food crops and to provide a food supply to meet the needs of an expanding population. New crop cultivars with improved agronomic traits have been the major factor contributing to increased crop production, and there will be a need for continued emphasis in improving these traits. Therefore, compositional traits that are consistent with good nutritional quality can be selected in new plant cultivars. Even selection for a simple chemical trait is a complex and expensive undertaking in a plant breeding program. Desired nutritional characteristics must occur in a cultivar that farmers will grow or the nutritional value will never get to consumers. That is, the specific nutritional trait must occur in a cultivar that gives a high and stable yield, has disease and lodging resistance, cold tolerance, and the like. It must be clearly established that a compositional trait will benefit nutrition before a plant breeder will be willing to add this trait to the long list of traits for which he or she selects already.

Breeding for Nutritional Traits

The explosive yield increases in crop production during recent decades indicate how successfully genetic resources can be exploited. Conscious screening for better nutritional value has been started only recently, so the available genetic diversity is far from having been evaluated. Röbbelen (1977) proposed three requisites for drafting effective strategies for modern "quality breeding" programs: (1) clear-cut information on the nature and priorities of the various nutritional criteria, (2) powerful analytical methods for their quantitative determination, and (3) sufficient access to genetic variability.

Cereal Protein Improvement

Frey (1977) estimated that on the average, cereal grains provide 70 percent of the calories consumed in human diets. The world's protein needs come approximately 50 percent from cereals, 20 percent from grain legumes, and 30 percent from animal products (Oram and Brock, 1972). In developing countries, animal products contribute only 10 percent and cereals 70 percent. Thus, in many parts of Asia, the Near East, Africa, and Latin America, people are dependent directly on cereals for approximately two-thirds of their dietary protein requirements.

Most rationale behind cereal protein improvement research can be understood by examining the major classes of storage proteins in cereal seeds. Osborne (1897), who initiated systematic studies of seed proteins in 1891, classified them according to their solubilities into albumins (water soluble), globulins (soluble in saline solutions), prolamins (soluble in relatively strong alcohol), and glutelins (soluble in dilute alkaline solutions). Prolamines from all cereals are relatively rich in proline and glutamine but low in basic amino acids including lysine. Miflin and Shewry (1979) have recently written an excellent review of cereal seed prolamins which clarifies the biology and biochemistry of this important class of proteins. Osborne and Mendel (1914) showed that rats of all ages went into rapid decline and eventually died if placed on a diet in which zein (maize (*Zea mays*) prolamin) was the sole source of dietary protein. The prolamin contents of the major cereal species fall into three rather distinct groups (Fig. 10.1). Because the protein qualities of cereals, in general, are inversely related to their prolamin contents, these groups of cereals increase in protein quality as the decrease in prolamin content. It is interesting to note that man domesticated major cereal species with a wide range in prolamin contents, whereas no such range in carbohydrate composion exists among cereal species. It is a challenge to deduce the biological significance of this diversity in prolamin contents among the cereals. Jones and Tsai (1977) have shown that zein is more rapidly mobilized than other proteins

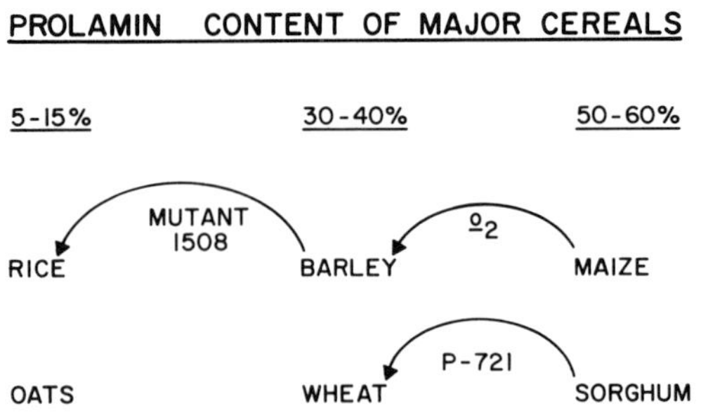

Fig. 10.1. Relationships among cereal grains based upon prolamin percentages in the seed proteins from normal and high-lysine mutant types.

during seed germination, and this may account for opaque-2 kernels germinating somewhat slower than normal ones of isogenic lines. In other words, the high prolamin content of maize and sorghum (*Sorghum bicolor*) may be advantageous for rapid stand establishment under competitive conditions in the tropics.

Major cereal species (Fig. 10.1) also differ in the photosynthetic pathway used to fix CO_2. Maize and sorghum use the "C-4 pathway" which is the more efficient CO_2 fixation package leading to greater potential productivity under tropical conditions. So the paradox exists that no currently available cereal species combines the nutritional advantage of low endosperm prolamin content and the high potential productivity characteristic of C-4 species. Two obvious approaches to solving this problem are: (1) the C-4 pathway advantages of CO_2 fixation could be incorporated into cereal species with lower prolamin content, namely, rice (*Oryza sativa*), oats (*Avena sativa*), wheat (*Triticum aestivum*), and barley (*Hordeum vulgare*); or (2) the low prolamin characteristics of rice and oats could be incorporated into maize and sorghum. Would alternative (1) or (2) be more easily accomplished? The answer will depend on which biochemical process is easiest to manipulate genetically, but unfortunately, today we do not understand the genetic control of either of these biochemical pathways in cereal grains.

Actually, alternative (2) is being researched in a sense. Mertz et al. (1964) discovered that the opaque-2 mutant of maize has an amino acid profile differing from that of common maize, being significantly higher in lysine and tryptophan. The opaque-2 gene drastically reduces the proportion of zein and increases the proportions of the more nutritious water and salt-soluble proteins. Similar changes in protein fractions account for nutritional improvements found in Mutant 1508 in barley (Ingverson et al., 1973) and the high lysine mutants in sorghum (Singh and Axtell, 1973; Mohan and Axtell, 1975) (Fig. 10.1).

The opaque-2 and high-lysine endosperm types in maize and sorghum, respectively, reduce the content of prolamin in these species to about the content of normal barley. The high-lysine mutant in barley reduces the prolamin content to that of normal rice and oats. Rice and oats may already have incorporated mutants homologous to those recently detected in maize, sorghum, and barley. Indeed, normal barley is intermediate between normal maize and rice in prolamin content. Mutant 1508 in barley has a prolamin content characteristic of rice and oats. These changes appear to occur in two steps from higher to lower prolamin content which suggests that the regulation of prolamin synthesis in these cereals may be controlled by two major genes (Nelson, 1974). If this interpretation is correct, it should be possible to find a "second step" mutation in sorghum and maize which will reduce the prolamin content to approximately 10 percent with a concomitant increase in overall protein quality. This would provide a cereal grain species which combines the photosynthetic efficiency and stress tolerance of sorghum with the protein and nutritional quality of oats or rice. The improved biological values associated with reduced prolamin and increased lysine contents in maize, barley, and sorghum are illustrated by the protein efficiency ratios (PER) values presented in Table 10.1.

Table 10.1. Lysine content and protein efficiency ratios (PER) for cereal grains when tested in isonitrogenous (8.8% protein) rat feeding trials. (Adapted from Mertz et al., 1975.)

Cereal	Lysine (% of protein)	PER (% of casein control)
Casein	8.6	100
Soft o_2	4.4	90
Hard o_2	4.5	87
Sorghum (*hl*)	3.4	80
Triticale	3.4	62
Normal Maize	2.6	50
Wheat	3.8	50
Sorghum (low tannin)	2.5	41
Sorghum (high tannin)	2.3	22

YIELD VS. PROTEIN QUALITY

Very significant progress has been made in improving the protein quality in barley, maize, and sorghum, but the question remains whether the productivity of high-lysine lines in these cereals can be improved to the point that the improved biological value is made available as improved cultivars.

Sorghum

Mohan (1975) utilized chemical mutagenesis to induce a high lysine mutation in sorghum. The parent line used for the mutagen treatment was photoperiod insensitive, three-dwarf, and relatively broadly adapted. It had a colorless pericarp and translucent (vitreous) endosperm so progeny from the mutagen treatments could be screened easily for mutant opaque kernels by using a light box. Selfed seed was treated with diethyl sulfate (DES), and heads of M_1 plants were bagged to insure self-fertilization. Heads on M_2 plants, grown in Puerto Rico, also were bagged, and approximately 23,000 bagged M_2 heads were harvested. M_3 seeds from each head were examined for opaque kernel segregants, and 445 M_2 progenies were identified as being putative opaque mutants. Seeds from each segregating head were separated into vitreous and opaque classes, and both seed classes from each putative mutant head were analyzed for protein and lysine concentration. Only 33 of the mutants had an increase in lysine concentration greater than 50 percent. Plants from 33 opaque and normal sib seed lots were grown in paired rows to permit their evaluation for morphological differences. Most opaque mutants had drastically changed plant or seed development. One (P-721) produced normal appearing plants and seeds, and it produced a 60 percent increase in lysine concentration. The high-lysine characteristic is conditioned by a single gene that is partially dominant. The biological value of P-721 grain is significantly higher than grain from its normal sib when tested in feeding experiments with monogastric animals. Note that extensive agronomic and biochemical selection occurred in identifying the particular high-lysine mutant to be used before the breeding program was initiated.

Van Scoyoc (1979) has examined dry matter accumulation during grain development to determine what effect the P-721 mutant has on grain yield potential in the original genetic background. No difference was apparent in dry matter accumulation until ca 31 days after pollination, but after that date, it leveled off in the P-721 opaque line, whereas in the normal sib line, dry matter accumulated until 38 days after pollination (Fig. 10.2).

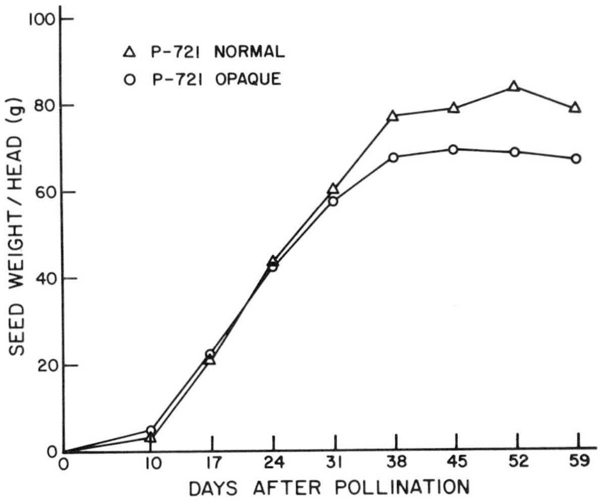

Fig. 10.2. Mean increase in seed weight/head of P-721 opaque and P-721 normal sorghum during grain development in a space planted population. (Van Scoyoc, 1979.)

The next phase of this sorghum improvement program involved the making of hundreds of crosses of the P-721 opaque mutant with high yielding entries from the World Sorghum Collection, with elite lines from the Purdue/AID sorghum breeding materials, and with individual plants selected from genetically heterogeneous random mating populations. Emphasis was put on incorporating the P-721 opaque gene into many and diverse genetic backgrounds to enhance the probability for identifying a genetic background which was optimal for expression of the P-721 gene. The pedigree breeding procedure was used in handling progenies from these crosses. Early generation selections were evaluated for agronomic desirability and yield potential at Lafayette, Indiana, USA, and for tropical adaptability in Puerto Rico. All segregating lines which lacked promising agronomic potential were discarded without attention to chemical evaluation because the major objective was to derive high-yielding, agronomically desirable sorghum lines in which the P-721 gene had survived. Some 197 homozygous opaque F_6 lines survived and after a final screening against lodging, stalk rot, and foliar diseases in Puerto Rico, Van Scoyoc (1979) tested the best 158 lines, 11 elite normal cultivars from international trials, and RS 671 for yield (Fig. 10.3). Several of the elite normal lines (the P-954 series) have yielded very well in Africa. Yields of the 158 P-721 lines and the 11

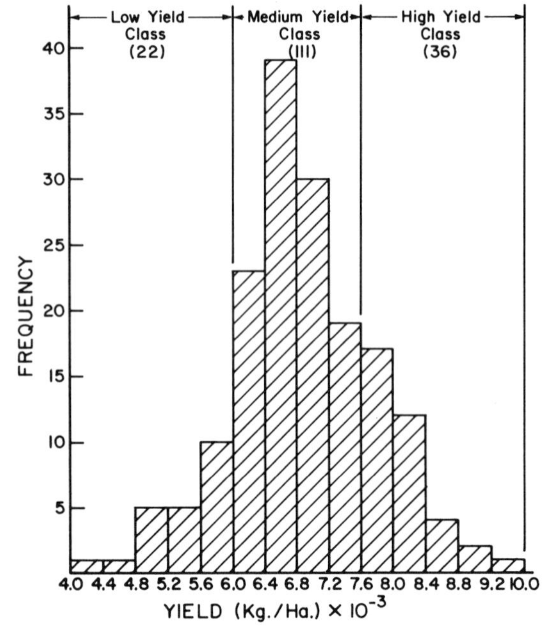

Fig. 10.3. Frequency distribution of grain yields of F_7 P-721 opaque-derived lines and elite normal cultivars of sorghum. (Van Scoyoc, 1979.)

elite normal lines divided into three classes: 22 with low yield, 111 with intermediate yield, and 36 with high yield. All check lines were in the high-yield class. Among entries that yielded above 8.0 t/ha, the 12 opaque lines and the seven vitreous controls both gave mean yields of 8.5 t/ha. Among entries yielding above 7.6 t/ha, the mean for 24 opaque endosperm lines was only slightly less than that for the 12 checks (8.1 vs. 8.2 t/ha, respectively). These data indicate that lines with the P-721 opaque gene can yield as well as the best normal sorghum cultivars if it occurs in the proper genetic background. Earlier, Christensen (1978), by studying a subset of the P-721 lines in F_5 breeding lines, also showed that the P-721 opaque gene when placed in an appropriate genetic background would not reduce grain yield potential (Table 10.2). Because seed weight was reduced about 15 percent by the P-721 opaque gene, we speculate that selection for grain yield in P-721 lines must have resulted in an increase in the number of seeds per panicle and/or the number of panicles per unit area in order to have maintained a good yield level (Axtell et al., 1979). It is likely that variation in sorghum

Table 10.2. Means of chemical and agronomic traits for opaque, heterozygous, and normal grain types in P-721 derived lines and high yielding checks of sorghum. (Adapted from Christensen, 1978.)

Genotype or Cultivar	No. of Entries	Dye Binding Capacity	Protein	Yield	Seed Weight
			%	*t/ha*	*g/100*
P-721 Genotype					
Opaque	300	49.3	12.5	4.7	2.50
Heterozygous	73	42.5	13.3	4.5	2.89
Normal	5	39.8	12.9	4.2	2.87
Checks					
954063 (cultivar)	4	34.0	11.4	5.7	2.58
RS-671 (hybrid)	4	34.3	11.7	5.6	2.30
NK-300 (hybrid)	4	33.8	10.9	6.7	2.18

panicle morphology allows compensation for reduced seed weight by increasing seed numbers per panicle.

Acceptance of high-lysine sorghum cultivars by growers and consumers presumably has been limited by problems associated with the opaque kernel phenotype; so Ejeta (1979) tried and was successful in identifying several lines with vitreous endosperm and high lysine content. Subsequently, these proved to be stable for vitreous endosperm phenotype and high lysine concentration. Also, seed treatments of P-721 opaque, high-lysine sorghum lines with DES resulted in mutants with vitreous endosperm and high lysine concentration (Porter, 1977; Ejeta, 1979). In general, the lines with modified vitreous endosperm from both sources had higher kernel weight and lower percentages of protein and lysine (Table 10.3). Also, the most vitreous types had the highest test weight.

Table 10.3. Protein and lysine contents and 100-kernel weight for vitreous (modified) and opaque endosperm kernels from lines derived from P-721 high-lysine sorghum. (Adapted from Ejeta, 1979.)

Endosperm type	Protein (%)	Lysine (% of protein)	100-kernel weight (g/100)
Modified	11.3	2.77	2.56
Opaque	12.6	2.83	2.27
Modified as % of opaque	90.0	97.9	112.8

Table 10.4. Seed, protein, and lysine yields and contents for high-lysine barley lines (mean of 6) derived from the cross hiproly × normal. (Perssons, 1975.)

Entry	Kernel Yield	Protein	Protein Yield	Lysine (g/100 g protein)	Lysine Yield
	t/ha	%	*kg/ha*		*kg/ha*
High lysine	4.2	11.8	493.4	4.11	20.3
Normal (Mona)	4.3	10.4	446.5	3.54	15.8
% of normal	97.3	113.6	110.5	116.1	128.2

Barley

Research on protein quality improvement in barley has centered on two sources of high-lysine germplasm. The 'Hiproly' source, identified by Munck et al. (1971), has its high-lysine content controlled by one major gene (*lys*) located on chromosome 7 and a number of minor genes (Persson, 1975). The yield of Hiproly is about one-third that of contemporary cultivars, but high-lysine lines derived from backcrossing to 'Mona' cultivar yielded from 95-100 percent as high as the recurrent parent (Table 10.4). However, under unfavorable growing conditions, the high-lysine lines were more inferior to the commercial cultivars. It is anticipated by the Svalov breeders that the use of high-lysine barley lines could reduce protein supplementation in pig rations by 35 percent. Small seed size and low grain yield continue to be major problems with high-lysine barley lines, but much breeding effort is being expended in Sweden to overcome these weaknesses by creating 12 new populations obtained by crossing Hiproly with lines that have desirable agronomic characteristics.

In Denmark, several high-lysine barley mutants were obtained via mutagenesis (Ingverson et al., 1973; Doll and Koie, 1975). Mutant 1508 has a single recessive gene that gives a 40 percent increase in lysine content of the protein. When compared with the parent cultivar, 'Bomi,' the mutant has a small increase in the protein content per seed, a 10 percent reduction in seed size, and a 17 percent reduction in kernel yield per unit area (Table 10.5).

Table 10.5. Productivity and seed composition of barley mutant 1508 and the parent cultivar Bomi. (Doll and Koie, 1975.)

Trait	Bomi	Mutant 1508	Mutant as % of Bomi
Lysine content (g/16 gN)	3.75	5.2	139
Lysine yield (g/plot)	10.4	13.7	132
Protein content (6.25 × N%)	9.7	11.0	113
Protein yield (g/plot)	278.0	263.0	95
Seed size (mg)	40.8	37.5	92
Number of seeds/plot × 10^{-3}	70.7	63.4	90
Grain yield (kg/plot)	2.88	2.38	83

Maize

Glover (1976) found that an opaque-2 hybrid of maize yielded from 86 to 92 percent of its normal counterpart over a four-year

Table 10.6. Grain yields of a commercial normal maize hybrid and opaque-2 counterpart grown in four years. (Glover, 1976.)

Year	Number of Tests	Genotypes	Yield	
			t/ha	*% of normal*
1	4	Normal	11.3	
		Opaque-2	10.3	91.4
2	4	Normal	10.8	
		Opaque-2	9.9	91.5
3	1	Normal	9.2	
		Opaque-2	8.2	89.5
4	1	Normal	12.4	
		Opaque-2	10.6	86.0
	Mean of 10 tests			90.7

period (Table 10.6). At the Center for Maize and Wheat Improvement (CIMMYT), Vasal et al. (1979) showed that some hard endosperm opaque-2 populations and their normal counterparts are nearly equal in yield, probably due to the selection of favorable modifiers (Table 10.7). Hard endosperm opaque-2 cultivars have been distributed internationally by CIMMYT for yield comparisons with the best normal cultivars in each country (Table 10.8). The best opaque-2 entry yielded equal to or better than the best normal check in seven countries. The philosophy at CIMMYT is to place major emphasis on breeding agronomically superior cultivars that are high yielding to maintain calorie production, with improved protein quality as a bonus. Their results indicate that significant progress toward that objective has been achieved. The question

Table 10.7. Grain yields (t/ha) of normal and opaque-2 types in two maize cultivars. (Adapted from Vasal, Villegas, and Bauer, 1979.)

Endosperm Type	Cultivar	
	Eto Blanco	LaPosta
Normal	4.1	5.3
Opaque-2	3.9	5.2
Opaque-2 as % of normal	95.0	97.0

Table 10.8. Grain yields (t/ha) of best opaque-2 maize cultivar and check cultivar in CIMMYT Experimental Variety Trial 15 grown at 11 locations in 1977. (Vasal, Villegas, and Bauer, 1979.)

Location	Best Opaque	Check
San Jeronimo (Guatemala)	5.0	3.6
Panama	2.9	2.6
Obregon (Mexico)	2.3	1.5
Costa Rica	3.5	2.9
Cotaxtla (Mexico)	4.4	3.9
Ludhiana (India)	6.0	5.5
Pirsabak (Pakistan)	4.4	2.7
Nicaragua	3.9	4.4
El Salvador	3.4	4.0
Jamaica	2.9	3.8
Suwan (Thailand)	5.6	6.7

remains, however, whether consistent grain yields comparable to those of the best yielding normal cultivars can be obtained with the opaque-2 gene system.

All prolamins appear to function as storage proteins. The concept of a "storage protein" implies a protein whose only function is to act as a storehouse for nitrogen. Miflin and Shewry (1979) suggest that a storage protein (1) is unlikely to have a metabolic function, (2) is likely to be formed relatively late in seed development, (3) is increased preferentially by increased levels of N nutrition, and (4) is likely to be stored in a package (namely, a protein body). The synthesis of the prolamin components is under genetic control and information is available regarding the chromosomal location of the structural genes that control its inheritance. The sequence of amino acids in several individual prolamins has been studied, and they show a high degree of homology indicating their closely related nature.

Tsai et al. (1978) and Tsai (1979, pers. commun.) have proposed a very intriguing hypothesis on the effect of N sink capacity on maize productivity. Genetic control of zein and glutelin synthesis may play an important role in promoting the growth and development of maize kernels. These two proteins may contain more than 90 percent of the total N in kernels and 60 percent of the total N in the maize plant at maturity. They appear to serve as an N sink in kernels and thereby prevent the accumulation of deleterious levels of free amides and ammonia in plant tissues. Both ammonium and nitrate ions can be taken up by maize roots, but the assimilation of ammonia and the subsequent organic N interconversion in developing kernels require organic acids, such as α-ketoglutaric acid and the like, which are derived from sucrose (Fig. 10.4). Thus the greater amount

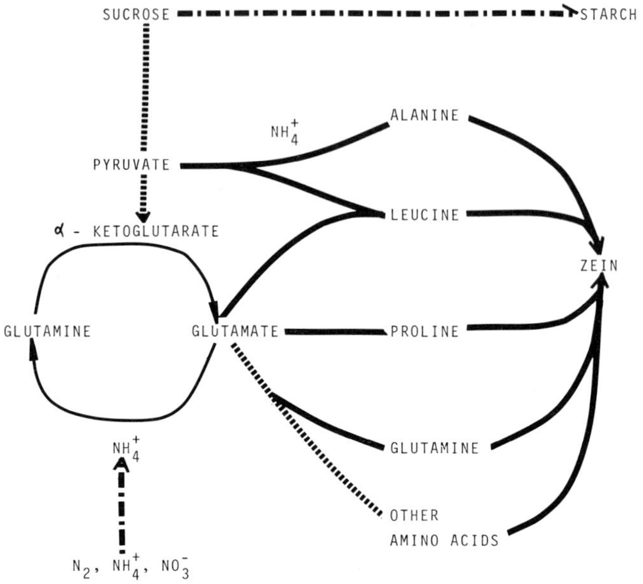

Fig. 10.4. Schematic pathway for the assimilation of ammonia via α-ketoglutaric acid through amino acids to zein to show how this prolamine may serve to cause greater CO_2 fixation, and thus, high yield in cereals. (Tsai et al., 1978.)

of sucrose translocated from leaves under N-rich soil conditions provides energy and essential carbon skeletons for ammonia assimilation and organic N interconversions. At the same time, the movement of more sucrose to N-rich tissues may enhance CO_2 fixation in leaves. The increase in photosynthetic efficiency, and the translocation of nitrogenous compounds and sucrose into the kernel which is considered a sink for photosynthetic assimilates, should further promote the synthesis of starch and thereby increase yield. Because nitrate, unlike ammonia, may accumulate without utilization, the mixture of nitrate and ammonium ions in plants may function as a "buffer" for optimizing N utilization. Thus, while ammonia enhances the movement of sucrose from leaves to kernels and increases photosynthetic efficiency, nitrate functions as an N reserve. The effectiveness of the ammonium ion in enhancing this pathway requires that the ammonia be readily assimilated and that the amides and other ammonium storage amino acids be deposited in some tissue. The kernel functions as this storage tissue after vegetative growth has ceased. About 60 percent of the final N in maize kernels is present in the shoot at pollination. While both zein and glutelin appear to function as an N sink in the kernel, zein appears to be the most effective sink in this regard, because first, it occurs in the greatest quantity, and second, its synthesis can be manipulated by N fertilization and genetic means (Fig. 10.5). Positive associations occur between zein content, kernel weight, and grain yield (Fig. 10.6), so selection of hybrids containing a large functional N sink (capable of producing the maximum amount of zein) might provide a way to increase grain and protein yields per ha.

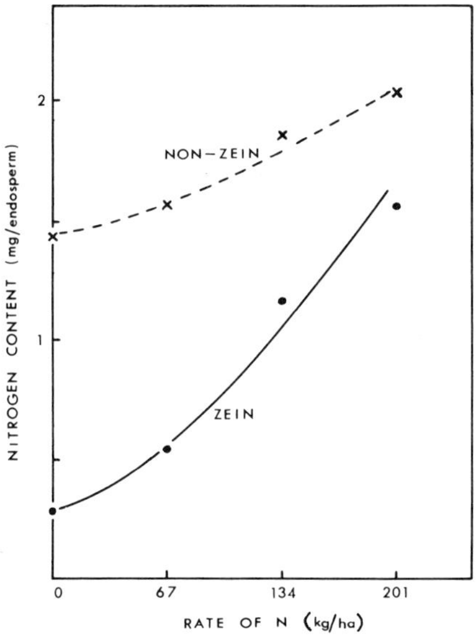

Fig. 10.5. Rates of zein and nonzein (albumin, globulin, and glutelin) accumulation with increasing rates of N fertilization. (Tsai et al., 1978.)

Because zein is low in lysine and tryptophan, essential amino acids for monogastric animals, the synthesis of maximum amounts of zein under high rates of N fertilization may increase yield of protein but reduce its nutritional quality. Maize, grown under N stress, produced lower kernel weight and zein content but lysine content in the protein was increased. Lysine concentration in the protein de-

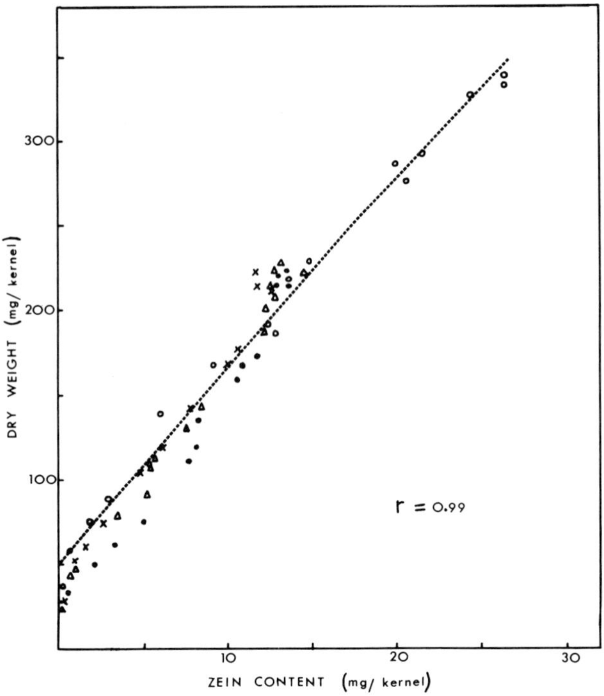

Fig. 10.6. Scatter diagram of coordinates for zein contents and dry weight in the developing kernels of inbred A545 (△) and B37 (×) of maize and their reciprocal crosses, A545 × B37 (ɼ) and B37 × A545 (o). (Tsai et al., 1978.)

creased as the level of N fertilizer increased up to 134 kg/ha, but no further reduction in the percentage of this amino acid occurred at 201 kg/ha of N fertilization. At the highest rate of N, zein accounted for only 40 percent of the total kernel protein.

Tsai's hypothesis suggests the need for a closer look at the role of storage proteins in endosperm development. Perhaps high prolamin genotypes should be sought in rice, wheat, barley, and oats as Frey (1977) has suggested. In fact, his "yield genes" in oats may represent high-prolamin oat genotypes. Genetic engineering techniques for transferring the high-zein gene(s) from maize to other cereals may be possible in the future. The question would arise of what trade-off in protein quality of rice, for example, could be accepted for significant increases in responsiveness to N fertilization and the resulting higher grain yields? This illustrates the importance of considering the N sink as well as the carbohydrate sink in cereals, and the interrelationship between them.

Table 10.9. Means and ranges of protein percentages in commercial lots of grain from cereals. (Miller, 1958.)

Cereal	Protein Percentage		Number of Samples
	Mean	Range	
Corn	10.4	7.5–16.9	1875
Wheat	12.0	8.1–18.5	309
Sorghum	12.5	8.7–16.8	1160
Barley	13.1	8.5–21.2	1400
Oats	13.3	7.4–23.2	1850
Rye	13.4	9.0–18.2	112

BREEDING FOR PROTEIN CONCENTRATION

Means and ranges for protein percentages of grains of the cereals as compiled by Miller (1978) are shown in Table 10.9. Commercial corn averaged only 10.4 percent whereas barley, oats, and rye averaged over 13 percent. The degree of genetic modification of chemical composition that can be accomplished by plant breeding is amply demonstrated by the classic selection experiments for protein and oil percentages carried out on maize at the Illinois Experiment Station (Dudley et al., 1974). In 70 generations of selection, breeders have lowered and raised the protein percentage from 10.9 percent in the original sample to 4.0 percent and 23.5 percent, respectively (Fig. 10.7). There is ample evidence in other cereals as well that demonstrates substantial genetic variation for protein concentration.

Frey (1977) has shown that in several cereal species, grain yield and grain protein content are negatively correlated almost universally (Table 10.10). Apparently, it is much easier to find genes that reduce starch accumulation in the grain, than it is to find genes that increase the accumulation of protein. Two examples illustrate, however, that it is possible to breed for higher grain protein concentration while at the same time maintaining very acceptable grain yields.

Wheat

At Nebraska, known wheat genes from 'Atlas 66' have been utilized to elevate protein concentration in 'Lancota' cultivar by 1 to 2 percent without a reduction in grain yield (Table 10.11). Middleton et al. (1954) reported that two wheat cultivars, 'Atlas 50'

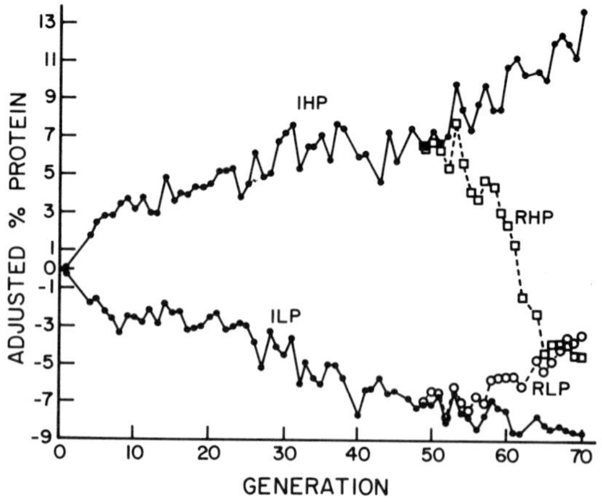

Fig. 10.7. Mean adjusted percent protein for IHP (Illinois high protein), ILP (Illinois low protein), RHP (reverse high protein), and RLP (reverse low protein) plotted against generations of selection. (Dudley et al., 1974.)

Table 10.10. Correlations between grain yield and protein percentages in the grain as reported by several researchers. (Frey, 1977.)

Crop	Correlation	Reference
Barley	-0.79**	Grant and McCalla, 1949
	-0.24*	Zubriski et al., 1970
Corn	-0.48**	Frey et al., 1951
	-0.33*	Dudley et al., 1971
Oats	-0.45**	Jenkins, 1969
	-0.59**	Sraon et al., 1975
Sorghum	-0.85**	Worker and Ruckman, 1968
	-0.26*	Malm, 1968
Wheat	-0.56**	Waldon, 1933
	-0.80**	Grant and McCalla, 1949
	-0.25**	Stuber et al., 1962

*, ** denotes significance at .05 and .01 levels, respectively.

and Atlas 66, were equally as productive as older cultivars and yet had from 0.9 to 3.2 percent higher protein in their grain (Table 10.12). Genes that produced this increase in protein content came from 'Frondosa,' a cultivar from Brazil. Atlas 50 and Atlas 66 were not consciously selected for protein, but rather they were selected for leaf-rust resistance obtained from the Frondosa parent. Their high protein content resulted from close linkage of the high-protein

Table 10.11. Yield and protein content of Lancota and several other cultivars of wheat when tested in the International Winter Wheat Performance Nursery grown at 25 sites in 1972 and 1973. (Johnson, Mattern, and Kuhr, 1979.)

Cultivar	Average Yield	Protein
	t/ha	%
Lancota	4.1	15.5
Zenith	4.0	14.4
Centurk	4.4	14.0
TAM 102	4.0	13.5
Maris Nimrod	4.3	13.2

Table 10.12. Grain yields and protein percentages of wheat cultivars tested for three years in Southeastern Wheat Nursery, USA. (Middleton, Bode, and Bayles, 1954.)

Cultivar	Grain Yield	Protein
	t/ha	%
Hardired 47-12	1.8	10.1
Chancellor	1.8	10.9
Purcam	1.8	11.4
Coker 47-27	1.9	11.6
Taylor	1.9	11.9
Atlas 50	1.9	12.8
Atlas 66	1.9	13.3

gene with the leaf-rust resistance gene in Frondosa (Johnson et al., 1968). According to Wilhelmi (1974), Lancota absorbed more soil nitrogen than did 'Lancer,' a check cultivar. Lancota also exhibited higher nitrate reductase activity (NRA), maintained NRA longer during the growing season, and translocated a higher percentage of absorbed N to its grain than did Lancer (Johnson et al., 1979). This is an example where a known "protein-content gene," that is independent of any effect on grain yield, has been successfully utilized in a long-term breeding program.

Oats

Oat grain protein has three unique features relative to other cereals: (1) its balance of essential amino acids is excellent, especially for a cereal with a high protein content (Robbins et al., 1971); (2)

the biological value of oat protein does not deteriorate as the protein percentage in the grain increases (Frey, 1951); and (3) the protein percentage of oat grain probably can be elevated to very high levels by genetic means. Generally, whole grain of cultivated oats ranges from 9 to 16 percent protein (Frey and Watson, 1950), and the maximum protein in groats (naked seeds) of commercial cultivars is 20 percent (Briggle, 1971). Groat-protein contents in *Avena sterilis*, a weedy oat type collected near the Mediterranean Sea, have been reported up to 27.3 percent by Ohm and Patterson (1973), to 28 percent by Campbell and Frey (1972), and to 35 percent by Frey et al. (1975). Efforts are currently under way to transfer genes for high groat protein from *A. sterilis* to commercially acceptable cultivars of oats. Recently, Frey (1975) reported the discovery of genes from *A. sterilis* that increase the yield of experimental lines of cultivated oats by 20 to 30 percent. These yield increases were accomplished without depressing groat-protein percentage. This discovery is at the opposite end of the spectrum from that made by the wheat breeders, but the two discoveries have a common focal point with respect to breeding cereal grains for higher grain-protein content. The wheat researchers discovered genes for high grain protein that have no effect on yield, and the oat researchers have discovered genes for high yield that have no effect on groat-protein content. It will be most interesting to study the protein fractionation patterns of the oat lines with "yield genes," especially in light of the hypothesis of Tsai et al. (1978) which suggests that some storage proteins may serve as important N sinks and prolong the grain-filling period of cereals.

Four high protein oat cultivars are currently being widely grown in Wisconsin (Forsberg, pers. commun.). 'Dal,' 'Goodland,' 'Marathon,' and 'Wright' cultivars have a 2 to 3 percent increase in groat protein, and still maintain acceptable yield levels (Table 10.13). These cultivars resulted from the accumulation of protein genes present in the adapted oat germplasm pool. Parenthetically, an extensive mutation breeding program by Frey (1977) was unsuccessful in inducing major mutations for groat-protein content in commercial oat cultivars. Forsberg et al. (1974) noted that protein percentage was significantly and negatively correlated with yield in three series of oat tests in 1971, whereas a nonsignificant positive correlation occurred in 1970. This relationship between yield and protein percentage is influenced by the amount of nitrogen available and removed from the soil, as is degree of lodging. There was considerably more lodging in 1970 than 1971 (data not shown) but the

Table 10.13. Grain and groat-protein yields and groat and groat-protein percentages for four oat cultivars tested in Wisconsin from 1973-1978. (Forsberg, Youngs, and Pendleton, 1979.)

Cultivar	Grain Yield	Groat-Protein	Groat	Groat-Protein
	t/ha	*kg/ha*	%	*kg/ha*
No. of Tests	39	30	30	30
Marathon	2.7	363	68.0	19.3
Dal	2.5	355	68.3	20.0
Froker	2.4	311	69.7	18.3
Lodi	2.5	307	65.9	18.3

correlation between protein and lodging was only -0.06 in 1970.

Rice

Much effort has been invested in increasing the protein concentration of rice grain. This is certainly justified since milled rice provides 40 to 80 percent of the calories and at least 40 percent of the protein in Asian diets. Rice has a high quality protein, but its protein concentration is low (7 to 14 percent).

Coffman and Juliano (1979), in summarizing 25 years of breeding effort, concluded that there is little prospect of improving the lysine concentration in rice protein using present techniques, but it should be possible to improve the protein concentration in the grain. The rice protein story demonstrates the difficulty of incorporating a biochemical phenotype, such as protein concentration, into a very dynamic crop improvement program. One frustration is the fact that environment contributes a large proportion of the total variability in protein concentration in rice, and variation due to environmental causes is exaggerated at high fertility levels. However, true genetic differences for protein concentration exist among rice cultivars. The inheritance of protein concentration is complex with low content being dominant.

Breeding for improved protein concentration in rice was begun in 1950, but it has not been particularly fruitful as witnessed by the lack of "high protein" cultivars in agricultural production today. At the International Rice Research Institute (IRRI), there has been some general progress in improving protein concentration in advanced breeding lines. They have exhibited superior protein concentration over several seasons, at yield levels comparable to check cultivars (Table 10.14). In 1976, 'IR2153-338-3,' an advanced high-

Table 10.14. **Protein content of brown rice and yield of rough rice for two high protein lines and two check cultivars in agronomy trials over six seasons and three nitrogen levels at the International Rice Research Institute. (Coffman and Juliano, 1979.)**

Cultivar or Line	Protein		Yield	
	Range	Mean	Range	Mean
	%		t/ha	
IR8 (check)	7.2 - 8.4	7.8	3.5 - 5.5	4.6
IR26 (check)	7.7 - 8.5	7.9	4.1 - 7.3	5.4
IR480-5-9	8.4 - 9.3	8.8	3.0 - 5.2	4.0
IR2153-338-3	8.7 - 9.1	8.9	3.2 - 6.8	4.6

protein line, had 1 percent higher protein concentration than 'IR8' and their yields were the same. However, 'IR26' was the best improved cultivar, and although IR2153-338-3 had 1 percent higher protein than IR26, its yield was not comparable. During 1977, IR2153-338-3 was seriously damaged by ragged stunt disease, which invalidated comparative testing of this line for yield and protein concentration (Coffman and Juliano, 1979). This illustrates the dilatory effects that environments can have on developing improved rice cultivars with high levels of protein. Because of environmental effects, extended periods of testing are required to verify the advantages in protein concentration, and by then, the protein-rich line is no longer suitable for release to farmers because the situation with prevalent diseases and insect pests changes so rapidly.

Marwaha and Juliano (1976) have been unsuccessful in finding a seedling or vegetative stage that is an indicator of high grain protein content for rice. Protein content differences are mainly due to variations in efficiency of translocation of foliar N to the developing grain after anthesis rather than to differences in total plant N (Perez et al., 1973). Until the mechanism(s) of N metabolism and translocation into the developing grain are more clearly understood, the IRRI researchers concluded that improving the level of protein concentration cannot be justified as a high research priority in rice. Note that in the IRRI rice program, as in the case of oats, the greatest progress in increasing genetic factors for high protein concentration has been made by using natural variability rather than induced mutants. Apparently, induced mutants that tend to be rich in protein result from reduced accumulation of starch in their endosperm. Probably, this will be true until the biochemical basis of high-protein content is clearly understood, and until efficient screening techniques designed to identify specific metabolic changes become available.

Maize

In maize the relationships between chemical composition and grain yield have been elegantly demonstrated by Dudley et al. (1977), who reported a unique study of the interrelationships among oil, protein, intrinsic energy contents, grain yield, and other agronomic traits. They studied the performance of crosses among nine strains, Illinois High Oil (IHO), Reverse High Oil (RHO), Switchback High Oil (SHO), Illinois Low Oil (ILO), Reverse Low Oil (RLO), Illinois High Protein (IHP), Reverse High Protein (RHP), Illinois Low Protein (ILP), and Reverse Low Protein (RLP) that represent the widest range of oil and protein percentages presently known (Table 10.15). Grain yield was negatively correlated with both percent oil and percent protein. The two highest yielding crosses, RLP × ILO and RLP × LRO, had almost normal levels of protein (approximately 10 percent) but slightly lower than average levels of oil. Grain calorie production per ha closely paralleled grain yield, and production of oil or protein per ha was influenced by both grain yield and concentration of oil or protein in the grain (that is, IHO had the highest general combining ability effect for kg oil/ha and IHP had the highest general combining ability for kg protein/ha). The caloric content of the grain as cal/g dry matter, on the other hand, was correlated with percent oil and to a lesser degree with percent protein and negatively correlated with grain yield.

This study shows that the optimum combination of protein, oil,

Table 10.15. Correlations among traits related to calorie production in maize crosses made between high oil and high protein strains. (Dudley, Lambert, and de la Roche, 1977.)

	Trait					
Trait	Cal/ Kernel	k cal/ ha	Percent Oil	Percent Protein	g/100 Kernels	Grain Yield
Cal/g dry matter	-.74*	-.45*	.94*	.51*	-.22	-.61*
Cal/kernel		.47*	-.78*	-.30	.99*	.58*
k cal/ha			-.32	-.67*	.49*	.98*
Percent oil				.23	-.85*	-.49*
Percent protein					-.36	-.70*
g/100 kernels						.62*

Table 10.16. Percentages of various fatty acids in the edible oils from various crops. (Adapted from Weiss, 1970.)

Crop	Palmitic	Stearic	Oleic	Linoleic	Linolenic	Eicosenoic	Erucic
Soybean	10.5	3.2	22.3	54.5	8.3	<1.0	...
Cotton seed	25.0	2.8	17.1	52.7	...	...	...
Corn	11.5	2.2	26.6	58.7	<1.0	...	...
Peanut	11.0	2.3	51.0	30.9	...	...	...
Safflower	6.7	2.7	12.9	77.5	...	<1.0	...
Sunflower	7.0	3.3	14.3	75.4	...	...	...
Rapeseed	4.0	1.3	17.4	12.7	5.3	10.4	45.6
"Tower" rapeseed[a]	4.3	1.7	59.1	22.8	8.2	<1.0	<1.0
Palm	46.8	3.8	37.6	10.0	...	...	...

[a]Adapted from Slinger (1977).

and grain yield depends on the objectives of a breeding program. For example, where production of protein or oil per ha is a primary consideration, intermediate levels of protein or oil with highest yields would be preferred. For maximum production of calories per ha, selection based primarily on grain yield with some increase in percent protein is appropriate.

BREEDING FOR OIL QUALITY AND CONTENT

The economic value of plant oils and special uses for oils with aberrant composition have caused breeders to examine the genetics of edible oil production in soybean (*Glycine max*), sunflower (*Helianthus annuus*), cotton (*Gossypium* spp.), peanut (*Arachis hypogaea*), safflower (*Carthamus tinctorius*), maize, and rape (*Brassica* spp.) (Table 10.16).

Variability for oil content in seeds is high in soybeans (Wilson et al., 1976), sunflowers (Fick, 1975), maize (Weber and Alexander, 1975), and safflower (Ashri et al., 1977). Russian breeders have increased oil content in sunflower seed from 30 percent to almost 50 percent in 50 years, using modified recurrent selection. Oil content of the safflower seed has been increased from 37 percent to 50 percent by breeding for reduced hull content (Knowles, 1975).

The classic recurrent selection experiment at the University of Illinois increased oil content in maize from 4.7 percent to 17 percent after 70 cycles of selection, but grain yield was reduced materially (Fig. 10.8). Maize hybrids with acceptable yields and up to 8 percent oil are being produced commercially (Weber and Alexander, 1975).

Oil content is partly determined by the maternal parent and partly by seed genotype in maize and rape (Grami and Stefansson, 1977b). Genes for oil content are additive in maize, sunflower, and safflower (Yermanos et al., 1967), and mainly additive with partial dominance for low oil content in sesame (*Sesamun indicum*) (Grami and Stefansson, 1977a).

Oil quality is determined by its fatty acid composition and the positional distribution of the fatty acids within the triglyceride. The genotype of the maternal parent influences the fatty acid composition in flax (*Linum usitatissimum* L.), maize, soybeans, and rape, but the genotype of the seed has an effect in flax, rape, and safflower. Cytological effects have been found in maize but not in rape (Thomas and Kondra, 1973).

Fatty acid composition of oils affects the nutritional and proc-

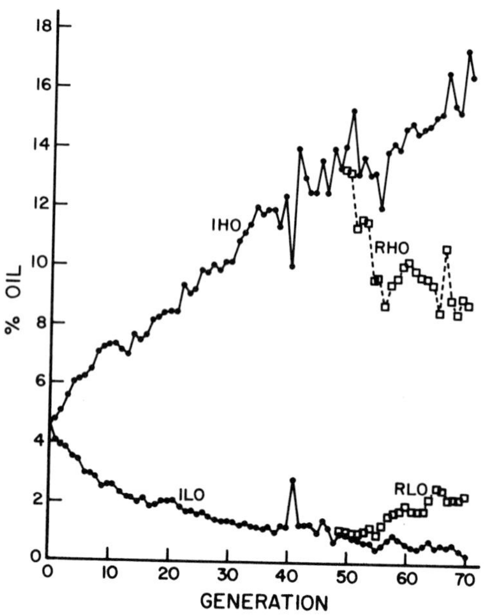

Fig. 10.8. Mean percent oil for IHO (Illinois high oil), ILO (Illinois low oil), RHO (reverse high oil) and RLO (reverse low oil) plotted against generations of selection. (Dudley et al., 1974.)

essing values of the oils. Polyunsaturated fatty acids (linoleic and linolenic) reduce the level of serum cholesterol in mammals, and saturated fatty acids raise it. Oleic acid, a monounsaturated fatty acid, is considered neutral, neither raising nor lowering serum cholesterol levels. Soybean oil quality could be improved by reducing the proportion of linolenic acid in the oil. This 18:3 (that is, 18 carbon atoms and three double bonds) fatty acid accounts for 7 to 8 percent of the total oil content, and it is responsible for lowered stability and poor flavor. Mechanical methods have been used to improve the quality of soybean oil, but generally, the best way to improve its quality would be through breeding. The level of linolenic plus linoleic acid in soybean oil is inversely proportional to the level of oleic acid. Selection for increased oleic acid lowered linolenic acid (37 percent), total polyunsaturated fatty acids, and total oil, but concentration of saturated fatty acids was unaffected (Table 10.17) in a study by Wilson et al. (1976). The inheritance of these fatty acids is not understood, but Howell et al. (1972) did suggest at least three additive genes sequentially desaturate a fatty acid.

Oil from rape and other *Cruciferae* species differs from most

Table 10.17. Percentages of fatty acids in the oils from a commercial cultivar and an improved soybean line. (Adapted from Wilson, Rinne, and Brim, 1976.)

Line or Cultivar	Palmitic + Stearic	Oleic	Linoleic + Linolenic
N70-3436	11.5	40.1	48.4
Dare	12.5	18.1	69.3

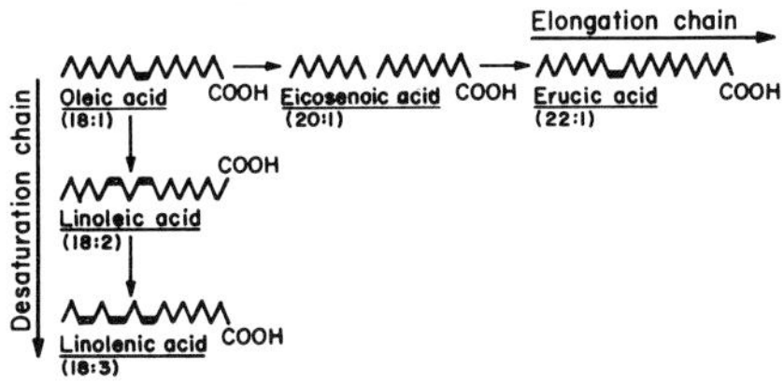

Fig. 10.9. Biosynthetic pathway of the primary fatty acids in rapeseed (Röbbelen, 1972.)

vegetable oils in that they contain high amounts of long chain monoenoic fatty acids, eicosenoic acid (20:1) and erucic acid (22:1) (Fig. 10.9). High levels of erucic acid make rapeseed oil a good quality industrial oil, but it may cause fatty deposition and myocardial lesions when fed to mammals. The discovery and use of erucic acid-free rape oil (Stefansson et al., 1961; Downey, 1964) has led to greatly increased production of rape in cooler climates.

Erucic acid and eicosenoic acid contents are controlled by multiple alleles at two loci in rape, and one locus in turnip rape. These alleles are additive for erucic acid content up to 30 percent, partially dominant for higher levels, and dominant to highly overdominant for eicosenoic acid content. Individually, the multiple alleles have been shown to produce 0, 3.5, 4.0, 7.5, 9.0, 10.0, and 12.5 percent erucic acid (Jönsson, 1977). Jönsson (1977) has shown that at least ten "homozygous" levels exist for erucic acid in rapeseed oil. Homozygous "effective" alleles at one locus produce from 5 to 10 percent erucic acid, at one or two loci they produce 10 to 35 percent, and at two loci they produce more than 35 percent.

Fatty acid composition in safflower is determined by three major alleles at the *ol* locus (Table 10.18). Oil with high linoleic acid is used in soft margarine, whereas oil with high oleic acid is used as a cooking oil. The increase of the major fatty acid in these two oils can be achieved by the management of genes with small effects (Knowles, 1975).

Table 10.18. Ranges of oleic and linoleic acid percentages in the oils of safflower strains with different genotypes. (Knowles, 1969.)

Genotype	Oleic	Linoleic
OlOl	10 - 15	75 - 80
Olol′	15 - 20	70 - 75
Olol	18 - 35	60 - 75
ol′ol′	35 - 50	42 - 54
ol′ol	55 - 63	30 - 40
olol	64 - 83	12 - 30

Flora and Wiley (1972) showed the effect of various corn endosperm mutants on oil content and fatty acid distribution (Table 10.19). Among 12 mutants, opaque-2 reduced oil percentage most with little increase in polyunsaturated fatty acids. Brittle-2 increased oil percentages most and caused a decrease in polyunsaturates.

The positional distribution of the fatty acids within the triglyceride affects the quality of the oil. Raghuveer and Hammond (1967) showed that concentrating unsaturated fatty acids at the middle position of the triglyceride stabilizes the fat against autoxidation. Placement of the fatty acids is somewhat heritable in maize (Table 10.20); so Weber and Alexander (1975) suggested breeders could concentrate the unsaturated fatty acids at the middle position of the triglyceride, and thus reduce stability problems of maize oil.

Many researchers have studied the correlations of oil with fatty acid content and of component fatty acids with one another. Oleic and linoleic acids, major components of vegetable oils, are highly negatively correlated. Stefansson and Storgaard (1969), working with four populations of rape, showed that correlations between pairs of fatty acids, measured as percent of total oil, are reduced when measured as percent of seed weight.

Several methods have been suggested to increase the variability for fatty acid composition of oils. Interspecific hybridization is useful when related species have the desired fatty acid character as in the oil palm (Ng et al., 1976). The concentration of unsaturated fatty acids was increased in the cultivated species, *Elaeis guineensis*, by selecting within a cross to the related wild species, *E. oleifera* (Table 10.21). Interspecific hybridization cannot be used to reduce the linolenic acid concentration in soybeans because the cultivated species is lower in linolenic acid than the wild ones (Howell et al., 1972).

Table 10.19. Percentages of fatty acids in oils from various endosperm mutant types in maize. (Flora and Wiley, 1972.)

Mutant	Palmitic	Stearic	Oleic	Linoleic	Linolenic	Oil Percentage
Opaque-2	11.5	1.2	17.7	66.3	3.3	5.2
Horny	10.7	1.6	21.1	64.0	2.3	9.2
Dull	10.7	1.6	23.8	61.7	2.2	9.2
Normal	9.6	1.5	21.4	65.5	2.0	9.5
Shrunken-1	11.4	1.4	19.4	65.2	2.6	9.7
Floury-1	9.4	1.5	21.2	65.7	2.2	9.8
Sugary-2	10.1	1.8	31.5	54.5	2.2	10.2
Amylose extender	11.0	1.6	21.1	64.0	2.3	10.4
Waxy	15.0	1.6	22.3	58.9	2.2	11.2
Sugary-1	12.0	1.7	20.6	62.9	2.8	11.6
Shrunken-2	17.3	1.6	21.3	56.6	3.2	13.0
Brittle-1	13.8	1.8	26.3	55.6	2.5	13.2
Brittle-2	14.3	2.4	33.7	47.9	1.7	18.4

Table 10.20. Percentage frequencies with which various fatty acids from two inbreds of maize and their hybrid are attached at the three positions of the glycerol molecule. (Adapted from Weber and Alexander, 1975.)

Inbred or Cross	Glycerol Position	Fatty Acid Distribution (mole %)				
		Palmitic	Stearic	Oleic	Linoleic	Linolenic
C103	1	22.4	4.2	41.2	31.7	0.5
	2	1.0	0.3	40.4	57.5	0.8
	3	14.2	2.1	47.5	35.6	0.6
C103 x NY16	1	21.3	3.2	30.4	44.3	0.8
	2	0.6	0.2	27.6	70.8	0.7
	3	9.0	1.5	38.4	50.1	1.0
NY16	1	15.6	3.9	21.4	57.8	1.3
	2	0.7	0.2	21.6	76.6	0.8
	3	7.0	1.6	19.2	70.6	1.6

Table 10.21. Percentages of fatty acids in the oils from the palm oil species, *Eloeis oleifera*, *E. guineensis*, and their hybrids. (Adapted from Ng, Corley, and Clegg, 1976.)

Species or Hybrid		Palmitic	Stearic	Oleic	Linoleic
E. oleifera		30.2	1.0	53.5	14.9
E. oleifera x *E. guineensis*	(1)	44.6	1.0	44.4	9.6
	(2)	42.1	1.0	46.7	9.8
E. guineensis	(1)	50.6	1.7	41.6	5.3
	(2)	50.0	3.0	39.6	6.9

A spontaneous mutation in the German summer rape, 'Liho,' provided the first low erucic acid cultivar (Downey, 1964). Srinivasachar et al. (1972) used mutagens, ethyl methanesulfonate, and gamma-rays to increase oil content and degree of unsaturation in linseed.

Soybeans are unusual among the oilseed crops in that their protein content is high, marketable, and worthy of improvement. The soybean contains twice as much protein as oil (41 percent vs. 21 percent) and the total market value of the protein usually approaches that of the oil. Protein and oil contents are negatively correlated, so the breeder must set minimum standards for one while selecting for the other. Higher protein levels have been achieved with some reduction in yield and oil content.

Caldwell et al. (1966) evaluated selection gains in soybeans for protein, oil, and yield when the traits were assigned equal economic weights. Selection for either oil or protein increased the seed content of that component but greatly reduced yield. Only when the selection index included yield did this trait increase. Improvement of total oil and protein yield can be achieved best by selecting for yield, while maintaining or slightly increasing oil and protein percentages.

Grami and Stefansson (1977a), working with rape, suggest that selecting for the sum of oil and protein content is better than selecting for either component alone. Although these traits are negatively correlated, an increase in one is not necessarily accompanied by a decrease in the other (Table 10.22). The variance of the sum of oil

Table 10.22. Means and within-row variances for percent protein, percent oil, and their sum in two rapeseed cultivars and their F_1, F_2, and backcross generations. (Adapted from Grami and Stefansson, 1977a.)

	Protein		Oil		Sum	
Generation	Mean	σ^2	Mean	σ^2	Mean	σ^2
P_1 (Midas)	25.4	1.34	43.7	3.44	69.0	1.16
P_2 (Tower)	30.3	2.06	41.7	3.63	72.0	1.34
F_1	27.6	1.93	42.3	4.36	70.0	1.21
F_2	27.3	2.29	42.8	4.97	70.1	1.87
BC_1 (F_1 x Midas)	26.5	1.82	43.0	4.48	69.5	1.82
BC_2 (F_1 x Tower)	28.7	2.84	42.1	4.65	70.8	1.93

and protein is less than the sum of the oil and protein variances, namely, 1.34 vs. 5.69, and smaller significant differences can be detected for the sum.

BREEDING FOR VITAMIN CONTENT

Agronomists tend to overlook the important contributions of fruits and vegetables as sources of vitamins and minerals in the human food supply. In USA diets they supply about 90 percent of the vitamin C, 50 percent of the vitamin A, 30 percent of the B_6, 25 percent of the magnesium, 20 percent of the thiamin, and 18 percent of the riboflavin and niacin (Senti and Rizek, 1975). Oranges (*Citrus sininsis*), white potatoes (*Solanum tuberosum*), tomatoes (*Lycopersicon esculentum*), and cabbages (*Brassica oleracea*) are the major contributors of vitamin C to the USA diets, whereas carrots (*Daucus carota*), tomatoes, and sweet potatoes (*Ipomoea batatas*) are the

Table 10.23. Contribution of specified food crops to USA per capita availability of ascorbic acid. (Senti and Rizek, 1975.)

Food Crop	Percentage
Oranges	20.4
White potatoes	19.7
Tomatoes	12.2
Cabbage	5.1
Grapefruit	4.0
Green peppers	3.0
Onions	1.8
Strawberries	1.8
Cantaloupe	1.7
Bananas	1.4
Cucumbers	1.3
Broccoli	1.2
Corn	1.2
Peas	1.2
Snap beans	1.2
Lettuce	1.1
Lemons	1.1
Sweet potatoes	0.9

primary dietary sources of vitamin A (Tables 10.23 and 10.24). Fruits and vegetables are more important in diets in developing countries than in the USA, but relatively little financial support is available to plant breeders to improve the nutrient composition of fruit and vegetable crops in these countries.

Tomatoes

The tomato provides an excellent example where genetic studies of the control of carotene synthesis have been utilized by plant breeders to increase provitamin A concentration. Avitaminosis A, which causes blindness in severe cases, is endemic in Southeast Asia, India, and the Middle East, and occurs sporadically in Africa and Latin America (Scrimshaw, 1966). Tomes (1972), in a review of the research on breeding tomatoes for high provitamin A concentration, says that source materials for high provitamin A tomatoes were crosses initially made in search of resistance to certain tomato diseases. Red-fleshed garden tomatoes were crossed with the green-fruited wild South American species (*L. hirsutum*), and among the progenies, some orange and orange-red fruit occurred. They contained quantities of the orange pigment beta-carotene (Kohler et al., 1947; Lincoln et al., 1943).

Table 10.24. Contribution of specified food crops to USA per capita availability of vitamin A. (Senti and Rizek, 1975.)

Food Crop	Percentage
Carrots	13.9
Tomatoes	9.5
Sweet Potatoes	5.6
Cantaloupe	2.6
Spinach	2.2
Oranges	1.3
Peaches	1.3
Pumpkin and Squash	0.9
Lettuce	0.8
Watermelon	0.8
Corn	0.7
Snap beans	0.7
Peas	0.6
Apricots	0.6
Escarole	0.5

Red color in tomatoes is due primarily to the red pigment lycopene, but normal tomatoes also contain some beta-carotene and small quantities of gamma-carotene. Red flesh depends on a mixture of red lycopene and orange beta-carotene, usually in a ratio 12:1 to 18:1, depending on cultivar and environment. High beta-carotene content is controlled by the interaction of two genes derived from the wild parent and pigment-producing genes and modifiers from the red-fleshed parent (Tomes et al., 1954). In these genotypes, beta-carotene is produced at the expense of lycopene. By suitable genetic manipulation, it is possible to develop intermediate orange-red types that contain intermediate levels of lycopene and beta-carotene and orange-fleshed types in which lycopene is nearly replaced by beta-carotene. Both types are extremely rich in provitamin A. Next, tomato breeders developed an agronomically acceptable cultivar rich in provitamin A which was named 'Caro-Red' (Tomes and Quackenbush, 1968). Caro-Red contains ten times more beta-carotene than normal tomatoes. Orange garden tomatoes such as 'Golden Jubilee,' 'Sunray,' and 'Penn Orange' contain a mixture of carotenes and have no more provitamin A than red types.

The fate of Caro-Red cultivar on the commercial market is instructive for breeders interested in developing strains with improved nutritional value. Caro-Red was the wrong color for USA consumers. Some said that the tomatoes tasted bad, but this must have been related to color differences. Indeed, in color masked experiments, 46 percent of the tasters preferred Caro-Red over a red-fleshed cultivar, whereas only 19 percent preferred it when color differences were not masked. As a result, Caro-Red is grown little, primarily by vitamin "faddists" or by some to add color variety for mixed salads.

The reaction of tomato breeders is predictable. The consumer prefers red tomatoes and USA grades for processing tomatoes are based largely on red color, so breeders pay little attention to developing high provitamin A cultivars.

A carotene mutant gene called crimson has been incorporated into some commercial strains recently. This gene enhances red color by lowering beta-carotene and increasing lycopene concentration. Though less nutritious, these tomatoes possess a bright red color.

The tomato represents a case where the plant breeder and nutritionist who work as a team need support from the tomato processing and marketing industry. Certainly, the tomato breeder should challenge human nutritionists to aid him in specifying areas of the world where tomatoes could be used to overcome vitamin A deficiencies in human diets, so that studies could assess the costs of solving this nutritional problem via using high provitamin A tomatoes vs. other nutritional intervention strategies. This problem requires interdisciplinary cooperation, but such studies are essential if plant breeders are to fulfill their role in supplying essential dietary nutrients for an expanding human population.

Carrots

The problem of market acceptability encountered with orange-pigmented tomatoes is an asset in carrot breeding programs. Some commercial cultivars of carrots have about 100 to 120 μg/g fresh weight of carotene (Gabelman, 1974). The carotenoids (C_{40}) can be divided into carotenes and xanthophylls. In commercial carrot cultivars (orange root), about 95 percent of the total carotenoids are carotenes, primarily beta- and alpha-carotene (Umiel and Gabelman, 1972). White color, which is dominant to orange, has only 1 μg/g of total carotenoids and no provitamin A potential.

In carrots, the intensity of orange color is associated with total carotenoid content, so carotene can be selected visually. Types containing from 1 μg/g to 400 μg/g total carotenoids have been isolated. In Wisconsin, breeding for color in the carrot research program includes two key procedures: (1) inbreeding and selecting for the dark orange types, and (2) hybridizing the dark orange inbreds to produce a vigorous and uniform F_1 hybrid of desirable color. Some carrot hybrids from this program that contain 140-160 μg/g of total carotenoids are approaching commercial production (Gabelman, pers. commun.). This represents an increase in provitamin A potential of 30 percent, which is substantial in view of the importance of carrots as a source of vitamin A in the USA diets (Table 10.24). Likely, future hybrids will have carotenoid contents up to 180 μg/g. Root color in carrots is inherited complexly; pigment inhibiting genes (Laferriere and Gabelman, 1968; Kust, 1970; Buishand and Gabelman, 1979), pigment enhancing genes (Kust, 1970), and genes that control the synthesis of specific pigments (Umiel and Gabelman, 1972) have been described. (A detailed literature review on the inheritance of carotenoid synthesis in carrots is in preparation by Gabelman for publication in Euphytica.) Studies of carotenoid synthesis in tissue cultures derived from carrots of red, dark orange, orange, light orange, yellow, and white root colors were conducted by Mok et al. (1976). Variation in culture pigmentation occurred, but plants regenerated from tissue cultures with variant colors all had the color of the original root. They proposed that if plantlets were regenerated only from diploid cells, callus color variations due to aneuploidy would never show as plants.

The carrot and tomato stories illustrate how plant breeders can contribute to improving the nutritional status of diets, and how genetic studies contributed to our understanding of carotenoid biosynthesis.

BREEDING FOR FORAGE QUALITY

Forage grasses and legumes are grown on over one-half of the USA land area (Hoveland, 1973). But, as pressure increases for more grain, oilseed, and fiber crops, forages likely will be relegated to land not suited to grain or seed production. Ruminant animals can utilize forage grown on less productive areas, and because they graze, no exogenous energy source is needed for harvesting. High-quality forage cultivars are needed to improve animal yield per ha. A review on the nutritive value of forage crops has been published by Raymond (1969).

Much progress has been made in breeding and management to improve forage quality. A premier example of what can be achieved by breeding for improved forage quality is that of Burton (1964), who bred Coastal bermudagrass (*Cynodon dactylon*), developed vegetative propagation methods for sowing large land areas to it, and set certification standards for it. Coastal bermuda is planted on more than four million ha in the southern USA, and it produces 110 more kg of beef per ha per year than the common bermuda it replaced.

'Coastcross-1' Bermudagrass

To improve the quality of 'Coastal' bermuda, it was hybridized with a high-quality introduction from Kenya (PI255445) to produce Coastcross-1 (CC1), the best of 385 F_1 hybrids. Compared with Coastal, CC1 establishes quicker, yields as much dry matter, makes more fall growth, is 12 percent more digestible, gives 30 and 50 percent better average daily gains (ADGs) when fed as hay and grazed, respectively, and is easier to control because it has no seed or rhizomes (Burton, 1972). CC1 bermudagrass is the first forage cultivar in the world bred for improved digestibility, and it is making a substantial contribution to livestock production in Mexico and Cuba (Burton, pers. commun.). Farmers in a Government Resettlement in Bolivia established excellent pastures of CC1 by planting a few stolons supplied free. In Spain, CC1 has survived, is spreading in southeastern Spain where only 10 to 20 cm of rainfall occurs per year, and is growing well on soils too high in salt content for other crops. CC1 is less winterhardy than Coastal, so it cannot be grown north of Florida. 'Coastcross-2,' a sister of CC1, is superior to CC1 in South Africa and is being extensively planted there.

'Tifton 44' Bermudagrass

To improve the quality and winterhardiness of Coastal bermuda, Burton (pers. commun.) made several thousand F_1 hybrids between Coastal and winterhardy bermudagrasses from Germany. These were screened for winterhardiness at the Georgia Mountain Experiment Station, and 70 were evaluated in clipping tests for yield, *in vitro* dry matter digestibility (IVDMD), and other important traits. The best nine were evaluated for winterhardiness, spring growth, and in a few cases, yield, in 14 states as far north as Oklahoma and Illinois. One, Tifton 44, (Burton and Monson, 1978; Utley et al., 1978) is as winterhardy as 'Midland,' is more disease resistant, and yields up to 3.3 tons more hay/ha/yr. When compared with Coastal, it yields as much dry matter, is 5 to 6 percent more digestible, has finer stems—which makes it easier to cure for hay, initiates spring growth earlier, and gives 19 percent better ADGs when grazed or fed as pellets.

Foundation stock of Tifton 44 was distributed to farmers in 14 states in 1978. It can be grown dependably in a belt 500 miles wide from the Atlantic to Central Texas. The 400 thousand ha of

Tifton 44 that will be planted by 1982 will produce 23 million kg more beef than would be produced from a similar area of Coastal bermuda.

'Tifleaf 1' Pearl Millet

Burton et al. (1969) showed that the d_2 gene, which reduced stem length of pearl millet (*Pennisetum typhoides*) by 50 percent increased leaf percentage of the forage from 54 to 81 percent. The d_2 gene reduced dry-matter yields, but it increased animal intake of dehydrated forage by 20 percent and ADGs from 20 to 50 percent. Later, Burton (1979) developed a d_2 short male parent to mate with 'Tift 23DA' to produce Tifleaf 1 hybrid pearl millet. The discovery of cytoplasmic male sterility (CMS) in pearl millet and its transfer to 'Tift 23' made possible the development of Tifleaf 1 and all other commercial millet hybrids in the world (Burton, 1958; Burton, 1969). ADGs and live weight gains per ha (LWGs/ha) for dairy heifers were 33 percent greater for Tifleaf 1 than for 'Gahi 1,' the check (Johnson et al., 1979).

Improving forage quality increases yields of animal product per unit of input, and thus cuts production costs. It reduces the time required by animals to reach a designated weight by increasing ADGs, and it reduces the amount of grain required to finish livestock or produce milk. The increasing world population, because of its growing demand for grain for direct human consumption, will eventually leave little grain for animal production. And, when that day arrives, quality forages that produce meat and milk with little or no grain supplementation will have increasing value. Truly, the future of the livestock industry depends on man's ability to improve forage quality.

These examples clearly demonstrate that the quality of forages can be improved by breeding, and the resulting cultivars with improved quality can significantly improve the production from animals consuming them. Further, they show that plant breeders can integrate nutritional objectives into an overall crop improve-provement program without sacrificing excellence in either area.

Antinutritional Components

Another important area in forage quality improvement is the selection against specific antinutritional components present in some forage species (Matches, 1973). One group of antinutritional compounds is the allelochemical substances (secondary metabolites) such as: terpenes, steroids, acetogenins, phenylpropanes, and alkaloids. Allelochemicals in forages may affect animal response directly or indirectly to serve as antinutritional components. Specific examples of forage antinutritional factors are alkaloids in reed canarygrass (*Phalaris arundinacea*), saponins in alfalfa (*Medicago sativa*), tannins in sorghum and lespedeza (*Lespedeza cuneata*), cyanogenic glycosides in sorghum and sudangrass (*Sorghum sudanense*) and white clover (*Trifolium repens*), plant estrogens in alfalfa and clovers, cyclopamine in veratrum (*Veratrum californicum*), and coumarin in sweetclover (*Melilotus alba*). Allelochemical substances affecting plant growth, such as phytotoxins and phytoalexins, may affect animal response indirectly. The effects of allelochemicals upon animals need to be assessed and understood by plant breeders and forage and livestock management specialists (Barnes and Gustine, 1973).

As in any area of nutritional quality improvement, having simple, quick, reliable, and inexpensive methods for laboratory prediction of forage quality is vital (Barnes and Marten, 1978).

THE NUTRITIONIST'S VIEWPOINT

In 1976, a workshop was convened in Boulder, Colorado, to assess the progress in the fields of breeding and food fortification for improving the protein quality and quantity of food staples (Wilcke, 1976). The major, general recommendations from this joint workshop are presented here along with my comments on what has been accomplished by breeders.

1. All improvements in the quality or quantity of cereal grains, whether by genetic manipulation or by fortification, should be directed toward the use by the intended populations or specific target groups.

The nutritionist must clearly define the nutritional constraints for specific populations of humans, and which group of individuals within each population is most vulnerable. For example, opaque-2 maize may provide an appropriate and cost-effective solution to a nutritional problem for the rural population in Guatemala, but not for the USA feedgrain industry. This poses a dilemma for the plant breeder because current markets do not recognize differences in nutritional quality as a basis for pricing. Occasionally, there is a negative incentive as when bird-resistant sorghum grain commands a lower price because it has a high tannin content. A clear-cut exception is the case of rape cultivars with zero erucic acid. The nutritionist delineated the nutritional problem in older cultivars as due to high erucic acid content in rapeseed oil, and plant breeders were able to overcome this problem. The lesson here is that a cooperative effort by all scientists is required if the problem is to be solved.

2. Continuous communication and cooperation among all individuals concerned with the improvement of the nutritional value of cereals, and particularly among the scientists, will expedite the success of such attempts in terms of time, the efficient use of scientific manpower and facilities, and cost.

All will agree with the wisdom of this recommendation, but the opportunities for communication between plant breeders and nutritionists are often very limited. We need the same kind and degree of communication as has developed between plant pathologists, entomologists, and breeders. The addition of this new dimension to plant breeding programs will certainly require substantial increases in funding.

3. Recommendations to the effect that no decreases in yield be accepted must be considered on a cost-benefit basis. Obviously, if improvements in nutrient level or availability are of sufficient magnitude, some decrease in total yield would be accepted.

To illustrate this point the recent and comprehensive nutritional and economic assessment of opaque-2 maize conducted by the Indian Agricultural Research Institute in New Delhi (Singh and Jain, 1977) is recommended as a model for breeders and nutritionists who are interested in protein quality improvement (Table 10.25). It represents the kind of team effort required for a meaningful assessment of the problem.

In this study, 18- to 30-month old children from low income families who habitually were underfed by 400 calories per day were given a midday supplementary diet of 400 calories and protein supplied from either milk, normal maize, or opaque-2 maize for 182 days. A control group received no supplementary food. The children fed opaque-2 maize had weight gains and weight to height ratios equal to those fed milk and superior to those fed normal maize or the check group (Fig. 10.10). The net protein utilization (NPU) was similar for milk protein and opaque-2 maize protein, but the utilizable protein of opaque-2 maize was supplied at 1/5 the cost of the utilizable protein in milk. More studies of this kind are needed to guide plant breeders.

4. The question of bulk of foods must be considered and evaluated. There is the distinct possibility that certain segments of the population, particularly the infant and young child, might

Table 10.25. Prices for protein from milk and opaque-2 maize in India. (Singh and Jain, 1977.)

Item	Protein Source	
	Milk	Opaque-2 maize
	%	%
Protein	3.7	10.0
Net protein utilization	75.0	72.0
Utilizable protein	2.8	7.2
Price	Rs. 2/liter	Rs. 1/kg
Price/g utilizable protein	7.2 paisa	1.4 paisa

not be able to consume sufficient total food, even though available, to supply nutrient requirements.

This is especially important in countries where maize and sorghum are staple food grains because infants frequently cannot tolerate consumption of the quantity of grain legumes required to balance the protein of cereals.

5. Consideration must be given to the possibility of providing special foods for special purposes. The use of high-lysine, sorghum

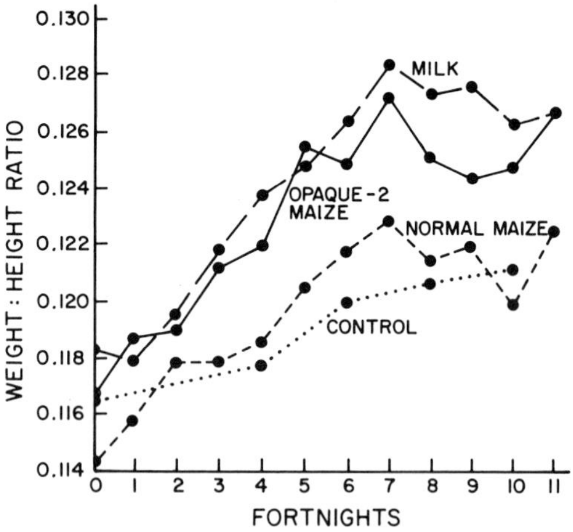

Fig. 10.10. Weight:height ratios of the experimental and control groups of 18- to 30-month old children given no supplemental diet or fed supplemental daily diets of 400 calories and protein from milk, opaque-2 maize, and normal maize for 182 days. (Singh and Jain, 1977.)

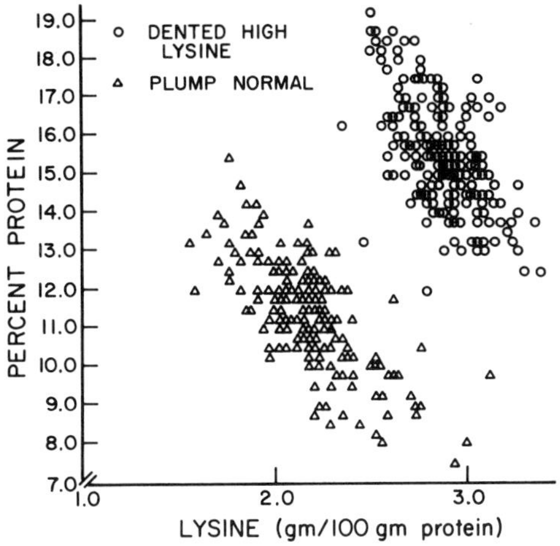

Fig. 10.11. Scatter diagram of coordinates for protein and lysine concentrations in normal and high-lysine sorghum lines collected from the same area in Ethiopia. (Ejeta, 1976.)

produced in Ethiopia as a weaning food is a good example, since the conventional foods used for this purpose would probably not be available to much of the population.

Harper (1976) and other nutritionists see a unique opportunity to develop the Ethiopian high-lysine sorghum cultivars as a special purpose crop for utilization as a weaning food of high nutritional quality for young children. These high-lysine lines identified by Singh and Axtell (1973) continue to be grown by Ethiopian farmers in Wollo province in mixed plantings with normal sorghum cultivars. Ejeta (1976) evaluated the protein and lysine contents of grain from high-lysine and normal cultivars grown in Ethiopia (Fig. 10.11). The mean lysine concentration, expressed as percent of protein, was 2.88 for the high-lysine entries and 2.17 for the normal ones and protein values were 15.7 and 11.4 percent, respectively. Likely, the high-lysine gene has been present in Ethiopia for a long time because there is great diversity in panicle morphology, maturity, and plant height among the high-lysine collections. Farmers roast the heads of the high-lysine and normal cultivars in the late dough stage and eat the grain as a mixture. Farmers recognize that these cultivars yield significantly less than normal ones (Table 10.26), but they continue to grow them because the grain has superior flavor and gives improved palatability when consumed in mixtures with normal sorghum grain. High-lysine strains could be utilized in African countries

Table 10.26. **Mean grain yields of 12 high lysine sorghum cultivars and two check cultivars evaluated at Alemaya, Ethiopia, (Gebrekidan, 1977.)**

Sorghum Type	Grain Yield 1975	1976	Mean
		t/ha	
High lysine	2.0	2.5	2.3
Normal	2.9	3.5	3.2

as high protein, special purpose cultivars for feeding people who have a high protein requirement. The protein concentration is increased by about 30 percent, and protein quality is increased significantly. Grain from high-lysine cultivars is recognizable in the marketplace because the kernel is somewhat dented. Farmers could produce high-lysine sorghum grain as a protein source for weaning children, pregnant women, and nursing mothers. And if a proper marketing system could be developed to include a premium for the grain, much of it could be sold in the cities. But a word of caution about these findings. Very recent studies by G. Graham at The Johns Hopkins University (pers. commun. to E. T. Mertz) suggest that sorghum flour is quite indigestible by infants. Since this is contradictory to results with laboratory animals, more information is needed before high-lysine sorghums are utilized as weaning foods.

6. In general, attempts to increase the vitamin or mineral content of cereals by genetic means offer prospects of only limited success. Fortification with either the individual vitamins or minerals known to be deficient, or with foods known to be a good source of the nutrients, should receive primary consideration.

Senti and Rizek (1975) clearly point out the need for expanded research emphasis on fruits and vegetables as important sources of nutrients in our diets. Food crops that provide major sources of important vitamins and minerals in the USA diet are shown in Table 10.27.

7. The nutrient contribution of all foods should be evaluated with due consideration for nutrient losses in processing and cooking.

This point could be especially important if a breeder was to use a tissue culture selection system designed to identify calluses with elevated levels of free amino acids in cereals. It is not clear whether free amino acids will be retained and remain stable during processing and/or cooking. Under any circumstances, however, such mutants

Table 10.27. Food crops identified as candidates for monitoring for one or more nutrients by the US Food and Drug Administration. (Senti and Rizek, 1975.)

Food crop	Nutrient				
Wheat	Protein	B_6	Thiamin	Niacin	Magnesium
White potatoes		B_6	Thiamin	Niacin	C
Carrots		A			
Tomatoes		A			C
Sweet potatoes		A			
Dry beans			Thiamin		Magnesium
Oranges					C
Cabbages					C
Peanuts				Niacin	

would be valuable in improving feedlot performance in animal rations.

8. Without reducing the importance of dietary protein, there has been increased recognition of the importance of the total diet, its total energy content and all of its composite nutrients, in determining nutritional health status.

ROLE OF BASIC RESEARCH

Two examples illustrate the potential benefits from basic research for improving protein quality in the future. Green and Phillips (1974) have described a system for isolating feedback inhibition-resistant mutants in maize which can be used to isolate mutants that overproduce lysine, threonine, or methionine, three amino acids that share a common biosynthetic pathway in maize. Feedback regulation of this pathway functions such that exogenously supplied lysine and threonine cause cessation of growth of calluses. Methionine added to normally inhibitory lysine and threonine levels allows normal growth. Two inbred lines, 'B37' and 'B76,' and several random-line extractions from Iowa Stiff Stalk Synthetic (BSSS) possess partial resistance to the lysine-threonine medium. Mutagenesis induced eight dominant mutants resistant to lysine-threonine (Phillips et al., 1978). The resistant lines, B37, 'B73,' and 'BSSS53' have higher total methionine relative to lysine on a whole-kernel basis than other sensitive lines (Table 10.28; Phillips, Gengenback, and Green, pers. commun., 1979). The methionine values for BSSS53 ranged from 3.6 percent to 4.8 percent of total amino acid content indicating that this line might be useful as a high methionine

Table 10.28. Lysine and methionine contents in seeds from maize lines regenerated from tissue cultures that were resistant and inhibited in growth by these amino acids. (Phillips et al., 1979.)

Source of Line	LT Effect	No. Analyses	Amino Acid (g/100 g protein)± S.D.		M/L Ratio
			Methionine	Lysine	
B37	Res	13	3.1 ± .5	2.1 ± .6	1,5 ± .5
B37o_2	Inh	8	3.2 ± .3	3.5 ± .6	0.9 ±. .2
B76	Res	15	3.6 ± .8	2.1 ± .4	1.8 ± .5
BSSS53	Res	9	4.7 ± .5	2.1 ± .4	2.2 ± .5
W23	Inh	16	3.0 ± .5	2.7 ± .7	1.1 ± .3
W23o_2	Inh	6	2.6 ± .4	3.7 ± .7	0.7 ± .2

source for cultivar development. The resistance of BSSS53 may be conditioned by one major recessive gene. Embryos isolated from the various resistant lines were inhibited by lysine and threonine; thus, the endosperm may furnish a stimulus to the embryo upon germination that confers resistance to the germinating seedling. Phillips et al. (pers. commun., 1979) hypothesize that resistant lines possess more available methionine relative to lysine in their endosperm; the methionine furnished to the embryo upon germination results in resistance. So the resistant lines probably are not feedback resistant mutants but may possess altered proportions of storage proteins. This basic research has provided a powerful selection technique as an additional tool when breeding for improved protein quality.

A second example is provided by basic research being done at Purdue in cooperation with the USDA on protein quality improvement in soybeans. This work gave high priority to increasing the sulfur amino acid contents of the seed protein. Progress toward achieving this goal will be slow until the structural genes that specify the important polypeptides have been identified and the mechanisms that operate during their expression have been elucidated. Only then can a rational selection be made between alternative genes that produce peptides with desirable nutritional characteristics and can conditions be manipulated to maximize their expression. Similar considerations apply to genes constructed via recombinant DNA techniques introduced into the genome, since these new genes must be expressed effectively *in vivo* to be useful.

In this program, the subunits of glycinin, a predominant storage protein in soybeans, have been purified and characterized (Moreira et al., 1979). Six polypeptides with acidic isoelectric points and four with basic ones were isolated from glycinin of the high protein cultivar 'CX635-1-1-1.' The acidic subunits had phenylalanine, leucine, isoleucine, and arginine at the N-termini, whereas the basic ones all had glycine at their N-termini (Fig. 10.12). N-terminal sequence analysis of the purified subunits revealed considerable homology (as indicated by lines between identical sequences) between members of individual families of the acidic and basic subunits, indicating that members of each family arose from a common ancestral gene. These data show that the glycinin subunit composition is complex.

The purified subunits were further characterized by amino acid analysis, and it was found that some subunits contained substantially more methionine than others (e.g., A_1 and $A_2 > A_3$ and A_4; B_1 and $B_2 > B_3$ and B_4) (Table 10.29). This observation may have

Fig. 10.12. N-terminal amino acid sequences of the 10 glycinin subunits aligned for maximum homology. (Moreira et al., 1979).

```
Acidic
Subunits
                          5               10                    15                     20             25
A4      NH2-Arg Arg Gly Ser Arg Ser Gln Lys Gln Gln Leu Gln  Asp (Ser) His Gln Lys Ile (Arg) His Phe Asn Glu Gly Asp  Gly

                             5                  10                   15                    20             25
A1         NH2-Phe Ser Ser Arg Glu Gln Pro Gln Gln Asn Glu  Cys  Gln  Ile Gln Lys Leu  Asn  Ala Leu Lys Pro Asp Asn (   ) Ile

                             5                  10                   15
F2(1)      NH2-Phe Ser Phe Arg Glu Gln Pro Gln Gln Asn Glu (Cys) Gln  Ile Gln

                                  5                  10                   15                    20
A2               NH2-Leu Arg Glu Gln Ala Gln Gln Asn Glu  Cys  Gln  Ile Gln Lys Leu  Asn  Ala Leu Lys Pro Asp Asn (Arg) Ile

                          5                                   10                  15                   20
A3      NH2-Ile Thr Ser Ser Lys Phe-----------------Asn Glu  Cys  Gln  Leu Asn Asn Leu  Asn  Ala Leu Glu Pro Asp His  Arg  Val Glu (    ) Glu Gly

                          5                                   10                  15
F2(2)   NH2-Ile Ser Ser Ser Lys Leu-----------------Asn Glu  Cys  Gln  Leu Asn Asn Leu  Asn  Ala Leu

Basic
Subunits
                          5                   10                    15
B1      NH2 Gly Ile Asp Glu Thr Ile Cys Thr Met Arg Leu Arg  Gln (Asn) Ile Gly Gln

B2      NH2 Gly Ile Asp Glu Thr Ile Cys Thr Met Arg Leu Arg  His   Asn  Ile Gly Gln
             *           *       *   *   *           *         *    *
B3      NH2 Gly Val Glu Glu Asn Ile Cys Thr Leu Lys Leu His  Glu   Asn  Ile Ala Arg

B4      NH2 Gly Val Glu Glu Asn Ile Cys Thr Met Lys Leu His  Glu   Asn  Ile Ala Arg
```

Table 10.29. Amino acid composition of the glycinin subunits in soybean protein[a]. (Moreira et al., 1979.)

Amino Acid	Subunit									
	A1	F2(1)	A2	A3	F2(2)	A4	B1	B2	B3	B4
Asx	36.8	34.8	42.1	45.5	9.0	50.8	25.5	24.3	19.2	20.7
Thr	12.0	12.4	12.3	15.5	3.9	11.8	8.1	9.1	6.2	5.4
Ser	18.3	19.0	16.4	27.1	7.7	23.5	13.5	12.4	12.1	12.4
Glx	85.3	85.3	86.4	91.6	14.9	92.6	22.5	22.7	24.8	21.0
Pro	24.0	25.0	21.3	33.9	8.1	27.3	10.5	10.8	10.2	9.1
Gly	31.0	27.3	29.9	29.5	7.9	22.4	11.1	10.4	13.4	16.1
Ala	14.4	15.9	18.1	10.9	5.0	6.2	15.6	14.3	12.4	11.2
Val	11.9	12.4	15.3	17.4	3.8	12.1	11.4	10.8	17.0	19.2
Met	3.6	4.1	5.8	2.4	1.1	1.4	2.3	2.7	0.0	1.3
Ile	17.6	16.6	15.3	12.2	5.1	10.4	9.2	9.8	7.0	7.3
Leu	20.1	18.7	20.0	21.8	10.7	14.0	17.9	17.4	18.1	18.1
Tyr	7.3	8.7	6.6	5.6	2.0	4.4	2.8	2.5	5.8	8.4
Phe	12.2	17.9	12.3	12.0	0.9	7.7	8.6	9.1	6.0	5.7
His	6.0	4.8	2.6	14.1	2.8	9.5	2.1	2.7	4.8	4.2
Lys	21.2	15.9	14.9	14.8	3.9	18.8	5.9	5.9	7.0	6.5
Arg	18.1	21.2	22.7	22.2	3.1	28.4	8.9	9.9	10.9	12.5
Cys	4.5	...	4.3	3.6	...	0.7	1.7	1.5	0.2	1.5

[a]Duplicate samples of each S-pyridylethyl-subunit were hydrolyzed in 6N Hcl at 110° C for 24 hours and the amino acids determined using standard techniques. The number of residues per subunit were calculated assuming an average molecular weight of 110 for each amino acid, and 340 amino acids per subunit for A1, F2 (1), A2 and A4; 380 residues for A3; 90 for F2 (2); and 175 for each of the basics. The values shown are averages of two determinations on each duplicate.

important practical significance in view of the fact that a sizable proportion of the world population is dependent upon grain legumes for dietary protein. If it can be established that genetic polymorphism exists to cause different relative amounts of the various glycinin subunits, and if the inheritance patterns of the various subunits are simple, perhaps subunit composition can be manipulated genetically to obtain more acceptable methionine levels. This is another example that illustrates basic research which might contribute greatly to our ability to design rational genetic improvement programs to improve protein quality.

CONCLUSIONS

Plant breeders have always shown an interest in the nutritional quality of the food crops with which they work. Yet, they are realistic scientists who understand the constraints imposed by the economics of breeding for a chemical phenotype, the necessity of handling large populations of plants, and the eccentricities of the marketplace.

Röbbelen (1977) stated the problem most clearly as follows:

> Plant breeding is the cheapest way for the nutritional improvement of cereal (food) grains (if such can be accomplished without loss of yield or acceptability). The new variety produces the better quality automatically without further expenditures. But it is generally accepted that quality breeding can only be successful when conducted as part of an overall breeding program in which yield factors always have top priority. The nutritionist who requests changes in nutritional values of cereals from the plant breeder must keep in mind that the development of a variety takes 10 or even 15 years. By then the plant breeder has invested around one million of U.S. dollars; he has planted tens of thousands of progenies and has run similar numbers of various determinations and analyses before the first kilogram of sold seed gives him the first financial return. Deciding upon the breeding aims 15 years in advance is a high risk and needs utmost knowledge and responsibility. It is indispensable that the breeding priorities are still accepted and valid after completion of the breeders program, thus securing a wide cultivation and market for the new variety. With the permanently expanding knowledge in the field of nutrition, and the rapid economic fluctuations, this basic requirement is not always fulfilled in quality breeding. This may explain why quantity still has priority in agricultural production and why internal quality of the product normally receives much less attention. In the European Economic Community (as well as in the U.S.) the marketing system of cereals (but not necessarily of forages or vegetables) is almost exclusively based on yield units. Financial stimuli for better qualities are insufficient or lacking. Yet the potential of plant breeding to improve nutritional values is at least as high as the chances for yield increases.

Röbbelen concludes that: ". . . it should be the continued effort of all those involved in agricultural production to realize this challenge." In my opinion Röbbelen has stated the realities of the dilemma faced by plant breeders very well.

In spite of these difficulties, however, significant progress has been made in breeding crop cultivars with improved nutritional quality. Analytical techniques available for rapid determination of quality parameters have improved greatly and will continue to improve. Certainly a most exciting area of plant genetics research is the study of the genetic control of storage protein biosynthesis in grain legumes (Hall and Bliss, 1978; Boulter, 1979; and others) and cereals (Larkins, 1979; Burr, B., 1979; Burr, F., 1979; Nelson, 1979; Miflin and Shewry, 1979; Brock and Langridge, 1975). The seed storage proteins provide one of the best "handles" for studying genetic regulation in eucaryotic systems.

Much has been and can be accomplished in improving nutritional quality of food and feed crops without a complete understanding of the basic mechanisms of how a change is manifest biologically. Achievements in improving forage quality have been remarkable, and recent research on reducing toxins or antinutritional factors in reed canarygrass, tall fescue, sorghum, and alfalfa is promising. Plant breeders have solved a potentially serious nutritional problem by developing zero erucic acid rape cultivars and perhaps this research has saved an important agricultural industry. High-protein wheat and oat cultivars are already on the market, contributing more protein to human diets. Special purpose high-lysine sorghum cultivars which could be used as a protein source for weaning children and pregnant women are being grown in Ethiopia.

Breeding crops for oil content and oil quality has been successful, and prospects for additional improvement of oil quality are good. Genes that control fatty acid composition in plant oils are still being discovered and genetic manipulation is a way to concentrate the polyunsaturated fatty acids in the middle position of the triglyceride to protect them with stable saturated and monounsaturated fatty acids.

Protein quality can be significantly improved in maize, barley, and sorghum. Quality of sorghum protein can be raised to that of wheat protein without loss of grain yield by substituting the high-lysine gene mutation in a germplasm background that can tolerate it. Barley and maize breeders have developed high-lysine lines that yield nearly as high as normal cultivars. Seemingly, opaque-2 could have nutritional advantages in developing countries where maize is a staple food. Development and use of opaque-2 maize is a form of nutritional intervention, and it has a cost in reduced grain yields. However, Singh and Jain (1977) showed that opaque-2 protein was a bargain relative to milk protein, even at lowered yield levels. In Guatemala, rats and children were used to show that the maximum PER value for protein in mixtures of normal corn and beans and of opaque-2 corn and beans occurred when the crop ratios were 50:50 (Fig. 10.13). Furthermore, the protein PER value for high-lysine maize ration did not decrease as the proportion of maize in the ration increased, whereas the ration with normal maize did.

There must be close cooperation between breeders and nutritionists as new opportunities for nutritional improvement via breeding are explored. It would be ludicrous for a plant breeder to attempt to improve disease resistance without the cooperation of a plant pathologist. It is just as ludicrous for a plant breeder to attempt to improve nutritional quality without the cooperation of a nutritionist. The objectives of plant breeders and nutritionists are the same, namely, to provide a better diet for all the people in the world.

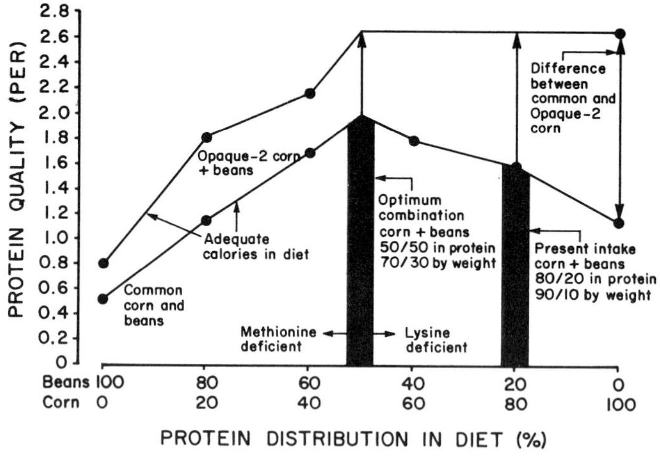

Fig. 10.13. Protein efficiency ratios (PER) for the proteins in mixtures with varying ratios of bean protein with normal and opaque-2 maize proteins. (Bressani, 1976.)

Discussion

W. H. GABELMAN
A. W. HOVIN
W. MARTINEZ

1. W. H. GABELMAN. Certainly, opportunity exists to change nutritional composition of cereals and vegetables via plant breeding. But sometimes the desire to affect such genetic changes is frustrated by human, political, and economic considerations. In the early 1960s, the FDA was empowered to monitor and enforce certain standards with respect to changes in composition of food brought about by breeding. These standards were referred to as the GRAS (generally regarded as safe) regulations. Personnel in the USDA advised the FDA on what crops and nutrient factors should be monitored.

Wheat was the only cereal recommended to the FDA for monitoring, primarily because it contributes so much protein, B6 vitamins, thiamine, niacin, and magnesium to USA diets. All other monitored crops were horticultural. This FDA list provided a summary that showed that a large number of vegetable crops were significant contributors to dietary needs for special factors. For example, take the cases of vitamin A in carrots, tomatoes, and sweet potatoes, and vitamin C in white potatoes, tomatoes, oranges, and cabbage.

The FDA list showed what everyone knows, that oranges contribute 20 percent of the vitamin C in USA diets, but few know that the white potato also contributes 20 percent. Tomatoes and cabbage each supply more than 5 percent. This threat by the FDA to monitor the composition of new vegetable cultivars, and perhaps apply the GRAS standards, shocked breeders into asking whether they should be concerned with improving ascorbic acid levels in vegetables.

Ascorbic acid content in potatoes was shown at Cornell University in the 1950s to be under genetic control. In fact, from a cross of two low vitamin C lines, 'Mohawk' and 'B355-24,' F_2 progeny were selected that had almost double the ascorbic acid level of the parents. Upon selfing and continued selection, a line with almost three times the parent level was found. So here is good evidence that ascorbic acid levels in potatoes can be changed via breeding, but, to date, no major potato breeding program has a primary objective of breeding for vitamin C content. In other words, the barrier to having a higher level of ascorbic acid in the potato is not genetic. Ascorbic acid content in cabbage was shown to be heritable in the 1940s, but a study showed that cabbage cultivars in 1970 had the same mean value for ascorbic acid content as did cultivars from 1945. Again, this is evidence of genetic opportunity not exploited for some reason.

Probably, increasing ascorbic acid content in potatoes is of more significance for countries like China, Poland, Korea, and others because the vitamin C sources such as tomato and citrus are available for only a few months each year. In these countries, the primary sources of ascorbic acid must be potatoes, cabbage, and onions. So much can be done to improve the vitamin C content in the diets for people worldwide if we increase the genetic potential for components in vegetable crops and especially in potatoes.

The FDA list shows that carrots, tomatoes, and sweet potatoes each contribute more than 5 percent of vitamin A to the USA diet. The carotenes are the source of vitamin A and about 900 carotenes are known. Of these 900, only eight have vitamin A potency. Beta carotene is a symmetrical molecule that, when cleaved enzymatically, gives two vitamin A units, so it is said to have vitamin A potency of 100 percent. All other effective carotenes have a vitamin A potency of 50 percent. For carrots, the predominant carotenes are the alpha and beta types. The tomato has beta carotene and lycopene, but lycopene has no vitamin A potency. White carrots have from 0 to 2 ppm of total carotenes, whereas orange roots have over 100 ppm. Commercial cultivars of carrots that are orange have 120 ppm while dark orange types have from 140 to 180 ppm total carotenes. In fact, carrot color can be used effectively in selecting for total carotenoid content until reaching the dark orange types. It is difficult to distinguish carotenoid content visually between 140 and 200 ppm. Visual selection is for total carotenoids, but carotenes increase proportionately.

The wild carrot is white. The orange and red carrots with carotene have been developed within the last millenneum. White, and therefore low carotene, is completely dominant to orange in the F_1. Practically all genes that enhance carotene content in carrots are recessive, so a carrot with intense orange color almost certainly is recessive for seven to nine carotene genes. Most carrot cores are a lighter orange color than is the cortex, and this core is low in carotenes. Thus, it should be possible to materially improve the carotene content by increasing the content of carotene in the core.

'Caro-Red,' a tomato cultivar developed at Purdue, has a very high beta carotene content; however, Caro-Red has never been accepted commercially. Organoleptic tests tend to downgrade Caro-Red tomatoes unless the tasters are blindfolded, because they discriminate against the orange-red color.

Red carrots have both beta carotene and lycopene, but no alpha carotene. The difference in carotene and lycopene contents in red and orange carrots is due to two gene pairs, one for the synthesis of alpha carotene and one for lycopene. The homozygous recessive at both loci gives only beta carotene, but the level of beta carotene is not enhanced. Tomatoes and carrots, in terms of qualitative and quantitative composition, are very similar. In tomato, however, an increase in lycopene occurs at the expense of beta carotene.

A breeder can set his objectives and establish his breeding priorities, but unless the marketplace pays a premium for nutritional improvement, foods with enhanced dietary value will not be bred. Of the foods listed on the GRAS list of the FDA, only wheat with higher food value, that is, increased protein, commands a higher price. Of all other crops listed, only carrots are apt to be selected either by growers or by processors with improved dietary quality, and this by virtue of the correlation of color and vitamin A potency. In some countries, selection of crop cultivars for improved nutritional quality may be dictated by governments, but in the USA, for as long as vegetables are purchased for supermarkets by so few wholesalers, nutritional quality will remain secondary to appearance.

2. A. W. HOVIN. There are three problem areas relative to forage breeding for improved quality: (1) digestible dry matter, (2) voluntary intake by ruminant animals, and (3) antinutritional quality constituents.

A good, rapid *in vitro* procedure is widely used in the laboratory to predict forage digestibility. Actually, the cellular content is highly digestible, so the real problem in digestibility is associated with the cell wall constituents, including cellulose, hemicellulose, lignin, and silica. In some species the presence of phenolic compounds and alkaloids affect digestibility. Plant breeders have access to various laboratory procedures for estimating these constituents as well as total digestibility. However, they are slow and expensive, and the concentrations of some constituents, such as lignin and silica, are relatively low. Narrow-sense heritabilities for several forage quality constituents in reed canarygrass are in the same range as those for yield; so it is necessary to make many evaluations of individual progenies, and to evaluate many progenies. As a consequence, newer and more rapid procedures are being sought for measuring forage quality constituents. The infrared reflectant instruments in use for analyzing grain protein and oil concentration are inadequate for measuring forage quality constituents at this time, but an experimental infrared spectra-computer system instrument built at Pennsylvania State University will measure the various forage constituents. Very good correlations have been obtained between readings from this instrument and those from standard laboratory procedures for all constituents except lignin.

Traits that affect voluntary intake potential of a forage are leafiness, cell wall constituents, and factors associated with pest resistance. Good examples of increasing uptake potential through breeding are 'Tiftlate-1' pearl millet, 'Kenhy' tall fescue, and 'Morpa' weeping lovegrass. Some factors, because we do not understand them, must be lumped together as taste-related factors.

Variation among genotypes for antinutritional constituents can be demonstrated in reed canarygrass. Cattle and sheep show decided preferences for certain strains of this species, and the preferred genotypes have low levels of indole alkaloids. If ingested in large quantities, they may cause disorders in livestock. The close negative correlation between indole alkaloid concentration and the palatability ratings for reed canarygrass prompted a breeding program at the University of Minnesota and included genetic studies on alkaloid concentration. Heritability estimates suggested the appropriate breeding procedure would be recurrent selection. By using thin layer chromatography we could demonstrate that certain genotypes carried one indole alkaloid, gramine, and no others.

From an analysis of data from 1700 reed canarygrass plants, we found that this indole alkaloid series is controlled by genes at two loci. All indole alkaloids are derived from tryptophan. Differences in concentration occur for methoxylated tryptamine and each of the nine other indole alkaloids. Gramine, the simplest indole alkaloid, results from the double recessive. Tryptamines are controlled by a single dominant *T* gene at one locus and methoxylate tryptamine and its derivatives by *M*. The *Mm* locus is epistatic to the *Tt* locus and closely linked to it. The genetics of the indole alkaloids has been elucidated through the cooperation with animal scientists, biochemists, forage physiologists, and management scientists, but the real test came when the biological significance of these reed canarygrass lines was tested via grazing trials with animals. In the 1978 feeding trial, which covered two periods of 28 days each, live weight gain was twice as great (120 grams per lamb per day) for reed canarygrass strain 'MN76' as for two commercially available cultivars. During these periods, the indole alkaloid concentration of MN76 was about 1/3 that of the commercial cultivars. This example shows that indole alkaloids in reed canarygrass are responsible for palatability differences and may cause real differences in live weight gain. Antinutritional quality constituents in other forage crops may also respond to selection.

3. W. MARTINEZ. Attention to compositional quality will be an increasingly demanding source of challenge to the plant breeders, but it is necessary to recognize that nutritional quality is but one of several quality characteristics in food. The nutrient composition of food is of vital importance to people. People, however, do not eat for nutritional satisfaction. They eat for the pleasure provided by the flavor, color, and texture of the food. The problem of nutrient delivery, therefore, only begins with the breeding and production of the commodity and is not accomplished until the commodity is consumed. In many instances, this requires processing or conversion of the new raw material into acceptable food forms. So the commodity must not only contain the desired nutrients; it also must meet the processing and fabrication requirements of the food industry and the sensory requirements of the consumer. Nutrition and quality must go hand in hand. For example, in countries where people consume wheat in the form of bakery goods, protein is not protein. Protein is gluten, that is, those wheat proteins with specific texture-forming characteristics. Any improvement in the amino acid balance of wheat protein that affects the baking characteristics would be highly unacceptable to the food industry and to the consumer. This is equally true when the food processor and the consumer are one and the same. High-lysine maize will not be acceptable until it closely approximates the handling characteristics and the color, flavor, and texture of regular maize.

The counterpart to nutritional quality resides in antinutritional factors. The current, highly appropriate emphasis on breeding for pest resistance as part of integrated pest management could increase the known and unknown chemical constituents, such as antinutritional factors, in plant cultivars and have a serious impact on the safety of our food supply. The increase in solanine content in potatoes was an early example of this problem. Another example was the failure to monitor the concentration of gossypol, a known antimetabolite, in the seed in new high-yielding, wilt-resistant cotton cultivars. The resulting high gossypol content has reduced the use of cottonseed meal by 50 percent in layer-hen rations in California, an area deficient in low-cost protein feed ingredients.

The problem occurs for several reasons. The number of samples to be assayed, the cost of analytical monitoring, and a lack of awareness concerning the antinutritional constituents can each contribute. The analytical chemist and instrumentation developer have failed to meet the critical need of the plant breeder. The plant physiologist and biochemist also have a responsibility to identify those enzymes and mechanisms that may provide a simpler, more rapid analytical alternative to direct compositional analysis. Much of the rapid success in producing the double O cultivars of rape was due to the availability of simple, quick tests for erucic acid and glucosinolate. Another factor in the success of the rapeseed breeding program was the identification of well-defined nutritional goals. Often, however, nutritional goals are hard to define. First, because the science of human nutrition is in its infancy and restricted by the moral problems involved in research with humans; much of what is accepted on human nutritional requirements is poorly quantified and represents a documented consensus based on the best available information. Requirements, therefore, are subject to change. Second, nutrient requirements differ with age group and health status, so the group for which the nutritional improvement is intended must be specifically targeted. Third, food habits do change; political, economic, social and regulatory forces all have an impact. As the level of income increases, people shift from cereals to animal products. In urban communities where both members of the household work, home preparation decreases and consumption of food from restaurants and fast food establishments increases. The recent emphasis on the relationship between diet and health, particularly with respect to heart disease, has resulted in significant changes in food habits. Regulatory policies, such as the removal of all snack foods and soft drinks from schools, also have an effect on food habits. Regulations on nutritional labeling of foods will tend to restrict food composition to the range established for the particular nutrient in the specific commodity or product. This regulation may also present the plant breeder with an opportunity of providing specific cultivars that are enriched in the nutrient of interest.

Therefore, the identification of nutritional goals for the plant breeder is a multifaceted problem. There will be no single, clear-cut directive from the nutritionist, but rather several. Sometimes overlapping objectives will be identified, depending upon the country, the segment of the population, the age of the population, and their economic and cultural status. The plant breeder must seek out the nutritionist, the food technologist, and the food industry, to acquaint each with the potential and limitations of plant breeding, and to establish the essential lines of communication and cooperation. For in plant breeders' hands reside both the food supply and the nutrition of the world.

4. R. A. JOHNSON. The task small grain breeders have in synthesizing cultivars that have favorable combinations of yield, disease resistance, and agronomic and quality traits is a formidable one. In light of this, would you recommend early generation screening for quality traits?

5. J. D. AXTELL. Yes, plant breeders favor early generation testing for quality traits. However, in the case of wheat breeding programs, baking quality characteristics are fairly well stabilized in the germplasm pools being used, so selection for these traits can be deferred to later generations in this specific case.

6. G. LESTER. Why is it more informative nutritionally to express protein as a biological value instead of as a percent of total?

7. J. D. AXTELL. In general, protein quality is a species characteristic. However, variations in protein quality may occur within a species as shown in opaque-2 maize, hiproly barley, and high-lysine sorghum. The problem really is—How can the high-protein-quality types be identified in the marketplace and how can the producer be compensated for the better protein quality in his product?

8. W. MARTINEZ. With regard to biological value of protein, it must be recognized that humans eat a variety of proteins and that these proteins are complementary. Therefore, it is not essential

for each ingested protein to have optimal biological value. Government regulations also have an impact upon the problem of biological quality because they specify feeding trials with rats, not humans, as the assay for the determination of protein quality. This anomaly has a real impact on the perceived quality of vegetable proteins since they are generally low in the sulfur amino acids and rat requirements for these amino acids are greater than human requirements. Recent data indicate that the sulfur amino acid requirements of humans can be adequately met with vegetable proteins.

9. D. WELLS. Is the basic nutritional problem in the world due to a lack of calories or a lack of protein?

10. W. MARTINEZ. Dialogue on this question has been continuing for 10 to 15 years. And its resolution is that both are important. Again, it depends upon the targeted group of humans to which the question is applied.

11. C. RODERUCK. The quality and amount of protein required by a human depends upon whether the individual is growing, lactating, or pregnant, and not so much on expenditure of energy for work. In the developing countries, there is often a shortage of food energy. But the shortage of energy and the shortage of protein are different in proportion to each other, depending on the physiological state of the population. The real difficulty in nutrition is that one cannot generalize from infants to adult men in terms of the need for protein and energy.

12. K. RAWAL. Most of the world's problems with undernutrition, malnutrition, poverty, high density population, and the like, occur between 20° north and 20° south latitude. In this band, however, cereals and such staples with high calorie value are not consumed by themselves. Almost always, they are consumed concurrently with grain or food legumes that supply protein. For example, in India, rice and wheat are eaten with chickpeas and pigeonpeas; in Latin America, beans are consumed with maize and potatoes; and in Africa, cowpeas and yams are eaten together. In the last decade, much effort has been spent on trying to improve the protein quality and quantity in cereals; yet, so little effort has been given to increasing the productivity of the high-protein grain legumes. Actually, the problems associated with the improvement of quality and quantity of food legumes have not been addressed.

13. J. D. AXTELL. No one would argue with the fact that the optimal dietary protein situation for a human diet can be obtained through a good cultivar of grain legumes. But right now, primary emphasis is being put on changing morphological types so as to increase yield and disease resistance so that grain legumes will be profitable to produce.

14. S. GALAL. It is possible to improve food and feed quantity and quality through breeding pairs of crop cultivars for use in mixed crop culture.

In Egypt, a mixed culture of maize and soybeans produced 1,040 kg of protein per ha, whereas maize alone produced only 500 kg of protein and soybeans alone only 850 kg of protein. And by grinding the maize and soybeans together, the biological value of the protein in the mixture was better than either the maize or soybean protein alone.

15. A. QUINN. A negative correlation usually occurs between protein percentages of seeds and seed yields. A similar association exists in the sugar beet industry where selections with increased sugar content usually have a low root yield. As pointed out, yield of protein can be increased more easily by selecting for grain yield than for high protein percentage. This also occurs in sugar beet breeding, and growers are well aware that growing a high tonnage-low sugar content cultivar of beets will give more sugar per ha than growing a low tonnage-high sugar content cultivar. Industry biases the scales a bit by paying a small premium for high sugar content beets. But, in fact, cultivars that have a higher sugar content are often so much lower in yield that the grower actually gets penalized. Many breeding companies compete for the same grower in the seed market, and the company with cultivars that have high tonnage and low sugar often gets the market. The processors ultimately suffer because they receive a low quality raw product. Does this happen with other crops, and if so, how does the industry attack it?

16. J. D. AXTELL. This problem exists for all crops, and if the sugar beet processing industry really wants to have a higher sugar concentration in the beet, they may have to pay a higher premium for it, because the farmer knows that sugar yield is correlated with beet yield. So the sugar beet industry may need to handle this problem through the marketing system. Another example occurs with wheat where the millers are requesting higher protein wheat

from the breeders to insure better baking quality. And yet, the industry in general does not pay a premium for higher protein.

17. A. THOMPSON. The crimson gene in tomato improves visual color by increasing lycopene, but concurrently reduces the beta carotene, and consequently, nutritive value of the tomato. There are other recessive high pigment genes in tomatoes that significantly increase the levels of both lycopene and beta carotene. The high pigment and crimson genes together usually increase the total pigment above that found with either gene by itself. It brings the total level of beta carotene back up to that usually found in the normal tomato. Thus, nutrition is not sacrificed when visual quality is enhanced. The combined product does have a very attractive color. However, the high pigment gene reduces germination, stand establishment, and seedling vigor, causes brittle stems, increases susceptibility to foliage diseases, and generally reduces yield. Considerable research effort has gone into trying to adapt this character, but this has resulted in little success. This example highlights the difficulty of incorporating improved nutritional and quality factors into cultivars that have satisfactory yield and required horticultural and agronomic characteristics.

18. G. LESTER. Nothing tastes worse than a bitter off-flavored carrot, so what work is being done to combine good flavor and good nutritional quality in a common carrot genotype?

19. W. H. GABELMAN. A major problem in carrots is high oil content, which causes a foul tasting product. But, in general, breeders are selecting for low oils and higher nonreducing sugars. High carotene carrots do not have to carry off-flavors.

20. T. S. COX. In cereals, high protein percentage is generally related to lateness, tallness, and especially low harvest index. That is, it is related to greater vegetative growth. This is usually explained by noting that most of the grain nitrogen is taken up before anthesis and later transported to the grain. Therefore, a small plant cannot provide as much nitrogen for grain filling and still maintain the leaves for starch production. It has been suggested that further yield increases in rice and wheat may come from breeding tall dwarfs. Would this approach give a good chance for maintaining or improving grain protein percentage?

21. J. D. AXTELL. In grain sorghum, three-dwarf, that is, very short types, produce very well under dry conditions or under conditions where lodging is a problem. Now, in general, sorghum hybrids are becoming taller as the breeder obtains better stalk-rot and lodging resistances. Whether this will increase the potential capability for protein content is unknown today. However, nitrogen translocation into the grain is a function of harvest index, so it is difficult to see how tall and short plants would differ in nitrogen translocation into the grain.

22. R. FINCHER. A number of maize and sorghum genotypes have been developed with high protein quality and adequate yielding ability. Are any of these genotypes being used for commercial or subsistence production anywhere in the world?

23. J. D. AXTELL. A strain of high-lysine sorghum is being grown in Ethiopia, and it has been grown there for centuries. Primarily, farmers mix it with normal sorghum grain. They know that it yields lower, but they grow it in mixtures because when roasted in the late dough stage, it improves the flavor of the mixture.

If our strains of sorghum with 30 percent higher protein and better protein quality would be accepted for special purpose foods, such as for recently weaned children, it would provide a great nutritional improvement in rural areas where there is no access to other sources of protein supplements. A farmer could grow a small section of his field to these high lysine-high protein types, and use this grain for a specific target population such as preschool children, pregnant women, and nursing mothers. This is a very viable concept. Where indigenous high-lysine cultivars already exist, farmers accept them. Therefore, a number of them have been collected and ICRISAT breeders are working on their agronomic improvement, with the idea of re-releasing them back into the areas where they are already being grown and accepted as cultivars.

Much work is being done at CIMMYT toward developing high-lysine improved cultivars and synthetics of maize. The actual hectarage of high-lysine maize in production is very small in the USA, and even in countries like Brazil. CIMMYT has developed hard endosperm opaque-2 cultivars with rather good yields, but it is a little early to judge what the ultimate result of that effort will be. If opaque-2 maize is judged on a cost-benefit basis, it provides a very cheap way to solve a nutritional problem relative to fortification.

By simply altering dietary habits, opaque-2 will solve that nutritional problem.

24. G. PERSSON. In Sweden, we have found a variation of about 10 percent in the digestibility of barley protein, and a high correlation exists between animal digestibility ratings from experiments and *in vitro* analysis.

25. J. D. AXTELL. No results on variation in protein digestibility are available for sorghum.

The high tannin trait in sorghum is an antinutritional factor, but it has a simple inheritance. High-lysine lines of sorghum do not contain tannin compounds, so there should be no interference.

26. D. NANDA. In your view, what is the future of improved quality protein corn and sorghum in the USA and what should the breeders of seed companies in the USA be doing about it?

27. J. D. AXTELL. First, there may be naturally occurring mutants, and if so, these should be used. For example, at the University of Minnesota, scientists are using a scheme for selecting maize tissue cultures that are resistant to lysine and threonine as an indirect way of selecting high-methionine cultivars. They have identified some cultivars that have resistance to these two amino acids. This may be an approach to getting lysine overproducers that have satisfactory grain yield. If such a type could be found in sorghum, the gene would need to be incorporated into advanced breeding materials to see if it could survive. This is the approach we have used with the P-721 gene for high lysine.

Right now, no one knows what the future of improved protein quality in maize or sorghum will be in the USA. It probably is needed more for people in developing countries that depend on a diet of cereals.

28. K. H. QUESENBERRY. Most forage breeders now have genotypes in their breeding programs that approach 70 percent digestibility at a reasonable stage of maturity, so probably this is near the upper level of digestibility that can be obtained in forages. If so, it seems that identification of and breeding against antinutritional components will be the area where forage breeders may make the important contribution in the next decade.

29. A. W. HOVIN. Of course digestibility and intake potential both go down with advanced maturity in forages. Some crops, like alfalfa, may have too high quality at times, and bloat may occur because the mesophyll membranes rupture too easily in the rumen resulting in a sudden release of soluble protein. Breeders of alfalfa are even thinking about selecting for higher tannin in alfalfa, so that it will behave more like the nonbloating legumes.

REFERENCES

Ashri, A., P. F. Knowles, A. L. Urie, D. E. Zimmer, A. Cahaner, and A. Marani. 1977. Evaluation of the germplasm collection of safflower, *Carthamus tinctorius*. III. Oil content and iodine value and their associations with other characters. Econ. Bot. 31:38-46.

Axtell, J. D., S. W. Van Scoyoc, P. J. Christensen, and G. Ejeta. 1979. Current status of protein quality improvement in grain sorghum, pp. 357-63. *In* Proc. Symp. on Seed Protein Improv. in Cereals and Grain Legumes, Neuherberg, Fed. Repub. Ger., 4-8 Sept. 1978.

Barnes, R. F., and D. L. Gustine. 1973. Allelochemistry and forage crops, pp. 1-15. *In* Matches, Arthur G. (ed.), Anti-quality components of forages. Crop Sci. Soc. Am., Madison, Wis.

Barnes, R. F., and G. C. Marten. 1978. Recent developments in predicting forage quality. Presented at Jt. Am. Dairy Sci. Assoc.-Am. Soc. of Animal Sci. Meet., Michigan State Univ., East Lansing, Mich., 9-12 July 1978.

Boulter, D. 1979. Structure and biosynthesis of legume storage proteins. pp. 125-34. *In* Proc. Symp. on Seed Protein Improv. in Cereals and Grain Legumes, Neuherberg, Fed. Repub. Ger., 4-8 Sept. 1978.

Bressani, Richardo. 1976. Productivity and improved nutritional value in basic food crops, pp. 265-87. *In* Wilcke, H. L. (ed.), Improving the nutrient quality of cereals. II. Rep. Second Workshop on Breed. and Fortification. U. S. Agency Int. Dev., Washington, D.C.

Briggle, L. W. 1971. Improving nutritional quality of oats through breeding. Agron. Abstr., p. 53.

Brock, R. D., and J. Langridge. 1975. Prospects for genetic improvement of seed protein in plants, pp. 3-13. *In* Breeding for seed protein improvement using nuclear techniques. Int. Atom. Energy Agency Publ. STI/PU8/400, Vienna.

Buishand, J. G., and W. H. Gabelman. 1979. Studies on the inheritance of root color and carotenoid content in red × yellow and red × white crosses of carrot, *Daucus carota* L. Euphytica (submitted).

Burr, B. 1979. Identification of zein structural genes in the maize genome, pp. 175-77. *In* Proc. Symp. on Seed Protein Improv. in Cereals and Grain Legumes, Neuherberg, Fed. Repub. Ger., 4-8 Sept. 1978.

Burr, F. A. 1979. Zein synthesis and processing on zein protein body membranes, pp. 159-62. *In* Proc. Symp. on Seed Protein Improv. in Cereals and Grain Legumes, Neuherberg, Fed. Repub. Ger., 4-8 Sept. 1978.

Burton, G. W. 1958. Cytoplasmic male-sterility in pearl millet (*Pennisetum glaucum*) (L.) R. Br. Agron. J. 50:230-31.

Burton, G. W. 1964. Coastal covers more than six million acres. Livest. J. Breed.

Burton, G. W. 1969. Registration of pearl millet inbreds Tift $23B_1$, Tift $23A_1$, Tift $23DB_1$, and Tift $23DA_1$. Crop Sci. 9:397-98.

Burton, G. W. 1972. Registration of Coastcross-1 Bermudagrass. Crop Sci. 12:125.

Burton, G. W. 1979. Registration of pearl millet inbred Tift 383 and Tifleaf 1 pearl millet. Crop Sci. (submitted).

Burton, G. W. 1979. (Pers. commun.)

Burton, G. W., and W. G. Monson. 1978. Registration of Tifton 44 Bermudagrass. Crop Sci. 18:911.

Burton, G. W., W. G. Monson, J. C. Johnson, Jr., R. S. Lowrey, H. D. Chapman, and W. H. Marchant. 1969. Effect of the d_2 dwarf gene on the forage yield and quality of pearl millet. Agron. J. 61:607-12.

Caldwell, B. E., C. R. Weber, and D. E. Byth. 1966. Selection value of phenotypic attributes in soybeans. Crop Sci. 6:249-51.

Campbell, A. R., and K. J. Frey. 1972. Inheritance of groat-protein in interspecific oat crosses. Can. J. Plant Sci. 52:735-42.

Christensen, P. J. 1978. Selection for yield and lysine concentration among opaque P-721 derived grain sorghum lines. Ph.D. thesis, Purdue Univ., Lafayette, Ind.

Coffman, W. R., and B. O. Juliano. 1979. Seed protein improvement in rice: Status Report, pp. 261-75. *In* Proc. Symp. on Seed Protein Improv. in Cereals and Grain Legumes, Neuherberg, Fed. Repub. Ger., 4-8 Sept. 1978.

Doll, H., and B. Køie. 1975. Evaluation of high lysine barley mutants, pp. 55-59. *In* Breeding for seed protein improvement using nuclear techniques. IAEA, Vienna.

Downey, R. K. 1964. A selection of *Brassica campestris* L. containing no erucic acid in its seed oil. Can. J. Plant Sci. 44:295.

Dudley, J. W., R. J. Lambert, and D. E. Alexander. 1974. Seventy generations of selection for oil and protein concentration in the maize kernel, pp. 181-212. *In* Dudley, J. W. (ed.), Seventy generations of selection for oil and protein in maize. Crop Sci. Soc. Am., Madison, Wis.

Dudley, J. W., R. J. Lambert, D. E. Alexander, and I. A. de la Roche. 1977. Genetic analysis of crosses among corn strains divergently selected for percent oil and protein. Crop Sci. 17:111-17.

Ejeta, G. 1976. Evaluation of high lysine and normal Ethiopian sorghum varieties for protein quality, carbohydrate composition, and tannin content and an assessment of nutritional value at various stages of grain development. M.S. thesis, Purdue Univ., Lafayette, Ind.

Ejeta, G. 1979. Selection for genetic modifiers that improve the opaque kernel phenotype of P-721 high lysine sorghum (*Sorghum bicolor* [L.] Moench). Ph. D. thesis, Purdue Univ., Lafayette, Ind.

Fick, G. N. 1975. Heritability of oil content in sunflowers. Crop Sci. 15:77-78.

Flora, L., and R. C. Wiley. 1972. Effect of various endosperm mutants on oil content and fatty acid composition of whole kernel corn (*Zea mays* L.). J. Am. Soc. Hort. Sci. 97:604-7.

Forsberg, R. A. 1979. (Pers. commun.)

Forsberg, R. A., V. L. Youngs, and J. W. Pendleton. 1979. (Pers. commun.)

Forsberg, R. A., V. L. Youngs, and H. L. Shands. 1974. Correlations among chemical and agronomic characteristics in certain oat cultivars and selections. Crop Sci. 14:221-24.

Frey, K. J. 1951. The relation between alcohol-soluble and total nitrogen content in oats. Cereal Chem. 28:506-9.

Frey, K. J. 1975. Heritability of groat-protein percentage of hexaploid oats. Crop Sci. 15:277-78.

Frey, K. J. 1977. Protein of oats. Z. Pflanzenzüchtg 78:185-215.

Frey, K. J., T. McCarty, and A. Rosielle. 1975. Straw-protein percentages in *Avena sterilis* L. Crop Sci. 15:716-19.

Frey, K. J., and G. I. Watson. 1950. Chemical studies on oats. I. Thiamine, niacin, riboflavin, and pantothenic acid. Agron. J. 42:434-36.

Gabelman, W. H. 1974. The prospects for genetic engineering to improve nutritional values, pp. 147-55. *In* White, Phillip (ed.), Nutritional quality of fresh fruits and vegetables. Futura Publ. Co., Mt. Kisco, N.Y.

Gebrekidan, Brhane. 1977. Unpublished data (pers. commun.).

Glover, D. V. 1976. Improvement of protein quality in maize, pp. 69-97. *In* Wilcke, H. L. (ed.), Improving the nutrient quality of cereals II. Rep. Second Workshop on Breed. and Fortification. U.S. Agency for Int. Dev., Washington, D.C.

Grami, B., and B. R. Stefansson. 1977a. Gene action for protein and oil content in summer rape. Can. J. Plant Sci. 57:625-31.

Grami, B., and B. R. Stefansson. 1977b. Paternal and maternal effects on protein and oil content in summer rape. Can. J. Plant Sci. 57:945-49.

Green, C. E., and R. L. Phillips. 1974. Potential selection system for mutants with increased lysine, threonine, and methionine. Crop Sci. 14:827-30.

Hall, T. C., and F. A. Bliss. 1978. Molecular biological approaches for improving bean seed proteins. Presented at Am. Assoc. Adv. Sci., Dec. 1978, Houston, Tex.

Harper, A. E. 1976. Widespread malnutrition: A protein problem, a food problem, or a population problem? pp. 255-64. *In* Wilcke, H. L. (ed.), Improving the nutrient quality of cereals II. Rep. Second Workshop on Breed. and Fortification. U.S. Agency Int. Dev., Washington, D.C.

Hoveland, C. S. 1973. Introduction. *In* Matches, Arthur G. (ed.), Anti-quality components of forages. Crop Sci. Soc. Am., Madison, Wis.

Howell, R. W., C. A. Brim, and R. W. Rinne. 1972. The plant geneticist's contribution toward changing lipid and amino acid composition of soybeans. J. Am. Oil Chem. Soc. 49:30-32.

Ingverson, J., B. Køie, and H. Doll. 1973. Induced seed protein mutant of barley. Experientia 29:1151-52.

Int. Rice Research Inst. 1978. Annu. Rep. for 1977, Los Baños, Philipp. (in press).

Johnson, V. A., P. J. Mattern, and S. L. Kuhr. 1979. Genetic improvement of wheat protein, pp. 165-78. *In* Proc. Symp. on Seed Protein Improv. in Cereals and Grain Legumes, Neuherberg, Fed. Repub. Ger., 4-8 Sept. 1978.

Johnson, V. A., J. W. Schmidt, and P. J. Mattern. 1968. Cereal breeding for better protein impact. Econ. Bot. 22:16-25.

Jones, R. A., and C. Y. Tsai. 1977. Changes in lysine and tryptophan content during germination of normal and mutant maize seed. Cereal Chem. 54:565-71.

Jönsson, R. 1977. Erucic-acid heredity in rapeseed (*Brassica napus* L. and *Brassica campestris* L.). Hereditas 86:159-70.

Knowles, P. F. 1969. Modification of quantity and quality of safflower oil through plant breeding. J. Am. Oil Chem. Soc. 46:130-32.

Knowles, P. F. 1975. Recent research on safflower, sunflower, and cotton. J. Am. Oil Chem. Soc. 52:374-76.

Kohler, G. W., R. E. Lincoln, J. W. Porter, F. P. Zscheile, R. M. Caldwell, R. H. Harper, and W. Silver. 1947. Selection and breeding for high β-carotene content (provitamin A) in tomato. Bot. Gaz. 109:219-25.

Kust, A. F. 1970. Inheritance and differential formation of color and associated pigments in xylem and phloem of carrot, *Daucus carota*, L. Ph.D. thesis, Univ. of Wisconsin, Madison, Wis.

Laferriere, L., and W. H. Gabelman. 1968. Inheritance of color, total carotenoids, alpha-carotene, and beta-carotene in carrots, *Daucus carota*, L. Proc. Am. Soc. Hort. Sci. 93:408-18.

Larkins, B. A. 1979. Seed storage proteins: Characterization and biosynthesis. *In* The biochemistry of plants: A comprehensive treatise. Vol. 6. Academic Press (in press).

Lincoln, R. E., F. P. Zscheile, J. W. Porter, G. W. Kohler, and R. M. Caldwell. 1943. Provitamin A and vitamin C in the genus *Lycopersicon*. Bot. Gaz. 105:113-15.

Marwaha, R. S., and B. O. Juliano. 1976. Aspects of nitrogen metabolism in the rice seedling. Plant Physiol. 57:923-27.

Matches, Arthur G. (ed.). 1973. Anti-quality components of forages. Crop Sci. Soc. Am., Madison, Wis.

Mertz, E. T., L. S. Bates, and O. E. Nelson. 1964. Mutant gene that changes protein composition and increases lysine content of maize endosperm. Science 145:279-80.

Mertz, E. T., R. Jambunathan, E. Villegas, R. Bauer, C. Kies, J. McGinnis, and J. S. Shenk. 1975. Use of small animals for evaluation of protein quality in cereals, pp. 306-29. *In* High Quality Protein Maize: Proc. CIMMYT-Purdue Symp. on Protein Quality in Maize. Dowden, Hutchinson, and Ross, Stroudsburg, Pa.

Middleton, G. K., C. E. Bode, and B. B. Bayles. 1954. A comparison of the quantity and quality of protein in certain varieties of soft wheat. Agron. J. 46:500-502.

Miflin, B. J., and P. R. Shewry. 1979. The biology and biochemistry of cereal seed prolamins, pp. 137-58. *In* Proc. Symp. on Seed Protein Improv. in Cereals and Grain Legumes, Neuherberg, Fed. Repub. Ger., 4-8 Sept. 1978.

Miller, D. F. 1958. Composition of cereal grains and forages. NAS-NRC Publ. 385, Washington, D.C.

Mohan, D. P. 1975. Chemically induced high lysine mutants in *Sorghum bicolor* (L.) Moench. Ph.D. thesis, Purdue Univ., Lafayette, Ind.

Mohan, D. P., and J. D. Axtell. 1975. Diethyl sulfate induced high lysine mutants in sorghum. Pap. presented at Ninth Biennial Grain Sorghum Res. and Util. Conf., Lubbock, Tex., 4-6 Mar. 1975.

Mok, M. C., W. H. Gabelman, and F. Skoog. 1976. Carotenoid synthesis in tissue cultures of *Daucus carota* L. J. Am. Soc. Hort. Sci. 101:442-49.

Moreira, M. R., M. A. Hermodson, B. A. Larkins, and N. C. Nielsen. 1979. Partial characterization of the acidic and basic subunits of glycinin. J. Biol. Chem. 254:9921-26.

Munck, L., K. E. Karlsson, and A. Hagberg. 1971. Selection and characterization of a high protein lysine variety from the world barley collection, pp. 544-58. *In* Nilan, R. (ed.), Barley genetics II. Pullman, Wash.

Nelson, Jr., O. E. 1974. Interpretive summary and review. Presented Workshop on Genetic Improv. of Seed Proteins. NAS, Washington, D.C.

Nelson, Jr., O. E. 1979. Inheritance of amino acid content in cereals, pp. 79-87. *In* Proc. Symp. on Seed Protein Improv. in Cereals and Grain Legumes, Neuherberg, Fed. Repub. Ger., 4-8 Sept. 1978.

Ng, B. H., R. H. V. Corley, and A. J. Clegg. 1976. Variation in the fatty acid composition of palm oil. Oleagineux 31:1-6.

Ohm, H. W., and F. L. Patterson. 1973. A six-parent diallel cross analysis for protein in *Avena sterilis*, L. Crop Sci. 13:27-30.

Oram, R. N., and R. D. Brock. 1972. Prospects for improving plant protein yield and quality by breeding. J. Aust. Inst. Agric. Sci. 38:163-68.

Osborne, T. B. 1897. The amount and properties of the proteids of the maize kernel. J. Am. Chem. Soc. 19:525-32.

Osborne, T. B., and L. B. Mendel. 1914. Nutritive properties of proteins of the maize kernel. J. Biol. Chem. 18:1-16.
Perez, C. M., G. B. Cagampang, B. V. Esmama, R. U. Monserrate, and B. O. Juliano. 1973. Protein metabolism in leaves and developing grains of rices differing in grain protein content. Plant Physiol. 51:537-42.
Persson, G. 1975. The barley protein project at Svalöv, pp. 91-97. *In* Breeding for seed protein improvement using nuclear techniques. IAEA, Vienna.
Phillips, R. L., B. G. Gengenbach, and C. E. Green. 1978. Genetic methods for improving protein quality in maize, p. 116. *In* Proc. 14th Int. Congr. Genet. Part II, Moscow.
Porter, K. S. 1977. Modification of the opaque endosperm phenotype of the high lysine sorghum line P-721 (*Sorghum bicolor* (L.) Moench). using the chemical mutagen diethyl sulfate. Ph.D. thesis, Purdue Univ., Lafayette, Ind.
Raghuveer, K. G., and E. G. Hammond. 1967. The influence of glyceride structure on the rate of autoxidation. J. Am. Oil Chem. Soc. 44:239-43.
Raymond, W. F. 1969. The nutritive value of forage crops. Adv. Agron. 21: 1-108.
Röbbelen, G. 1972. Selection for oil quality in rapeseed, pp. 215-19. *In* Lupton, F. G. H., G. Jenkins, and R. Johnson (eds.), The way ahead in plant breeding. Proc. Sixth Congr. Eucarpia, Cambridge Univ. Press.
Röbbelen, G. 1977. Possibilities and limitations of breeding for nutritional improvement of cereals, pp. 47-57. *In* Nutritional evaluation of cereal mutants. IAEA, Vienna.
Robbins, G. S., Y. Pomeranz, and L. W. Briggle. 1971. Amino acid composition of oat groats. Agric. Food Chem. 19:536-39.
Scrimshaw, N. S. 1966. Applications of nutritional and food sciences to meeting world food needs, pp. 48-55. *In* Prospects of the World Food Supply: A Symp. NAS.
Senti, F. R., and R. L. Rizek. 1975. Nutrient levels in horticultural crops. Hort. Sci. 10:243-46.
Singh, J., and H. K. Jain. 1977. Studies on assessing the nutritive value of opaque-2 maize. Indian Agric. Res. Inst., New Delhi.
Singh, R., and J. D. Axtell. 1973. High lysine mutant gene (hl) that improves protein quality and biological value of grain sorghum. Crop Sci. 13:535-39.
Srinivasachar, D., A. Seetharam, and R. S. Malik. 1972. Combination of the three characters (high oil content, high iodine value, and high yield) in a single variety of linseed *Linum usitatissimum* L. obtained by mutation breeding. Curr. Sci. 41:169-71.
Stefansson, B. R., F. W. Houghen, and R. K. Downey. 1961. Note on the isolation of rape plants with seed oil free from erucic acid. Can. J. Plant Sci. 41:218-19.
Stefansson, B. R., and A. K. Storgaard. 1969. Correlations involving oil and fatty acids in rapeseed. Can. J. Plant Sci. 49:573-80.
Tomes, M. L. 1972. Breeding for improved nutritional value. *In* Proc. Symp. on the Role of Hortic. in Meet. World Food Requir. Hort. Sci. 7:154-56.
Tomes, M. L., and F. W. Quackenbush. 1968. Caro-Red, a new provitamin A rich tomato. Econ. Bot. 12:256-60.
Tomes, M. L., F. W. Quackenbush, and M. McQuistan. 1954. Modification and dominance of the gene governing formation of high concentrations of β-carotene in the tomato. Genetics 38:117-27.
Thomas, P. M., and Z. P. Kondra. 1973. Maternal effects on the oleic, linoleic, and linolenic acid content of rapeseed oil. Can. J. Plant Sci. 53:221-25.
Tsai, C. Y. 1979. (Pers. commun.)

Tsai, C. Y., D. M. Huber, and H. L. Warren. 1978. Relationship of the kernel sink for N to maize productivity. Crop Sci. 17:399-404.
Umiel, N., and W. H. Gabelman. 1972. Inheritance of root color and carotenoid synthesis in carrot, *Daucus carota*, L.: Orange vs. red. J. Am. Soc. Hort. Sci. 97:453-460.
Utley, P. R., W. G. Monson, G. W. Burton, R. E. Hellwig, and W. C. McCormick. 1978. Comparison of 'Tifton 44' and Coastal Bermudagrasses as pastures and as harvested forages. J. Animal Sci. 47:800-804.
Van Scoyoc, S. W. 1979. Comparison of yield and of dry matter and endosperm protein accumulation during seed development of P-721 opaque and its normal sib line, and yield evaluation of F-7 P-721 opaque high lysine derivatives with elite normal varieties of *Sorghum bicolor* (L.) Moench. Ph.D. thesis, Purdue Univ., Lafayette, Ind.
Vasal, S. K., E. Villegas, and R. Bauer. 1979. Present status of breeding quality protein maize, pp. 127-48. *In* Proc. Symp. on Seed Protein Improv. in Cereals and Grain Legumes, Neuherberg, Fed. Repub. Ger., 4-8 Sept. 1978.
Weber, E. J., and D. E. Alexander. 1975. Breeding for lipid composition in corn. J. Am. Oil Chem. Soc. 52:370-73.
Weiss, T. J. 1970. Food oils and their uses. Avi Publ. Co., Westport, Conn. 224 pp.
Wilcke, H. L. (ed.). 1976. Improving the nutrient quality of cereals II. Rep. Second Workshop on Breed. and Fortification 1976. U.S. Agency Int. Dev., Washington, D.C.
Wilhelmi, K. D. 1974. Effect of variety, cultural practices, and nitrogen fertilizer regime on protein content of wheat, p. 362. *In* Proc. Lat. Am. Wheat Conf., Porto Alegre, Brazil.
Wilson, R. F., R. W. Rinne, and C. A. Brim. 1976. Alteration of soybean oil composition by plant breeding. J. Am. Oil Chem. Soc. 53:595-97.
Yermanos, D. M., S. Hemstreet, and M. J. Garber. 1967. Inheritance of quality and quantity of seed-oil in safflower (*Carthamus tinctorius* L.). Crop Sci. 7:417-22.

CHAPTER 11

Meeting Human Needs Through Plant Breeding: Past Progress and Prospects for the Future

G. W. BURTON

WHAT ARE HUMAN NEEDS and how many of them have been satisfied in part at least by plant breeding? How shall I answer these questions?

Suppose we imagine a young woman with a baby in some remote area of the Earth—an area untouched by and unacquainted with civilization and domesticated agriculture. What are her needs?

She and her baby need food that she must gather from the natural environment. Plant breeding has contributed nothing to her food supply. Certainly plant breeding can claim no credit for the animal skins that clothe this young woman and her baby. Her housing is probably a cave or a shelter made from tree boughs and an abundance of dead twigs and branches in the woods can satisfy her need for fuel. Man has not tampered with her environment, nor endangered its species. She can live in "perfect harmony with the environment." She has to. She has no choice.

It is reasonable to assume that this young woman's world consists of the area in which she lives. It contains neither books, radio, TV, nor great works of art. Its music is limited to the songs of the creatures in her natural environment. Her world contains no comfortable cottages, beautiful mansions, great cathedrals, churches,

Research Geneticist, Agricultural Research, Science and Education Administration, USDA, and the University of Georgia, College of Agricultural Experiment Stations, Coastal Plain Station, Agronomy Department, Tifton, Georgia.

Cooperative investigations of Agricultural Research, Science and Education Administration, USDA, and the University of Georgia, College of Agriculture Experiment Stations, Coastal Plain Station, Agronomy Department, Tifton, Georgia.

schools, hospitals, supermarkets, department stores, or transportation except her legs. It has neither well-kept gardens, lawns, nor parks.

We can only imagine the fear and distress that can plague this young woman on those days when she can find no food, her clothing and shelter are inadequate, she or her baby is sick, or they are threatened with physical violence. Life expectancy for mother and child is probably less than 35 years.

Your needs and life today might be similar to those described for this young woman but for the contributions of plant breeding.

PAST PROGRESS

Plant breeding has contributed directly to your need for food, clothing, shelter, recreation, conservation, environmental enrichment, peace, and peace of mind. Indirectly, plant breeding and the agricultural development associated with it have made possible your way of life today. Without plant breeding that converted weedy plants into our basic food crops, every one of us would spend most of our time hunting food, wondering each day what, if anything, we could eat tomorrow. No one would have had time to develop the level of culture that we enjoy today.

Without cultivated crops, most of us would not exist because the world's native vegetation could feed less than 5 percent of today's world population. The wars between Indian tribes over hunting grounds are evidence that in terms of Indian cultures and land use the United States was "overpopulated" when the white man arrived. Estimates of the Indian population in the United States during the era of initial colonization by Western Europeans are less than three million, or about 1 percent of our present population. Yet the population size would have been much smaller had the Indians been forced to find all of their food in the natural vegetation. Most American Indian tribes grew some crops.

Directly or indirectly, plant breeding deserves credit for most of man's food today. Peter Jennings came close to the truth when he entitled his 1976 *R. F. Illustrated* article on plant breeding, "No Plant Breeders, No Crops: No Crops, No Food." Without the primitive plant breeders who fashioned our food crops from the weeds about them, we would have none of the crops that supply most of our food today. What kind of people were they? Probably most of them were women who gathered seeds and fruits while the men hunted. They knew nothing about genetics, sex in plants, metabolic pathways, DNA, RNA, and the like. They were without microscopes, computers, or any sharply honed tools. They had neither written language nor libraries, and probably no word for the introgression and recurrent selection that they practiced.

What did they have? They had their plants, probably with a more restricted germplasm base than most of us have at our disposal. They lived with their plants and they knew them as few of us know ours today. They had hands and used them. They were motivated; their life depended on their own success. Finally, they had time on their side—more time than the hungry world can give us today.

Past progress in meeting human needs through plant breeding has been greater by far than most people realize. A few specific examples will serve to substantiate this claim.

In the past 45 years, crop yields in the United States have increased from 50 percent for oats (*Avena sativa*) to more than 300 percent for maize (*Zea mays*) and sorghum (*Sorghum bicolor*). Yields of major food crops such as wheat (*Triticum aestivum*), rice (*Oryza sativa*), soybeans (*Glycine max*), and potatoes (*Solanum tuberosum*) have more than doubled in this period. A combination of better cultivars and improved crop management made these increases possible. How much of it can be credited to plant breeding?

Russell (1974), comparing the performance of corn hybrids representing different eras of corn breeding, estimated that 63 percent of the advance in Iowa's grain yield in the past 48 years could be credited to plant breeding.

Duvick (1977) compared the yield performance of double and single cross corn hybrids between inbreds used in central Iowa in the past 40 years at plant densities of 32M, 44M, and 66M per ha. In both experiments, maximum yields were obtained at the higher populations. He concluded that 57 percent and 60 percent of the yield gain in central Iowa in the past 48 years can be credited to genetics. He also recognized "that the improved yields of the new hybrids can be expressed only with modern cultural techniques."

From 1936 to 1975, wheat yields in New York State doubled. Jensen's (1978) research indicates that 49 percent of this advance can be credited to wheat breeding.

Luedders (1977) compared the yield performance of 21 soybean cultivars from 0, 1, and 2 selection cycles for three years in Missouri. The cultivars from the 0 cycle represent those generally grown in 1930. Cycle 2 cultivars are those grown in the USA today. His studies revealed that plant breeding in 47 years has increased yield 45 percent; plant height, 7 percent; and reduced lodging, 31 percent.

Continuous well-supported plant breeding coordinated with other agricultural research has increased the yield of every food crop so treated. Neither plant breeding nor other agricultural research alone could have made such remarkable progress. Together they have met man's basic need for food so effectively that less than 5 percent of our people can produce enough food for all of us. This has freed the rest of us to develop the culture and way of life that prevails today.

Plant breeding has made today's fruits and vegetables more productive, dependable, beautiful, and nutritious. My father would never have believed that plant breeders working with agricultural engineers would be able to modify the tomato (*Lycopersicon esculentum*) and its culture so that the crop could be grown and harvested mechanically. At Tifton, Georgia, in 1978, one of the new disease- and nematode-resistant indeterminate hybrid tomatoes tied to 10-foot poles gave us lovely ripe fruits from June 1 until frost in mid-December. Twenty years ago we rarely had tomatoes for more than one month.

Forage breeding has not received the attention or the support given our crop and food plants. However, a few well-supported programs continuous for over 40 years have made great improvements in the yield, dependability, quality, and overall performance of several forage species. The result of this effort has been more meat and milk at a lower cost to the consumer.

What has plant breeding done to help satisfy man's other needs?

Breeders of cotton (*Gossypium hirsutum*) and flax (*Linum usitatissimum*) as they have improved the yields and quality of their crops have reduced the cost and improved the quality of man's cotton and linen clothing. Forage breeders, who have bred better forages and increased the efficiency of the livestock consuming that forage, have lowered the cost of the wool and leather products that we use.

A century ago most of man's shelter was made of wood. As the population explosion and man's affluence have increased by manyfold the demands for building materials, the world's supply of timber has rapidly disappeared. Today, cutover timberlands are being replanted with seedlings from selected trees. The increased growth rate and production efficiency of certain tree hybrids have been demonstrated. Plant breeding projects designed to develop superior cultivars of our timber species are receiving modest support and will one day produce the same improvement in timber production that has been achieved in crop production. When one considers the years required to complete a breeding cycle and grow a tree, neither you nor I can expect to reap much benefit from this effort. It is gratifying to know, however, that plant breeding will help to satisfy the lumber needs of our grandchildren.

For most of his existence on earth, man has depended on wood to supply his fuel. More than a billion people still depend on wood as their principal energy source. Millions of people unable to get wood now rely on cow dung as fuel to cook their meals. Wood for fuel is a necessity in the world. However, plant breeding has contributed little to meeting this need to date, but the poplar (*Populus* spp.), sycamore (*Planatus* spp.), sweet gum (*Liquidambar styraciflua*), and other species with fuel potential exhibit the same variability and response to plant breeding as our crop plants. Cloning outstanding trees selected from the wild is feasible with most of these species and is being used to plant improved forests for fuel. Plant breeding will help to satisfy our children's need for fuel.

Dwindling oil supplies and man's increasing demand for energy are stimulating an assessment of biomass as a possible energy source. Preliminary studies suggest that man's need for land and the cost of biomass production will keep it from supplying more than 5 percent of man's total energy needs. In Georgia, 'Merkeron' napiergrass (*Pennisetum purpureum*), the best vegetatively propagated F_1 hybrid, set on 0.9 m centers in the field in May, had reached a height of 4 m by November and produced 28.5 T/ha of dry matter. With a full season to grow, perennial grass probably can produce 50 T/ha. The total production was harvested with the same forage chopper used to fill silos with corn. Dry matter contained 1 percent of N and K and 0.12 percent P. Such hybrids can be fertilized with sewage effluent and sludge, disposing of these waste products as they produce biomass for energy. Thus, the plant breeding that developed Merkeron napiergrass may help to meet two of man's needs, energy and waste disposal.

In the last 30 years, plant breeding has produced more than 50 improved turfgrass cultivars. The 'Tif-bermudagrass' (*Cynodon dactylon*) hybrids, for example, have revolutionized games such as golf, bowling, and football played on turf in the subtropics of the world. Fifty-three of Hawaii's 55 golf courses now use 'Tifdwarf' or 'Tifgreen' bermudagrass instead of bentgrass (*Agrostis* spp.) on their greens. Miami's Orange Bowl has replaced its artificial football field turf with 'Tifway' bermudagrass. Plant breeding has helped to meet the recreational needs of millions of people by giving them better turfgrasses.

Sod-forming grasses such as tall fescue (*Festuca elatior*), bermuda, and bahiagrass (*Paspalum notatum*) are conserving millions of acres of erodable land that would be full of gullies. The soil under these grasses is producing feed for livestock instead of polluting rivers and destroying reservoirs. Plant breeding that develops improved cultivars of these grasses increases man's food supply today and protects his soil for posterity.

The contribution that plant breeding has made in beautifying our environment with thousands of new ornamental plants defies any written assessment. Seventeen hundred plant breeders in The Netherlands have developed 4,000 gorgeous named cultivars of tulips (*Tulipa* spp.), species of which grow wild in Turkey.

Finally, plant breeding, by increasing food production in the world, has brought peace to some of mankind: peace in the form of freedom from riots by the hungry; peace by reducing pressures that lead to war; peace in the form of freedom from the fear of starvation; and peace of mind for those who have shared their food with the hungry.

This week we have heard 11 excellent papers describing the problems of the plant breeder and suggesting methods for solving them. These papers have been discussed by highly competent scientists, so there is no justification for reconsidering what has been discussed already. It is better to delineate the task that lies ahead and suggest emphases that will deal with the problems successfully.

PROSPECTS FOR THE FUTURE

The world's population will increase; nothing short of a holocaust can stop it. Programs designed to slow its rate of growth can only delay for a few years the time required to double the number of mouths to be fed. Whether it takes 25 or 40 years, the world's agriculture must double its food production for man to eat as well as he eats today. If man is to live as well as he has in the past, agriculture must expand its production of clothing, shelter, fuel, and energy, and promote soil conservation, recreation, and environmental enrichment. Most of the world is demanding more than it has today. The task seems almost insurmountable. How can the challenge be met?

Agricultural productivity and income must be increased. Wortman and Cummings (1978) tell us that farmers, regardless of the size of landholding, generally will increase their productivity provided the following four requisites are met:

1. *An improved farming system.* A combination of materials and practices that is clearly more productive and profitable, with an acceptably low level of risk, than the one he currently uses must be available to the farmer.

2. *Instruction of farmers.* The farmer must be shown on his own farm or nearby how to put the practices into use, and he should understand why they are better.

3. *Supply of inputs.* The inputs required, and, if necessary, credit to finance their purchase must be available to the farmer when and where he needs them, and at reasonable cost.

4. *Availability of markets.* The farmer must have access to a nearby market that can absorb increased supplies without excessive price drops.

In fact, Wortman and Cummings (1978) have described a basic strategy that will feed the world for the next 50 years. The evidence for my optimism can be found in the story of peanut production in Georgia (Burton, 1976).

From 1920 to 1949, "neglect" (little support from research, extension, or industry) was the principal constraint that kept Georgia's peanut yields around 800 kg/ha. Government price supports ranging from 75 to 90 percent of parity had supplied price incentives since 1941, but with low yields, the incentive was small.

In the early 1950s, research pathologist L. W. Boyle discovered that the contraint imposed on peanut yields by southern blight (*Sclerotium rolfsii* Sacc.) could be removed by complete deep burial of crop residue and controlling weeds without throwing soil on the peanut plant. Chemical weed. control methods developed by weed research specialist Ellis Hauser supplied weed control without throwing soil on the plant and increased flowers per plant and yield. In 1954, J. Frank McGill, an outstanding county agent who became a full-time extension peanut specialist, included these research findings in a production package and set about to overcome the human constraint that resisted change. Result and method demonstrations, demonstration clinics, annual peanut production schools in each peanut-growing county, and ton-per-acre clubs were a few of the strategies that helped sell the program. McGill's expertise and untiring efforts earned the growers' confidence that resulted in rapid adoption of constraint-removing practices coming from research. Georgia's peanut yield in 1960 was 1,340 kg/ha.

Intense research efforts to remove constraints (often identified by McGill and growers) developed better cultivars and improved insect and disease control. Industry added new pesticides and equipment to the ever-changing peanut production package. By 1970, Georgia produced 2,486 kg/ha of peanuts on 213,000 ha of land.

'Florunner,' a new cultivar with greater yield potential, was introduced in 1970 and occupied 93 percent of the hectarage by 1975. The constraint imposed by limited water on 20 percent of the sandiest peanut soils was removed by irrigation during this five-year period. These two breakthroughs added to the peanut production package and applied by growers highly receptive to new technology set a five-year world record for yield (increasing from 2,790 kg/ha in 1971 to 3,700 kg/ha in 1975). In just 25 years, Georgia's peanut yields have been increased 4.6 times.

The strategy is simple. It depends on people—dedicated people who care. There must be a leader—one who is energetic, knowledgeable, enthusiastic, and capable of gaining cooperation and respect. He must be anxious to devote his talents and energy to feeding people. In the Georgia peanut program, the leader was an extension specialist. In the Mexican wheat improvement program, the leader was a plant breeder, Norman Borlaug. Tomorrow's leader could be you.

There must be a team of capable research workers—specialists who make their top priority the discovery of ways to overcome biological constraints. There must be a reward for such efforts. Otherwise, they may be tempted to devote their time and talents to research that may contribute little or nothing to eliminating food production constraints. Too many scientists in some of the world's hungriest nations spend their time on such research.

There must be continued interchange between the farmer who applies the new technology to grow more food and the scientist who creates it. Laboratory solutions will probably need modification on the farm. The scientist must be made aware of new constraints as soon as they appear. Working closely with the farmer may be the best way for the scientist to discover the nature and scope of each new constraint.

The inputs required to overcome the constraint must be supplied to the farmer. Discovery of a chemical that will protect a crop from the ravages of a disease or some other pest will not save the crop. The chemical must be available, growers must be induced to use it, and someone must show them how. In countries like the USA industry assumes this responsibility. In the hungry nations, governments must accept this role.

Overcoming food production constraints costs money; hence financial resources must be supplied. If the private sector cannot provide financing at a reasonable rate of interest, governments must.

There must be incentives and there must be rewards for everyone involved in overcoming constraints. For the farmer, particularly, there must be financial rewards. He cannot be expected to assume the extra financial risk without being assured of a reasonable profit. The magnitude and certainty of this profit will determine the rate at which the new practice will be adopted. In most of the world, governments must supply the fiscal incentive; yet there are other incentives. For many, the satisfaction of helping to feed the hungry will be reward enough. Recognition by government or peers as the best or one of the best workers in a particular discipline can be a strong motivating force. Production contests, citation of winners, and an analysis of how they succeeded can be a great incentive and an educational tool in any nation or society.

In the strategy just described for feeding the world in the next 50 years, plant breeding was mentioned only once, and for good reason. The plant breeder working alone cannot develop the improved farming systems required to feed the world. Neither can these systems be developed without him. The plant breeder must produce cultivars capable of meeting man's needs in environments with stress situations seldom experienced today. He cannot do it alone. He must become a part of a team that includes farmers and members from every discipline involved in developing the improved farming system. He must be able to work with people. Plant breeding candidates who cannot should be encouraged to find another profession.

At Plant Breeding Symposium I, my prediction was that "The successful plant breeder and his profession will increasingly enjoy one of the most honored positions in society" (Burton, 1966). The selection of Norman Borlaug as recipient of the 1970 Nobel Peace Prize substantiates that statement. However, the reduction in plant breeding positions and support, and the marked increase in genetic engineering research, casts doubt on the validity of this statement today. Budget builders responsible for this shift in support need to know that genetic engineering is only a potentially useful tool. It cannot feed people except as an adjunct to conventional crop improvement. Jennings (1976) expressed his concern: "Just when he is most needed, the plant breeder working in the field has been sidelined by 'falsely sophisticated' work in computerized laboratories." Certainly, Jennings would endorse strong support for "genetic engineering" and basic research of the type reported at this symposium but not at the expense of plant breeding.

Administrators and economy-minded leaders who advocate the reduction of support for plant breeding should realize that:

1. The limits to crop improvement by conventional plant breeding have not been reached.
2. Only conventional plant breeding can produce the cultivars required to feed the world.
3. Effective plant breeding requires continuous support adjusted for inflation. An effective plant breeding program cannot be built and maintained with intermittent funding.
4. Strong continuous plant breeding programs offer the best solution to the genetic vulnerability problem.
5. Breeding better cultivars takes time. We cannot wait until we are hungry to start.
6. Years are required to build an effective plant breeding program.

Plant breeders helping to develop improved farming systems for the next 50 years will be some of the busiest people in the world. The needs and potentials will provide a tremendous challenge. The plant breeders' opportunity to serve millions of people by creating better cultivars can bring them much satisfaction and widespread recognition. The following remarks are presented primarily for them.

Tomorrow's plant breeders must keep up to date. They cannot spend their lives using exactly the same procedures they learned as students. Neither can they read all of the cytogenetic and plant breeding literature. If they do, they will have no time to add their contributions. Review articles, symposia, and discussions with leaders in the field will help to supply useful ideas. Plant breeders do not need to know everything in the literature. They do need to know how to reach their breeding objectives most efficiently. Properly designed experiments to test new breeding procedures can improve their efficiency and provide research papers needed for promotion without detracting from the breeding program. Staff scientists and technicians should continually bear in mind that "we have not found the best way to do anything." It has improved our breeding and testing procedures and paid real dividends in the form of better cultivars and publications.

Tomorrow's plant breeders should be endowed with an insatiable curiosity. If they are, plant breeding questions for which there are no immediate answers will be raised. When this happens, I think breeders must provide energetic and flexible programs to try to find the answers. It has been my privilege to visit a number of young plant breeders in recent years. When they ask me a question I cannot answer (and they do this frequently), I reply, "I don't know. I'd like to know. Why don't you try it?" And so to tomorrow's plant breeders who must feed the world, let me say: "When you get an idea you think may work, don't just wonder—TRY IT."

Efficiency in plant breeding may be defined in more than one way. The commercial plant breeder may measure it in terms of cost per unit of advance. The hungry world will be more interested in gain per year. Hallauer (Chap. 1) presented an excellent, in-depth discussion on recurrent selection. The pros and cons of phenotypic and progeny selection were set forth. Tomorrow's plant breeder must constantly consider how these ideas can be used to increase the efficiency of his plant breeding operation. How can his objective be reached in less time? We have raised these questions as we have tried to improve mass selection with our recurrent restricted phenotypic selection (RRPS) method (Burton, 1974). Perhaps a discussion of its development will indicate how plant breeding research and the development of superior cultivars can be combined.

We chose mass selection as a method of improving forage yield in Pensacola bahiagrass, a perennial pasture grass, because we believed it had potential. We knew heritabilities on a single plant basis for yield in forages are usually low and suggest breeding methods other than mass selection. We believed, however, that regular mass selection, depending on open-pollinated seed for the next cycle, could be improved, and we determined to try it. Yield potential of spaced plants in the field must be evaluated under the most uniform conditions possible. This meant planting polycross seed of each selection at uniform rates in flats of sterile soil in the greenhouse in December so seedlings could be transplanted to 5-cm clay pots in late January and set in the field in April. Careful transplanting in a 0.9 × 0.9 m spacing in a uniformly fertilized field with adequate irrigation for establishment was also required.

Because winterhardiness is important, we decided that all selected plants should have survived one winter.

Two populations were studied. The wide-gene population

(WGP) consisted of a mixture of equal quantities of seed from 39 south Georgia farms. The narrow-gene population (NGP) consisted of seed harvested from 75 F_1 plants of 'Tifhi-2,' a high-yielding 2-clone hybrid comparable to a double cross in maize. Remnant seed of these populations was placed in cold storage along with seed of each cycle for future testing to measure progress. We have usually grown 1,000 spaced plants of each population in each cycle.

The restrictions imposed to improve efficiency were: (1) selecting the five highest-yielding plants in each 25-space plant square block to reduce soil heterogeneity effects; (2) intermating the selected plants in isolation to double the rate of advance over open-pollinated mass selection, (3) placing two culms (ready to flower) from each selection together in water under one large bag with daily agitation to insure intermating of all selections; and (4) avoiding unequal representation of the parents in the polycross seed by intermating two flowering heads from each selection.

Correlation coefficients between first-year spaced plant yields of 25 selections from NGP and WGP and their five-year yields in replicated clonal plots were 0.43 and 0.39. This test proved that our space-planting method would permit screening Pensacola bahiagrass for forage yield potential in the first year.

Forage yield of Pensacola bahiagrass has increased at a uniform rate of 2.5 percent per cycle in seeded plots and 8.7 percent per cycle as spaced plants. The top three spaced plants in cycle 6 yielded 40 percent more than the top three spaced plants in the original populations. The coefficients of variability for forage yields of 125 spaced plants of cycles 4, 5, and 6 grown in 1978 were 33.0, 34.7, and 31.3, respectively, indicating that continued advance from RRPS may be expected. Means and CVs for NGP and WGP have not been different in the last three cycles. The space planting of cycle 6 in which blocks of NGP and WGP were interplanted is the source of breeders' seed for a potential new cultivar.

A diallel among nine of the best clones from cycle 5 has given 3-year average forage yields ranging from 9.6 to 15.9 T/ha. Cycle 5 of NGP and WGP yielded 11.3 and 12.1 T/ha, respectively. Self-sterile cross-fertile clones 15-21 and 17-37 (parents of the top-yielding hybrid) planted vegetatively in alternate strips in seed fields should produce F_1 hybrid seed for a number of years by harvesting all seed produced.

Studies designed to improve the efficiency of RRPS in breeding Pensacola bahiagrass revealed that 90 percent of the five better plants in a 25-plant block at the end of the first year were also better in two-year performances. Further study showed that a visual selection of the five better plants in each 25-plant block made in late July of the first year were almost always the higher-yielding plants when cut and weighed in October of the first year. In mild winters, no plants were lost because of winter injury, but in the severe 1977-1978 winter, 21.7 percent of the better first year plants were discarded because of winter injury. Plants discarded for winter injury had yielded 680 grams per plot compared with 653 grams for all plants saved.

Ideally, the intermating in each RRPS cycle should produce a diallel so all recombinations could be tested for yield. For the 200 plants selected, 19,900 crosses would be required. How could one test the yield performance of these 19,900 diallel crosses assuming the polycross seed from our RRPS intermating system represents them? Well, research demonstrated the existence of a small positive correlation between seedling vigor and yield of mature plants. So we planted enough polycross seeds to give 100 seedling plants from each selection in a 45-cm row in a flat in the greenhouse and selected the five most vigorous seedlings to transplant to 5-cm clay pots. This permitted the screening of 20,000 seedlings from each cycle for yield.

By starting each cycle earlier in the greenhouse and carefully transplanting them to the field in fertilized soil fumigated with methyl bromide for weed control, all good plants can be made to flower in sufficient quantity by late July of the first season to permit making an RRPS polycross.

It costs $750/acre to fumigate soil with methyl bromide. It is expensive but no other treatment will control weeds so effectively and allow the good plants to flower well enough to make a selected polycross in the first year. A closer spacing than 0.9 × 0.9 m would cut fumigation costs. Tests of 0.6 × 0.6 m and 0.75 × 0.75 m spacings proved the 0.6 m spacing to be inadequate, but 0.75m was satisfactory.

Thus, through continuous study, we have doubled the efficiency of the RRPS breeding method applied to Pensacola bahiagrass. Modifications responsible for improving the RRPS efficiency include:

1. Screening for seedling vigor as well as mature plant yield to increase the population screened from 200 to 20,000.
2. Spacing seedlings in the field 0.75 × 0.75 m apart instead of 0.9 × 0.9 m to reduce land and fumigation costs by 30 percent.

3. Improving cultural procedures to permit making the RRPS polycross in July of the year planted.

4. Retaining the selection number with its polycross progeny so progeny of selections with poor two-year performance or inadequate winterhardiness may be eliminated from the field planting and their maternal effect on the population be removed. This procedure permits one cycle per year instead of one cycle every two years. Progeny discarded have ranged from 5 to less than 25 percent.

5. Using visual selection in late July of the first season to select the four top-yielding plants (instead of five) to produce the RRPS polycross for the next cycle.

Have you ever wondered how many intermated selections fail to contribute to the next cycle in a recurrent selection program? By keeping the mother plant number with each polycross plant set at random in the field, we were able to answer this question. The test involved 165 and 160 selections of NGP and WGP Pensacola bahiagrass, respectively, each thoroughly intermated the RRPS way. Seven of the most vigorous seedlings in 5-cm pots from each mother plant were placed individually in numbered paper bags and randomized in seven groups for each population before they were space planted in the field.

In late July, the top five plants in each 25-plant square block were selected to produce the next polycross. Later, when we checked the maternal parent of these selections, we discovered that 27.9 percent and 25.8 percent of the mother plants in the NGP and WGP populations had been omitted. The paternal influence of these plants on the next cycle could not be ascertained. Eliminating approximately one-fourth of the maternal effect in each cycle of improvement must be narrowing the gene base. One can only wonder how long it will be before inbreeding effects will offset any increases in yield from applying RRPS to these populations.

Elmer Johnson, maize breeder with CIMMYT in Mexico, has reduced the height of Tuxpeno maize 50 percent with 13 cycles of recurrent selection. Lonnquist and others have used recurrent selection to develop earlier maturing populations in maize. The success of such efforts may have caused some to overlook other breeding tools that may be more efficient for certain objectives.

The recessive d_2 gene in pearl millet (*Pennisetum glaucum*) when homozygous reduces the length of all internodes but the peduncle and reduces plant height 50 percent (Burton et al., 1969). A single recessive gene e_1 in pearl millet that occurred as a natural mutant, when homozygous in 'Tift 23DB,' reduces days to flowering of May and early June planted millet from 80 to 45. Three other single recessives introduced by mutation, when homozygous, hasten flowering of their normal inbreds by three weeks when planted in May and June. The recessive *tr* gene in pearl millet that removes all trichomes when homozygous alters the cuticle to reduce water loss from the leaves up to 35 percent (Burton et al., 1977). During the drought at Tifton, Georgia, in 1978, 'Tift 23DBS,' homozygous for the *tr* gene, was more drought-tolerant and made more growth than its isogenic counterpart Tift 23DB. Do similar genes occur in other cereal grasses? Have we really looked for them? Could they be induced by mutation breeding?

Reducing the height of an inbred or a cultivar by backcrossing can be accomplished much faster and with less likelihood of character loss or shift than by recurrent selection. Three to five years are required to complete a backcrossing operation. Can it be done in less time? Table 11.1 shows the minimum time and effort required to place a single gene into a recurrent parent.

Five backcrosses plus one generation to segregate out recessive or homozygous dominants should combine a donated gene with 98.4 percent of genes from the recurrent parent. Linkages cannot reduce this percentage much. All but the last generation could be carried out in a greenhouse. A minimum of ten hybrid seeds of each back-

Table 11.1. Generations required to rapidly combine a donor (Dd) gene with various percentages of the recurrent parent (RP) genes with a success of P = .999.

Generations	Percent RP Genes Recovered	Manipulation
1	50.0	Dd x RP
2	75.0	BC Dd to RP
3	87.5	BC Dd to RP
4	93.7	BC Dd to RP
5	96.9	BC Dd to RP
6	98.4	BC Dd to RP

Self 16+ plants of the final backcross, grow 24-plant progenies of each and intermate the recessives or homozygous dominants. Backcross RP to dominant F_1.

Self and backcross 10 F_1s in each generation and grow 24-plant progenies of each self at the same time as the F_1s to discover heterozygous-recessive-carrying F_1s for the next backcross.

Mixed pollen from a number of plants of a recurrent variety backcrossed to more F_1s will be required in the later generations to retain the variation in a highly variable variety.

cross will be adequate for inbreds or self-pollinated species. The years required will be determined by the generations per year. How many generations of your crop do you grow? How many could you grow? For pearl millet, a summer growing annual, we used to say two generations. Now we can say four generations per year, thanks to plant physiology research. Physiological research revealed that millet that flowers in 80 days in midsummer will flower in 45 days if subjected to 8-hour days and temperatures above 32°C (Hellmers and Burton, 1972). An inexpensive automatic dark box provides the 8-hour days when needed (Burton and Stansell, 1971). A 1-hour soak in a water solution of 1 percent 2-chloroethanol plus 0.5 percent sodium hypochlorite breaks seed dormancy in pearl millet, permitting immediate germination of freshly harvested seed (Burton, 1969).

We are presently following the procedure outlined in Table 11.1 to transfer the *tr*, d_2, and e_1 genes to several inbred lines and two variable African cultivars of pearl millet. There are several questions we hope to answer from this research. Do five backcrosses that can recover up to 98 percent of the recurrent parent's genes provide the same quality and performance as the recurrent parent and thus eliminate or greatly reduce costly time-consuming testing? We think it should. How many plants of a variable cultivar must be involved in each backcross to recover its characteristics? Must our theoretical estimates be modified?

Can the backcrossing procedure outlined in Table 11.1 be used to transfer monogenic pest resistance to outstanding widely adapted F_1 hybrids or cultivars? With an adequate screen, the answer should be "yes." Much of the pest resistance that has protected our crops to date has been monogenic in its inheritance. Parlevliet (Chap. 9.) has told us that monogenic disease resistance tends to be race specific but may be nonrace specific.

Protecting the world's crops from new pests will be a major assignment for tomorrow's plant breeders. The pressures for food and other plant products will require the breeder to protect each victimized crop as rapidly as possible. The new protected cultivar should be as productive as the victim and will be accepted most readily if it is like the victim in quality and most other traits. How better can these objectives be realized than via backcrossing?

Increasing genetic variability in a cultivar usually reduces vulnerability to pests. It can rarely be achieved, however, without a loss in performance and acceptability. It has been estimated that 30 percent of the maize acreage in the USA is presently planted to slight modifications of a single genotype, B73 × Mo17. It is more vulnerable than the double crosses it replaced, but farmers will continue to grow it or a better single cross until losses due to pests greatly exceed those presently realized.

A number of strategies for protecting tomorrow's crops from new pests have been suggested this week. Here is another strategy that involves the backcross.

1. Learn how to maximize the number of generations of each crop that can be grown in a year as soon as possible.

2. Locate active breeding programs with capable pathologists, entomologists, nematologists, and others, strategically throughout the world.

3. Develop a thorough pest monitoring system at the breeding sites and an intercommunication system to report the occurrence of a new pest to all breeders of the crop.

4. When a new pest appears, screen the world's germplasm for sources of resistance as rapidly as possible. If feasible, do this at the plant breeding station discovering the new pest to slow its rate of spread. Because resistance genes may be recessive as well as dominant, the world's germplasm (seed) collections of cross-pollinated crops should be in an S_1 condition (if possible) to facilitate immediate screening for both dominant and recessive genes. Logical mixtures of several sources will reduce the labor involved in screening.

5. Start immediately to transfer the resistant genes to the best cultivars or hybrids by backcrossing.

6. At the same time, study the new pest—its effect on the host and the inheritance of resistance.

7. Make the source(s) of resistance available to other plant breeders so they can start a backcrossing program with their superior cultivars or hybrids.

The next century will see F_1 hybrids of every major crop in general use in most of the world. The greater yield potential, efficiency, and ability of the outstanding F_1 hybrid to withstand stress when compared with cultivars will be needed. The uniformity in maturity to reduce losses caused by birds, in lodging resistance to facilitate harvest, and in quality to satisfy man's tastes will appeal to all farmers just as the outstanding maize single cross appeals to those who grow this crop today. Discovering practicable methods of putting the F_1 hybrid on the farm is a major challenge for those who work with autogamous crops such as soybeans and rice.

Can F_1 hybrids be used in developing agriculture? Many

say "no," but the answer can be "yes" if the hybrids are good enough. When testing throughout the pearl millet belt of India had shown HB1 (Tift 23A × Bil 3B) outyielded open-pollinated checks by 88 percent, many said, "It won't go. Pearl millet is a minor crop, grown without fertilizer by small farmers using oxen for power. They won't pay the high price for seed." But these farmers paid the high price for hybrid seed, and pearl millet grain production in India increased from 3.5 million tons in 1965 to 8.0 million tons in 1970.

Discovering usable apomixis in nonhybrid crops would offer an ideal method of putting the F_1 hybrid on the farm. Chromosomal and cytoplasmic manipulations of the type described here by Peloquin (Chap. 4) will surely help to develop methods for commercial propagation of F_1 hybrids. Innovative cytogeneticists and plant breeders will find a way, and the world's increased demands will readily cover any extra propagation costs.

Those breeding perennial plants should not overlook vegetative propagation as a practicable method of making the F_1 hybrid available to the public. Thirty-six years ago the leading pasture specialist in the USA advised us to discard Coastal bermudagrass because it could not be propagated by seed. Pastures were not propagated vegetatively in the USA in 1943. A conviction that vegetative propagation was possible and subsequent research have developed methods that have planted more than 4 million ha of 'Coastal' bermudagrass across the South. With today's bermudagrass sprig harvesters and planters, selective herbicides to control weeds, and a little know-how, it is possible to plant bermudagrass vegetatively on an extensive scale at a very reasonable price.

Most wide crosses are sterile. Some would have economic potential if they could be propagated vegetatively. Bob Buckner tells me that some ryegrass × tall fescue F_1 hybrids exhibit considerable heterosis and have better quality than tall fescue. These hybrids are sterile, and enough backcrossing to tall fescue to give good seed production loses most of the heterosis and quality in the F_1 hybrid. It is conceivable that pastures planted to a good ryegrass × tall fescue F_1 hybrid could produce 50 percent more animal product per ha/yr than has tall fescue.

Machines and methodology to successfully and economically plant outstanding genotypes of perennial bunch grasses and stoloniferous legumes would be a great boon to the forage breeder. He would no longer be concerned with progeny testing, seed yields, and commercial seed production of new cultivars. He would only need to produce and test a superior genotype. Sterility would generally enhance its value. Increase and distribution could be easier than you may think. Perhaps a description of the steps involved in the increase and distribution of 'Tifton 44' bermudagrass will be helpful.

Tifton 44 bermudagrass is the best of several thousand F_1 hybrids between Coastal and a winter hardy bermuda available from Berlin, Germany, in 1966. Six years of testing had shown that Tifton 44 was earlier, finer stemmed, 6 percent more digestible, more winterhardy, and equal to Costal in yield and other characteristics. In two years of grazing trials with steers and feeding pellets to steers, Tifton 44 had given 19 percent better average daily gains than Coastal. Thirty agronomists in 14 southeastern states had reported excellent agronomic performance and winterhardiness for Tifton 44.

A 1/4 ha breeder's nursery was planted to Tifton 44 and its description was given to the agricultural press. Letters requesting planting stock were answered with a form letter stating that only people qualifying as certified growers could get sprigs. To qualify, their land for growing sprigs had to be inspected by their state crop improvement association and found free of bahiagrass, bermudagrass, and nut sedge. A letter from their crop improvement association to the Georgia Seed Development Commission in Athens, Georgia, was evidence that they had qualified.

On April 15, those who had qualified were notified that breeder's sprigs of Tifton 44 would be available at Tifton for them to pick up on May 3. On May 2, we dug and packaged (in white plastic garbage bags) 500 bushels of sprigs from our nursery. By 2:00 P.M. on May 3, 122 farmers had paid $20 for four bushels of sprigs and were on their way home, some as far as central Texas. Most growers planted about 0.5 ha with their four bushels of sprigs. By fall, one grower had increased his area planted to 3 ha.

Sprigs were distributed to more qualified growers on June 21, to bring the total to 263. These growers are now producers-sellers. Many are also custom planters. Lists of these growers have been prepared and sent to extension specialists in each state. They are also used to answer inquiries for sources of sprigs.

Tomorrow's plant breeders will need to use all their tools. One tool overlooked by many today is mutation breeding. It is not a panacea for all plant breeding problems as many once thought. But it has a place that should not be ignored. Parlevliet (Chap. 9) has mentioned its use in creating a disease-resistant mutant of a special genotype of Mitcham mint. At CIMMYT, sorghum breeder Vartan Guiragossian showed me induced mutants flowering much earlier than the earliest material in his germplasm collection. Our three induced pearl millet mutants that flower three weeks earlier than their parent inbreds were mentioned earlier.

'Tiffine,' 'Tifgreen,' and 'Tifway,' widely used for turf, are sterile triploids. 'Tifdwarf,' the best of the lot for golf and bowling greens, is a natural mutant. Why not step up the natural mutation rate of these sterile heterozygous hybrids with a mutagenic agent? By exposing dormant rhizomes to gamma rays from a Cobalt 60 source, we produced 158 mutants of Tifdwarf, Tifgreen, and Tifway (Powell et al., 1974). Much testing has reduced this number to nine that appear to be equal to their parents in most characteristics and superior in at least one.

Our research that proved pollen shedders in our cytoplasmic male sterile (cms) Tift 23DA were normal true breeding Tift 23DB maintainers caused us to try to reverse the process (Burton, 1977). Seed of Tift 23DB inbred treated with ethidium bromide (EB) gave rise to cms mutants in the M_1 generation (Burton & Hanna, 1976). Later EB induced cms mutants in sorghum, cereal rye (*Secale cereale*), and soybeans. One sorghum cms mutant appears to be stable. All EB-induced cms pearl millet mutants, like many naturally occurring cms mutants, throw too many pollen shedders for commercial hybrid seed production. However, we now have eight cms mutants of Tift 23DB, induced by soaking seeds in water solutions of streptomycin and mitomycin. These were as stable as Tift 23DA in the summer of 1978.

Successful mutation breeding requires that many thousands of mutagen-exposed seeds or plant parts be screened. The better the screen, the greater the success.

Axtell (Chap. 10) has given a good resume of the successes and the problems associated with breeding for improved quality. I believe it has a tremendous potential if it can reach the consumer. The world needs protein. Nebraska's 'Lancota' wheat bred by a team made up of a breeder, physiologist, and chemist contains 10 to 20 percent more protein than cultivars generally grown. If other wheat cultivars grown in Nebraska were replaced with Lancota or a similar cultivar, the state would produce an extra 45 million kg of protein at no extra cost. But the market generally pays no premium for extra protein, and Lancota is not quite as winterhardy as the cultivars generally grown. As a consequence, it occupies only 2.5 percent of Nebraska's wheat acreage.

Frequently it seems that yield, dependability, or an acceptable set of food characteristics are sacrificed for improvements in quality such as increased lysine content. The consumer then refuses the high-lysine cultivar because he can see no need to sacrifice taste for a characteristic he cannot see or feel. The challenge is to improve one characteristic without altering others that the market demands. Can such an objective be realized? Mutation breeding and repeated backcrossing are two tools that might help to meet such a challenge.

Tomorrow's plant breeders must be sure that their objectives are sound. As they weigh their objectives, they must ask: Is this objective important? What economic weight does it have? Is it compatible with maximum performance? An example of a mistaken objective was the 1920 wheat breeder's ideal of awnlessness. Those of us worn sore by awns inside our clothes as we worked in a header barge had created the demand. None of us knew that awnlessness was incompatible with maximum yield. Thirty-five years passed before Atkins and Norris (1955) demonstrated with 10 isogenic lines that removing awns reduces wheat yields an average of 4.1 percent and over 8 percent in dry years. How much this mistake cost USA wheat growers, no one will ever know.

Wilson (1980) concluded his analysis by indicating the need for more isogenic material in studying the importance of morphological and physiological traits. The plant breeder could produce isogenic material and would probably do so if the physiologist would make him a cooperator and a coauthor of publications resulting from its use.

In Plant Breeding Symposium I (Burton, 1966), "near-isogenic populations" which could be created in less time and with less effort than isogenic lines were suggested. We used near-isogenic populations to learn that by delaying maturity in pearl millet 25 days, we could increase yield from 9 to 19 percent and increase persistence, leafiness, protein content, and digestibility of the forage produced (Burton, et al., 1968). "Near-isogenic populations" could enjoy wider use than they have received to date.

Plant breeding is a numbers game. Other things being equal, success will be directly proportional to the volume of material handled. A new assembly-line technique that will double the number of plants or hybrids that can be screened and evaluated can contribute more to the development of a superior cultivar than can many of the research papers being published today.

The importance of increasing the precision of plant breeding and quantitative genetic research cannot be overemphasized. Well chosen experimental designs improve experimental precision. The 9×9 lattice square design that we have used for many years to evaluate pearl millet single crosses has lowered CVs to an average of 6.6 percent. Perhaps more important, they have saved an average of 1.4 of the top five and 2.4 of the top ten crosses assuming that lattice square adjusted yields are more nearly "true" yields of the entries tested.

Fields with good reasonably uniform soil are one of the most important of all tools used by the plant breeder. Quantitative genetic relationships cannot be accurately measured in fields previously used for fertilizer investigations, crop-rotation studies, and the like. Land close enough to terraces to differ in topsoil depth, water accumulation, or runoff is not suitable for plant breeding or quantitative genetic studies.

Fields to be used for plant breeding and genetic research should be uniformly prepared, fertilized, and cropped. Alternating fields used for grass research with a well-adapted annual legume, such as the velvet bean (*Stizolobium deeringianum*), will help control grass pests and leave the soil more uniform for the next grass crop.

A plant breeder needs hands—many more than his own—to carry a full program. Some hands can belong to high school and college students. More than 200 have helped me in the past 43 years. Other hands, invaluable to me, have belonged to technicians—honest, dependable, capable, hardworking high school graduates trained by me. Two have been with me more than 20 years, another, 10 years. These men are more than technicians—they are thinking assistants. Their help is essential.

I have yet to learn how to breed better grasses in my air-conditioned office. Attempts to do so when the temperature and relative humidity in the millet fields have been in the 90s have never been successful.

In January, 1979, I spent three interesting days in a conference room of the Office of Technology Assessment (OTA), Washington, D.C. Four plant physiologists and I were there to consider limits to photosynthetic production and synthetic biological systems as a part of OTA's assessment of energy from biological processes. The experts on photosynthesis expressed doubt that the basic photosynthetic process could be improved. The relationships between yield and photosynthesis, light respiration, and dark respiration in C_3 and C_4 plants were discussed. The failure of any one of the processes to be highly correlated with yield was reported. During their discussions, I asked the following questions: Why should a process as complex as growth (producing yield) be closely correlated with one or two of its contributing processes? If the correlation were perfect, could these processes be measured as precisely and as cheaply as yield itself?

If yield is important, and it usually is, few tools are as efficient as the trained eye for selecting the better plants among thousands of individuals. To isolate the best from the better plants, study the activities of men like Orville Vogel. Recognizing that plant breeding is a "numbers game," Vogel developed special machines and techniques to plant and take yields from many times the number of plots most plant breeders handle. His success is an excellent endorsement for his methods.

This discussion cannot close without asking, "Who will train tomorrow's plant breeders?" Jennings (1976) tells us, "For better or worse, we export not only our goods, but our values. It is disheartening to see, in developing countries, young graduates who are experts in the mysteries of genetics and computer science but inexperienced in old-fashioned field work, trying to raise crop yields." He also says "excellence in graduate training in applied breeding exists in only a few unitversities." I do not know how many of you are faculty members in one of those excellent programs. But if you are, you can make it better if you try. If your university does not warrant a rating of excellence in graduate training in applied plant breeding, please do one of two things. Either make the program excellent or send your applied plant breeding candidate to a graduate school that has such a program. The future of a hungry world is at stake.

May I summarize my remarks with six words that should describe a plant breeders's activities for a long time? They are:

Variate ←
Isolate → Intermate
Evaluate →
Multiplicate
Disseminate

Forty years ago, the five words at the left would have been considered adequate. The word "intermate" that briefly describes recurrent selection, has been set off to the right to give it emphasis. The arrows show how it fits into a cyclic process. For those characteristics with good heritability on a single plant basis, intermating can involve superior phenotype isolated from a variable population grown in a suitable environment. For characteristics such as grain yield, only those selections proven superior in evaluation tests should be intermated to start another improvement cycle.

You will note that all of these words are verbs. They are action words. It will always take action-work-to breed better plants. The subject for these verbs is only implied and if spelled out would be you—every person who is a plant breeder.

You will, no doubt, find assistance in the multiplication and dissemination of your improved cultivars. You may argue that these phases of the operation are no concern of yours. In reply, I would ask, Who has more at stake than you? How can the investment in you and your program be paid if the product of your work never reaches the ultimate consumer?

May I remind us again that primitive plant breeders had no knowledge of genetics, sex in plants, genes, DNA, and RNA. They had no wirtten language, no libraries, and probably no words for the ideas expressed in the six words above. Their germplasm was probably more restricted than ours. They had no microscopes, computers, nor any sharply honed tools.

What did they have? They had their plants. They lived with them and knew them as few of us know ours. They had their hands and they used them. They worked as if their very life depended on it. It did—and we might get more done if we were similarly motivated. Finally they had time—more time than the hungry world will give us. We must be more efficient than they.

Primitive plant breeders took what they had, worked hard, and made tremendous advances in plant improvement by our profession—plant breeding. We can and must do likewise.

Discussion

W. L. BROWN
S. FONSECA MARTINEZ
B. J. ZOBEL

1. W. L. BROWN. The role of plant breeding in meeting human needs must vary because human needs vary, so the breeder's goals and procedures are quite different, depending upon the particular segment of the population he is trying to serve. For example, the state of agriculture in most of the developing world is such that the primary needs are greatly increased yields, adequate disease resistance, and a product that conforms to long established local food standards, if the crop is used directly for human food. The farmer operating in a state of primitive agriculture is not impressed by a 10 percent increase in yield. Such farmers, however, will respond to yield increases of 50 to 100 percent, values which are still attainable in parts of the developing world.

Conversely, the farmer of the developed world, who is accustomed to a highly sophisticated agriculture, will recognize the value of a 10 percent yield increase, providing that increase is reflected in a harvestable yield. On the other hand, such farmers would not be impressed with yield increases of 25 percent or more if at harvest most of the increase is on the ground, or if for any reason, the improved yielding cultivar is unadapted to their highly mechanized system of farming. My point is that breeding objectives and procedures designed to achieve those objectives must be flexible and vary greatly depending upon the state of agriculture in which a breeder is operating.

Earlier, it was suggested that 30 percent of the maize acreage of the USA presently is planted to the single cross 'B73 × Missouri 17,' or slight modifications of it. An estimate from a 1975 survey showed that this hybrid was produced in sufficient quantities to plant about 7 percent of the USA maize hectarage. Today, this value is probably not greater than 10 percent.

2. S. FONSECA. The past progress of meeting human needs through plant breeding can be summarized in three words, "The Green Revolution." And, the prospects for the future also can be summarized in three words, "The Green Evolution." Plant breeders have been very successful in dramatically contributing to increased crop production as has been amply shown by the bountiful 1977 and 1978 crop harvests. Production in these two years has brought world grain stocks to near adequate levels; yet millions of people will die prematurely in the next decade from malnutrition and starvation. Therefore, careful consideration should be given to improved food production and distribution in the countries where deficits exist.

As much as 45 percent of the total cereal consumption in Central America and the Caribbean is imported now, and with an average annual increase of 2.7 percent in production, the deficit could reach 4.5 million tons by 1985. This will be equivalent to nearly half of the food needs in those countries. In Central America alone, population will more than double before 2000, and over 70 percent of the staple food is produced by farmers with less than 35 ha. This traditional farmer generally uses primarily his own energy, or in the best instances, that of an animal, for farming activities. His farm is located on a hillside, and he usually cultivates more than one crop in the same field during the growing season.

Multiple cropping in time and/or space is a rule rather than an exception. Limited research efforts have been devoted to understanding the multiple cropping systems that have been developed by the small farmers over the centuries. Uncertainty of the environment, limited availability of outside support, and the need for family sustenance compel the farmer to prefer stable yields and low-risk cropping systems.

Studies on some agronomic aspects of the cropping systems used by the small farmers in Central America have shown that minimal tillage gives a canopy that protects the delicate balance between soil yield potential and the environment as well as saving hand labor at critical periods. Modifying crop arrangements in the traditional farming systems has increased yield significantly. For example, two systems under study at Turrialba, Costa Rica, show that with intermediate technology, yields of 843 kg of protein and 31,000 Mcal per ha per year could be obtained. However, the real breakthrough in improving crop production of the traditional farmer of the tropics is still to come, and it is in hands of the plant breeders.

The plant breeder, specifically, can stimulate the collection and evaluation of locally grown potential food crops, develop cultivars for highly intensive cropping patterns, develop genotypes to withstand stress conditions, and produce cultivars with a high nutritional value. How quickly the plant breeders can achieve the great challenge of meeting human needs in the tropics will depend on how well they can evolve the technology already available to confront the current factors and constraints of the tropics for improving crop production.

3. B. J. ZOBEL. I want to emphasize that the good life is more than food. What value is it to have a full stomach, if you don't have a place to live, if you can't cook your food, if you don't have clothing? There is often an overemphasis on food, but there are other considerations. What are you going to wear? How are you going to keep warm? What do you use for toilet paper? These are the types of things that go to make up a complete good life, and we must keep this in mind. Tree breeding is geared to help the good life.

Trees are long-lived plants, and breeding them requires much space and money for experimentation. Actually, the initial methods of breeding applied to trees are very simple. First, variations are discovered in the natural populations and assessed genetically. Then the useful variations are combined into genotypes which are reproduced through seed production into millions of propagules. The success of this simple methodology is shown by the 20 percent increase in productivity achieved in one cycle of breeding and planting of 200,000 ha per year of improved genotypes. Second generation gains are 30 to 50 percent.

Forest trees are very heterozygous and this aids in making trees adaptable to environmental changes. Another great advantage in breeding forest trees is the fact that once a genotype is developed, it can be reproduced through vegetative propagation. Not many realize that characteristics for adaptability are genetically independent from characteristics of economic importance in forest trees, so breeders can develop a broad adaptive base into a forest tree cultivar simultaneously with a narrowing of the base for economic traits. So breeders are producing improved lines without resorting to the narrow genetic base that often occurs in agriculture.

Where will emphasis be put in future forest tree breeding?

Because forest production will be relegated to marginal lands and sites, breeders must develop trees that will grow on these sites. Already, breeders have special strains for wet sites, dry sites, cold areas, disease areas, and the like; so they are familiar with developing trees for adverse conditions. Second, trees will continue to be tailor-made for special uses, that is, one strain of trees for newsprint, a second for tissues, a third for writing papers, and the like. A new objective will be to breed trees for maximum BTU production and for organic chemicals. In America, the use of wood to heat homes is becoming popular again, and in Brazil, most energy needed in the steel industry is supplied by charcoal.

The job of the forest tree breeder cannot be taken lightly. He is responsible for breeding trees that provide the raw material for a multibillion dollar industry. Once trees are planted, the die is cast for 20 to 30 years. Mistakes cannot be erased by discing up a crop. His decision will make or break a whole industry throughout the world, so he had better be right.

4. D. WOOD. It was suggested that budget makers should provide adequate resources to support plant breeding at a highly productive level, but it is doubtful that budget makers will permit plant breeders to proceed in the manner to which they have become accustomed. So perhaps breeders need to consider plant breeding programs in terms of modern management techniques that permit a degree of accountability between the breeder and the budget maker. What are the latest insights on the applications of modern management techniques to the plant breeding enterprise?

5. P. BUSEY. There are two ways in which management science can be applied to plant breeding. First, the vast amount of quantitative data available could be used to construct utility functions based on cost-benefit analysis, similar to ones used in modern decision analysis. Second, control systems analysis could be applied to plant breeding. This type of work involves flow charting which has already been done to make selection criteria more efficient. Both of these require cooperative research between various disciplines.

6. G. W. BURTON. Cooperation with others is very important. The number of persons who have coauthored papers with me exceeds 100. And these people were in many different disciplines. This cooperative effort has been tremendously helpful to me, it has been

helpful to them, and it has been a very rewarding experience for me to be able to solve problems more effectively than I could have done alone. Much of my cooperation is with animal scientists, because when working with forages the final judge of your success is the cow. If the animal scientists will say a good word for a new forage cultivar, it will carry more weight in getting that cultivar accepted than anything a plant breeder says.

The attitude a plant breeder must take is that he or she is willing to do more than his or her share without reservation. Make sure that everyone is acknowledged, that appreciation is expressed, that something worthwhile comes out of the effort, and most researchers will cooperate. Many of the problems that need solving are interdisciplinary in nature, so it will take teams of researchers to solve them.

7. W. L. BROWN. Probably most plant breeders utilize funds allocated for plant breeding research as efficiently as possible, but it is likely to become more important to use those funds even more efficiently. However, we must hope that the situation never deteriorates to the point where the breeder has to spend as much time determining how best to use funds as he spends at plant breeding.

8. K. RUSSELL. Dr. Burton stressed that there is a crucial need for funds for plant breeding research to be allocated in an intelligent and wise manner. The persons responsible for allocating funds for research do not seem convinced that what a plant breeder does is very important relative to other scientific endeavors. How do we as plant breeders make our voices heard and convice the policymakers that sound, well-developed breeding programs are not a luxury, but a necessity?

9. G. W. BURTON. Plant breeders do not do a good enough job of letting humankind or budget planners know what plant breeding can do and what plant breeding must do in order to meet the needs of the world. Dr. Borlaug has said that people who draw up the budgets take their information from the *Wall Street Journal* rather than from anything that plant breeders do.

10. R. BARHAM. My questions are many. When we reconvene in fourteen years, will we be able to assess that agricultural production has stayed ahead of population growth? If yes, do we again embark on another round of research to stay ahead of increasing popula-

tions? Or, by supporting increasing populations, are we creating conditions for even more massive hunger and starvation? But do we have a choice but to increase production to buy time for the world to work out its political, economic, and moral problems?

11. S. FONSECA. Certainly, humans must be fed now and in the future to the best of our ability to do so. Food production problems must be solved where they are, and the problem now is in many of the developing countries. Therefore, we must make an effort to use our knowledge to increase the production and productivity in these countries.

12. N. BORLAUG. We have an obligation as scientists to try to point out to our political and religious leaders what the balance is between available food and the human population on earth today and what it will be in the year 2015. There will be another doubling in population by that year if something is not done to stabilize it in a humanitarian way. This planet Earth has only certain carrying capacity if all humans are to have a reasonable standard of living. However, even if we are good salesmen of this message, increased population size of some magnitude will still occur; in the developing countries, where most of the overpopulation problems exist, the only reasonable approach to feeding these persons is through increasing yield on the land already under cultivation. If adequate food is not available in those countries it can and will lead to legal or illegal movements of people, which ultimately will lead to political chaos and upheavals.

13. E. A. CLARK. Sensational yield advances achieved in major food and feed crops in past years are attributable to crop improvement and to increased energy use in such forms as mechanization, fertilizers, pesticides, and irrigation. Now if we assume that our energy base is finite, what types of modifications will need to occur in cropping systems and how can plant breeders contribute to an agriculture that must operate within an increasingly energy-limited future?

14. G. W. BURTON. A group of five people in the OTA in Washington has estimated that biomass probably can supply only about 5 percent of USA energy needs because its production must compete with land and fertilizer use for production of food, clothing, and shelter. The time may come when people will be

willing to walk so they do not have to produce their food with hand tools. In other words, as long as energy is available at all, I believe it will be available to agriculture.

We desperately need to learn how to utilize fusion energy. Without it, there is little hope for humankind in 100 or 200 years. This generation will be looked upon as a generation that was extremely extravagant and wasteful of energy. For example, tremendous quantities of natural gas are being flared off of oil wells, never to be available again. This is a crime against our children and our children's children.

15. E. AYEH. At this symposium, it has been stressed continually that crop improvement is needed to aid in solving the world food problem, and further, this improvement needs to be done in the developing countries. Admittedly though, the state of agriculture, the population, and the state of technology in the various areas of the world differ; so does this mean that plant breeders in different areas of the world have different responsibilities?

16. S. FONSECA. Most plant breeders in the developing countries have been trained in developed countries; so most newly educated plant breeders have much the same skills and attitudes. However, they have the obligation when returning to their home countries to adapt to the real problems that confront them in those countries. They must take the technology already developed into account when breeding plants to feed the country's people.

17. R. MILLER. Can mutation induction provide any variability not present in our natural germplasm? Would a more efficient way of breeding involve screening and utilization of the vast variation we presently have in our germplasm?

18. G. W. BURTON. Mutations, artificially produced, will be no different from natural variants. Mutation breeding merely increases the frequency of mutations. It is a useful technique in the bag of techniques that plant breeders can use. Let me illustrate a case where mutation breeding was the obvious breeding procedure to use. Tifdwarf and Tifgreen are triploid turf bermudagrasses that cover golf course greens across the southern USA and the tropics around the world. They are sterile, so the only methods for improving them were mutation breeding or making more triploid hybrids. We made many more triploids, but none was any better than we already had.

So the logical approach was to use mutation breeding. Because these bermudagrasses were highly heterozygous, we were able to induce 158 mutants in the M_1 generation. Nine of these seem to be equal to their parent in most traits and superior in at least one. Certainly, mutation breeding was the best breeding procedure in this case.

19. R. BAKER. Dr. Burton, will you elaborate on where isogenic lines are most useful in plant breeding research?

20. G. W. BURTON. Isogenic lines can be tremendously useful in developing appropriate breeding objectives. Where increased yield is a major objective, and it usually is, the effect of other objectives on yield must be understood. Objectives that reduce yield cannot be accepted unless they are more important than yield. Yields of isogenic lines or populations that differ in a characteristic thought to be important can indicate if it should become a breeding objective.

21. A. LEFFLER. Dr. Borlaug spoke about the tremendous vulnerability of pure-line cultivars of self-pollinated crops to disease epidemics. Doesn't the same degree of vulnerability apply to hybrids of allogamous crops?

22. W. L. BROWN. No, hybrids are not as vulnerable to disease epidemics as are pure-line cultivars. Multilines of small grains provide a type of insurance against disease epidemics, and somewhat the same thing occurs through the use of hybrids, in that each single, 3-way, and double cross has more than one parent. Also, there is no reason the multiline approach cannot be extended to hybrids in totality.

REFERENCES

Atkins, I. M., and M. J. Norris. 1955. The influence of awns on yield and certain morphological characters of wheat. Agron. J. 47:218-20.

Burton, G. W. 1966. Plant breeding: Prospects for the future, pp. 391-402. *In* Frey, K. J. (ed.), Plant breeding. Iowa State Univ. Press, Ames, Ia.

Burton, G. W. 1969. Breaking dormancy in seeds of pearl millet, *Pennisetum typhoides*. Crop Sci. 9:659-64.

Burton, G. W. 1974. Recurrent restricted phenotypic selection increases forage yields of Pensacola bahiagrass. Crop Sci. 14:831-35.

Burton, G. W. 1976. Overcoming constraints and realizing potentials in the physical and biological aspects of feeding people, pp. 71-86. *In* Proc. World Food Conf. of 1976, Iowa State Univ. Press, Ames, Ia.

Burton, G. W. 1977. Fertility sterility maintainer mutants in cytoplasmic male sterile pearl millet. Crop Sci. 17:635-37.

Burton, G. W., and J. R. Stansell. 1971. An automatic darkbox to induce flowering in short-day plants in midsummer. Crop Sci. 11:595-96.

Burton, G. W., and Wayne W. Hanna. 1976. Ethidium bromide induced cytoplasmic male sterility in pearl millet. Crop Sci. 16:731-32.

Burton, G. W., Joel B. Gunnels, and R. S. Lowrey. 1968. Yield and quality of early and late-maturing near-isogenic populations of pearl millet. Crop Sci. 8:431-34.

Burton, G. W., W. W. Hanna, J. C. Johnson, Jr., D. B. Leuck, W. G. Monson, J. B. Powell, H. D. Wells, and N. W. Widstrom. 1977. Pleiotropic effects of the *tr* trichomeless gene in pearl millet on transpiration, forage quality, and pest resistance. Crop Sci. 17:613-16.

Burton, G. W., W. G. Monson, J. C. Johnson, Jr., R. S. Lowrey, Hollis D. Chapman, and W. H. Marchant. 1969. Effect of the d_2 dwarf gene on the forage yield and quality of pearl millet. Agron. J. 61:607-12.

Duvick, D. N. 1977. Genetic rates of gain in hybrid maize yields during the past 40 years. Maydica 12:187-96.

Hellmers, Henry, and G. W. Burton. 1972. Photoperiod and temperature manipulation induces early anthesis in pearl millet. Crop Sci. 12:198-200.

Jennings, Peter R. 1976. No plant breeders, no crops: No crops, no food. Rockefeller. Illus. vol. 3, no. 1, July.

Jensen, Neal F. 1978. Limits to growth in world food production. Science 201:317-29.

Luedders, Virgil D. 1977. Genetic improvement in yield of soybeans. Crop Sci. 17:971-72.

Powell, Jerrel B., G. W. Burton, and J. R. Young. 1974. Mutations induced in vegetatively propagated turf bermudagrasses by gamma radiation. Crop Sci. 14:327-32.

Russell, W. A. 1974. Comparative performance for maize hybrids representing different eras of maize breeding. Twenty-ninth Annu. Corn and Sorghum Res. Conf. 29:81-101.

Wortman, S., and R. W. Cummings, Jr. 1978. To feed this world. The Johns Hopkins Univ. Press, Baltimore, Md.

CHAPTER 12

Increasing and Stabilizing Food Production

N. E. BORLAUG

AN ADEQUATE FOOD SUPPLY is the first necessity for a decent life. It directly affects the stability of both nations and world society. Nonetheless, the interest of political leaders and the non-rural general public in agriculture waxes and wanes directly with scarcity or abundance of the food supply. Consumers everywhere want cheap food. Few realize the costs in investments and labor and the risks, both biologic and economic, that are involved in producing the world's food supply.

Recognizing that we are increasingly a part of an ever more interdependent world, it is astonishing that so few people give much thought to the importance and difficulties of increasing agricultural output sufficiently to provide for the increased food and fiber demands that result from the combined effect of population growth and increased affluence. Moreover, it never occurs to them that provisions must also be made, insofar as possible, to fulfill these requirements, despite poor harvests from adverse weather and crop losses from diseases and pests.

When listening to the nightly weather forecasts on radio or television in the affluent nations, one is struck by what seems to be the public's primary concern: Will the weather be pleasant and comfortable for outdoor recreation? Seldom, except on farm programs, is any mention made of the effect of weather on agriculture and food production. Yet weather is a principal influence on crop production. Even though weather control and effect lie largely beyond the management of man, its year-to-year variations

Director, Wheat Improvement Program, International Maize Improvement Center, Mexico City, Mexico.

must be considered in planning food production.

Although man has little or no direct regulation over weather, we should not ignore the indirect positive effects that have accrued from better cultural practices that improve moisture utilization and minimize yield reductions in dry seasons. Similarly, the development of drought-tolerant and frost-tolerant cultivars has been important in minimizing effects of adverse weather in certain areas.

During the past five decades, great progress has been made in developing disease-resistant cultivars of crop plants. However, despite the overall benefits that such cultivars have contributed to greatly stabilized yield and production, the results are far from perfect. Disease-resistant crop cultivars become susceptible as the disease-causing organisms mutate or hybridize to give rise to new races capable of attacking them. Unless the once resistant cultivars are promptly replaced by newer resistant ones, the stage is set for the development of a serious epidemic whenever weather conditions become favorable for the pathogen (Johnson, 1961). This frustrating problem is most serious in autogamous crop species such as wheat (*Triticum aestivum*), rice (*Oryza sativa*), oats (*Avena sativa*), barley (*Hordeum vulgare*), and the like, which are attacked by one or more airborne obligate pathogens. Recurring disease epidemics contribute to instability in world crop production from year to year and sometimes directly result in hunger and famine.

Certainly, the current methods of relying on pure-line disease-resistant cultivars of autogamous crop plants can be improved upon. More than 25 years ago, my colleagues and I began to explore the feasibility of using multiline cultivars to control rusts of wheat (Borlaug, 1953; Borlaug and Gibler, 1953). For various reasons, mostly related to the development and subsequent improvement of high-yielding semidwarf wheat cultivars, the research on multiline cultivars was discontinued for more than a decade, not because the concept and its application were unworkable, but rather because of the higher priority of other research.

Currently though, the time is opportune for the development and use of multiline wheat cultivars in developing nations. Such cultivars can help to minimize losses from rusts and, thereby, contribute to stabilizing year-to-year yields and production.

Much of the remainder of this paper will be devoted to a discussion of the development and use of multiline wheat cultivars in developing nations. However, before doing so, let me present my perceptions of present and short-run future world food production problems.

FACTORS INFLUENCING AGRICULTURAL PRODUCTION AND FOOD AVAILABILITY

The abundance and scarcity of food supply is cyclic. It probably has been so from earliest history (Borlaug, 1978a). It is affected by political stability, area of arable land under cultivation, availability of agricultural inputs, prices of inputs and products, weather, crop yields, and the increased demands for food resulting from population growth.

During the past three decades, three distinct cycles have occurred. The period 1950-1971 was characterized by food surpluses in the principal food-exporting nations giving rise to relatively cheap food worldwide. Widespread food aid programs during the 1950s and 1960s assisted many newly independent nations with food deficits to meet their food requirements. Unfortunately, all too often this led to neglect of agricultural improvements in these countries while overemphasis was given to the development of other sectors of the economy. Such policies contributed to, and made many developing nations highly vulnerable to, food deficits during the subsequent food shortage years of the mid-1970s.

The period from 1972-1975 was one of widespread food shortages and soaring food prices everywhere. Poor harvests in 1972, 1973, and 1974 resulted from the interaction of many adverse factors. Unfavorable weather—droughts, floods, untimely frosts—losses from pests, shortages of fertilizer, and soaring prices for energy contributed to food production shortfalls. As reserve grain stocks in the exporting nations fell below a critical level, food aid programs were reduced and food prices soared worldwide. Severe food shortages developed in many food-deficit nations and famine prevailed in some. One additional year of bad harvest would have plunged the world into very serious turmoil. The world narrowly escaped a major disaster (Borlaug, 1978a).

At the height of the shortages in 1974, a United Nations World Food Conference was held in Rome, Italy. The conference was characterized by an excess of rhetoric from political leaders from around the world. Irresponsible statements claiming that no child would go to bed hungry within the next decade and that hunger would be banished from the face of the earth by the turn of the century were made. Such oratory provides great hope but no sustenance.

Assuring the production and equitable distribution of an adequate supply of food for the world population of four billion appeared a simple task to political leaders and social planners. Moreover, little was said about the relentless growth in human numbers, which adds about 78 million people annually to the world population. The world conference did, however, agree to (1) establish an international grain reserve for use in emergencies anywhere in the world, (2) increase food aid for food-deficit developing nations, and (3) establish a fund to assist the food-deficit developing nations to increase their agricultural production.

What has happened in the past five years toward achieving these lofty goals? Despite four years of discussions and debates by the World Food Council, which was created by the World Food Conference to develop a workable international grain reserve, no effective agreement has been approved, nor has an internationally financed reserve been established. And an internationally financed reserve is not likely to be established in the foreseeable future. Actually, many nations of the world have already forgotten the desperate years of the mid 1970s and appear to be drifting back to complacency about the need for an international grain reserve.

During the past three years, with bountiful harvests almost everywhere in the world, grain reserves have been rebuilt rapidly in the grain-exporting nations, as well as in some developing nations that embarked on building modest national reserves a few years ago.

The grain stocks in the USA have already reached levels where acreage allotments have been reestablished to prevent a further drop in grain prices, as well as to prevent the accumulation of stocks that may become unbearable economically. If stocks continue to accumulate, almost certainly there will again be an expansion of the PL 480 type of concessional sales to developing nations, which if improperly handled by the recipient nations as in the 1960s, will again become a disincentive to progress in agricultural production in the developing nations.

Is the increased agricultural production real or is it an illusion? Many will attribute most of the improvement to favorable weather, but is it likely that more favorable weather is involved? Skeptics believe that this is a temporary period of abundance and that the world will soon again be faced by another period of severe and worsening food crisis unless a drastic improvement in agricultural production takes place in the developing nations.

This point of view is clearly and forcefully brought out in a study undertaken by the International Food Policy Research Institute (IFPRI) entitled *Food Needs of Developing Countries: Projections of Production and Consumption to 1990.* This study indicates that the heart of the world food problem is, and will continue to be, in the low-income and food-deficit countries concentrated in Asia and sub-Sahara Africa. The study excludes the People's Republic of China and other countries with centrally planned economies. It points out that the food deficit, expressed in grain equivalents is projected to increase from 13 million tons in 1975 to 70-85 million tons by 1990 in the low-income developing countries. To maintain consumption at the inadequate 1975 per capita level would require 35 million tons more than the projected production. The anticipated shortfalls in production are shown in Figures 12.1 and 12.2. The plight of 11 of the countries that are foreseen to be confronted with the most severe food shortages is presented in Table 12.1. India and Nigeria will have the largest deficits in tons. India's grain deficit of 1.4 million tons in 1975 will increase to a deficit of 17.6 to 21.9 million tons by 1990. Nigeria's deficit will increase from 0.4 million tons in 1975 to 20.5 million in 1990, whereas Bangladesh's deficit will increase from 1.0 million tons to 6.4 to 8.0 million tons in the same period. Such a deficit would be more serious for India and Bangladesh than for Nigeria because of their shortages of foreign exchange for financing imports, whereas the

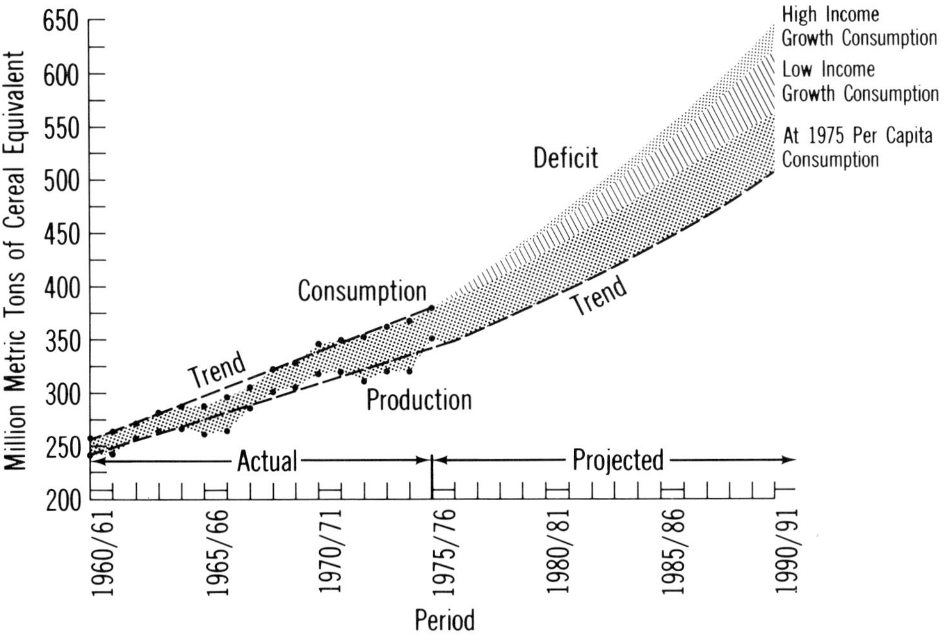

Fig. 12.1. All food deficit developing market economies: Production and consumption of major staples, 1960-1975 and projected 1990. (International Food Policy Research Institute, 1977.)

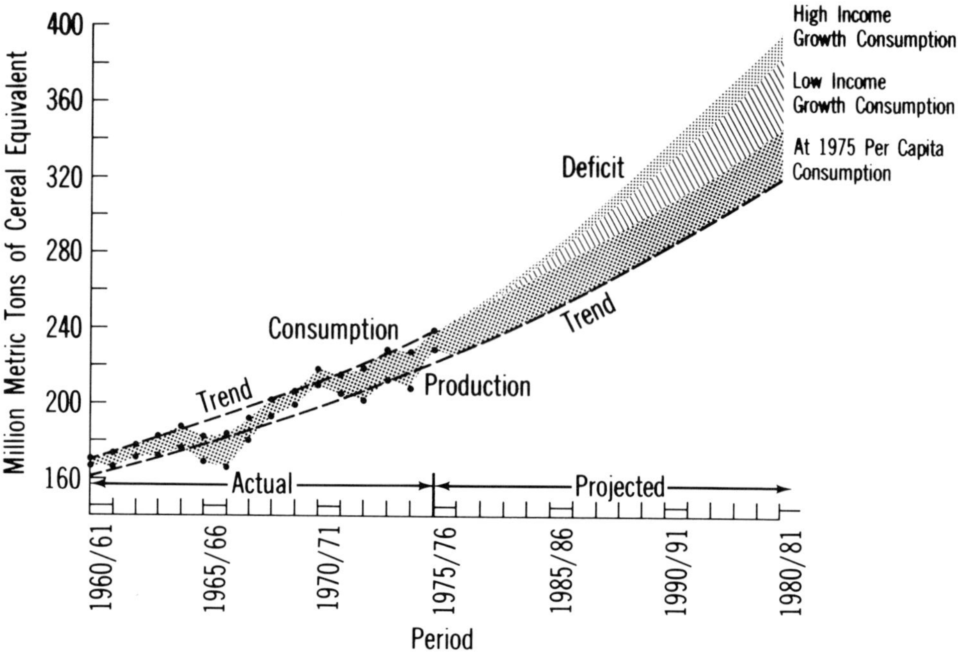

Fig. 12.2. Low income food deficit developing market economics: Production and consumption of major cereals, 1960-1975 and projected 1990. (International Food Policy Research Institute, 1977.)

latter country has large foreign exchange earnings from petroleum exports.

There is considerable doubt about the validity of the projections for India because the 1960-1975 base period used for making extrapolations was probably abnormal in terms of the situation today. The revolution in Indian wheat culture did not significantly affect production until 1968, and the progress in production was interrupted in 1974 and 1975 by shortages of fertilizer and diesel oil for pumping irrigation water. The dynamic upward increase in Indian production has been reestablished during the past three years with three successive record-breaking crops. Moreover, very significant increases in rice production were just beginning in 1975. The outlook for substantial increases in sorghum and maize production in the next decade is excellent also. Consequently, the 1960-1975 trend may not accurately represent the current production trends that appear to be considerably higher. In the case of Bangladesh, recent increases in wheat and rice production, if continued, may reduce the projected deficit also.

Table 12.1. 1975 actual and projected 1990 cereal grain equivalent deficiency in low-income developing market economy countries. (International Food Policy Research Institute, 1977.)

Country	Actual 1975 Millions of Tons	Actual 1975 Percent of Consumption	Projected 1990 Millions of Tons	Projected 1990 Percent of Consumption
India	1.4	1	17.6 to 21.9	10 to 12
Nigeria	0.4	2	17.1 to 20.5	35 to 39
Bangladesh	1.0	7	6.4 to 8.0	30 to 35
Indonesia	2.1	8	6.0 to 7.7	14 to 17
Egypt	3.7	35	4.9	32
Sahel Group	0.4	9	3.2 to 3.5	44 to 46
Ethiopia	0.1	2	2.1 to 2.3	26 to 28
Burma	(0.4)[a]	(7)[a]	1.9 to 2.4	21 to 25
Philippines	0.3	4	1.4 to 1.7	11 to 13
Afghanistan	...	...	1.3 to 1.5	19 to 22
Bolivia and Haiti	0.3	24	0.7 to 0.8	35 to 38

[a]Surplus

Why may some of the developing food-deficit nations be increasing their agricultural production faster than is indicated by the IFPRI report? Let me discuss some pertinent points.

The adoption of high yield technology, dubbed the so-called *Green Revolution* by the popular press, had its first impact on cereal production in some densely populated food-deficit developing nations of Asia only a decade ago (Borlaug, 1976). India, Pakistan, and the Philippines made spectacular progress in increasing production of wheat and rice from 1968 through 1972. Though progress slowed or even retrogressed in some cases during 1974-1975 because of shortages of chemical fertilizer and the large increase in price of energy, the Green Revolution remained alive. While the world was focusing attention on the worldwide shortage of grain and soaring food prices, grain production in all three countries, as well as in the People's Republic of China, was making a steady comeback. Despite the shortage of fertilizer, the adoption of improved technology began to exert a very positive influence on wheat production in Turkey and Tunisia and on both soybeans and wheat in Argentina and Brazil.

The magnitude of changes in production that have occurred can be illustrated by wheat production in India, Pakistan, and Turkey. In 1966, before the introduction of the high-yield technology, India harvested 10.5 million metric tons of wheat. Two years later, largely as a result of the new technology, 16 million tons were harvested. By 1977, the harvest exceeded 28 million tons in India, and the 1978 harvest was estimated at 31.5 tons. If political stability prevails, if wise agricultural policies are continued, and if the rust diseases are kept under control, wheat production can be raised to 40 million tons and 16 million tons in India and Pakistan, respectively, by 1988. The technology is also available in both countries to continue to increase production of rice, maize, and sorghum at a rapid rate.

India has recently been confronted by a completely different kind of problem on the food front. During the past two years, stocks of grain—mostly wheat—have accumulated in government warehouses to more than 20 million tons. This is probably more than double the level required for an ample strategic reserve as protection against a poor harvest. The major problem for the government of India now is one of finding an effective way for transferring the grain from overflowing warehouses into the nearly empty stomachs of underemployed or unemployed people who still go hungry because they have little or no purchasing power. There is really no "overproduction" of food but rather a problem of "underconsumption" because of inequities in distribution. In the search for the past two years to alleviate this dilemma, we encouraged the government of India to embark on a large "public food for work program." This is visualized as a challenge similar to the emergency public work programs (Civilian Conservation Corps and the Work Projects Administration) undertaken to provide employment and food during the economic depression of the 1930s in the USA. Such "food for work" programs, besides providing employment and food to the unemployed and underemployed, can be very effective in improving the rural agricultural infrastructure by the building of grain warehouses, farm-to-market roads, and rural schoolhouses; expanding and improving the networks of irrigation canals and drainage ditches; and reforesting and reseeding denuded and eroded areas unsuitable for agriculture.

About a year ago, a sizable food for work program was launched in several Indian states supported in part by the transfer of wheat from the grain reserves of the national government to state governments undertaking such programs. More than a million metric tons were transferred to such programs during the first eight months of its operation. How effectively these programs are being organized remains an unanswered question, but the challenges and opportunities are obvious. It will be interesting to see what political leaders, economists, sociologists, social anthropologists, and planners can do to utilize the accumulating stocks of grain effectively in food for work programs or in other ways to serve the twofold purpose of improving the rural agricultural infrastructure and equitably transferring needed food to the unemployed and under-employed landless peasants in the process.

My personal experience and observations in a number of key agricultural countries including Turkey, Tunisia, Pakistan, the People's Republic of China, South Korea, Taiwan, Argentina, Brazil, South Africa, and Rhodesia indicate that agricultural production is increasing much faster than was seemingly possible a decade ago. But these indications do not leave room for complacency in food production. In effect, the IFPRI report is a clear signal to government planners, political leaders, and agricultural scientists that there is no place for relaxation on the agricultural production front.

An additional development from the 1974 World Food Conference recommendations that should accelerate agricultural expansion and production in the next decade is the establishment and funding of the International Fund for Agricultural Development (IFAD). To date, ca. one billion dollars have been pledged for the next three years to finance projects which will assist developing nations to improve their agriculture. Of the funds pledged, 60 and 40 percent come from the Organization for Economic Cooperation and Development (OECD) nations and the OPEC countries, respectively. Five projects have been approved under this fund.

Right now, my greatest concern is avoiding, as far as possible, severe crop losses. Reduction of losses from bad weather, droughts, untimely frosts, and hail are factors largely beyond the control of man; but losses caused by disease epidemics and insect outbreaks are another matter (Stakman, 1966). The strategy that has been used with considerable success to control plant diseases, namely, the development and widespread use of pure-line disease-resistant cultivars of autogamous crop species, is far from perfect. It is especially deficient for Third World nations where government seed organizations have the exclusive rights and responsibilities for multiplying and distributing the new disease-resistant cultivars to farmers. All too often they are bureaucratic inefficient organizations that move their seed multiplication programs with the velocity of a tortoise. Moreover, they usually produce and distribute seed of poor quality.

RECENT CASE HISTORIES OF DISEASE EPIDEMICS

Twice within the past two years, in two widely separated areas of the world, leaf rust (caused by *Puccinia recondita*) epidemics of wheat have caused yield and production losses. These two case histories will illustrate the difficulties that agricultural scientists, and especially plant breeders, encounter in trying to avoid epidemics, particularly in developing nations.

Generally, certain factors responsible for efficient control of diseases in the USA are lacking in developing countries. The first handicap is the lack of an effective monitoring system to identify the appearance of new races of rusts capable of destroying the principal cultivars; thus, plant breeders and government officials are not alerted to the new danger. The second is the slowness of cultivar release committees in approving for multiplication new cultivars with genetic resistance that would protect against the new pathogen races. Thirdly, the organizations responsible for the multiplication and distribution of new cultivars are usually bureaucratic and inefficient government monopolies. For these reasons, the commercial crop remains vulnerable to losses from rust during the long lag period (five to seven years) from the time a new cultivar is released until there is sufficient seed to sow the area that is prone to the disease. During this long lag period, if weather conditions are favorable for the pathogen, epidemics may develop and cause serious crop losses.

The first case history involves the leaf rust epidemic that occurred in Sinaloa and Sonora, Mexico, in the 1976-1977 crop season. Previously, there had been no appreciable losses in wheat production from either stem or leaf rust in this region for 25 years (since 1951). Government researchers, seed organizations, scientists, and even most farmers—a new generation who had never experienced the destructiveness of a rust epidemic—had become rather complacent.

Never in the previous 33 years of my experience have so many seemingly unrelated factors—biologic, weather both favorable (for the rust) and unfavorable for agriculture (shortage of irrigation water), land tenure conflicts, political instability, organizational ineptitude and complacency—come together so perfectly to favor a new race of rust and generate an epidemic. But it was heartening to see how the rust pathologists, with the decisive support and leadership from top government officials, took the initiative away from the rust pathogen. For the first time in history, anywhere in the world, with the use of chemical sprays properly and timely applied, they stopped a widespread rust epidemic with only modest yield losses.

The sequence of events that set the stage for this epidemic were:

1. During 1975, 50 percent of the leaf rust races were capable of attacking 'Jupateco,' the most widely grown and popular cultivar of bread wheat. Four other commercial cultivars retained their resistance, and three new ones with different genes for leaf rust resistance were released.

2. Invasion of private agricultural properties by landless peasants during the fall and winter (1975-1976), created both confusion and conflict.

3. Uncertainty about future ownership of much of the remaining private farm holdings at time of harvest of the wheat crop in April 1976 added further to the confusion.

4. The general political instability had its effect.

5. There was a shortage of irrigation water that prevented the planting of summer soybeans or maize. These crops normally are sown immediately following the harvest of the wheat crop. But since there was much doubt about who the owners or users of the land would be in November 1976—the time of wheat sowing—no land was fallowed and the wheat stubble remained throughout the summer. Consequently, volunteer wheat seedlings that emerged after light rains in July were infected with leaf rust spores from the wheat stubble and the rust grew throughout the summer.

6. The failure of both the government and farmers' seed organizations to multiply the newly released cultivars was a factor.

7. With the shortage of irrigation water, a few farmers planted Jupateco cultivar in early October, two months before the normal dates of planting, to take advantage of rain brought by Hurricane Liza on September 29, 1976. These plantings soon became heavily infected from inoculum carried over the summer on volunteer plants.

8. Due to political uncertainty, most farmers did not sow wheat until mid-December or January, thus providing an opportunity for the heavy infection over all of its vegetative cycle.

9. Many farmers who had never seen a severe epidemic of rust disregarded the scientists' warnings about the danger of planting Jupateco and ignored the recommendations to sow 'Torim,' 'Anahuac,' 'Cocorit,' and 'Mexicali' cultivars. They learned a sad lesson the hard way.

By early to mid-January, there was severe leaf rust infection on most Jupateco plantings.

Fortunately, the Mexican Ministry of Agriculture sprayed 70,000 hectares of Jupateco with Bayleton or Indar fungicides. It stopped the epidemic and saved the crop. Fields that were not sprayed produced 0.7 to 0.9 T/ha of unsaleable grain, whereas nearby fields that were timely and properly sprayed produced near normal yields of 5.0 to 5.5 T/ha.

Some items of scientific and practical value were learned from this epidemic. It was clearly demonstrated that a widespread leaf rust epidemic on commercial wheat could be brought under control economically by proper and timely spraying with fungicides. This demonstration was a landmark in rust control and added another weapon to our defense against airborne wheat diseases.

The second case of loss from leaf rust occurred in Pakistan in the 1977-1978 crop season. Its cause is much easier to explain.

The principal bread wheat cultivar, 'Chenab,' had become susceptible to one or more races of leaf rust that appeared in 1972. Three new cultivars resistant to the new races, as well as the older races, were released by the Punjab wheat breeders in 1973, but there was no effective seed organization to multiply and distribute them widely. Too much reliance was placed on the effectiveness of farmer-to-farmer spread of seed.

The winter crop season of 1977-1978, throughout most of Pakistan, was characterized by frequent rains and temperatures favorable for rust development. Early infection occurred and a rust epidemic built up, which cut yield and production seriously, necessitating the importation of approximately two million tons of wheat. Various explanations have been advanced to explain the production loss. One was that most of the reduction in yield and production resulted from high temperatures during the period of grain filling. But, there is both ample circumstantial and observational evidence to indicate that the reduction in yield and production was principally the result of rust. For example, during the 1977-1978 crop season, India harvested a record wheat crop while Pakistan's production fell two million tons. The largest wheat producing areas in India and Pakistan, respectively, are the East Punjab and West Punjab, and these areas are similar in climate and are not separated by any significant geographical features. Nevertheless, while East Punjab harvested a record crop, West Punjab experienced a poor harvest. Furthermore, India had opportunely developed, released, multiplied, and distributed new rust-resistant cultivars to replace those that had become susceptible, and by doing so, it avoided rust losses. Pakistan had developed and released new resistant cultivars, but the seed organization did not multiply and distribute them widely. The result was that Pakistan suffered severe losses from rust (Borlaug, 1978b).

The lesson that wheat breeders and pathologists in developing countries should learn from these two case histories is that we must find a more foolproof system of deploying and managing disease-resistance genes (cultivars) in autogamous crops so as to reduce the risk of losses from epidemic diseases. The greater the human population pressure on the land, the less scientists can professionally and morally accept crop losses.

Stakman (1966) succinctly described the dilemma as follows:

> In two successive years, 1953 and 1954, stem rust destroyed one-quarter of the bread wheat and three-fourths of the durum wheat in the principal spring wheat area of the United States. Losses of this magnitude are catastrophic in countries that are always hungry and often on the verge of famine, as is illustrated by the death of upward of a million people in India, about a decade earlier, because a *Helminthosporium* blight destroyed much of the rice crop in large areas of production. The control of the living enemies of crop plants often makes the difference between food and famine.

The time has arrived to develop and widely use multiline wheat cultivars in developing nations. If properly developed and utilized, they will reduce losses from diseases; hence they will help to bring greater stability to food production. Multiline wheat cultivars and multiline cultivars of other autogamous crop species may also have an important future role to play in the agricultural production of the developed nations, if their potential value is explored without undue bias toward pure-line cultivars.

THE EVOLUTION OF GENETIC HETEROGENIETY/ HOMOGENEITY IN AGRICULTURAL CROPPING SYSTEMS AND ITS IMPACT ON DISEASE EPIDEMICS

Today, many persons and groups assert that modern agricultural methods have made our crops and forests more vulnerable to losses from diseases and insect pests than they were in the past (Borlaug, 1977). Are these charges true? If true, or partially so, what can be done to correct the defects in agricultural production? Remember that these corrections must be compatible with feeding and clothing, not 15 million people who lived on earth at the dawn of agriculture 12,000 years ago, but 4.3 billion people who live on earth today.

Prior to domestication of today's principal cereal and grain legume species, and even during the early centuries of primitive agriculture, the so-called "crowd diseases" of small grains, such as rusts and mildews, seldom, if ever, caused severe epidemics. Quite likely, the wild cereal hosts and their obligate rust and mildew pathogens lived in relatively well balanced "peaceful equilibria."

Even though the obligate pathogens were always present, probably the cereal host populations were protected from severe epidemics by the interaction of spatial and genetic factors because: (1) there was spatial separation and low density of the susceptible and resistant cereal host plants among nonsusceptible or immune plants in the mixed populations of wild grasses; (2) plant populations were mixtures of species, many of which gave genetic resistance or immunity to the cereal host's pathogens; and (3) the cereal host itself was heterogeneous for genetic resistance to its pathogens.

Unfortunately, we have little recorded historical information concerning the nature and composition of the indigenous plant communities from which our present-day principal crop species evolved to use as a point of reference. For example, the wild progenitors of maize had disappeared before the earliest recorded history. Mangelsdorf (1966), on the basis of archeological evidence recovered in caves in Puebla, Mexico, and pollen data from Lake Texcoco, has established that wild maize was still in existence in 5000 B.C. The historical evolution of the small grain species is better known because most of their presumed ancestors and wild relatives still grow wild in limited areas of their original habitats. Nevertheless, the state of development and the distribution of these wild species in plant communities in present-day ecosystems undoubtedly is markedly different from what it was in preagricultural times. The adverse effects of human activities over the past eight to ten millennia, such as cultivation, fire, and grazing by domestic livestock, have certainly greatly modified the ecosystems in which we observe these wild forms today.

But one aspect is retained in present-day ecosystems of wild small grains. They grow in mixed populations with many other genera and species. Although these populations were infected with rusts and other obligate pathogens, the spread was slow and the disease intensity did not often, if ever, reach epidemic proportions. The result was that the host and pathogen generally coexisted in a peaceful "equilibrium" until man began to intervene.

Borlaug (1977) has said that forests were the midwives at the birth of agriculture. There is considerable circumstantial evidence that many of the significant crop plants of modern agriculture, except rice, grew wild at forest edges or in new openings in the forest canopy prior to domestication. This presumably was the pattern of evolutionary development of barley, rye, oats, wheat, and maize, as well as beans (*Phaseolus vulgare*), broad beans (*Vicia faba*), peas (*Pisum sativa*), lentils (*Lens esculenta*), and chickpea (*Cicer *arietinum*). These species had two characteristics in common: (1) they were shade intolerant species that did not prosper in the heavy shade of the forests; and (2) although they were sun-loving species, they were poor competitors with sod-forming grasses, which had spread across the well-watered temperate zones of the earth during the Miocene period. Wild progenitors of these crop plants had prospered primarily at the edges of the forests, in temporary openings in the forest canopy, or in semiarid areas, less favorable for sod-forming grasses.

Neolithic farmers certainly must have been frustrated by the inadequacy of their stone tools for preparing land for planting in areas covered by sod-forming grasses. Eventually they discovered a way to use the stone tools and fire to open small "patches" in the forest canopy for agriculture. Thus slash and burn agriculture came into being. It was effective because it opened the forest canopy so that sunlight could reach the ground. Organic materials were burned; mineral nutrients recycled; and a mineral soil seed bed was exposed, free of weeds and grasses. After two or three years in crops, soil fertility dropped, weed competition increased, and crop yields plummeted. Consequently, the clearings were abandoned and a new "patch" was cleared nearby by slashing and burning. Slash and burn migrant agriculture is still practiced in many areas of the world.

We can surmise, based in part on observation of present-day slash and burn migrant agriculture, that when neolithic farmers began to cultivate wheat, or perhaps barley, in small cleared patches in the forests, they did not encounter epidemics of rusts or mildews, despite having modified several factors that influence the development of rust epidemics. The seeds they planted were possibly mixtures of several to many different genotypes of wheats, or more probably mixtures of other cereals. Nevertheless, these plantings (populations) were much less diverse genetically from both an interspecies and intraspecies point of view, than were the naturally occurring wild plant communities from which they were selected. Also, they were shielded from inblown inoculum by the physical barrier of the surrounding forest. It probably was not until wheat and barley were grown extensively on the flood plains of the Nile, Tigris, Euphrates, and Indus rivers that ecological and epidemiological conditions became conducive to the early buildup of inoculum which occasionally resulted in epidemics.

As farmers migrated from the centers of origin of cereals in the Near East they carried with them seed of mixed populations of wheat (and/or other cereal grains). Thus, diverse populations were spread widely into the remote parts of Asia, Europe, and Africa, and millennia later into the Americas and Oceania. These mixed populations or land races continued to be the principal basis of small grain production worldwide until the early 1900s. The great genetic diversity between and within the land races aided in slowing the development and spread of rust epidemics. But genetic diversity alone was inadequate to prevent crop losses from disease when inoculum arrived early and ecological conditions remained favorable for the development of widespread epidemics. Under severe epidemic conditions, though individual plants remained free or nearly free of rust, most were severely damaged or killed.

By 1900, scientists in Europe, America, and Australia began to select from such land races, pure-line wheat cultivars with high grain yield and uniform resistance to stem rust (caused by *Puccinia graminis tritici*). Many pure-line cultivars were developed and grown successfully for a number of years. It soon became apparent, however, that these pure-line cultivars were not uniformly resistant throughout a country or region nor in the same region from one year to another. The effective usefulness of the resistance of many pure-line cultivars was short-lived and inexplicable, until Stakman and Piemeisal (1917) showed that the population of stem rust spores consisted of a mixture of pathogenic or physiologic races that differed in their ability to attack different wheat cultivars. Moreover, they showed that the races varied in different regions and their prevalence varied from year to year. This was a forewarning that the rust resistance genes of improved pure-line cultivars were not universally effective against all rust spore populations nor were they likely to be long lasting.

The wheat cultivar breeding programs in the USA, Canada, and Australia during the 1920s and 1930s concentrated on hybridization to combine stem and leaf rust resistances of two or more cultivars, while simultaneously increasing grain yield and improving agronomic type and milling and baking quality in the progeny. Again, many new cultivars were produced during the 1930s that resulted in higher yields and provided good protection against both rusts. There was confidence that the effectiveness of rust resistance genes would be longer with this approach. Wheat cultivars remained resistant to leaf rust for only three to six years, but they remained resistant to stem rust for ten years or longer.

During the 1930s and 1940s, pathologists in several countries developed a vast amount of basic information that demonstrated the great diversity of physiologic races that existed in both the stem and leaf rust pathogens. This clearly indicated the complexity of developing and maintaining rust resistance in commercial pure-line cultivars.

DEVELOPMENT OF INTEREST IN MULTILINE CULTIVARS

My interest in finding a different, more stable, and lasting method of controlling rust resulted from witnessing the disastrous losses from stem rust in South America in the late 1940s and in North America in the early 1950s, during my early and formative years in the wheat research program in Mexico.

A disastrous epidemic of stem rust destroyed the popular and previously highly resistant cultivar 'Klein Cometa,' which was grown on 70 percent of the wheat in Argentina in the 1947-1948 season, and it resulted in a serious shortage of wheat for domestic needs. The shortage was so great that it forced Argentinians to blend sorghum and wheat flour to meet their food needs.

In the USA and Canada, race 15B of stem rust became a major biotype in 1950 throughout the major wheat producing areas of the Great Plains. It has been found for many years in local areas near barberry (*Berberis vulgaris*) plantings. All commercial cultivars of wheat grown in the USA and Canada were susceptible to this race. All that was necessary to develop a stem rust epidemic was a source of inoculum of this race early in the season and weather favorable for the pathogen. No serious local losses from stem rust occurred in the USA in 1951 and 1952, but in 1953 and 1954, widespread epidemics developed in the northern part of the winter wheat belt of the USA and throughout the spring and durum wheat areas of the USA and Canada. The losses from rust amounted to hundreds of millions of dollars. These epidemics frightened me.

In 1950, in Mexico, we began to explore the feasibility of developing stem rust resistant multiline cultivars with the belief that they would provide a more flexible and lasting protection against rust. Initially we developed multiline cultivars based on the improved cultivars 'Yaqui 50' and 'Kentana 48.' This program, based on backcrossing, complemented conventional breeding programs for developing cultivars.

From 1955 through 1959, it became apparent that multiline cultivars could be suitable and acceptable to farmers, millers, bakers, and pathologists. A summary of this research was presented at the First International Wheat Genetics Symposium in Winnipeg, Canada (Borlaug, 1959).

Why were Yaqui 50 and Kentana 48 multiline cultivars not released? Several unforeseen developments in wheat research and production in Mexico precluded the release and distribution of these multiline cultivars. First, the pure-line cultivars 'Chapingo 53,' 'Lerma Rojo,' 'Yaqui 54,' and 'Nainari' maintained their resistance to stem rust; so there was no urgent need for releasing a multiline cultivar. Second and more decisive, however, was the 90 percent increase in yield potential, namely, from 4.5 to 8.5 T/ha, that occurred with the development and release of the semidwarf cultivars 'Pitic 62' and 'Penjamo 52.' These high-yielding wheats ushered in a new era in wheat production that made the tall multiline cultivars Yaqui 50 and Kentana 48 obsolete and unacceptable because of the vast difference in yield potential.

These events did not dampen my enthusiasm for multiline cultivars of autogamous crop plants as a means of reducing losses from airborne obligate pathogens. However, we were forced to abandon the multiline breeding project temporarily to concentrate on correcting several inherent weaknesses, that is, low test weight, weak gluten, and lodging susceptibility, present in the earliest semidwarf cultivars. Before this weakness-elimination work was completed, other changes delayed our return to developing multiline cultivars.

Among these changes was a metamorphosis in several national agricultural research organizations. The Mexican Office of Special Studies (a joint dependency of the Mexican Ministry of Agriculture and the Rockefeller Foundation) by 1960 had completed the task of training an excellent corps of young Mexican scientists to whom responsibility was given for research on wheat and other crops in the newly formed National Institute for Agricultural Research. Moreover, from 1962 onward there was a deeper involvement on my part in assisting the governments of India and Pakistan to renovate their wheat research programs to help alleviate the huge food shortages in those countries. India and Pakistan had twin problems: (1) stagnant agricultural production, and (2) relentless pressure of the human population growth. Additionally, in 1964, the International Center for Maize and Wheat Improvement (CIMMYT) was chartered and assigned worldwide responsibility for assisting the developing nations to improve their research and production of maize and wheat. Time was required to recruit a professional staff and organize a worldwide network of collaborative projects. Consequently, little time remained for the development of multiline wheat cultivars.

Finally, in 1970, CIMMYT again began to develop isolines

of the widely adapted wheat cross 8156 which is grown extensively under more than 20 names in 16 countries. These isolines are being tested in 20 countries in Latin America, Africa, and Asia. They are made available to national programs for use whether as pure-line cultivars or as isolines in multiline cultivars. Moreover, three Indian wheat breeding programs also are developing isolines of 'Siete Cerros' ('Kalyan Sona') and 'Sonalika' for use in multiline cultivars.

THE DEVELOPMENT AND USE OF MULTILINE OAT CULTIVARS

A pioneering role has been played in the development and use of multiline oat cultivars to control crown rust (*Puccinia coronata avenae*) by scientists at Iowa State University (Browning and Frey, 1969; Frey et al., 1977). Over the past ten years, this group has distributed a number of multiline oat cultivars that are widely grown in Iowa and neighboring states.

These scientists have developed imaginative and useful methods for identifying levels of tolerance to crown rust in the genotypes used as recurrent parents, as well as procedures for developing isolines. They have extensively evaluated the performance of multiline cultivars in the Corn Belt, where such cultivars are grown commercially, and in southern Texas, where severe epidemics occur regularly. Multiline cultivars have performed excellently under both sets of ecological conditions. Perhaps most significant of all is the research that these scientists have done on the epidemiology of the rusts in "natural (mixed) populations" of oats and other cereals in Israel where the alternate hosts of the rust pathogens occur in their natural habitats. These studies have provided insight into the "peaceful equilibrium" under which hosts and their obligate pathogens have existed over long periods of time.

During the past 80 years this "peaceful equilibrium" of pathogen and host has been upset repeatedly by man via the development and intensive cultivation of a single or few closely related pure-line cultivars. These have provided excellent short-term protection from rust losses, but eventually new rust races have evolved with the genetic capability to attack these resistant cultivars. At times, these new virulent races have caused serious economic losses before the newly susceptible cultivars could be replaced.

For the past 12 years, multiline cultivars have provided good protection against crown rust for the commercial oat crop in Iowa and surrounding states. The good performance of these multiline oat cultivars has ushered in a new era in plant protection.

Another milestone was the release in 1978 of the multiline wheat cultivar, 'KSML3' (Kalyan Sona Multiline 3) from the University of the Punjab in India. The cultivar Kalyan Sona is a reselection made in India from the CIMMYT cross 8156 which also gave rise to the Mexican cultivar Siete Cerros. Kalyan Sona and its sibs represent the principal cultivars that revolutionized wheat production in a number of Asian and African countries during the period from 1968 to 1978. Recently the area sown to Kalyan Sona has been decreasing due to its increasing susceptibility to leaf rust and/or stripe rust (caused by *Puccinia striaformis*). The isolines incorporated into KSML3 are resistant to the prevalent races of these two diseases also. KSML3 was distributed to 4000 small farmers in the zone of adaptation of the Kalyan Sona parent. Two more multiline cultivars of wheat, one from the Indian Agricultural Research Institute in New Delhi and the other from the Institute of Agricultural Sciences, Kanpur, U.P., are planned for release to Indian farmers for 1979 fall plantings. These two multiline cultivars, both built upon the Kalyan Sona background, will be distributed in other geographic zones where the parent cultivar is adapted. Another multiline cultivar based on Sonalika, an early maturing genotype sown after rice in India, is in the final stage of testing.

ADVANTAGES AND DISADVANTAGES OF MULTILINE CULTIVARS

There is some fear that widespread use of multiline cultivars will cause the evolution of a super rust race that can attack all cultivars of wheat grown. There is no reason to believe that this will happen. The evidence is reassuring. Open-pollinated maize cultivars have been grown in the semitropics and tropics in contact with its two obligate rust pathogens *Puccinia sorghi* and *P. polysora* for thousands of years, but no super race of these rusts has evolved. Moreover, various species of *Pinus* and their corresponding rust pathogens have been living together in "natural multiline open-pollinated systems" for 100 million years and yet no super races have evolved.

A multiline cultivar breeding program can and should be carried as only one segment of a general breeding program, especially in developing nations. It is not a total substitute for a dynamic general crop improvement program. Rather, it should be used as a part of a total breeding program in areas where the rust hazard is high. It should be used to broaden the disease resistance base of broadly adapted, high-yielding cultivars that have good acceptance by farmers, industry, and consumers. Multiline breeding, if properly conceived and organized, has great flexibility and can be applied with many different degrees of sophistication to fit the needs of countries whose agriculture is in different stages of development. Isolines developed in such a program can either be used commercially in a multiline cultivar or the best isoline can be multiplied and released as a pure-line cultivar.

Many plant breeders feel that a multiline cultivar soon becomes obsolete because of the rapid improvement in yield potential that occurs with the progressive development of newer and better pure-line strains. This may or may not be true. It was true in the case of Yaqui 50 multiline cultivar, which was ready for release in 1960 because of the 90 percent yield improvement by the first semidwarf cultivars, Pitic 62 and Penjamo 62. But such tremendous increases in yield happen once in a lifetime. In fact, in the 16 years since Siete Cerros and its sib cultivars (such as Kalyan Sona) were identified in 1962, there have been only marginal yield increases in newer cultivars. Cases where Siete Cerros has been outyielded 15 to 20 percent by the new variety 'Pavon' probably are due to the fact that Siete Cerros is susceptible—and Pavon is resistant—to leaf rust. Consequently, the high yield potential of Siete Cerros and Kalyan Sona are worth defending by making them into multiline cultivars.

Meanwhile, breeders must continue to use conventional breeding methods to develop pure-line strains with broad adaptation, significantly higher yielding ability, and superiority in other traits. Whenever a significantly worthwhile advance is achieved, another multiline cultivar should be developed to defend this more advanced genotype from diseases.

As originally perceived, two to four backcrosses were to be used to recover isolines sufficiently alike for use in multiline cultivars. It is now apparent, however, that isolines phenotypically similar to the basic parental cultivar can be recovered after only one or two backcrosses. At CIMMYT, it has been possible to recover isolines phenotypically similar to the multiline parent from double crosses of the type (A × B) × (A ×C), where A represents the recurrent parent. However, when a selection from a primitive land race is used as a donor parent, generally four or more backcrosses are necessary (Rajaram and Dubin, 1977; Breth, 1976).

The greatest advantages of multilines in developing countries where the seed organizations are weak is the protection they can provide to the commercial crop when pathogen races change. Many developing nations now have good-to-excellent wheat breeding programs, but few have good seed distribution organizations that can multiply and distribute new cultivars to farmers rapidly when changes in rust races require prompt replacement of the currently grown commercial cultivars to avoid production losses from disease. The multiline cultivar, by contrast, is made up of many isolines, and it is unlikely that all component isolines will become susceptible to a new race at once. Consequently, the multiline cultivar will continue to provide protection from the disease until such "slow moving" seed organizations can get a new cultivar distributed to the farmer. The buildup of inoculum and rate of spread of rust is very slow in a population where 50 percent of the population remains resistant. In fact, Frey et al. (1977) have shown that such partially susceptible or "dirty" multilines will provide economically adequate protection from rust for a number of years, or perhaps, indefinitely.

The distribution of a multiline of a popular cultivar is easy to handle in a developing country. The breeder simply needs to multiply the isolines separately to a quantity which, when blended, will give 50 tons of the multiline cultivar. The seed of the multiline can be put into 5-kg packets for distribution to 10,000 small farmers through extension workers, fertilizer distributors, or others in the area where the original cultivar is adapted. The farmers, because they are already familiar with the parental cultivar, will rapidly accept the new multiline. This approach to seed distribution can bypass seed distribution organizations and assure that good quality seed reaches the farmers where the cultivar is adapted. The seed will spread from farmer to farmer rapidly after the first harvest.

To take full benefit of multiline cultivars of wheat requires close cooperation between the plant pathologist, plant breeder, and cereal technologist. First, the cereal technologist must define what genotypes that might be used as recurrent parents have satisfactory milling characteristics. Then the pathologist and breeder must decide what isolines in the reserve bank should be committed for inclusion in the multiline cultivar to keep the rust populations off-balance. Initially, of course, the breeder and pathologist should have developed the isolines jointly, and they should maintain seed of the isolines until needed.

During each of the past five years, CIMMYT has sent 150 Siete Cerros isolines to 20 locations in the world for evaluation for rust resistance and adaptation. Many valuable lines have been identified in these tests; some have been utilized as parents in crosses in national programs, as commercial cultivars, and as isolines in multiline cultivars.

Additionally, the CIMMYT breeding program contributes greatly to adding general genetic variability, including rust resistance, to the gene pools of the national wheat programs around the world. During the past decade, the CIMMYT staff has made ca. 60,000 crosses among bread wheat strains from which 11,000 promising advanced lines have been evaluated in trials in Mexico. The 3300 best lines from these have been incorporated into uniform International Bread Wheat Screening Nurseries (IBWSN) and are grown at more than 120 locations in the world for evaluation of their adaptability and disease resistance.

It has become apparent from this widespread testing of wheats with a broad range of genetic diversity that this is a very valuable procedure for identifying unique combinations of genes for disease resistance. Some seemingly broad types of field resistance, presumably polygenic in nature, which convey a broad type of resistance to both stem and leaf rust, have been identified. Probably, if pathologists can identify such partial resistances with more exactitude, many more valuable resistances will be found. These types of resistance, if they can be identified, will likely improve the stability of resistance to rust pathogens. Such resistances could be used in either pure-line or multiline cultivars.

WHAT KIND OF WORLD DO WE WANT OUR GRANDCHILDREN TO LIVE IN?

For those involved in food and fiber production, there must be concern about the number of people the world's land base will feed and clothe. Approximately 98 percent of the world's food production in 1975 was produced on the land, and the land base available for agriculture is shrinking annually as arable land is diverted to other uses. Anyone engaged in increasing world food production soon comes to realize that human misery from world food shortages and world population growth are two sides of the same coin. Unless these two interrelated problems and the energy problem are brought into better balance within the next several decades, the world will become increasingly chaotic. The social, economic, and political pressures and tensions are building at different rates in different countries depending upon human population density, human population growth rate, and the natural resource base. Poverty in many developing nations may become unbearable. The USA and other developed nations import more and more of their raw materials with the result that all nations of the world become increasingly interdependent. Standards of living in affluent nations will likely stagnate, or in some cases, regress. Will the peoples of these nations have the fortitude to adjust their needs to more modest levels, or will these nations also be threatened by social and political chaos?

Even in affluent nations with well-educated people, such as the USA, people have begun to concern themselves with symptoms of the complex malaise that threatens civilization rather than with the basic underlying causes. People support new legislation relative to polluting the environment, or protecting species from extinction, or controlling carcinogens in food, and the like, but they fail to recognize that the major underlying cause of the malaise is the terrifying rate at which the human population is growing. After all, it is the huge human population and its rapid rate of growth that cause unacceptable levels of environmental pollution, the extinction of present day species of fauna and flora, and our food shortage.

But let me again focus briefly on the growth in human numbers and the task that lies ahead if we are to leave our grandchildren with some hope for a decent world. Evidence clearly indicates that man or "near man" appeared on the planet Earth at least three million years ago. About 12 millennia ago, when humans discovered agriculture and domesticated animals, world population was about 15 million. With a stable food supply, the population doubled four times to 250 million by the time of Christ. The next doubling to 500 million occurred in 1650 years, and the next doubling required only 200 years, to give a population of one billion in 1850. At that time, the nature and cause of infectious diseases were discovered, and the beginnings of modern medical practice occurred and soon resulted in a reduced death rate. The third doubling to two billion required only 80 years, that is, until 1930. Sulfa drugs, antibiotics, improved vaccines, and improved diets further reduced the death rate spectacularly, with the result that world population doubled again to four billion people by 1975, that is, in 45 years. In total, we have had an increase of 256 fold or eight doublings since the discovery of agriculture. Even with some headway in reducing population growth, another doubling to eight billion souls may occur by the year 2015.

For the future, this means that in the next 40 years world food production may have to increase from the 1975 record of 3.3 billion metric tons to more than 6.6 billion tons, namely, an increase in food production in the next four decades as great as the total that occurred in the 12,000 years from the discovery of agriculture to 1975. It also clearly indicates the urgency of dealing effectively and humanely with the Human Population Monster. For plant and food scientists, this is a tremendous undertaking of vital importance to the future of civilization. Failure will plunge the world into social, economic, and political chaos. Can the production of food and fiber be doubled in the next 40 years? It can, providing world governments have the political will to give high enough priority to support for research in agriculture and forestry. It cannot be achieved with a continuation of the miserly and sporadic support that has been given to world agriculture and forestry over the past 50 years.

We agricultural scientists, who have the experience and insight to comprehend the complexities of the food/population problem, can no longer remain mute on this issue if we are truly concerned about the welfare of our grandchildren and great grandchildren. We must play our part in convincing political and religious leaders to give immediate attention to restraining population growth. It is later than most of us believe.

REFERENCES

Borlaug, N. E. 1953. New approach to the breeding of wheat varieties resistant to *Puccinia graminis tritici.* Phytopathology 43:467.

Borlaug, N. E. 1959. The use of multilineal or composite varieties to control airborne epidemic diseases in self-pollinated crop plants. Proc. First Int. Wheat Genet. Symp., Univ. Manitoba, Winnepeg, Canada, pp. 12-28.

Borlaug, N. E. 1976. The green revolution: Can we make it meet expectations? Proc. Am. Phytopath. Soc. 3:6-21.

Borlaug, N. E. 1977. Forest for people: A challenge in world affairs—Why the challenge? Proc. Nat. Conv. Soc. Am. Foresters, Alberquerque, N.M., pp. 1-14.

Borlaug, N. E. 1978a. The magnitude and complexities of producing and distributing equitably the food required for a population of four billion which continues to grow at a frightening rate, pp. 1-59. *In* Ramanujam, S. (ed.), Proceedings fifth international wheat genetics symposium: Indian society genetic plant breeding, Indian Agric. Res. Inst., New Delhi, India.

Borlaug, N. E. 1978b. Pakistan's agriculture and enigma. Proc. 1978 Pakistan Wheat Workers Workshop. Pakistan Agric. Res. Counc., Islamabad, Pakistan.

Borlaug, N. E., and J. W. Gibler. 1953. The use of flexible composite wheat varieties to control the constantly changing stem rust pathogen. Agron. Abst., p. 81.

Breth, Steven A. 1976. Multilines: Safety in numbers. CIMMYT Today, No. 4.

Browning, J. A., and K. J. Frey. 1969. Multiline cultivars as a means of disease control. Annu. Rev. Phytopath. 7:350-81.

Frey, K. J., J. A. Browning, and M. D. Simons. 1977. Management systems for host genes to control disease loss. Ann. N.Y. Acad. Sci. 287:225-74.

International Food Policy Research Institute. 1977. Food needs in developing countries: Projections of production and consumption in 1980. Res. Rpt. 3, December, 1977.

Johnson, T. 1961. Man-guided evolution in plant rusts. Science 133:357-62.
Mangelsdorf, Paul C. 1966. Genetic potential for increasing yield of food crops and animals. Proc. Symp. on Prospects of World Food Supply. Nat. Acad. Sci. Publ., pp. 66-71.
Rajaram, S., and H. J. Dubin. 1977. Avoiding genetic vulnerability in semi-dwarf wheats. Ann. N.Y. Acad. Sci. 287:243-54.
Stakman, E. C. 1966. Pest, pathogen, and weed control for increased food production. Proc. Symp. on Prospects of World Food Supply. Nat. Acad. Sci. Publ., pp. 72-77.
Stakman, E. C. and R. J. Piemeisel. 1917. A new strain of *Puccinia graminis*. Phytopathology 7:73.

INDEX

Additive genetic effects, 8, 15
Additive genetic variance, 7
Agricultural production, 469-75. *See also* Food supply
Agrobacterium hybrid vector, 103
Alfalfa, insect resistance breeding, 292-93
Aluminum tolerance, 156, 157, 159, 165
Anther culture, 88-89
Antibiosis, 293, 294
Antinutritional components. *See* Forages
Apomixis, 140, 450
Artificial insect infestation, 351
Assimilate partitioning, 270
Associated cropping, 182
Autogamous crop species
  genetic variability, 10
  recurrent selection, 22
Avena sativa L. *See* Oats
Avena sterilis. *See* Oats
Avoidance mechanisms, 309, 310, 346

Backcross, 44, 447, 448, 453, 487
Barley
  leaf rust, 313, 322, 335
  powdery mildew, 322
  scald, 316
B chromosomes, 143
Bermudagrass, 399-400
Bermudagrass hybrids, 437
Beta-carotene, 397-98, 416-17
Biological yield, 234-55
Biomass production, 437
Biosphere reserves, 63
Breeding
  forage quality, 399-402
  insect resistance, 291-92, 298-301
  nutritional traits, 366-67
  oil, 389-95
  resistance, 331-46
  stress environments, 260-67
  vitamin content, 395-99
Breeding methodology
  host-plant resistance to insects, 298-301
  intercropping, 199-200
Breeding objectives, 4-6
Breeding populations, 6

Carotene, 398
Carrots, 398-99
C_4 plants, 236, 238, 250, 252, 266, 274, 368, 454
C_3 plants, 236, 238, 250, 252, 266, 274, 454
Cell line selection
  5-methyl tryptophan, resistance to, 87
  for urease, 88
  for valine resistance, 87
Centers of diversity, 58, 76
Centers of origin, 481
Cereals
  protein improvement, 367-70
Chromosome pairing
  B chromosomes, effect of, 123
  genetic control in, 121
  heterochromatin, effect of, 122
Cold tolerance, 158
Convergent breeding, 47
Convergent cross, 44

Cotton, insect resistance breeding, 295-96
Crop
digestibility, 259
growth rate, 235
ideotype, 233
respiration, 250
Cuticular resistance, 264
Cybrid, 95
Cynodon dactylon. *See* Bermudagrass
Cytogenetics, plant breeding, 117
Cytoplasmic male sterility, 140

Dark respiration, 252-55, 275
Daucus carota L. *See* Carrots
Diallel Selective Mating System, 23, 25, 45, 47
Dilatory resistance, 349, 350
Disease epidemics, 475-79
Disease tolerance, 358
Dominance effects, 8
Drought tolerance, 161, 167, 171, 172, 261-62

Eceriferum, 44
Economic yield, 255-60
animal trials, 259
from reserves, 256
Endosperm, 133-36
Endosperm heterosis, 134
Environmental stress, 152
Epilleles, 136
Erectoides, 44
Erucic acid, 391, 413, 420

First division restitution, 125, 128
Flax rust, 324
Food-deficit nations, 470, 471, 473
Food supply, 467
Forages
antinutritional components, 402
antinutritional quality constituents, 417
digestible dry matter, 417
voluntary intake, 417

Gametophytic modifications, 129-32
indeterminate gametophyte gene, 129
Gene-for-gene relationship, 326-30, 352, 356
Gene-for-gene system, 323,325
evolution of 328-31
Gene parks, 76
"General" resistance, 349
Genetic diversity, 57-58
Genetic dominance variances, 7
Genetic engineering, 162, 380
legal implications, 113
Genetic erosion, 58-60
Genetic gain formula, 13
Genetic gains for yield, 155
Genetic homogeneity, 479-83
Genetic resistance to stress, 164
Genetic resource utilization, 68-69
Genetics of disease resistance, 322-24
horizontal resistance, 322
vertical resistance, 322
Genetic variability, 6-11, 50
Genotype x environmental interaction, 167, 169
Genotype x system interaction
statistical measures of, 196-99
Germplasm
collecting, 61-62
collection, 78, 124
conservation, 64
ex situ and *in situ*, 62
data storage and retrieval, 66-68
evaluation, 66, 78, 80
exotic, 34
flint maize, 77
international coordination, 69-70
maize, 34, 71
preservation
active and base collections, 63-64
regenerative, 65
Resources Information Project, 76
tomatoes, 73
wheat, 79
Glycinin, 409
Gossypium hirsutum. *See* Cotton
Grain crop canopies, designs for, 245
Grain reserve, 469, 470
Gramine, 418
GRAS standards, 415
Green Revolution, 440, 473

Haploid methods, 48
Haploids, 118-21
Haploids in breeding, 89
barley, 120
maize, 120
Harvest index, 255, 256
Heat tolerance, 161
Hessian fly, 292
High-lysine barley, 369
High-lysine maize, 419
High-lysine sorghum, 369, 370-75, 404, 425

High-protein maize, 387-89
High-protein oats, 383-85
High-protein rice, 385-87
High-protein wheat, 381-83
Hiproly barley, 375
Hordeum vulgare. See Barley
Horizontal resistance, 309, 314, 318, 320-22, 324, 349, 351, 353, 356
Host-pathogen, co-evolution of, 311-13
Host-pathogen relationship, 309, 328
Host-pathogen system, 311-32
Host-plant resistance, 322
  to insects, 299-98, 306

Ideotype, 45, 276
Insect biotypes, 303-4
Insect control via insecticides, 348
Insect infestations, 299-300
Insect resistance, 291-92
  chemistry of, 296
  screening for, 300-1
Insect resistant cultivars, 301-2
Insect supply, 299
Integrated Pest Management, 297, 298, 301-2, 306
Intercropping, 182
  tolerance, 192
Intermating, 46
IPM. *See* Integrated Pest Management
Iron chlorosis, 158, 165
Isogenic populations, 453
Land equivalent ratio, 183, 219
Leaf area index, 235
Light transmission coefficient K, 241, 242
Linoleic acid, 390
Linolenic acid, 390, 392
Lycopersicon esculentum. See Tomato
Lycopersicon spp. *See* Tomato

Maize, 5, 7, 387
  additive effects, 9
  feed-back inhibition-resistant mutants, 407
  genetic gains, 5
  genetic variances, 7
  Helminthosporium leaf blight, 327
  "latente" trait, 170
  northern corn blight, 324
  recurrent selection, 14
  rusts, 313, 486
Maize, insect resistance breeding, 294
Maize ideotype, 246
Maize-sorghum hybrids, 139, 143, 145
Maize-*Tripsacum* hybrid, 138, 143, 145
Mass selection, 32, 49, 50
  in alfalfa, 28
  in oats, 22
  in onion populations, 9
  in soybeans, 22
  in sweetclover, 28
  in sweet potato, 33
Mating designs, 7, 9
Mechanisms of host-plant resistance to insects, 302-3
  antibiosis, 302-3
  antixenosis, 303, 304
  tolerance, 303, 304
Medicago sativa. See Alfalfa
Mentha piperita. See Mint
Metal toxicity, 161
Mint, Verticillium wilt, 332-33
Mitochondrial DNA, 141
Monoculture, 182
Monoculture-intercropping correlations, 192, 199
Monoploid methods, 48
Monosomics, 130
Multiline cultivars, 302, 305, 342, 349, 350, 355, 356, 357, 468, 479, 483-89
Multiple cropping, 182
  competitive ability, 220
  competitive ability and yield, 211
  cultivar choice, 189, 216, 221
  desirable plant traits, 202-5
  genotype by cropping system interactions, 190
  interacting factors, 187
  legume-grain combinations, 188
  productivity from, 210
  reviews, 181
  species choice, 188-90, 217
  species interaction and resource utilization, 212
  temperate zone, 218-19
  testing cultivars for, 201-2
  varieties for, 186
Multiple cropping research, 185
  breeding techniques, 206-8
  on-farm testing, 208
Multiple crossing, 423
Mutation breeding, 451, 453

Net assimilation rate, 235
Net photosynthesis
  leaf age, effect of, 239
Niacin, 395
Nitrogen input, 270-71
Nonspecific resistance, 318
Nuclear repetitive DNA, 141

Oats
  Victoria blight, 327
Oleic acid, 390
Opaque-2 maize, 368, 403, 414, 421, 425
Oryza sativa. See Rice
Ozone resistance, 157, 159, 161-62, 172
Parasexual hybrids, 100
Peanut production, 439
Pearl millet, 401, 418, 448
Pennisetum typhoides. See Pearl millet
Phenotypic plasticity, 172
Photoperiod response, 261
Photorespiration, 236, 250, 269, 274, 275
Photosynthesis
  awns, role of, 246
  of crop canopy, 241-50
  flag leaf, role of, 245
  individual leaf, 235-41
Photosynthetic adaptability, 239
Photosynthetic rate, 237
Photothermal induction, 48
Phylloxera, 292
Phytoalexins, 313, 402
Plant breeding, 434
  definition, 3
  goals, 4, 5
Plant cell culture, 85
Plant digestibility, 259
Plant diseases, 475
Plant exploration, 60-62
Plant regeneration, 86, 104-5
Plant roots, 262-63, 276
Plasmid vector, 101
Polyculture, 182
Polyploids, 119
Population growth, 490-91
Potato
  late blight, 316, 332
Prebreeding, 68
Production of 2n gametes
  bilateral sexual polyploidization, 127
  unilateral sexual polyploidization, 124
Production stability, 167
Prolamins, 367, 368, 369, 377, 380
Protein digestibility, 426
Protein efficiency ratio, 369
Protein quality, 402
Protoplast fusion, 92
Public works programs, 474-75
Pyramiding resistance genes, 354, 355

Recurrent restricted phenotypic selection, 443
Recurrent selection, 11-34, 39, 44, 45, 443, 447
  in alfalfa, 29, 36
  application, 34-41
  with autogamous crop species, 21
  for cantaloupes, 36
  for forages, 28
  goals, 42
  in horticultural crops, 33
  in maize, 14, 35
  in oats, 35
  in Pensacola bahiagrass, 31
  for pest resistance, 36
  phenotypic, 23, 32
  reciprocal, 15, 20, 47
  for resistance to *Diplodia zeae*, 18
  for resistance to European corn borer, 17
  in sorghum, 32
  in soybeans, 25, 35
  for specific combining ability, 15
  in tobacco, 22
Recurrent selection vs. pedigree selection, 37, 50
Resistance genes, strategies for use, 337-45
  absence of a strategy, 338-39
  development of resistant genes, 344
  mosaic patterns of cultivars, 341-42
  multiline cultivars, 342
  multiple genes, 340
  polygenic resistance, 340-41
  tolerance, 345
Resistance mechanisms, 309, 310, 346
  avoidance, 313, 323
Resistance to drought, 169
Resistant cultivars, 298, 301, 305-6
Restricted phenotypic selection, 31
Riboflavin, 395
Rice
  bacterial leaf blight, 315, 322
  rice blast, 322

Saccharum officianalis. *See* Sugar-cane
Salt tolerance, 159, 160, 166, 173, 174
Screening for resistance, 332-33
  under natural conditions, 335
Seed multiplication, 475
Selection
  for biotype, 304-5
  for disease resistance, 332-36, 336-37
  intensity, 48
  in tissue culture, 86-88
Semi-dwarf cultivars, 484, 487
Senescence patterns of leaves, 246
Sequential cropping, 182
Sexual polyploidization, 128
Shoot:root ratio, 258
Single-seed descent, 21, 27
Slash and burn agriculture, 481
Solanum tuberosum. *See* Potato
Somatic hybridization, 89, 92, 95
Somatic hybrids, 95, 145
Somatic instability, 132-33
  C-mutants, 132
  intraplant variations, 133
  via chromosome substitution, 133
Sorghum
  breeding for resistance to green-bug, 296
  conversion program, 32
  genetic variability in, 10
  Periconia circinata, 333
Sorghum bicolor. *See* Sorghum
Source and sink, 274
Sources of resistance, 332-33
S_1 progeny, 28
Spike photosynthesis, 246
Specific resistance, 314, 349, 356
Stabilizing selection, 338, 342, 349
Stomatal frequency, 264
Stomatal resistance to diffusion, 238
Stress environments
  direct breeding, 156-63
  field tests, 156
  indirect breeding, 154-56
  laboratory tests, 158
Stress resistance, 160, 170, 265
Sugar cane
  Helminthosporium sacchari, 326-27
Synthetic cultivar, 51

"Tall-dwarf" cereals, 249
Team research, 273
Temperature stress, 266
Thiamin, 395
Tillering, 258
Timing crop development, 262
Tissue culture, 85
  plant propagation, 85-86
  Douglas fir, 106
  loblolly pine 106
Tolerance, 294
  to insects, 302
  mechanisms, 346
Tomato, 396-98, 436
Transpiration, control of, 263
Triticum aestivum. *See* Wheat
Triticum spp. *See* Wheat
2n gamete, 143

Unit area of leaf, 237
U.S. Plant Germplasm System, 74-75

Vertical resistance, 309, 314, 319, 324, 349, 351, 353
"Vertifolia effect," 336
Vitamin A, 395. 416
Vitamin C, 395, 416

Water-use efficiency, 263
Wheat
  glume blotch, 323
  Hessian fly, 355, 360
  insect resistance breeding, 293-94
  leaf rust, 316, 322, 475, 478, 486, 488
  mildew, 320, 354
  powdery mildew, 322
  Septoria blight, 316
  stem rust, 322, 482, 483
  take-all, 328
Winter hardiness, 159
Woolly aphid, 292

Yield improvement, 435, 436
Yield plateau, 5
Yield stability, 222

Zea mays. *See* Maize